Investment in Electricity Generation and Transmission

Antonio J. Conejo · Luis Baringo
S. Jalal Kazempour · Afzal S. Siddiqui

Investment in Electricity Generation and Transmission

Decision Making under Uncertainty

Springer

Antonio J. Conejo
Departments of Integrated Systems Engineering, Electrical and Computer Engineering
The Ohio State University
Columbus, OH
USA

Luis Baringo
E.T.S.I. Industriales
Universidad de Castilla—La Mancha
Ciudad Real
Spain

S. Jalal Kazempour
Department of Electrical Engineering
Technical University of Denmark
Kongens Lyngby
Denmark

Afzal S. Siddiqui
Department of Statistical Science
University College London
London
UK

and

Stockholm University
Stockholm
Sweden

ISBN 978-3-319-29499-5 ISBN 978-3-319-29501-5 (eBook)
DOI 10.1007/978-3-319-29501-5

Library of Congress Control Number: 2016936287

Printed on acid-free paper

This Springer imprint is published by Springer Nature
The registered company is Springer International Publishing AG Switzerland

To Olaia, Mireia, and Núria
To María and Carmen
To Azita
To my parents, Shamim and Saeed, for taking risks

Preface

Rigorous analysis is essential for making informed decisions about the construction of new electricity generation plants or transmission lines. This book provides a number of relevant models that constitute a framework for such investment analysis.

Since electricity generation and transmission assets last for many years, investment decisions are challenging because they involve high costs, are often made under high uncertainty, and generally require risk control on cost variability. Moreover, given the required investment level, wrong decisions may prove catastrophic from a financial viewpoint. Also, investment decisions are dynamic in nature since they are made by practitioners throughout a long-term planning horizon as uncertainty unfolds. As a result of the uncertainty involved, the required risk control, and the dynamic nature of investment decisions, appropriate mathematical models are complex and often need tailored solution techniques.

Within a market framework, distinct agents with different and often conflicting objectives are generally involved in making investment decisions, and thus a multiobjective equilibrium approach is desirable. Moreover, market power is common in electricity markets, and its representation entails nontrivial modeling features. Consequently, the resulting game-theoretic models are complex enough to require carefully crafted solution techniques.

By focusing on the application of the state-of-the-art mathematical tools for decision-making, this book aims to convey the principles of investment analysis in the electricity industry to students and practitioners alike. Initially, a social planning viewpoint is adopted, and generation expansion, transmission expansion, and generation plus transmission expansion problems are considered. Subsequently, a market perspective is taken, and generation investment equilibria are analyzed.

This book consists of seven chapters and five appendices. Chapter 1 provides an introduction to both electricity transmission and electricity generation expansion planning problems, emphasizing their long-term nature, the high degree of uncertainty involved, and the market framework in which electricity is produced, transported, distributed, and supplied in most parts of the world. Next, the chapter describes specific decision-making problems involving electricity transmission and/or generation facilities and introduces the computational tools needed to tackle these problems. It concludes by stating the scope of the book.

Chapter 2 is devoted to electricity transmission expansion planning adopting the viewpoint of a social planner. The rationale for this perspective is that the electricity transmission network is undoubtedly a natural monopoly. A simple deterministic model is first introduced to clarify the elements of this important decision-making problem, followed by a detailed robust adaptive model to cope with uncertainty in a secure yet economical manner.

Chapter 3 considers generation expansion planning also from a social planning viewpoint. The optimal outcome is intended to guide private investment in generation facilities. The chapter describes a number of increasingly complex models. The first model involves a single decision point, includes no network representation, and is deterministic. Subsequently, the single decision point is substituted by multiple decision points, the network is represented, and short-term (demand and renewable production variability within one year) and long-term (changes in demand and investment/operational costs across years) uncertainties are incorporated via stochastic programming.

Chapter 4 similarly adopts a social planning viewpoint and considers the joint generation and transmission expansion planning problem. Addressing the expansion of generation and transmission facilities together yields a transmission–generation coordinated solution that is optimal. The transmission component of this solution is to be built by the transmission operator, while its generation component serves as a guide for private investment in generation facilities. The chapter introduces models of increasing complexity: first, a deterministic single decision-point model, followed by a deterministic multiple decision-point model. Uncertainty is then introduced via stochastic programming, and finally, risk control (on total cost variability) is incorporated into this stochastic model.

Chapter 5 considers a market viewpoint and describes the decision-making problem of a private investor seeking to build electricity generation facilities from which to sell its output in the market for a profit. We assume that this investor has the capability to alter market outcomes, i.e., it has market power. This requires a complementarity or bilevel model, which generally entails high modeling and computational complexity. A single decision-point model is first introduced, followed by a multiple decision-point model. Short-term uncertainty pertaining to demand levels and renewable (solar- and wind-based) production is then introduced, but for simplicity, long-term uncertainty (change in investment/fuel cost and demand growth) is not considered. The chapter concludes by discussing computational techniques to tackle this type of large-scale complementarity problem.

Chapter 6 likewise takes a market viewpoint and considers a number of private investors competing in building power plants and in selling their generated electricity in the market for a profit. We assume that these investors are able to exert market power. The chapter describes models to identify the equilibria that are eventually reached by these competing investors. For simplicity, short-term uncertainty is represented, but long-term uncertainty is not. A number of solution methodologies to tackle these resulting equilibrium problems are discussed. Identifying equilibria is also of interest to the industry regulator in order to ascertain ways to improve market design and market rules.

Chapter 7 describes the real options methodology for identifying the timing, sizing, and technological characteristics of a specific investment project in generation or transmission facilities. The uncertainty unfolding over time is carefully represented to enable sequential decision analysis comprising features such as operational flexibility, modularity, and capacity choice. Risk control via utility functions is naturally embedded in the analysis.

Appendix A reviews the fundamentals of engineering economics. Appendix B provides an introduction to optimization under uncertainty. Appendix C reviews complementarity analysis, including equilibrium and hierarchical (bilevel) problems. Appendix D introduces the fundamentals of risk management. Appendix E provides an introduction to dynamic programming.

The material in this book can be arranged in different ways to address the needs of graduate teaching in a one-semester course. Chapters 1–4 and Appendices A, B, and D constitute the core of a capacity expansion planning course with no market focus. Chapters 1, 5, and 6 and Appendices A, B, C, and D include fundamental material for a market-focused capacity expansion planning course. Chapters 1 and 7 and Appendices A, D, and E constitute the basis for a real options course.

The book provides an appropriate blend of theoretical background and practical applications. This feature makes the book of interest to practitioners as well as to researchers and students in engineering, operations research, and business. Practical applications are developed up to working algorithms (coded in the GAMS environment) that can be readily used.

Reading this book provides a comprehensive understanding of current investment problems in electric energy systems, including the formulation of decision-making models for both generation and transmission expansion planning, the familiarization with efficient solution algorithms for such decision-making models, and insights into these investment problems through the detailed analysis of numerous illustrative examples.

This book opens the door to analyzing investment decisions in electricity generation and transmission facilities using the most advanced models available. Such models are explained in a tutorial and simple manner with illustrations provided by many worked examples. Hence, the concepts and insights can be accessible to practitioners and students.

To conclude, we would like to thank our colleagues and students for insightful observations, pertinent corrections, and helpful comments.

December 2015

Antonio J. Conejo
Luis Baringo
S. Jalal Kazempour
Afzal S. Siddiqui

Contents

Chapter 1
Investment in Generation and Transmission Facilities

This introductory chapter provides an overview of the investment decision-making process in electric energy systems, and it spans both transmission and generation investment. Transmission investment is regulated because it critically influences the operation of the system as a whole, while generation investment is generally the concern of private entities. The distribution reinforcement and expansion planning problem is outside the scope of this book, which focuses on generation and transmission. We adopt an electricity market viewpoint since this is the prevalent paradigm throughout the world. This chapter seeks to provide an insightful general overview, while other chapters provide precise models and detailed descriptions.

We first describe the three main characteristics of the decision-making process for planning electric energy systems, namely, long-term view, uncertainty, and high dimensionality. We then summarize the basic features of current electricity markets since expansion planning decisions are made within a market framework. Next, we specifically focus on the transmission expansion planning problem, the generation investment problem, and the problem of valuation and timing of generation investment alternatives. Subsequently, we adopt the regulator's viewpoint to consider the joint transmission and generation expansion problem.

We conclude the chapter by clarifying what we do and what we do not do in this book.

1.1 Long-Term Decision Making Under Uncertainty

Electricity generation facilities have effective operation lives of 30–50 years, while transmission facilities stay in service even longer. Therefore, reinforcing or expanding the transmission system and investing in generation facilities are long-term exercises leading to decisions that influence the future operation of the system up to fifty years and beyond. These investments are capital intensive, which generally involves complex financial arrangements, and building periods range from six months to two years for transmission lines and from two to five years for generation facilities.

The models required to make informed decisions in transmission expansion and generation investment are necessarily large-scale since they should include

A.J. Conejo et al., *Investment in Electricity Generation and Transmission*,
DOI 10.1007/978-3-319-29501-5_1

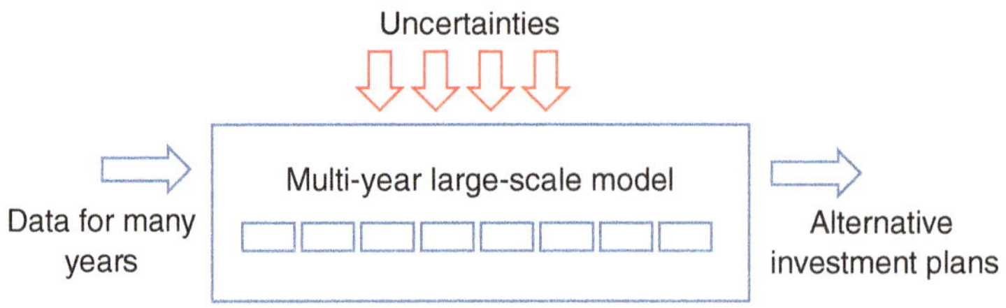

Fig. 1.1 Long-term decision-making under uncertainty

many variables and constraints to represent a large number of different operating conditions [7]. Moreover, the description of such conditions includes both continuous and binary variables. Binary variables are needed to represent the selection of discrete facilities, while continuous variables are needed to represent operating decisions. These models push available computational hardware and software to their limits since they typically entail millions of constraints and variables, many of them binary.

In summary, investment in electricity transmission and generation facilities requires:

- Taking a long-term view.
- Accounting carefully and comprehensively for uncertainty.
- Dealing with large-scale optimization problems.

Figure 1.1 illustrates the nature of long-term reinforcement and expansion problems in power systems along with the associated uncertainty.

In particular, modeling the uncertainty involved is critical to the decision-making process for investment in electric energy systems. Long-term expansion and reinforcement planning involve the following uncertainties:

1. Future evolution of the loads consumed by the demands located throughout the transmission network.
2. Future evolution of investment costs of the different production technologies considered as investment alternatives throughout the planning horizon, particularly renewable power units. This is contingent on the level of maturity of each technology.
3. Future evolution of the operation costs of the different production technologies throughout the planning horizon, particularly fuel costs. This is contingent on the time evolution of fuel costs.
4. Future investment decisions in production facilities, stochastic or not, made by producers. Such decisions are unknown to market agents other than the one involved in the specific investment decision.

Investment in electric energy systems is generally a multistage process. That is, expansion and reinforcement interventions are carried out sequentially at different

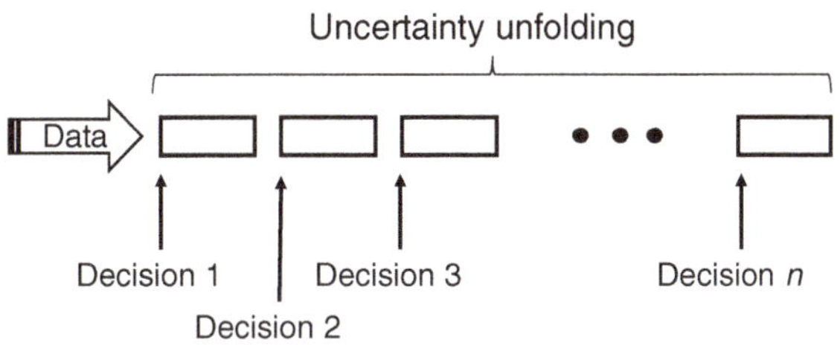

Fig. 1.2 Long-term decision making under uncertainty: multistage decision framework

points in time. Therefore, a full-fledged model to assist the decision maker in investment decisions needs to embody this dynamic multistage framework. Figure 1.2 illustrates this multistage decision framework.

However, embedding such a dynamic framework in a decision-making tool may render it computationally intractable and therefore useless. Therefore, if a dynamic framework is considered, then a number of simplifying assumptions in the description of the system are generally needed.

Alternatively, a rolling-window approximate static framework might be used, within which decisions are made at a given point in time, considering all future uncertainties but disregarding the fact that future investment decisions will be made afterward. In turn, the rolling window is moved forward to the next decision point in time and the process repeated. Such a simplified framework results generally in computational tractability even if a very detailed representation of the components of the system is incorporated. Figure 1.3 illustrates this rolling-window single-stage decision framework.

The decision to use a multistage or a rolling-window single-stage model is contingent on the size of the specific electric energy system under study, the required

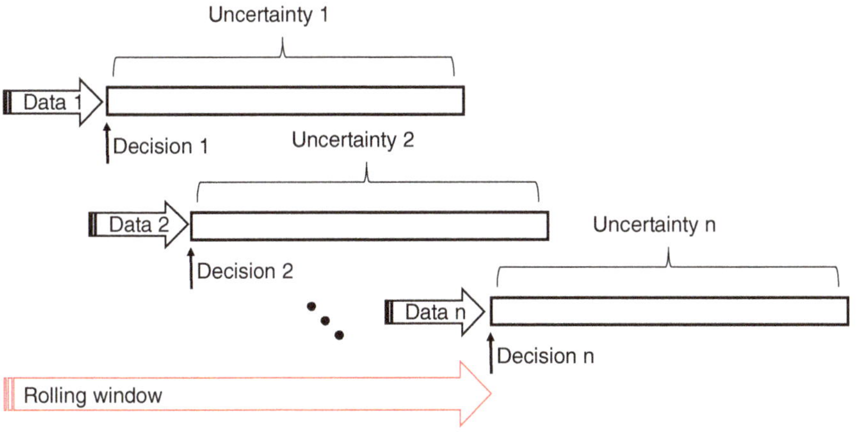

Fig. 1.3 Long-term decision making under uncertainty: rolling-window single-stage decision framework

detail in uncertainty description, and the software and computational machinery available. Decomposition techniques [13] that solve a given large-scale problem by decomposing it into smaller and/or simpler subproblems and using an iterative scheme are generally able to handle nearly intractable expansion planning problems.

1.2 Electricity Markets

In most parts of the world, the production, transmission, distribution, and supply of electric energy revolves around an *electricity market*, which typically includes several marketplaces, notably the *day-ahead market* and the *balancing market*. These two marketplaces constitute what is known as the *pool* [20, 31]. The day-ahead market closes at around noon and spans the 24 h of the next day, while the balancing market closes several minutes prior to power delivery on an hourly basis. Generally, a medium-term contracting framework is available as well and is typically known as the *futures market*. The futures market allows trading energy from one week to several years in advance at rather stable prices compared to day-ahead and balancing prices. Figure 1.4 illustrates the trading timeline.

In these marketplaces, *producers* submit offers in the form of production blocks (MWh) and their corresponding prices ($/MWh), while *consumers* submit bids in the form of consumption blocks (MWh) and their corresponding prices ($/MWh). In turn, the *independent system operator* (ISO), called the *market operator* in some jurisdictions, matches supply and demand using an appropriate *market-clearing tool* that results in accepted production levels per generating unit and hour, accepted consumption levels per demand and hour, and hourly clearing prices. Clearing prices

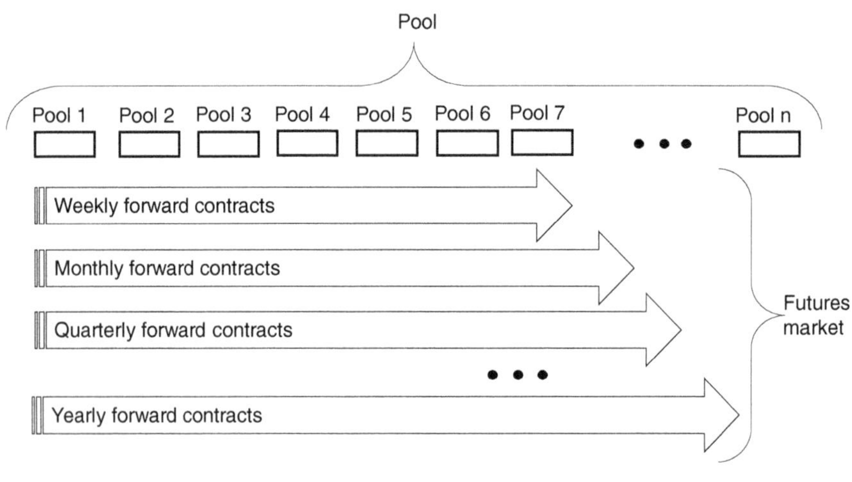

Fig. 1.4 Trading timeline

are generally marginal prices, which might be supplemented with uplifts if needed. Uplifts may be needed to ensure cost recovery to some critical production units, e.g., peakers.

Market-clearing tools in the pool generally include a network representation, which results in clearing prices differentiated by location if the network is congested, i.e., if some transmission lines are working at capacity. We call such prices *locational marginal prices*, or simply LMPs. Each consumer pays the LMP at its location multiplied by its consumption level, while each producer receives the LMP at its location multiplied by its production level. Some critical production units may not recover their corresponding costs with such a price scheme due to nonlinearities (e.g., lumped startup costs or minimum production levels) and market imperfections (e.g., the participation of a few large producers). If this is the case, then uplifts might be established to ensure cost recovery by these critical units.

The *regulator* is an independent entity in charge of ensuring an efficient and competitive working of the electricity market. The regulator in the United States (US) is the *Federal Energy Regulatory Commission* (FERC), while in the European Union (EU), regulators are country-based and have heterogeneous powers of oversight.

The key agents in an electricity market are the following:

1. Flexible producers with units that provide energy and *reserve* (backup power), e.g., combined-cycle gas turbines.
2. Inflexible producers with units that provide just energy, e.g., nuclear power units, which lack the flexibility to adapt their respective productions to real-time operating conditions.
3. Stochastic producers with units that provide energy on a stochastic basis, e.g., wind-power units.
4. Consumers that require energy and that may or may not have the capability of adapting to real-time operating conditions, e.g., a large aluminum factory.
5. Retailers that buy energy in the market to sell it to their respective clients for a profit.
6. The ISO that runs the market and uses an appropriate market-clearing procedure to derive reserve levels, production, and consumption quantities and market-clearing prices.

It is important to recognize that to make informed investment decisions, the operation of the market under many plausible future working conditions needs to be represented carefully. This is generally done using operating scenarios that capture most load/production market conditions.

A given operating scenario in the market can be represented by an optimization problem of the form [11]:

$$\begin{aligned} & \min_{\mathbf{x}} f(\mathbf{x}) \\ & \text{s.t.} \quad \mathbf{h}(\mathbf{x}) = \mathbf{0} : \boldsymbol{\lambda} \\ & \qquad \mathbf{g}(\mathbf{x}) \leq \mathbf{0} \\ & \qquad \mathbf{x} \in \mathscr{X}. \end{aligned} \tag{1.1}$$

Vector $\mathbf{x} \in \mathscr{X}$ includes the variables that characterize the market outcome, including the production allocated to each generating unit and the consumption level assigned to each demand, as well as reserves allocated to both production units and demands. Constraints $\mathbf{h}(\mathbf{x}) = \mathbf{0}$ describe the market-equilibrium conditions. If problem (1.1) is continuous, then vector $\boldsymbol{\lambda}$ contains the dual variables related to market-equilibrium conditions $\mathbf{h}(\mathbf{x}) = \mathbf{0}$, which are marginal prices. Constraints $\mathbf{g}(\mathbf{x}) \leq \mathbf{0}$ are used to state production and consumption bounds as well as equipment limits.

If vector $\mathbf{x}$ in (1.1) includes binary variables to represent the on/off status of production units, then problem (1.1) becomes a mixed-integer problem [11]:

$$\begin{aligned} \min_{\mathbf{x},\mathbf{y}} \; & f(\mathbf{x}, \mathbf{y}) \\ \text{s.t.} \quad & \mathbf{h}(\mathbf{x}, \mathbf{y}) = \mathbf{0} \\ & \mathbf{g}(\mathbf{x}, \mathbf{y}) \leq \mathbf{0} \\ & \mathbf{x} \in \mathscr{B} \\ & \mathbf{Y} \in \mathscr{Y}. \end{aligned} \tag{1.2}$$

In problem (1.2), vector $\mathbf{x} \in \mathscr{B}$ includes binary variables, while vector $\mathbf{y} \in \mathscr{Y}$ includes continuous ones.

In the EU, market-clearing algorithms are generally linear programming problems, while in the US they are mixed-integer linear programing problems. Therefore, in EU markets, marginal clearing prices are directly obtained from the solution of (1.1), while in US markets, a relaxation of (1.2) needs to be used to derive marginal clearing prices, which have to be supplemented with uplifts [34]. Long-term expansion planning models should embody all operational constraints that influence planning decisions. However, an excess of operational details may result in an intractable model. Thus, an appropriate tradeoff should be achieved between modeling accuracy and computational tractability.

Uncertainty plagues decision-making in electricity markets. Particularly, the following uncertainties are most likely to be present in short-term market operation:

1. Producer offering behavior in terms of both offered quantities and prices. Any particular market agent (producer, consumer, or retailer) has no precise knowledge of the offering/bidding behavior of other agents in the market.
2. Consumer/retailer bid behavior in terms of both bid quantities and prices. Any particular market agent has no precise knowledge of the offering/bidding behavior of other agents in the market.
3. Production levels by stochastic production units, such as wind power units or concentrating solar power units. Stochastic producers owning stochastic units, e.g., wind farms or solar power units, have no precise knowledge of the future production levels of such units.
4. Equipment failures involving both production and transmission facilities. Both generation facilities and transmission equipment fail with a nonnegligible failure rate, and thus their availabilities at a future point in time are uncertain.

Finally, it is important to state that as stochastic power units (wind- and solar-based) increasingly penetrate the generation mix, market-clearing procedures based

Fig. 1.5 Electricity market: agents, input, and outcomes

on stochastic programing might be advisable [30]. In such a case, instead of deterministic model (1.1) or (1.2), a stochastic model similar to (1.7) needs to be used.

Figure 1.5 illustrates the functioning of an electricity market.

1.3 Transmission Expansion Planning

The transmission network is a natural monopoly (i.e., an infrastructure not to be duplicated and over which no competition makes sense) and is used by all producers and consumers to trade electric energy. Its relative cost as compared to the cost of production facilities is small, but its impact on the functioning of the system is huge. Congestion prevents the transport of cheap energy from some production areas to areas of consumption, and line outages may endanger the stability of the system as a whole. It is therefore crucial to have a transmission network that in most cases is invisible to the system, i.e., energy trading is carried out without the presence of the network being noticed. Under these considerations, we adopt a social welfare and infrastructure viewpoint to plan the reinforcement and expansion of the transmission network.

It is important to realize that infrastructure is generally designed to be able to operate correctly under the worst plausible conditions, e.g., a bridge should withstand a maximum weight under adverse climate conditions, and a sea breakwater should withstand the strongest plausible wave. Thus, traditionally, every infrastructure is designed by considering the worst plausible conditions of future operation. In fact, this is also common practice regarding electricity transmission expansion planning.

However, the worst future condition is a priori unknown, and we have to cope properly with this uncertainty, which entails real reliability and economic consequences. If we are too conservative and design a transmission network around a worse-than-actual condition, then we end up building a stronger transmission system than needed at higher cost. If we, conversely, design a transmission system around a better-than-actual worst case, then we build a transmission system unable to withstand the actual worst condition, thereby leading to losses or collapse. To confront this challenge, we need to describe carefully the uncertain parameters involved in the transmission expansion planning problem.

Additionally, we need to ensure that the future transmission system will be able to withstand the worst operating condition. Consequently, we will work on robust optimization approaches, i.e., those that guarantee worst-case protection within a plausible uncertainty set. Moreover, since the uncertainty description that we use is parameterized through a set of *conservatism* parameters, we are able to characterize the critical tradeoff security versus cost and generate an array of efficient design solutions from which the decision-maker can choose.

Within the electric power industry, *transmission expansion planning* (TEP) refers to the decision-making process faced by a *transmission system operator* (TSO) to resolve the best way to expand or to reinforce an existing electricity transmission network. Reference [15] provides an industry perspective of this important decision-making problem. The TSO is the publicly controlled entity in charge of operating, maintaining, reinforcing, and expanding the electricity transmission network within a given jurisdiction and with the objective of maximizing social welfare. TSOs in the different European countries are coordinated through ENTSO-E [18]. In the US, TSOs have generally more limited roles than those in Europe and are generally referred to as *regional transmission organizations* (RTO). European TSOs generally own their respective transmission networks, which is not generally the case in the US. Prominent regional transmission organizations in the US include PJM [32] and the Midcontinent Independent System Operator, MISO [29].

Adaptive robust optimization (ARO) [9] allows modeling decision-making under uncertainty with recourse. For transmission expansion problems, ARO involves three steps:

1. Making investment decisions to maximize social welfare (or to minimize social cost).
2. Outlining the worst uncertainty realizations within a plausible uncertainty set that respects the physical constraints of the problem.
3. Making operational decisions to mitigate the negative effect of the uncertainty realization and to achieve maximum social welfare. These are *recourse* decisions.

Traditional robust optimization techniques [36] do not allow for controlling the level of robustness, i.e., conservatism, of the solution attained, which is a major drawback; however, the theoretical work in [9] introduces formulations that make it possible to control the level of robustness of the solutions attained. Such formulations allow developing valuable planning tools, which are relevant in practice.

ARO has two advantages with respect to stochastic programming models that generally require a large number of scenarios to represent the uncertainty involved [19]. First, scenarios do not need to be generated, and since generating scenarios, based on probability distributions or others, may entail an approximation of the description of the uncertain parameters, not needing scenarios is an advantage. Instead, robust sets are used in ARO models [10], and constructing such sets is generally simpler than generating scenarios. Second, an ARO model typically has a moderate size, which does not grow with the number of scenarios, and thus computational tractability is not generally at stake.

An ARO model has the general form below:

$$
\begin{aligned}
\min_{\mathbf{x}} \quad & \max_{\mathbf{u}} \quad \min_{\mathbf{y}} \quad f(\mathbf{x}, \mathbf{u}, \mathbf{y}) \\
& \qquad\qquad \text{s.t.} \\
& \qquad\qquad \mathbf{h}^{\mathrm{O}}(\mathbf{x}, \mathbf{u}, \mathbf{y}) = \mathbf{0} \\
& \qquad\qquad \mathbf{g}^{\mathrm{O}}(\mathbf{x}, \mathbf{u}, \mathbf{y}) \leq \mathbf{0} \\
& \qquad\qquad \mathbf{y} \in \mathscr{Y} \\
& \quad \text{s.t.} \\
& \quad \mathbf{u} \in \mathscr{U} \\
\text{s.t.} \\
& \mathbf{h}^{\mathrm{I}}(\mathbf{x}) = \mathbf{0} \\
& \mathbf{g}^{\mathrm{I}}(\mathbf{x}) \leq \mathbf{0} \\
& \mathbf{x} \in \mathscr{X}.
\end{aligned}
\tag{1.3}
$$

The objective function $f(\mathbf{x}, \mathbf{u}, \mathbf{y})$ represents the minimization of the total system costs, including transmission line investment costs and operating costs. The transmission investment decision variables are gathered in vector $\mathbf{x}$. Note that the entries of vector $\mathbf{x}$ are binary variables that represent the build/not-build nature of each investment decision. Similarly, the entries of vector $\mathbf{y}$ are the operating decision variables, which are assumed to be continuous. These operating variables include power production from each generating unit, load shedding per node, and power line flows. The worst-case realization of the uncertainty and the successive adaptive actions are considered in the max–min right-hand-side problem, while the min left-hand-side problem seeks minimum total social cost. Constraints $\mathbf{h}^{\mathrm{I}}(\mathbf{x}) = \mathbf{0}$ and $\mathbf{g}^{\mathrm{I}}(\mathbf{x}) \leq \mathbf{0}$ represent investment conditions and limits (budget, capacity, locations, and others). Constraints $\mathbf{h}^{\mathrm{O}}(\mathbf{x}, \mathbf{u}, \mathbf{y}) = \mathbf{0}$ and $\mathbf{g}^{\mathrm{O}}(\mathbf{x}, \mathbf{u}, \mathbf{y}) \leq \mathbf{0}$ include equality constraints related to the systems operations, e.g., the power balance at each node of the network, and inequality constraints, e.g., upper and lower power outputs for each generating unit, related to system limits, respectively. It should be emphasized that the operational constraints depend on both $\mathbf{x}$ and $\mathbf{y}$, which implies that the transmission investment decisions alter the configuration of the transmission network and thus the operation of the system. Finally, $\mathbf{u} \in \mathscr{U}$ defines the uncertainty set. Uncertainty parameters are modeled as decision variables so that the worst realization of such parameters can be represented.

Figure 1.6 illustrates the ARO methodology to address the transmission reinforcement and expansion planning problem.

1.4 Generation Investment

Investment in generating facilities is generally a competitive endeavor. Specifically, producers compete among each other to build and to operate their respective production facilities with the objective of maximizing profit. However, building electricity

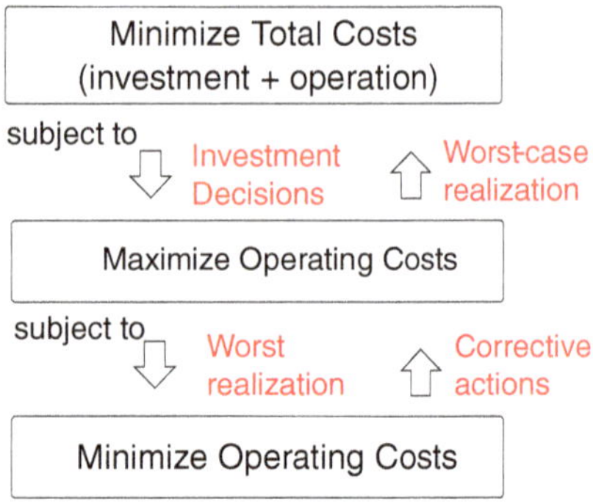

Fig. 1.6 Transmission reinforcement and expansion via ARO

production facilities is capital-intensive and plagued with a number of administrative "hurdles" (e.g., siting permissions, authorizations by different but overlapping administrative bodies with conflicting interests, and others), which highly complicates the actual construction of power units. As a result, and more often than not, different types of subsidies (e.g., a price guarantee or an increase in price over the market-clearing price) are available to encourage private investors to build electricity production facilities.

Among the array of technologies available to produce electricity, power units based on natural gas and renewable energy (solar- and wind-based) are nowadays most prominent [22]. Combined-cycle gas turbines, based on natural gas, are highly efficient, burn an environmentally friendly fossil fuel, and involve moderate investment and operational costs. Moreover, such units are highly flexible from the operational point of view and make possible the operation of systems even with a high level of penetration of stochastic renewable sources, such as wind- or solar-based power units. Mature renewable technologies include wind power, concentrated solar power, and photovoltaic power. Biomass units employ standard thermal-based technology, and thus they are technologically mature as well. Coal units with carbon capture and sequestration (CCS) are also considered as future alternatives as are a new generation of nuclear power units. Considering environmental impact, a possible and desirable (from the perspective of achieving a sustainable planet Earth) future generation mix will include natural gas units with high operating flexibility and renewable units including wind and concentrated solar units.

In deciding on investments in production facilities, two types of uncertain phenomena should be carefully modeled: short-term and long-term. Short-term uncertainty includes:

1. Demand uncertainty throughout the hours of the day and the days of the week.
2. Production uncertainty from stochastic (solar- and wind-based) units throughout the hours of the day.
3. Equipment failures including both production units and transmission lines.

Long-term uncertainty includes:

1. Yearly increase/decrease in demand.
2. Change in geographical demand distribution.
3. Yearly change in investment and operation costs.
4. Yearly change in the interest rate.

Mathematically, investment models in electricity production facilities are naturally formulated as optimization problems constrained by other optimization problems, which may also be reformulated as mathematical programs with equilibrium constraints (MPECs) [20]. A rather general instance of such a model is provided below:

$$\max_{\mathbf{x}} \; f(\mathbf{x}, \mathbf{y}^{(1)}, \ldots, \mathbf{y}^{(n)}, \boldsymbol{\lambda}^{(1)}, \ldots, \boldsymbol{\lambda}^{(n)}) \tag{1.4a}$$

s.t.

$$\mathbf{h}(\mathbf{x}, \mathbf{y}^{(1)}, \ldots, \mathbf{y}^{(n)}) = 0 \tag{1.4b}$$

$$\mathbf{g}(\mathbf{x}, \mathbf{y}^{(1)}, \ldots, \mathbf{y}^{(n)}) \leq 0 \tag{1.4c}$$

$$\left\{\begin{array}{l} \max_{\mathbf{y}^{(1)}} \; f^{(1)}(\mathbf{x}, \mathbf{y}^{(1)}, \ldots, \mathbf{y}^{(n)}) \\ \text{s.t.} \\ \quad \mathbf{h}^{(1)}(\mathbf{x}, \mathbf{y}^{(1)}, \ldots, \mathbf{y}^{(n)}) = 0 : \quad \boldsymbol{\lambda}^{(1)} \\ \quad \mathbf{g}^{(1)}(\mathbf{x}, \mathbf{y}^{(1)}, \ldots, \mathbf{y}^{(n)}) \leq 0 \end{array}\right. \tag{1.4d}$$

$$\vdots$$

$$\left\{\begin{array}{l} \max_{\mathbf{y}^{(n)}} \; f^{(n)}(\mathbf{x}, \mathbf{y}^{(1)}, \ldots, \mathbf{y}^{(n)}) \\ \text{s.t.} \\ \quad \mathbf{h}^{(n)}(\mathbf{x}, \mathbf{y}^{(1)}, \ldots, \mathbf{y}^{(n)}) = 0 : \quad \boldsymbol{\lambda}^{(n)} \\ \quad \mathbf{g}^{(n)}(\mathbf{x}, \mathbf{y}^{(1)}, \ldots, \mathbf{y}^{(n)}) \leq 0. \end{array}\right. \tag{1.4e}$$

Problem (1.4a)–(1.4c) is the upper-level problem, while problems (1.4d)–(1.4e) are the constraining lower-level problems. It is important to note that the objective functions of the upper-level problem and of the lower-level problems are different and generally conflicting; otherwise, problem (1.4) reduces to a single-level problem. The upper-level problem is the investment problem, which seeks to identify the investment decisions that maximize profit, while the lower-level problems represent market-clearing conditions. Vector $\mathbf{x}$ includes the set of optimization variables specifically belonging to the upper-level problem, i.e., investment decisions, while vector $\mathbf{y}^{(i)}$ includes the set of optimization variables of constraining lower-level problem (i), i.e., the market-clearing outcomes of market condition (i). Vector $\boldsymbol{\lambda}^{(i)}$ is the dual variable vector pertaining to the equality constraints of the constraining lower-level problem (i). Such a vector contains market-clearing prices.

Note the hierarchical structure of problem (1.4) above, i.e., an optimization problem (identifying the investment that results in maximum profit) constrained by its own constraints and a set of other optimization problems (market-clearing conditions). Note also that some dual variables, indicated following a colon after the

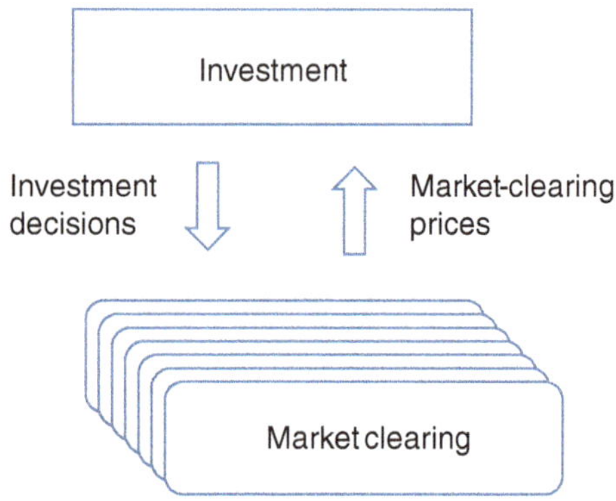

Fig. 1.7 Investment in generation facilities

corresponding constraints, of the constraining lower-level problems, which are clearing prices, affect the upper-level problem, but not the other way around. That is, the dual variables of the upper-level problem do not directly affect the lower-level problems. In a stochastic multistage decision-making framework, upper-level problem (1.4a)–(1.4c) should include *nonanticipativity* constraints, i.e., those that prevent anticipating future information.

Model (1.4) allows representation of strategic behavior from the power investor's point of view. This is realized by "manipulating" both offer power levels and offer prices, which are input to the lower-level problems (1.4d)–(1.4e), to alter market-clearing prices for the benefit of the producer, which are inputs to the upper-level problem (1.4a)–(1.4c). Model (1.4) also allows representation of the investment decision-making in stochastic (e.g., solar- and wind-based) power units. This requires including a large number of lower-level problems (1.4d)–(1.4e) representing many operational conditions describing different production levels and demand levels. Moreover, if a producer has a generating mix dominated by stochastic units, then model (1.4) can also be used to represent the investment strategies of such a strategic stochastic producer.

Figure 1.7 illustrates the investment problem in generation facilities.

1.5 Generation and Transmission Expansion Planning

Since investment in transmission facilities is the responsibility of the independent system operator (or the transmission system operator) and is carried out in the pursuit of social welfare maximization, while investment in generation facilities is a private profit-oriented endeavor, the following question arises: who is interested in the joint expansion planning of both generation and transmission facilities?

The regulator is the independent entity that pursues the harmonious expansion and reinforcement of both the transmission infrastructure and the generation facilities. Its objective is to maximize social welfare or to minimize social costs.

Regarding the transmission system, the outcomes of this planning exercise are transferred to the transmission system operator to be analyzed further and eventually implemented, while the outcomes pertaining to generation investment are provided to private investors as guiding plans. These generation investment guiding plans are often used by governmental agencies to design subsidies to promote investment in certain generation facilities, e.g., renewable power units.

A natural modeling approach to addressing investment problems in transmission and production facilities is stochastic programming [14]. Stochastic programming allows the representation of sequential decisions in time along with a precise description of uncertain phenomena and the unfolding of uncertainty over time. Moreover, in contrast to robust optimization, stochastic programming incorporates many of the future operating conditions of the transmission and production facilities under consideration, which is a requirement in making economically meaningful build/not-build decisions [7]. However, the size of a stochastic programming model grows with the number of scenarios, which may result in intractability, and a large number of scenarios is often needed to represent the uncertain parameters accurately. Therefore, it is important to reach an appropriate tradeoff between accuracy and computational complexity. Decomposition techniques [13] are helpful in achieving high modeling accuracy while preserving computational tractability.

With respect to a single investment point in time, the transmission and generation investment problem above can be represented by the two-stage *stochastic programming* problem below:

$$\begin{aligned} \min_{\mathbf{x}} \ & f^{\mathrm{I}}(\mathbf{x}) + \mathbb{E}_w\{z_w^{\mathrm{O}}\} \\ \text{s.t.} \ & \mathbf{h}^{\mathrm{I}}(\mathbf{x}) = \mathbf{0} : \boldsymbol{\lambda} \\ & \mathbf{g}^{\mathrm{I}}(\mathbf{x}) \leq \mathbf{0} \\ & \mathbf{x} \in \mathscr{X}, \end{aligned} \quad (1.5)$$

where

$$\begin{aligned} z_w^{\mathrm{O}} = \{\min_{\mathbf{y}_w} \ & f^{\mathrm{O}}(\mathbf{y}_w) \\ \text{s.t.} \ & \mathbf{h}_w^{\mathrm{O}}(\mathbf{x}, \mathbf{y}_w) = \mathbf{0} : \boldsymbol{\lambda}_w \\ & \mathbf{g}_w^{\mathrm{O}}(\mathbf{x}, \mathbf{y}_w) \leq \mathbf{0} \\ & \mathbf{y}_w \in \mathscr{Y}\} \ \forall w \in \mathscr{W}. \end{aligned} \quad (1.6)$$

Problem (1.5) seeks to minimize the total cost, including investment costs, $f^{\mathrm{I}}(\mathbf{x})$, pertaining to investment variables $\mathbf{x}$ in both transmission and generation facilities, and expected operational costs, $\mathbb{E}_w\{z_w^{\mathrm{S}}\}$, pertaining to operational variables $\mathbf{y}_w$, $\forall w$. Problem (1.6), which constrains problem (1.5), represents operational decisions per scenario $w \in \mathscr{W}$.

Under mild mathematical assumptions, the expectation and minimization operators can be interchanged in problems (1.5)–(1.6), thereby rendering the single-level problem below:

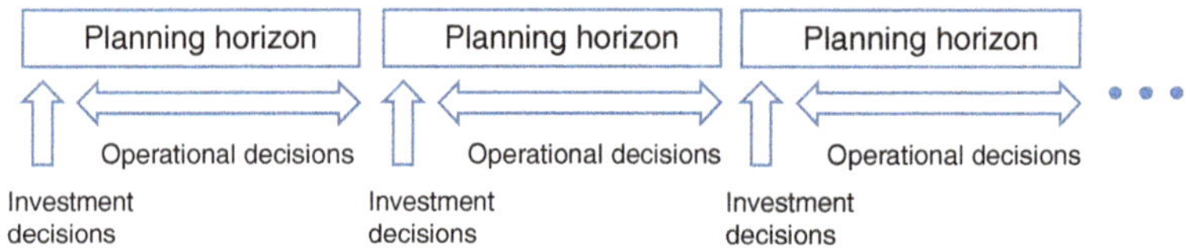

Fig. 1.8 Investment in both transmission and generation facilities

$$\begin{aligned}
&\min_{\mathbf{x};\mathbf{y}_w,\forall w} f^{\mathrm{I}}(\mathbf{x}) + \mathbb{E}_w\{f^{\mathrm{O}}(\mathbf{y}_w)\} \\
&\text{s.t.} \quad \mathbf{h}^{\mathrm{I}}(\mathbf{x}) = \mathbf{0} : \boldsymbol{\lambda} \\
&\quad \mathbf{g}^{\mathrm{I}}(\mathbf{x}) \le \mathbf{0} \\
&\quad \mathbf{x} \in \mathscr{X} \\
&\quad \mathbf{h}^{\mathrm{O}}_w(\mathbf{x}, \mathbf{y}_w) = \mathbf{0} : \boldsymbol{\lambda}_w, \ \forall w \in \mathscr{W} \\
&\quad \mathbf{g}^{\mathrm{O}}_w(\mathbf{x}, \mathbf{y}_w) \le \mathbf{0}, \ \forall w \in \mathscr{W} \\
&\quad \mathbf{y}_w \in \mathscr{Y}, \ \forall w \in \mathscr{W}.
\end{aligned} \tag{1.7}$$

In problems (1.5)–(1.7), the superscript "I" denotes the investment (first) stage, while the superscript "O" denotes the operation (second) stage. The subscript w denotes the scenario, and $\mathscr{W}$ represents the set of all possible scenarios. Constraints $\mathbf{h}^{\mathrm{I}}(\mathbf{x}) = \mathbf{0}$ and $\mathbf{g}^{\mathrm{I}}(\mathbf{x}) \le \mathbf{0}$ pertain to investment decisions, while constraints $\mathbf{h}^{\mathrm{O}}_w(\mathbf{x}, \mathbf{y}_w) = \mathbf{0}$ and $\mathbf{g}^{\mathrm{O}}_w(\mathbf{x}, \mathbf{y}_w) \le \mathbf{0}$ pertain to operational decisions in scenario w. Extending the above optimization framework to a multistage setting can be done in a straightforward manner, as explained in [14], but at the cost of greater computational complexity and potential intractability.

Figure 1.8 illustrates the investment problem in both transmission and generation facilities.

1.6 Investment Valuation and Timing

The investment decision pertaining to a particular facility, i.e., a power plant or a transmission line, can be deferred in time and may be subject to modification in its capacity. This happens frequently since power companies generally prefer to defer the refurbishment of existing production facilities or the building of new ones until it is strictly necessary. Moreover, they often try to identify the best point in time to undertake new construction or refurbishment.

Real options analysis (ROA) is a methodology tailored to this, i.e., to identifying the best time to undertake the building of a new facility or the refurbishment of an existing one, and/or to ascertain the best capacity (size) of such a facility. Moreover, ROA precisely represents the uncertainty plaguing this decision-making process.

ROA models are naturally formulated using a *dynamic programming* framework [17, 33], which has the form:

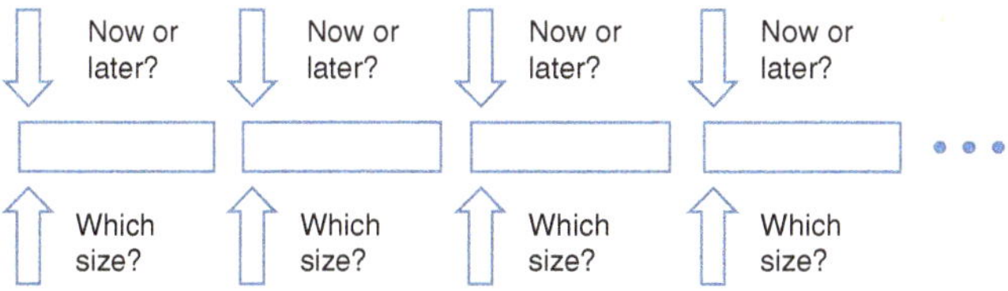

Fig. 1.9 Valuation and timing of investments

$$
\begin{aligned}
& f(\mathbf{s_0}) = \max_{\mathbf{x}_0,\mathbf{x}_1,\ldots} \quad \mathbb{E}_{\mathbf{u}} \left\{ \sum_{t=0}^{\infty} \beta^t g(\mathbf{s}_t, \mathbf{x}_t, \mathbf{u}) \right\} \\
& \text{s.t.} \\
& \mathbf{s}_{t+1} = \mathbf{h}(\mathbf{s}_t, \mathbf{x}_t, \mathbf{u}), \;\; \forall t = 0, 1, 2, \ldots \\
& \mathbf{x}_t \in \mathscr{X}(\mathbf{s}_t), \;\; \forall t = 0, 1, 2, \ldots
\end{aligned}
\tag{1.8}
$$

At time period t, vector $\mathbf{s}_t$ represents the state of the system, and $\mathbf{x}_t$ is the vector of investment decisions. Vector $\mathbf{u}$ contains the uncertain parameters. Function $\mathbf{h}(\mathbf{s}_t, \mathbf{x}_t, \mathbf{u})$ gives the state of the system at $t + 1$ if investment decision vector $\mathbf{x}_t$ is implemented at state $\mathbf{s}_t$, i.e., $\mathbf{s}_{t+1} = \mathbf{h}(\mathbf{s}_t, \mathbf{x}_t, \mathbf{u})$. The set of feasible investment decisions at state $\mathbf{s}_t$ is contained in set $\mathscr{X}(\mathbf{s}_t)$, which depends on the state $\mathbf{s}_t$, i.e., $\mathbf{x}_t \in \mathscr{X}(\mathbf{s}_t)$. Function $g(\mathbf{s}_t, \mathbf{x}_t, \mathbf{u})$ gives the profit at time period t for state $\mathbf{s}_t$ if investment decision $\mathbf{x}_t$ is implemented, while $f(\mathbf{s_0})$ is the discounted expected profit, where $\mathbf{s_0}$ is the initial state. Furthermore, $\mathbb{E}_{\mathbf{u}}\{\cdot\}$ is the expectation operator conditional on $\mathbf{u}$, and β is the discount factor. In addition to investment decisions, $\mathbf{x}_t$ can also include decisions on capacity sizing, operations, and abandonment of power plants. Thus, model (1.8) may be generalized to incorporate discretion over not only timing and sizing of the initial investment, but also subsequent operational levels and capacity modifications. Hence, (1.8) consists of nested (or compound) options, which may be valued quasi-analytically given assumptions about the underlying stochastic process and an infinite time horizon.

Figure 1.9 illustrates the valuation and timing of a specific investment project.

1.7 What We Do and What We Do Not Do

Investment decision-making requires identifying the best possible alternative plans using a suitable computational model and then testing those alternative plans, considering all practical details that are absent from the decision-making model but are relevant in practice. This book provides detailed computational models to identify alternative investments plans that are optimal with respect to the objectives and limitations of those models. This is what we specifically do in this book. Some of these models are reported as indicated below:

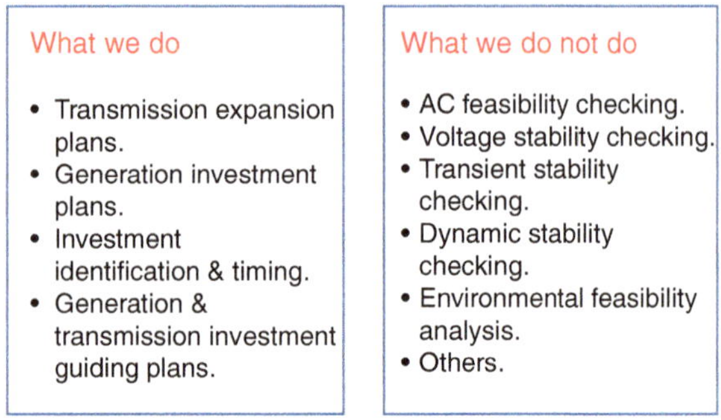

Fig. 1.10 What we do and what we do not do in this book

- Transmission reinforcement and expansion: [1, 2, 4, 12, 16, 21, 23, 35].
- Investment in conventional units: [24–28].
- Investment in renewable stochastic units: [3, 5, 6, 8].

In turn, alternative investment plans need to be tested for all the practical details not present in the models. This important second step, "testing for all the practical details not present in the models," is not addressed in this book. Thus, this is what we specifically do not do in this book. A description of this complementary analysis is provided in [15] for transmission expansion planning. In short, this book focuses on the development of mathematical models to produce investment plans that are optimal with respect to the objective of the models.

Regarding transmission investment planning, alternative investment plans in transmission lines need to be tested for their impact on system stability, voltage profile, and others factors [15]. Regarding investment in production facilities, the plans obtained need to be screened for environmental impact, appropriate location, network impact, and other factors.

Figure 1.10 illustrates what we do and what we do not do in this book.

1.8 End-of-Chapter Exercises

1.1 What are the main uncertainties that plague investment in electricity production units? Are all uncertainties of a long-term nature?

1.2 Why are electric energy investment problems large-scale? Do they include only continuous variables? Or do they also include binary variables? Do they include many constraints? If so, then why?

1.3 What is the main marketplace in which to trade electricity in most electricity markets around the world? Besides the main marketplace, list other other relevant markets for trading electricity.

1.4 Why would a mixed-integer linear programming model be needed to clear the day-ahead market? Can marginal prices be derived using such a clearing tool? If so, then how are the marginal prices obtained? Why might uplifts be needed?

1.5 Who is the entity in charge of reinforcing and expanding the transmission network within a given jurisdiction? What are the main differences between the US and the EU in this respect?

1.6 Why is adaptive robust optimization a good option for infrastructure expansion planning? Would it not be more appropriate to use stochastic programming? If not, then why?

1.7 Who is responsible for investing in power units? Has any public entity the obligation of investing in power units? How can we guarantee that enough generating capacity is added to the system? What is the role played by the regulator in ensuring that enough capacity is added to the system?

1.8 Why is a mathematical program with equilibrium constraints (MPEC) needed to represent decision-making problems pertaining to generation investment? What is the main difference between an optimization problem and an MPEC?

1.9 Which is the institution interested in the joint expansion of transmission and generation facilities? Why? Are the plans derived by such an institution directly implemented?

1.10 Why is stochastic programming appropriate for the joint expansion of the transmission and generation facilities? Does stochastic programming allow representation of multiple decision points in time? Are stochastic programming models easy to solve? If not, then why?

References

1. Aguado, J.A., de la Torre, S., Contreras, J., Conejo, A.J., Martinez, A.: Market-driven dynamic transmission expansion planning. Electric Power Syst. Res. **82**(1), 88–94 (2012)
2. Alguacil, N., Motto, A.L., Conejo, A.J.: Transmission expansion planning: a mixed-integer LP approach. IEEE Trans. Power Syst. **18**(3), 1070–1077 (2003)
3. Baringo, L., Conejo, A.J.: Wind power investment within a market environment. Appl. Energy **88**(9), 3239–3247 (2011)
4. Baringo, L., Conejo, A.J.: Transmission and wind power investment. IEEE Trans. Power Syst. **27**(2), 885–893 (2012)
5. Baringo, L., Conejo, A.J.: Wind power investment: a Benders' decomposition approach. IEEE Trans. Power Syst. **27**(1), 433–441 (2012)
6. Baringo, L., Conejo, A.J.: Risk-constrained multi-stage wind power investment. IEEE Trans. Power Syst. **28**(1), 401–411 (2013)
7. Baringo, L., Conejo, A.J.: Correlated wind-power production and electric load scenarios for investment decisions. Appl. Energy **101**, 475–482 (2013)
8. Baringo, L., Conejo, A.J.: Strategic wind power investment. IEEE Trans. Power Syst. **29**(3), 1250–1260 (2014)

9. Bertsimas, D., Brown, D.B., Caramanis, C.: Theory and applications of robust optimization. SIAM Rev. **53**(3), 464–501 (2011)
10. Bertsimas, D., Brown, D.B.: Constructing uncertainty sets for robust linear optimization. Oper. Res. **57**(6), 1483–1495 (2009)
11. Castillo, E., Conejo, A.J., Pedregal, P., Garcia, R., Alguacil, N.: Building and Solving Mathematical Programming Models in Engineering and Science. Wiley, New York (2001)
12. Chen, B., Wang, J., Wang, L., He, Y., Wang, Z.: Robust optimization for transmission expansion planning: minimax cost versus minimax regret. IEEE Trans. Power Syst. **29**(6), 3069–3077 (2014)
13. Conejo, A.J., Castillo, E., Minguez, R., Garcia-Bertrand, R.: Decomposition Techniques in Mathematical Programming. Engineering and Science Applications. Springer, Heidelberg (2006)
14. Conejo, A.J., Carrion, M., Morales, J.M.: Decision Making Under Uncertainty in Electricity Markets. Springer, New York (2010)
15. de Dios, R., Soto, F., Conejo, A.J.: Planning to expand? Looking at mainland Spain to see the importance of well-planned transmission expansion. IEEE Power Energy Mag. **5**(5), 64–70 (2007)
16. de la Torre, S., Conejo, A.J., Contreras, J.: Transmission expansion planning in electricity markets. IEEE Trans. Power Syst. **23**(1), 238–248 (2008)
17. Dixit, A.K., Pindyck, R.S.: Investment Under Uncertainty. Princeton University Press, Princeton (1994)
18. European network of transmission system operators for electricity, ENTSO-E (2015). www.entsoe.eu
19. Gabrel, V., Murat, C., Thiele, A.: Recent advances in robust optimization: an overview. Eur. J. Oper. Res. **235**(3), 471–483 (2014)
20. Gabriel, S., Conejo, A.J., Hobbs, B., Fuller, D., Ruiz, C.: Complementarity Modeling in Energy Markets. Springer, New York (2012)
21. Garces, L.P., Conejo, A.J., Garcia-Bertrand, R., Romero, R.: A bilevel approach to transmission expansion planning within a market environment. IEEE Trans. Power Syst. **24**(3), 1513–1522 (2009)
22. Gil, J., Caballero, A., Conejo, A.J.: Power cycling: CCGTs: the critical link between the electricity and natural gas markets. IEEE Power Energy Mag. **12**(6), 40–48 (2014)
23. Jabr, R.A.: Robust transmission network expansion planning with uncertain renewable generation and loads. IEEE Trans. Power Syst. **28**(4), 4558–4567 (2013)
24. Kazempour, S.J., Conejo, A.J., Ruiz, C.: Strategic generation investment using a complementarity approach. IEEE Trans. Power Syst. **26**(2), 940–948 (2011)
25. Kazempour, S.J., Conejo, A.J., Ruiz, C.: Strategic generation investment considering futures and spot markets. IEEE Trans. Power Syst. **27**(3), 1467–1476 (2012)
26. Kazempour, S.J., Conejo, A.J., Ruiz, C.: Generation investment equilibria with strategic producers—Part I: formulation. IEEE Trans. Power Syst. **28**(3), 2613–2622 (2013)
27. Kazempour, S.J., Conejo, A.J., Ruiz, C.: Generation investment equilibria with strategic producers—Part II: case studies. IEEE Trans. Power Syst. **28**(3), 2623–2631 (2013)
28. Kazempour, S.J., Conejo, A.J.: Strategic generation investment under uncertainty via Benders' decomposition. IEEE Trans. Power Syst. **27**(1), 424–432 (2012)
29. Midcontinent Independent System Operator, MISO (2015). www.misoenergy.org
30. Morales, J.M., Conejo, A.J., Perez-Ruiz, J.: Economic valuation of reserves in power systems with high penetration of wind power. IEEE Trans. Power Syst. **24**(2), 900–910 (2009)
31. Morales, J.M., Conejo, A.J., Madsen, H., Pinson, P., Zugno, M.: Integrating Renewables in Electricity Markets. Springer, New York (2014)
32. PJM Interconnection, PJM (2015). www.pjm.com
33. Powell, W.B.: Approximate Dynamic Programming: Solving the Curses of Dimensionality, 2nd edn. Wiley, New York (2011)
34. Ruiz, C., Conejo, A.J., Gabriel, S.A.: Pricing non-convexities in an electricity pool. IEEE Trans. Power Syst. **27**(3), 1334–1342 (2012)

35. Ruiz, C., Conejo, A.J.: Robust transmission expansion planning. Eur. J. Oper. Res. **242**(2), 390–401 (2015)
36. Soyster, A.: Convex programming with set-inclusive constraints and applications to inexact linear programming. Oper. Res. **21**(5), 1154–1157 (1973)

Chapter 2
Transmission Expansion Planning

A critical issue in the operation of electric energy systems is the capacity of the transmission lines that enable energy flows from generation nodes to demand nodes. In this chapter, we analyze the transmission expansion planning (TEP) problem, which allows a transmission planner to identify the optimal transmission reinforcements to be carried out with the aim of facilitating energy exchange among producers and consumers, e.g., by reducing generation or load-shedding costs. With this purpose, two models are described and analyzed: first, a deterministic model that solves the TEP problem based on a future demand forecast, and second, an adaptive robust optimization (ARO) model that takes into account the influence of different sources of uncertainty, such as future demand growth and the availability of generating units in the TEP problem. These two models are formulated using a static approach in which transmission expansion plans are made at a single point in time and for a future planning horizon.

2.1 Introduction

This chapter analyzes the transmission expansion planning (TEP) problem, which refers to the decision-making problem of determining the best transmission reinforcements to be made in an existing electric energy system. TEP is motivated, among other reasons, by the aging of the current infrastructure [20], expected demand growth, and the building of new renewable production facilities, usually located far away from demand centers. These issues make it essential to reinforce and to expand the existing transmission network in order to facilitate energy exchanges among producers and consumers as well as to guarantee supply–demand balance.

TEP is a critical issue in modern electric energy systems because transmission lines allow energy flows from generating to demand nodes. These demands must be supplied even in the worst situations, e.g., those corresponding to a peak load or the failure of a generating unit. Thus, transmission expansion plans should be decided in such a way that demands are efficiently supplied even if one of these situations

A.J. Conejo et al., *Investment in Electricity Generation and Transmission*,
DOI 10.1007/978-3-319-29501-5_2

occurs. Note that the transmission expansion has twin objectives: on the one hand, facilitating economic trade, and on the other hand, engineering reliability.

The TEP problem is generally tackled under two different frameworks: centralized and competitive. In a centralized framework, an entity controls both generation and transmission facilities and is in charge of performing generation and transmission planning jointly. In a competitive environment, an independent and regulated entity is generally in charge of operating and expanding the transmission network with the aim of maximizing energy trade opportunities among producers and consumers. There is a third option that consists in considering the scope for a merchant investor that is profit-motivated in expanding the transmission system.

The first view, i.e., a central planner deciding both generation and transmission expansion, is analyzed in Chap. 4 of this book. Here, we adopt the second view, i.e., we consider that a single regulated entity decides the transmission expansion plans. The existence of this entity, known as the transmission system operator (TSO), is common in most European countries [14, 25]. In the US, TSOs have comparatively more limited attributes, and they are usually in charge of a specific region and therefore, are referred to as regional transmission organizations (RTOs) [19, 24].

The TSO (or RTO) decides the best way to reinforce and expand the existing transmission network with the goal of facilitating energy trade opportunities of the market players, e.g., by minimizing generation costs or by reducing load-shedding costs. Note that a joint economic and engineering objective is considered. By expanding the transmission network we reduce the generation and load-shedding costs (economic objective), but at the same time, we improve the reliability in the supply of demands (engineering objective).

The relevance of the TEP problem in electric energy systems has motivated significant research effort in this area over the past few decades. Pioneering work is due to Garver [12], who in 1970 proposed a linear programming problem that determines the transmission expansion plans based on the location of overloads. Since then, many relevant contributions have been published based on mathematical programming [1, 21, 23].

TEP is a complex decision-making problem since it generally involves a multiattribute objective, nonlinear constraints, and a nonconvex feasible region. As a result, different approaches have been proposed to deal with this complexity. These approaches are based on the application of decomposition techniques such as Benders decomposition [5, 18, 26, 27] or on the use of heuristics [5, 9, 22, 28].

An important observation is that the TEP problem is usually considered for a long-term planning horizon. When the TSO decides about the transmission expansion plan to be carried out, it should take into account the future demand growth, the availability of existing generating units, and the building of new generating facilities. This means that transmission expansion decisions are made within an uncertain environment, and such uncertainties must be properly represented in order to achieve informed expansion decisions. To do so, stochastic programming [11, 30] and robust optimization (RO) [7, 16, 29] have been used in a TEP context. On the one hand, stochastic programming is based on the generation of scenarios that describe the uncertain parameters [8]. However, this scenario generation usually requires knowing

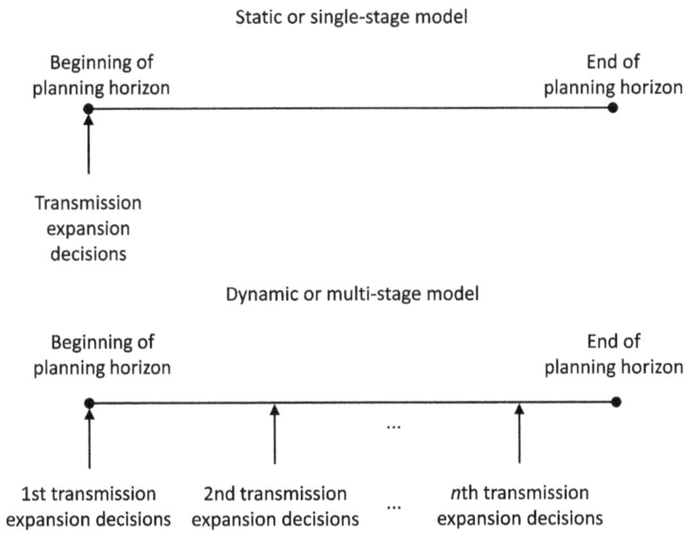

Fig. 2.1 Static and dynamic TEP models

the probability distribution functions of the uncertain parameters, which is generally a hard task. Moreover, a large enough number of scenarios must be generated to represent the uncertainty accurately, which increases the computational complexity of the problem. On the other hand, RO does not need scenarios to be generated but robust sets, which are generally simpler to obtain [2]. Additionally, RO models have a moderate size, which reduces the computational complexity compared with stochastic programming models. A general disadvantage of RO is that the results are usually too conservative. However, this is not a disadvantage for the TEP problem, in which a reliable supply of demands is required.

As previously mentioned, the TEP problem is generally considered for a long-term planning horizon, e.g., 30 years. In this sense, there are two expansion strategies. The first is to make the transmission expansion plans, i.e., to build the new transmission lines, at a single point in time (usually at the beginning of the planning horizon). The model that results from this strategy is known as a static or a single-stage model. The second one is to make the transmission expansion decisions at different points in time of the planning horizon. In this case, the model is known as a dynamic or a multistage model. This dynamic approach usually provides more accurate solutions since it allows the transmission planner to adapt to future changes in the system. However, it further increases the complexity of the TEP problem. Figure 2.1 illustrates the differences between these two expansion strategies. For the sake of simplicity, in this chapter we focus on a static approach. The use of dynamic models for making expansion decisions is described and analyzed in the following chapters of this book.

The remainder of this chapter is organized as follows. Section 2.2 provides a TEP-problem model considering a deterministic approach, in which transmission expansion decisions are made considering a future demand forecast. In this case, the transmission expansion plan is determined so that the transmission network is capable of dealing with the worst expected demand realization in the future. The model in Sect. 2.2 is extended in Sect. 2.3 to consider the uncertainties faced by the TSO in carrying out the TEP-problem exercise. The model is formulated in this case using an adaptive robust optimization (ARO) approach. Both Sects. 2.2 and 2.3 include clarifying examples. Section 2.4 summarizes the chapter and discusses the main conclusions of the models and results reported in the chapter. Section 2.5 proposes some exercises to enable a deeper understanding of the models and concepts described in the chapter. Finally, Sect. 2.6 includes the GAMS code for one of the illustrative examples.

2.2 Deterministic Approach

The transmission infrastructure is a critical point in electric energy systems. Demands should be supplied even in the worst possible situations, e.g., during a peak demand period or during the failure of an important generating unit. Thus, transmission expansion plans should take into account such situations. For the sake of simplicity, we consider as the worst case a single-load scenario that corresponds to the maximum load demand expected in the planning horizon for which the TEP analysis is carried out. This assumption is usually made in the technical literature [1, 5, 18, 31].

In order to formulate the TEP problem, it is necessary to use binary variables to model whether a prospective transmission line is built. In systems of hundreds or thousands of nodes, there is a large number of transmission expansion options that need to be considered. This requires the use of a very large number of binary variables, which increases the complexity of the problem. To avoid formulating a very complex problem, we consider a static approach in which TEP decisions are made now for a future long-term planning horizon, e.g., 20 years. This assumption is typical in TEP problems because it allows us to formulate a comparatively easier problem to solve. However, note that under a deterministic assumption, it is also generally possible to formulate the TEP problem considering a dynamic approach.

The problem is formulated from the perspective of a TSO that aims at facilitating energy trade opportunities among producers and consumers. Therefore, the objective function of the TSO's problem is the minimization of generation and load-shedding costs. Since the TSO is responsible for building transmission lines, we also include their construction costs in the objective function.

The following sections provide the formulation of the TEP problem under a deterministic approach. This problem can be formulated as a mixed-integer nonlinear programming (MINLP) problem since it includes the products of continuous and binary variables. These kinds of problems are generally hard to solve, and the convergence of the MINLP problem to the optimum is not guaranteed [6]. However, it is possible to formulate an exact equivalent mixed-integer linear programming (MILP) problem, which allows us to solve the TEP problem by applying conventional branch-and-cut solvers [6, 15].

2.2.1 Notation

The main notation used in this chapter is provided below for quick reference. Other symbols are defined as needed throughout the chapter.

Indices

d Demands.
g Generating units.
ℓ Transmission lines.
n Nodes.
ν Iterations.

Sets

$r(\ell)$ Receiving-end node of transmission line ℓ.
$s(\ell)$ Sending-end node of transmission line ℓ.
Ω_n^{D} Demands located at node n.
Ω_n^{E} Generating units located at node n.
$\Omega^{\mathrm{L}+}$ Prospective transmission lines.

Parameters

B_ℓ Susceptance of transmission line ℓ [S].
C_d^{LS} Load-shedding cost of demand d [\$/MWh].
C_g^{E} Production cost of generating unit g [\$/MWh].
$F_\ell^{\max}$ Capacity of transmission line ℓ [MW].
$\tilde{I}_\ell^{\mathrm{L}}$ Annualized investment cost of prospective transmission line ℓ [\$/MW].
$\tilde{I}^{\mathrm{L,max}}$ Annualized investment budget for building prospective transmission lines [\$].
$P_d^{\mathrm{D^{max}}}$ Load of demand d [MW].
$\underline{P}_d^{\mathrm{D^{max}}}$ Lower bound of the load of demand d [MW].
$\overline{P}_d^{\mathrm{D^{max}}}$ Upper bound of the load of demand d [MW].
$P_g^{\mathrm{E^{max}}}$ Production capacity of generating unit g [MW].
$\overline{P}_g^{\mathrm{E^{max}}}$ Upper bound of the production capacity of generating unit g [MW].
Γ^{D} Uncertainty budget for load demand.
Γ^{G} Uncertainty budget for production capacity.

Binary Variables

x_ℓ^{L} Binary variable that is equal to 1 if prospective transmission line ℓ is built and 0 otherwise.

Continuous Variables

p_g^{E} Power produced by generating unit g [MW].
p_ℓ^{L} Power flow through transmission line ℓ [MW].
p_d^{LS} Load shed by demand d [MW].

η Auxiliary variable to reconstruct objective function of the ARO problem gradually [$].

θ_n Voltage angle at node n [rad].

2.2.2 MINLP Model Formulation

The deterministic TEP problem can be formulated as the following MINLP problem:

$$\min_{\Delta} \quad \sum_{\ell \in \Omega^{L+}} \tilde{I}_{\ell}^{L} x_{\ell}^{L} + \sigma \left[\sum_{g} C_{g}^{E} p_{g}^{E} + \sum_{d} C_{d}^{LS} p_{d}^{LS} \right] \tag{2.1a}$$

subject to

$$\sum_{\ell \in \Omega^{L+}} \tilde{I}_{\ell}^{L} x_{\ell}^{L} \leq \tilde{I}^{L,\max} \tag{2.1b}$$

$$x_{\ell}^{L} = \{0, 1\} \quad \forall \ell \in \Omega^{L+} \tag{2.1c}$$

$$\sum_{g \in \Omega_n^E} p_g^E - \sum_{\ell | s(\ell)=n} p_{\ell}^L + \sum_{\ell | r(\ell)=n} p_{\ell}^L = \sum_{d \in \Omega_n^D} \left(P_d^{D^{\max}} - p_d^{LS} \right) \quad \forall n \tag{2.1d}$$

$$p_l^L = B_{\ell} \left(\theta_{s(\ell)} - \theta_{r(\ell)} \right) \quad \forall \ell \setminus \ell \in \Omega^{L+} \tag{2.1e}$$

$$p_l^L = x_{\ell}^L B_{\ell} \left(\theta_{s(\ell)} - \theta_{r(\ell)} \right) \quad \forall \ell \in \Omega^{L+} \tag{2.1f}$$

$$-F_{\ell}^{\max} \leq p_{\ell}^L \leq F_{\ell}^{\max} \quad \forall \ell \tag{2.1g}$$

$$0 \leq p_g^E \leq P_g^{E^{\max}} \quad \forall g \tag{2.1h}$$

$$0 \leq p_d^{LS} \leq P_d^{D^{\max}} \quad \forall d \tag{2.1i}$$

$$-\pi \leq \theta_n \leq \pi \quad \forall n \tag{2.1j}$$

$$\theta_n = 0 \quad n\text{: ref.}, \tag{2.1k}$$

where variables in set $\Delta = \left\{ x_{\ell}^L, p_{\ell}^L, p_g^G, p_d^D, \theta_n \right\}$ are the optimization variables of problem (2.1).

The aim of the TSO is to facilitate energy trading and at the same time, to minimize the costs incurred in building new transmission lines. On the other hand, the TSO is constrained by the requirements for maintaining grid reliability.

Therefore, objective function (2.1a) comprises the three terms below:

1. $\sum_{\ell \in \Omega^{L+}} I_{\ell}^L x_{\ell}^L$ is the annualized cost of building new transmission lines.
2. $\sum_{g} C_g^E p_g^E$ is the operating cost of generating units.
3. $\sum_{d} C_d^{LS} p_d^{LS}$ is the load-shedding cost.

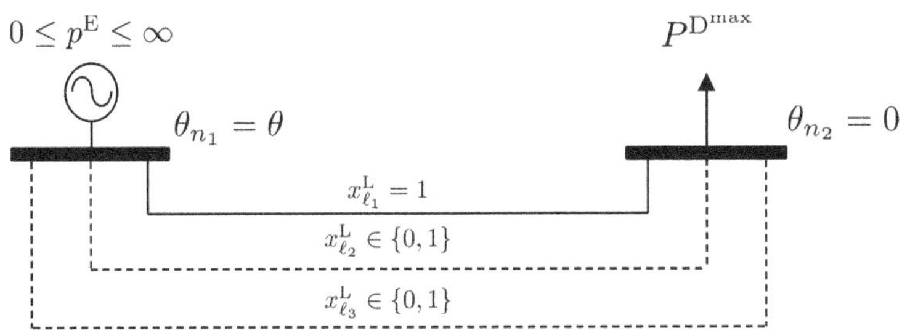

Fig. 2.2 Illustrative Example 2.1: two-node system (one stage)

The terms in 2 and 3 above are multiplied by the factor σ to make them comparable with investment costs. Since we consider that $\tilde{I}^{\mathrm{L}}_{\ell}$ are annualized investment costs, then σ is equal to 8760 h, i.e., the total number of hours in a year.

Objective function (2.1a) is constrained by Eqs. (2.1b)–(2.1k). Constraint (2.1b) ensures that the cost of building new transmission lines is below the available budget when it exists. Constraints (2.1c) define binary variables x^{L}_{ℓ} that indicate whether a prospective line is built ($x^{\mathrm{L}}_{\ell} = 1$) or not ($x^{\mathrm{L}}_{\ell} = 0$). Constraints (2.1d) impose the generation–demand balance at each node of the system. Equations (2.1e) and (2.1f) define the power flows through existing and prospective transmission lines, respectively, which are limited by the corresponding capacity limits by constraints (2.1g). Note that subscripts $s(\ell)$ and $r(\ell)$ denote the sending-end and receiving-end nodes of transmission line ℓ, respectively. Constraints (2.1h) and (2.1i) impose bounds for the power produced by generating units and the unserved demand, respectively. Finally, constraints (2.1j) and (2.1k) impose bounds on voltage angles and fix to zero the voltage angle at the reference node, respectively.

The network is represented using a dc model without losses. This assumption is usually made in the technical literature and consists in considering that voltage magnitudes are approximately constant in the system and that voltage angle differences are small enough between two connected nodes [13]. This allows us to formulate the power-flow Eqs. (2.1e)–(2.1f) using linear expressions.

Illustrative Example 2.1 *Deterministic TEP: Static solution*

The deterministic TEP model (2.1) is applied to the two-node system depicted in Fig. 2.2. There is a generating unit located at node 1. Its production cost is C, and it has a very large capacity (for the sake of simplicity, we consider that its capacity is infinite). There is also a demand located at node 2 with a peak load equal to $P^{\mathrm{D}^{\max}}$ and a very large load-shedding cost (also considered infinite, i.e., load shedding is not possible).

Nodes 1 and 2 are connected through a transmission line ℓ_1 of capacity equal to $F^{\max}$ and susceptance equal to 1 p.u. It is possible to build two additional transmission lines (ℓ_2 and ℓ_3) between these nodes with the same characteristics as the existing one and an annualized investment cost equal to I with an unlimited budget. Node 2 is the reference node.

Considering the above data, model (2.1) results in the following optimization problem:

$$\min_{x^{\mathrm{L}}_{\ell_2}, x^{\mathrm{L}}_{\ell_3}, p^{\mathrm{E}}, \theta} \quad I x^{\mathrm{L}}_{\ell_2} + I x^{\mathrm{L}}_{\ell_3} + \sigma C p^{\mathrm{E}}$$

subject to

$$\begin{aligned}
& x^{\mathrm{L}}_{\ell_2}, x^{\mathrm{L}}_{\ell_3} = \{0, 1\} \\
& p^{\mathrm{E}} - p^{\mathrm{L}}_{\ell_1} - p^{\mathrm{L}}_{\ell_2} - p^{\mathrm{L}}_{\ell_3} = 0 \\
& p^{\mathrm{L}}_{\ell_1} + p^{\mathrm{L}}_{\ell_2} + p^{\mathrm{L}}_{\ell_3} = P^{\mathrm{D}^{\max}} \\
& p^{\mathrm{L}}_{\ell_1} = \theta \\
& p^{\mathrm{L}}_{\ell_2} = x^{\mathrm{L}}_{\ell_2} \theta \\
& p^{\mathrm{L}}_{\ell_3} = x^{\mathrm{L}}_{\ell_3} \theta \\
& - F^{\max} \leq p^{\mathrm{L}}_{\ell_1} \leq F^{\max} \\
& - F^{\max} \leq p^{\mathrm{L}}_{\ell_2} \leq F^{\max} \\
& - F^{\max} \leq p^{\mathrm{L}}_{\ell_3} \leq F^{\max} \\
& 0 \leq p^{\mathrm{E}} \leq \infty \\
& - \pi \leq \theta \leq \pi .
\end{aligned}$$

Note that if $0 \leq P^{\mathrm{D}^{\max}} \leq F^{\max}$, the capacity of the existing transmission line allows the power to flow from the generating unit at node 1 to the demand at node 2. Therefore, we do not need to build any additional transmission lines. However, for values of $P^{\mathrm{D}^{\max}} > F^{\max}$, the capacity of the existing transmission line ℓ_1 is not enough. If $F^{\max} < P^{\mathrm{D}^{\max}} \leq 2F^{\max}$, then it is necessary to build one additional transmission line between nodes 1 and 2, while if $2F^{\max} < P^{\mathrm{D}^{\max}} \leq 3F^{\max}$, then we need to build two additional transmission lines. Note that for $P^{\mathrm{D}^{\max}} > 3F^{\max}$, the problem is infeasible, and additional expansion options must be considered.

That is, the optimal solution of the TEP problem for this illustrative example depends on the value of $P^{\mathrm{D}^{\max}}$: we build zero, one, or two additional transmission lines depending on the value of the expected largest demand in the planning horizon. □

Illustrative Example 2.2 *Deterministic TEP: Dynamic solution*

In Illustrative Example 2.1, we solve the TEP problem considering a static approach, in which transmission expansion plans are decided and made now for a future planning horizon. For illustration purposes, let us now consider that this planning horizon is divided into two time periods, t_1 and t_2. The expected largest demands in these two time periods are different and equal to $P_{t_1}^{\mathrm{D^{max}}}$ and $P_{t_2}^{\mathrm{D^{max}}}$, respectively. Instead of considering a static approach, we consider that the transmission planner can build additional lines at the beginning of each of the two considered time periods, i.e., we consider a dynamic approach. Figure 2.3 illustrates the TEP problem in this case. Note that the remaining data are obtained from Illustrative Example 2.1.

Considering the above data, the TEP problem considering a dynamic approach results in the following optimization problem:

$$\min_{x_{\ell_2 t_1}^{\mathrm{L}}, x_{\ell_3 t_1}^{\mathrm{L}}, x_{\ell_2 t_2}^{\mathrm{L}}, x_{\ell_3 t_2}^{\mathrm{L}}, p_{t_1}^{\mathrm{E}}, p_{t_2}^{\mathrm{E}}, \theta_{t_1}, \theta_{t_2}} \quad 2 I x_{\ell_2 t_1}^{\mathrm{L}} + 2 I x_{\ell_3 t_1}^{\mathrm{L}} + I x_{\ell_2 t_2}^{\mathrm{L}} + I x_{\ell_3 t_2}^{\mathrm{L}} + \sigma C \left(p_{t_1}^{\mathrm{E}} + p_{t_2}^{\mathrm{E}} \right)$$

subject to

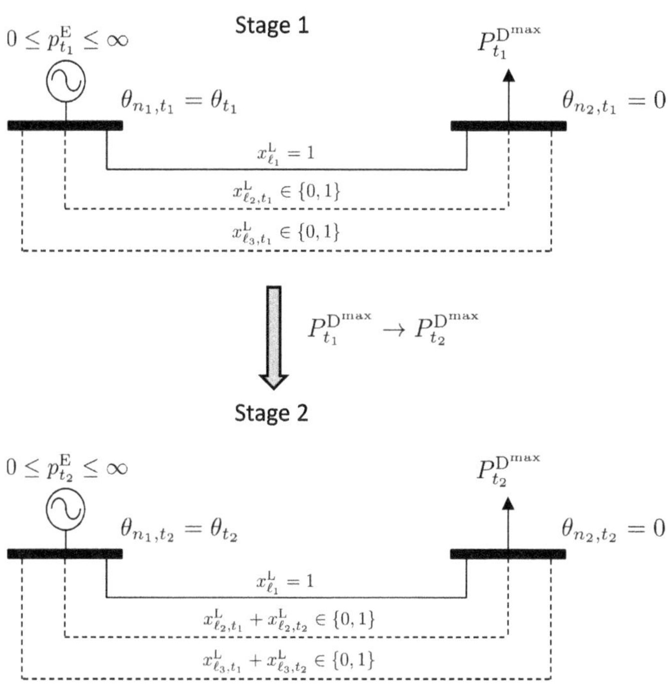

Fig. 2.3 Illustrative Example 2.2: two-node system (two stages)

$$
\begin{aligned}
& x^{\mathrm{L}}_{\ell_2 t_1}, x^{\mathrm{L}}_{\ell_3 t_1}, x^{\mathrm{L}}_{\ell_2 t_2}, x^{\mathrm{L}}_{\ell_3 t_2} = \{0, 1\} \\
& x^{\mathrm{L}}_{\ell_2 t_1} + x^{\mathrm{L}}_{\ell_2 t_2} \leq 1 \\
& x^{\mathrm{L}}_{\ell_3 t_1} + x^{\mathrm{L}}_{\ell_3 t_2} \leq 1 \\
& p^{\mathrm{E}}_{t_1} - p^{\mathrm{L}}_{\ell_1 t_1} - p^{\mathrm{L}}_{\ell_2 t_1} - p^{\mathrm{L}}_{\ell_3 t_1} = 0 \\
& p^{\mathrm{L}}_{\ell_1 t_1} + p^{\mathrm{L}}_{\ell_2 t_1} + p^{\mathrm{L}}_{\ell_3 t_1} = P^{\mathrm{D^{max}}}_{t_1} \\
& p^{\mathrm{L}}_{\ell_1 t_1} = \theta_{t_1} \\
& p^{\mathrm{L}}_{\ell_2 t_1} = x^{\mathrm{L}}_{\ell_2 t_1} \theta_{t_1} \\
& p^{\mathrm{L}}_{\ell_3 t_1} = x^{\mathrm{L}}_{\ell_3 t_1} \theta \\
& - F^{\max} \leq p^{\mathrm{L}}_{\ell_1 t_1} \leq F^{\max} \\
& - F^{\max} \leq p^{\mathrm{L}}_{\ell_2 t_1} \leq F^{\max} \\
& - F^{\max} \leq p^{\mathrm{L}}_{\ell_3 t_1} \leq F^{\max} \\
& 0 \leq p^{\mathrm{E}}_{t_1} \leq \infty \\
& - \pi \leq \theta_{t_1} \leq \pi \\
& p^{\mathrm{E}}_{t_2} - p^{\mathrm{L}}_{\ell_1 t_2} - p^{\mathrm{L}}_{\ell_2 t_2} - p^{\mathrm{L}}_{\ell_3 t_2} = 0 \\
& p^{\mathrm{L}}_{\ell_1 t_2} + p^{\mathrm{L}}_{\ell_2 t_2} + p^{\mathrm{L}}_{\ell_3 t_2} = P^{\mathrm{D^{max}}}_{t_2} \\
& p^{\mathrm{L}}_{\ell_1 t_2} = \theta_{t_2} \\
& p^{\mathrm{L}}_{\ell_2 t_2} = \left(x^{\mathrm{L}}_{\ell_2 t_1} + x^{\mathrm{L}}_{\ell_2 t_2}\right) \theta_{t_2} \\
& p^{\mathrm{L}}_{\ell_3 t_2} = \left(x^{\mathrm{L}}_{\ell_3 t_1} + x^{\mathrm{L}}_{\ell_3 t_2}\right) \theta_{t_2} \\
& - F^{\max} \leq p^{\mathrm{L}}_{\ell_1 t_2} \leq F^{\max} \\
& - F^{\max} \leq p^{\mathrm{L}}_{\ell_2 t_2} \leq F^{\max} \\
& - F^{\max} \leq p^{\mathrm{L}}_{\ell_3 t_2} \leq F^{\max} \\
& 0 \leq p^{\mathrm{E}}_{t_2} \leq \infty \\
& - \pi \leq \theta_{t_2} \leq \pi .
\end{aligned}
$$

The main differences between this model and that used in Illustrative Example 2.1 are as follows:

1. The subscripts t_1 and t_2 are used to denote the values of variables and parameters at time periods 1 and 2, respectively.
2. Prospective transmission lines built in the first time period remain built in the second time period.
3. Prospective transmission lines can be built only once.
4. The investment costs of transmission lines built in the first time period are twice the investment costs of transmission lines built in the second one since they are used (and thus amortized) twice.
5. Constraints are considered for both time periods.

As in Illustrative Example 2.1, the optimal transmission expansion plan depends on the value of $P_{t_1}^{D^{max}}$ and $P_{t_2}^{D^{max}}$:

1. In the first time period:
    - If $0 \le P_{t_1}^{D^{max}} \le F^{max}$, then we build no prospective transmission line.
    - If $F^{max} < P_{t_1}^{D^{max}} \le 2F^{max}$, then we build one additional transmission line.
    - If $2F^{max} < P_{t_1}^{D^{max}} \le 3F^{max}$, then we build two additional transmission lines.
2. In the second time period, the expansion plan depends on $P_{t_2}^{D^{max}}$ and also on the expansion plan considered in the first time period. Assuming that $P_{t_2}^{D^{max}} \ge P_{t_1}^{D^{max}}$ (which is generally true since the demand growth in a system is usually positive), the transmission expansion plan in the second time period is as follows:
    - If no prospective transmission line is built in the first time period, then we build no additional transmission line if $0 \le P_{t_2}^{D^{max}} \le F^{max}$, we build one additional transmission line if $F^{max} < P_{t_2}^{D^{max}} \le 2F^{max}$, and we build two additional transmission lines if $2F^{max} < P_{t_2}^{D^{max}} \le 3F^{max}$.
    - If one prospective transmission line is built in the first time period, then we build no additional transmission line if $F^{max} < P_{t_2}^{D^{max}} \le 2F^{max}$, and we build one additional transmission line if $2F^{max} < P_{t_2}^{D^{max}} \le 3F^{max}$.
    - If two prospective transmission lines are built in the first time period, then no additional transmission line is built.
3. If $P_{t_1}^{D^{max}} > 3F^{max}$ or $P_{t_2}^{D^{max}} > 3F^{max}$, then the problem is infeasible, and additional expansion options should be considered.

Note that Illustrative Example 2.2 is comparatively more complex than Illustrative Example 2.1 since the number of binary variables and constraints is approximately twice. However, considering a dynamic approach as in Illustrative Example 2.2, we can adapt to future changes in the system, and therefore, its solution is usually better than that obtained considering a static approach. □

Illustrative Examples 2.1 and 2.2 are very simple, and their solutions are trivial. However, as the number of nodes of the system under study and the number of prospective transmission lines increase, the TEP problem becomes complex, and its solution is no longer trivial. Note also that problem (2.1) includes binary variables, as well as products of binary and continuous variables in constraints (2.1f), i.e., problem (2.1) is an MINLP problem. These problems are usually hard to solve. Nevertheless, it is possible to remove the nonlinearities, as explained in the following section.

2.2.3 Linearization of Products of Binary and Continuous Variables

MINLP model (2.1) provided in the previous section is nonlinear, due to the products of binary (x_ℓ) and continuous (θ_n) variables in constraints (2.1f). However, it is possible to replace these nonlinear constraints by the following sets of exact equivalent mixed-integer linear constraints:

$$-x_\ell^{\mathrm{L}} F_\ell^{\max} \le p_\ell^{\mathrm{L}} \le x_\ell^{\mathrm{L}} F_\ell^{\max} \quad \forall \ell \tag{2.2a}$$

$$-\left(1-x_\ell^{\mathrm{L}}\right) M \le p_\ell^{\mathrm{L}} - B_\ell \left(\theta_{s(\ell)} - \theta_{r(\ell)}\right) \le \left(1-x_\ell^{\mathrm{L}}\right) M \quad \forall \ell, \tag{2.2b}$$

where M is a large enough positive constant [5, 31].

The working of Eqs. (2.2) is explained below.

On the one hand, let us consider that prospective transmission line ℓ is built, i.e., binary variable x_ℓ^{L} is equal to 1. In such a case, Eqs. (2.2) impose that $-F_\ell^{\max} \le p_\ell^{\mathrm{L}} \le F_\ell^{\max}$ and $p_\ell^{\mathrm{L}} - B_\ell \left(\theta_{s(\ell)} - \theta_{r(\ell)}\right) = 0$. Note that these equations are equivalent to constraints (2.1f) and (2.1g) when $x_\ell^{\mathrm{L}} = 1$.

On the other hand, let us consider that prospective transmission line ℓ is not built, i.e., binary variable x_ℓ^{L} is equal to 0. In such a case, Eqs. (2.2) impose that $p_\ell^{\mathrm{L}} = 0$ and $-M \le p_\ell^{\mathrm{L}} - B_\ell \left(\theta_{s(\ell)} - \theta_{r(\ell)}\right) \le M$. First, we impose that the power flow through this transmission line is null. Second, we consider large enough bounds on the difference between the voltage angles at two nodes that are not connected through the disjunctive parameter M. These equations are equivalent to constraints (2.1f) and (2.1g) when $x_\ell^{\mathrm{L}} = 0$. The interested reader is referred to [5, 31], which provide a discussion on how to select the value of parameter M effectively.

2.2.4 MILP Model Formulation

Using the linearization procedure described in the previous section, it is possible to reformulate the TEP problem considering a deterministic static approach as in the MILP problem below:

$$\min_{\Delta} \quad \sum_{\ell \in \Omega^{\mathrm{L}+}} \tilde{I}_\ell^{\mathrm{L}} x_\ell^{\mathrm{L}} + \sigma \left[\sum_g C_g^{\mathrm{E}} p_g^{\mathrm{E}} + \sum_d C_d^{\mathrm{LS}} p_d^{\mathrm{LS}} \right] \tag{2.3a}$$

subject to

$$\sum_{\ell \in \Omega^{\mathrm{L}+}} \tilde{I}_\ell^{\mathrm{L}} x_\ell^{\mathrm{L}} \le \tilde{I}^{\mathrm{L,max}} \tag{2.3b}$$

$$x_\ell^{\mathrm{L}} = \{0, 1\} \quad \forall \ell \in \Omega^{\mathrm{L}+} \tag{2.3c}$$

$$\sum_{g\in\Omega_n^{\mathrm{G}}} p_g^{\mathrm{G}} - \sum_{\ell|s(\ell)=n} p_\ell^{\mathrm{L}} + \sum_{\ell|r(\ell)=n} p_\ell^{\mathrm{L}} = \sum_{d\in\Omega_n^{\mathrm{D}}} \left(P_d^{\mathrm{D^{max}}} - p_d^{\mathrm{LS}}\right) \quad \forall n \tag{2.3d}$$

$$p_\ell^{\mathrm{L}} = B_\ell \left(\theta_{s(\ell)} - \theta_{r(\ell)}\right) \quad \forall \ell \setminus \ell \in \Omega^{\mathrm{L+}} \tag{2.3e}$$

$$-F_\ell^{\max} \le p_\ell^{\mathrm{L}} \le F_\ell^{\max} \quad \forall \ell \setminus \ell \in \Omega^{\mathrm{L+}} \tag{2.3f}$$

$$-x_\ell^{\mathrm{L}} F_\ell^{\max} \le p_\ell^{\mathrm{L}} \le x_\ell^{\mathrm{L}} F_\ell^{\max} \quad \forall \ell \in \Omega^{\mathrm{L+}} \tag{2.3g}$$

$$-\left(1 - x_\ell^{\mathrm{L}}\right) M \le p_\ell^{\mathrm{L}} - B_\ell \left(\theta_{s(\ell)} - \theta_{r(\ell)}\right) \le \left(1 - x_\ell^{\mathrm{L}}\right) M \quad \forall \ell \in \Omega^{\mathrm{L+}} \tag{2.3h}$$

$$0 \le p_g^{\mathrm{E}} \le P_g^{\mathrm{G^{max}}} \quad \forall g \tag{2.3i}$$

$$0 \le p_d^{\mathrm{LS}} \le P_d^{\mathrm{D^{max}}} \quad \forall d \tag{2.3j}$$

$$-\pi \le \theta_n \le \pi \quad \forall n \tag{2.3k}$$

$$\theta_n = 0 \quad n\text{: ref.} \tag{2.3l}$$

Illustrative Example 2.3 *Deterministic TEP: Six-node system*

The deterministic TEP model (2.3) is applied to the six-node system depicted in Fig. 2.4. This system comprises six nodes, five generating units, four demands, and three transmission lines. The system is divided in two zones: region A (nodes 1–3) and region B (nodes 4–6), which are initially not connected. Note also that node six is initially isolated, and thus the demand at this node can be supplied only by generating unit g_5.

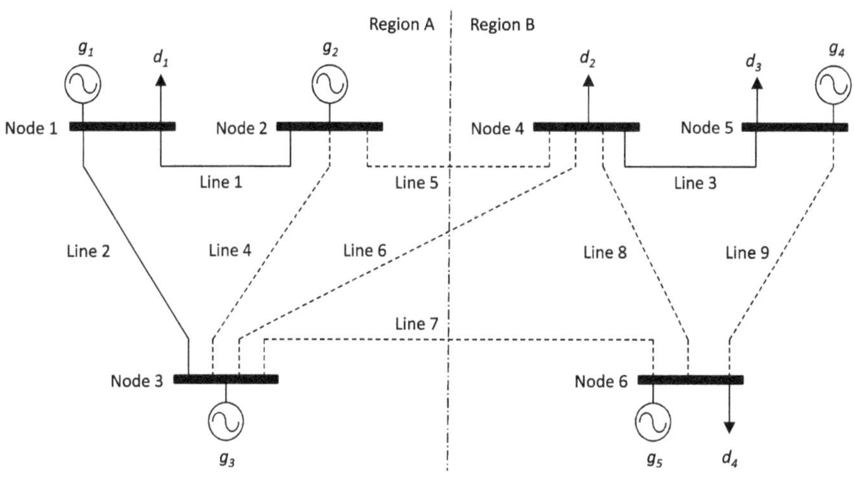

Fig. 2.4 Illustrative Example 2.3: six-node system

Table 2.1 Illustrative Example 2.3: data for generating units

Generating unit	Node	$P_g^{\mathrm{E^{max}}}$ [MW]	C_g^{E} [\$/MWh]
g_1	n_1	300	18
g_2	n_2	250	25
g_3	n_3	400	16
g_4	n_5	300	32
g_5	n_6	150	35

Table 2.2 Illustrative Example 2.3: data for demands

Demand	Node	$P_d^{\mathrm{D^{max}}}$ [MW]	C_d^{LS} [\$/MWh]
d_1	n_1	200	40
d_2	n_4	150	52
d_3	n_5	100	55
d_4	n_6	200	65

Table 2.3 Illustrative Example 2.3: data for existing transmission lines

Line	From node	To node	B_ℓ [S]	$F_\ell^{\max}$ [MW]
ℓ_1	n_1	n_2	500	150
ℓ_2	n_1	n_3	500	150
ℓ_3	n_4	n_5	500	150

Table 2.1 provides data for the generating units. The second column identifies the node allocation, while the third and fourth columns provide the capacity and production cost of each generating unit, respectively.

Table 2.2 provides data for the demands. The second column identifies the node allocation, while the third and fourth columns provide the maximum load demand and the load-shedding cost of each demand, respectively. Note that the optimal expansion plan is obtained based on a future demand forecast. Thus, these demands represent the worst realization of the demand at each node in the future, which corresponds to the largest expected demand at each node in the considered planning horizon.

Table 2.3 provides data for the existing transmission lines. The second and third columns identify the sending-end and receiving-end nodes, respectively, while the fourth and fifth columns provide the susceptance and capacity of each existing transmission line, respectively.

We consider that it is possible to build six different transmission lines, whose data are provided in Table 2.4. The second and third columns identify the sending-end and receiving-end nodes, respectively, while the fourth and fifth columns provide the susceptance and capacity of each prospective transmission line, respectively. The sixth column gives the annualized investment cost. The annualized investment budget is considered equal to \$3,000,000, which limits the number and type of prospective lines to be built.

Table 2.4 Illustrative Example 2.3: data for prospective transmission lines

Line	From node	To node	B_ℓ [S]	$F_\ell^{\max}$ [MW]	$\tilde{I}_\ell^{\mathrm{L}}$ [\$]
ℓ_4	n_2	n_3	500	150	700,000
ℓ_5	n_2	n_4	500	200	1,400,000
ℓ_6	n_3	n_4	500	200	1,800,000
ℓ_7	n_3	n_6	500	200	1,600,000
ℓ_8	n_4	n_6	500	150	800,000
ℓ_9	n_5	n_6	500	150	700,000

Investment costs and investment budget are provided as annualized values. Therefore, factor σ is equal to 8760 (i.e., the number of hours in a year) to make the annualized costs and the load-shedding/generating costs comparable.

Finally, the reference node is node 1, the base power is 1 MW, and the base voltage is 1 kV.

The above data are used to solve the TEP problem (2.3). The optimal solution consists in building prospective lines ℓ_5 and ℓ_7, i.e., lines connecting nodes 2–4 and 3–6, respectively. Note that the system is divided into two regions, A and B, which are originally independent. While most of the generation capacity is located in region A, most of the demand is located in region B. Moreover, the cheapest generating units are located in region A. This means that without building any prospective line, demands in region B are supplied by expensive generating units, and load-shedding occurs, resulting in load-shedding costs. Therefore, it is optimal to build prospective lines ℓ_5 and ℓ_7, which connect regions A and B, and thus part of the demand in region B can be supplied by the cheap generating units in region A, which contributes to reducing generation and load-shedding costs. □

Illustrative Example 2.4 *Deterministic TEP: Impact of investment budget*

The expansion decisions obtained in Illustrative Example 2.3 are conditioned by the available investment budget. Note that building transmission lines ℓ_5 and ℓ_7 requires an annualized investment cost of \$3,000,000, i.e., the considered investment budget. A larger investment budget may allow the TSO, i.e., the planner, to reduce further the generation and load-shedding costs, as analyzed next.

The lower plot of Fig. 2.5 depicts the investment and load-shedding costs, while the upper plot of Fig. 2.5 depicts the generation and the total, i.e., the value of objective function (2.3a), costs for different values of the investment budget. On the other hand, the expansion decisions for different investment budgets are provided in Table 2.5.

For values of the annualized investment budget lower than \$700,000, it is impossible to build any transmission lines since the cheapest transmission line has an annualized cost of \$700,000. In such a case, demands in region B are supplied by expensive generating units, and/or load-shedding occurs with the corresponding load-shedding cost, as can be observed in the lower plot of Fig. 2.5. As the investment

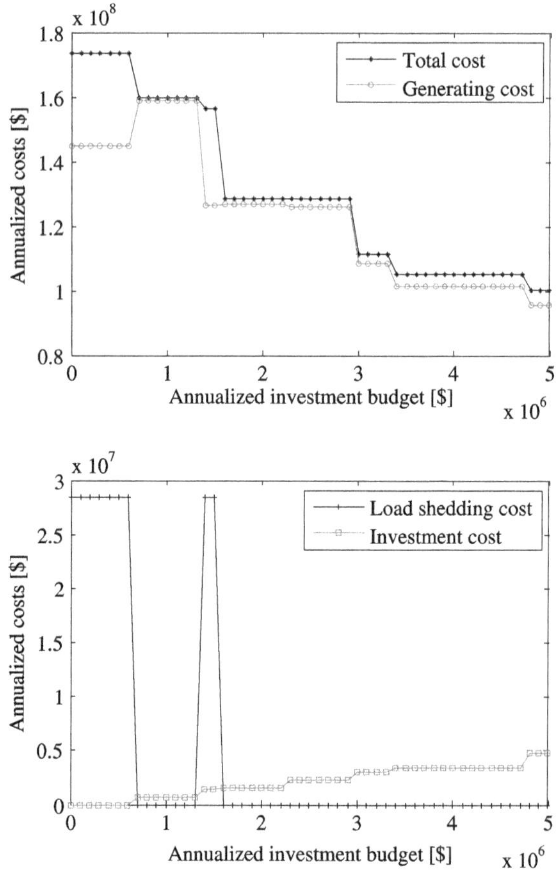

Fig. 2.5 Illustrative Example 2.4: costs versus investment budget

budget increases, it becomes possible to build some of the prospective transmission lines. Moreover, building some of the prospective transmission lines is most appropriate since this contributes to reducing the total costs by reducing the generation and/or the load-shedding costs. Note that for some values of the investment budget, the cost of load-shedding increases, and the generation cost decreases, i.e., load-shedding substitutes generation, and vice versa. Note also that the cost of building new transmission lines is significantly lower than the generation costs. Finally, we observe that there is no incremental progression in the building of new transmission lines as the investment budget increases, i.e., the optimal solution of the TEP problem does not consist in building additional transmission lines as the investment budget increases, but in considering different transmission expansion plans. □

Table 2.5 Illustrative Example 2.4: expansion decisions for different investment budgets

Annualized investment budget [M$]	Transmission lines built[a]
[0, 0.7)	–
[0.7, 1.4)	ℓ_9
[1.4, 1.6)	ℓ_5
[1.6, 2.3)	ℓ_7
[2.3, 3.0)	ℓ_4 and ℓ_7
[3.0, 3.4)	ℓ_5 and ℓ_7
[3.4, 4.8)	ℓ_6 and ℓ_7
[4, ∞)	ℓ_5, ℓ_6 and ℓ_7

[a] ℓ_4: 2–3, ℓ_5: 2–4, ℓ_6: 3–4, ℓ_7: 3–6, ℓ_8: 4–6, ℓ_9: 5–6

Table 2.6 Illustrative Example 2.5: expansion decisions for different values of the total demand in the system

Total demand in the system [MW]	Transmission lines built[a]
100	ℓ_5 and ℓ_9
200	ℓ_7 and ℓ_9
More than 300	ℓ_5 and ℓ_7

[a] ℓ_4: 2–3, ℓ_5: 2–4, ℓ_6: 3–4, ℓ_7: 3–6, ℓ_8: 4–6, ℓ_9: 5–6

Illustrative Example 2.5 *Deterministic TEP: Impact of demand*

Illustrative Example 2.3 considers a deterministic approach in which expansion decisions are obtained for a given value of the demand. The problem is solved for the worst case, i.e., the largest expected demand in the considered planning horizon. However, expansion decisions are conditioned by this demand level, as can be observed in Table 2.6, which provides the transmission expansion decisions for different values of the demand in the system and a fixed annualized investment budget of $3,000,000.

We observe that no matter what the demand in the system is, it is optimal to connect regions A and B using some of the prospective transmission lines. Moreover, initially isolated node 6 is always connected to the system since the generating unit and demand at this node are the most expensive (in terms of generation and load-shedding costs, respectively). However, note that there is no incremental progression in the expansion decisions.

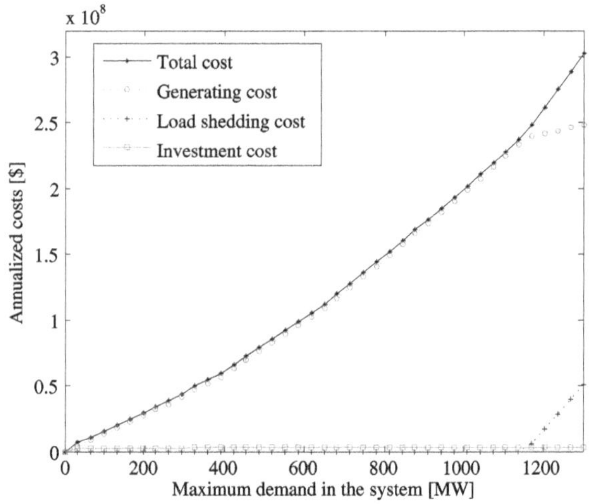

Fig. 2.6 Illustrative Example 2.5: costs versus total demand

On the other hand, as the future demand forecast increases, expansion decisions change, as do the investment, load-shedding, and generation costs as well. These costs for different values of the total demand in the system are depicted in Fig. 2.6. □

2.3 Robust Approach

In the previous section, we described a deterministic model in which the TEP is parameterized based on the largest future demand forecast. We assume that we know this demand as well as the remaining data of the system, i.e., given all the information, it is possible to find the optimal transmission expansion plan that minimizes objective function (2.3a). However, the future is uncertain. It is difficult to forecast the maximum demand in the future if the planning horizon is long enough. Moreover, there are other sources of uncertainty that could contribute to the so-called *worst case*, e.g., availability of generating units, availability of transmission lines, or newly built generating units. Thus, it is necessary to consider an appropriate approach that takes into account the influence of these uncertainties on decision-making in the TEP problem. This is analyzed in this section.

As explained in the introductory chapter, there are generally two ways of handling this uncertainty. The first is to use stochastic programming, which requires generating scenarios of uncertain parameters whose probability distribution functions we need [8], which are generally not available. Additionally, stochastic programming usually leads to computationally complex problems. Therefore, we use in this chapter the second option, which is based on RO [2]. RO allows us to represent the uncertain parameters through robust sets, which are generally easier to obtain than probability distribution functions. Moreover, RO models are comparatively less complex than stochastic programming models.

2.3.1 Adaptive Robust Optimization Formulation

For the sake of simplicity, we assume that the uncertain parameters in the TEP problem (2.3) are (i) demand–load levels and (ii) available generating resources, i.e., we assume that uncertainty affects only parameters $P_d^{\mathrm{D}^{\max}}$, $\forall d$, and $P_g^{\mathrm{E}^{\max}}$, $\forall g$, respectively. Given this, we aim to determine the optimal transmission expansion plan, i.e., the optimal values of variables x_ℓ^{L}, $\forall \ell$, which optimizes objective function (2.3a), but anticipating the worst possible realizations of the uncertain parameters. To do so, we formulate an ARO problem, whose main characteristics are summarized below:

1. The optimal transmission expansion plan is sought by minimizing objective function (2.3a).
2. This optimal transmission expansion plan is sought by anticipating that once transmission expansion decisions are made, the worst uncertainty case will occur, i.e., assuming a given transmission expansion plan, uncertain parameters will take the values that maximize objective function (2.3a).
3. The worst-case realization of uncertain parameters is considered by anticipating that once this worst case is realized, the system adapts to it. That is, assuming that the transmission expansion decisions and uncertain parameters are fixed, we select the optimal values of the remaining variables (i.e., the operating decision variables) that minimize objective function (2.3a).

Note that the above decision sequence is consistent with reality. First, the transmission planner (the TSO) decides the transmission expansion plan to be implemented. Then, a *worst case* occurs, e.g., an unexpected peak demand in the system and/or the failure of some generating units. Finally, the system operator decides the best actions in order to minimize the operating costs.

The hierarchical structure described above can be represented using the three-level optimization problem below:

$$\min_{x_\ell^{\mathrm{L}}} \quad \sum_{\ell \in \Omega^{\mathrm{L}+}} \tilde{I}_\ell^{\mathrm{L}} x_\ell^{\mathrm{L}}$$

$$+ \max_{P_d^{\mathrm{D}^{\max}}, P_g^{\mathrm{E}^{\max}} \in \Xi} \quad \min_{\Delta \backslash x_\ell^{\mathrm{L}} \in \Omega(x_\ell^{\mathrm{L}}, P_d^{\mathrm{D}^{\max}}, P_g^{\mathrm{E}^{\max}})} \quad \sigma \left[\sum_g C_g^{\mathrm{E}} p_g^{\mathrm{G}} + \sum_d C_d^{\mathrm{LS}} p_d^{\mathrm{LS}} \right] \tag{2.4a}$$

subject to

$$\sum_{\ell \in \Omega^{\mathrm{L}+}} I_\ell^{\mathrm{L}} x_\ell^{\mathrm{L}} \leq I^{\mathrm{L,max}} \tag{2.4b}$$

$$x_\ell^{\mathrm{L}} = \{0, 1\} \quad \forall \ell \in \Omega^{\mathrm{L}+} \tag{2.4c}$$

In problem (2.4) we include set Ξ, which defines the uncertainty set and set $\Omega\left(x_\ell^{\mathrm{L}}, P_d^{\mathrm{D}^{\max}}, P_g^{\mathrm{E}^{\max}}\right)$, which ensures the feasibility of the operating decision variables given the expansion decisions and the realizations of the uncertain parameters. Further details of these two sets are provided in the sections below.

2.3.2 Definition of Uncertainty Sets

In order to represent the uncertainty that appears in the TEP problem effectively, it is necessary to have an accurate definition of the uncertainty set Ξ. To do so, we consider a polyhedral uncertainty set, such as the one used in [2, 3, 29]. This uncertainty set is characterized by the following equations:

$$P_g^{\mathrm{E}^{\max}} \in \left[0, \overline{P}_g^{\mathrm{E}^{\max}}\right] \quad \forall g \tag{2.5a}$$

$$\frac{\sum_g \left(\overline{P}_g^{\mathrm{E}^{\max}} - P_g^{\mathrm{E}^{\max}}\right)}{\sum_g \overline{P}_g^{\mathrm{E}^{\max}}} \leq \Gamma^{\mathrm{G}} \tag{2.5b}$$

$$P_d^{\mathrm{D}^{\max}} \in \left[\underline{P}_d^{\mathrm{D}^{\max}}, \overline{P}_d^{\mathrm{D}^{\max}}\right] \quad \forall d \tag{2.5c}$$

$$\frac{\sum_d \left(P_d^{\mathrm{D}^{\max}} - \underline{P}_d^{\mathrm{D}^{\max}}\right)}{\sum_d \left(\overline{P}_d^{\mathrm{D}^{\max}} - \underline{P}_d^{\mathrm{D}^{\max}}\right)} \leq \Gamma^{\mathrm{D}}. \tag{2.5d}$$

Constraints (2.5a) and (2.5c) impose upper and lower bounds for $P_g^{\mathrm{E}^{\max}}$ and $P_d^{\mathrm{D}^{\max}}$, respectively. We consider that the lower bound of the available generating capacity is zero to represent the uncertainty in the availability of a generating unit or the uncertainty in building new generating units. On the other hand, constraints (2.5b) and (2.5d) limit the variability of uncertain variables $P_g^{\mathrm{E}^{\max}}$ and $P_d^{\mathrm{D}^{\max}}$, respectively, through the so-called uncertainty budgets Γ^{G} and Γ^{D}, as explained below.

Uncertainty budget Γ^{G} can take values between 0 and 1. If Γ^{G} is chosen equal to 0, then $P_g^{\mathrm{G}^{\max}} = \overline{P}_g^{\mathrm{E}^{\max}}$, $\forall g$, i.e., uncertainty in the available capacity of generating units is not considered. On the other hand, if Γ^{G} is chosen equal to 1, then $P_g^{\mathrm{E}^{\max}}$, $\forall g$, can take any value within the interval $\left[0, \overline{P}_g^{\mathrm{E}^{\max}}\right]$. This can be seen as the case of maximum uncertainty. Similarly, Γ^{D} can also take values between 0 and 1. If Γ^{D} is chosen equal to 0, then $P_d^{\mathrm{D}^{\max}} = \underline{P}_d^{\mathrm{D}^{\max}}$, $\forall d$, i.e., uncertainty in demand is not considered. However, if Γ^{D} is chosen equal to 1, then $P_d^{\mathrm{D}^{\max}}$, $\forall d$, can take any value within the interval $\left[\underline{P}_d^{\mathrm{D}^{\max}}, \overline{P}_d^{\mathrm{D}^{\max}}\right]$. By choosing different values of Γ^{G} and Γ^{D}, we can analyze the impact of different levels of uncertainty on transmission expansion decisions.

2.3.3 Feasibility of Operating Decision Variables

Given the expansion decision variables, x_ℓ^{L}, and the realizations of the uncertain parameters, $P_d^{\mathrm{D}^{\max}}$ and $P_g^{\mathrm{E}^{\max}}$, we define set $\Omega\left(x_\ell^{\mathrm{L}}, P_d^{\mathrm{D}^{\max}}, P_g^{\mathrm{E}^{\max}}\right)$ to ensure the feasibility of the operating decision variables $\Delta \setminus x_\ell^{\mathrm{L}}$ as follows:

$$\Omega\left(x_\ell^{\mathrm{L}}, P_d^{\mathrm{D}^{\max}}, P_g^{\mathrm{E}^{\max}}\right) = \{\Delta \setminus x_\ell^{\mathrm{L}} :$$

$$\sum_{g \in \Omega_n^{\mathrm{E}}} p_g^{\mathrm{E}} - \sum_{\ell | s(\ell)=n} p_\ell^{\mathrm{L}} + \sum_{\ell | r(\ell)=n} p_\ell^{\mathrm{L}} = \sum_{d \in \Omega_n^{\mathrm{D}}} \left(P_d^{\mathrm{D}^{\max}} - p_d^{\mathrm{LS}}\right) \ : \lambda_n \quad \forall n \tag{2.6a}$$

$$p_\ell^{\mathrm{L}} = B_\ell\left(\theta_{s(\ell)} - \theta_{r(\ell)}\right) \ : \phi_\ell^{\mathrm{L}} \quad \forall \ell \setminus \ell \in \Omega^{\mathrm{L}+} \tag{2.6b}$$

$$p_\ell^{\mathrm{L}} = x_\ell^{\mathrm{L}} B_\ell\left(\theta_{s(\ell)} - \theta_{r(\ell)}\right) \ : \phi_\ell^{\mathrm{L}+} \quad \forall \ell \in \Omega^{\mathrm{L}+} \tag{2.6c}$$

$$- F_\ell^{\max} \le p_\ell^{\mathrm{L}} \le F_\ell^{\max} \ : \phi_\ell^{\mathrm{L,min}}, \phi_\ell^{\mathrm{L,max}} \quad \forall \ell \tag{2.6d}$$

$$0 \le p_g^{\mathrm{E}} \le P_g^{\mathrm{E}^{\max}} \ : \phi_g^{\mathrm{E,min}}, \phi_g^{\mathrm{E,max}} \quad \forall g \tag{2.6e}$$

$$0 \le p_d^{\mathrm{LS}} \le P_d^{\mathrm{D}^{\max}} \ : \phi_d^{\mathrm{D,min}}, \phi_d^{\mathrm{D,max}} \quad \forall d \tag{2.6f}$$

$$- \pi \le \theta_n \le \pi \ : \phi_n^{\mathrm{N,min}}, \phi_n^{\mathrm{N,max}} \quad \forall n \tag{2.6g}$$

$$\theta_n = 0 \ : \chi^{\mathrm{ref}} \quad n\text{: ref.} \, \} . \tag{2.6h}$$

Note that the dual variable associated to each constraint is provided following a colon.

Equations (2.6a) ensures the generation–demand balance at each node of the system. Equations (2.6b) and (2.6c) define the power flows through existing and prospective transmission lines, respectively, which are bounded by the corresponding capacity limits by Eqs. (2.6d). Equations (2.6e) and (2.6f) impose bounds on the power of generating units and on the load-shedding, respectively. Finally, Eqs. (2.6g) and (2.6h) define bounds on voltage angles and fix to zero the voltage angle at the reference node, respectively.

2.3.4 Detailed Formulation

Given the definitions of the uncertainty sets and the operating feasibility region provided in the previous sections, the TEP problem considering an ARO approach can be formulated using the following model:

$$\min_{x_\ell^{\mathrm{L}}} \quad \max_{P_g^{\mathrm{E^{max}}}, P_d^{\mathrm{D^{max}}}} \quad \min_{p_g^{\mathrm{E}}, p_d^{\mathrm{D}}, p_\ell^{\mathrm{L}}, \theta_n} \quad \sum_{\ell \in \Omega^{\mathrm{L+}}} \tilde{I}_\ell^{\mathrm{L}} x_\ell^{\mathrm{L}} + \sigma \left[\sum_g C_g^{\mathrm{E}} p_g^{\mathrm{E}} + \sum_d C_d^{\mathrm{LS}} p_d^{\mathrm{LS}} \right] \tag{2.7a}$$

subject to

$$\sum_{g \in \Omega_n^{\mathrm{E}}} p_g^{\mathrm{E}} - \sum_{\ell | s(\ell)=n} p_\ell^{\mathrm{L}} + \sum_{\ell | r(\ell)=n} p_\ell^{\mathrm{L}} = \sum_{d \in \Omega_n^{\mathrm{D}}} \left(P_d^{\mathrm{D^{max}}} - p_d^{\mathrm{LS}} \right) \quad \forall n \tag{2.7b}$$

$$p_\ell^{\mathrm{L}} = B_\ell \left(\theta_{s(\ell)} - \theta_{r(\ell)} \right) \quad \forall \ell \setminus \ell \in \Omega^{\mathrm{L+}} \tag{2.7c}$$

$$p_\ell^{\mathrm{L}} = x_\ell^{\mathrm{L}} B_\ell \left(\theta_{s(\ell)} - \theta_{r(\ell)} \right) \quad \forall \ell \in \Omega^{\mathrm{L+}} \tag{2.7d}$$

$$-F_\ell^{\max} \le p_\ell^{\mathrm{L}} \le F_\ell^{\max} \quad \forall \ell \tag{2.7e}$$

$$0 \le p_g^{\mathrm{E}} \le P_g^{\mathrm{E^{max}}} \quad \forall g \tag{2.7f}$$

$$0 \le p_d^{\mathrm{LS}} \le P_d^{\mathrm{D^{max}}} \quad \forall d \tag{2.7g}$$

$$-\pi \le \theta_n \le \pi \quad \forall n \tag{2.7h}$$

$$\theta_n = 0 \quad n\text{: ref.} \tag{2.7i}$$

subject to

$$P_g^{\mathrm{E^{max}}} \in \left[0, \overline{P}_g^{\mathrm{E^{max}}} \right] \quad \forall g \tag{2.7j}$$

$$\frac{\sum_g \left(\overline{P}_g^{\mathrm{E^{max}}} - P_g^{\mathrm{E^{max}}} \right)}{\sum_g \overline{P}_g^{\mathrm{E^{max}}}} \le \Gamma^{\mathrm{G}} \tag{2.7k}$$

$$P_d^{\mathrm{D^{max}}} \in \left[\underline{P}_d^{\mathrm{D^{max}}}, \overline{P}_d^{\mathrm{D^{max}}} \right] \quad \forall d \tag{2.7l}$$

$$\frac{\sum_d \left(P_d^{\mathrm{D^{max}}} - \underline{P}_d^{\mathrm{D^{max}}} \right)}{\sum_d \left(\overline{P}_d^{\mathrm{D^{max}}} - \underline{P}_d^{\mathrm{D^{max}}} \right)} \le \Gamma^{\mathrm{D}} \tag{2.7m}$$

subject to

$$\sum_{\ell \in \Omega^{\mathrm{L+}}} I_\ell^{\mathrm{L}} x_\ell^{\mathrm{L}} \le I^{\mathrm{L,max}} \tag{2.7n}$$

$$x_\ell^{\mathrm{L}} = \{0, 1\} \quad \forall \ell \in \Omega^{\mathrm{L+}}. \tag{2.7o}$$

Note that the above problem (2.7) has a three-level structure whose three optimization problems include as optimization variables the expansion decision variables x_ℓ^{L}, the worst realizations of uncertain parameters $P_g^{\mathrm{E^{max}}}$, $P_d^{\mathrm{D^{max}}}$, and the operating decision variables p_g^{E}, p_d^{D}, p_ℓ^{L}, θ_n, respectively. Constraints (2.7b)–(2.7i) define the feasibility of the operating decision variables. Constraints (2.7j)–(2.7m) define the uncertainty sets. Finally, constraints (2.7n) and (2.7o) define the investment budget and the binary variables that represent which transmission lines are built, respectively.

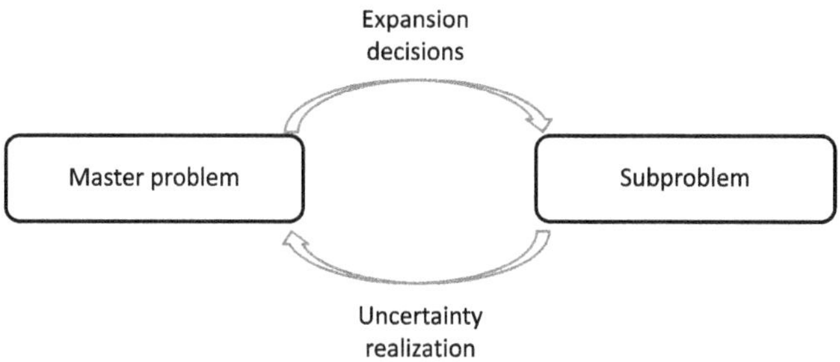

Fig. 2.7 Adaptive robust optimization approach: interactions between master problem and subproblem

2.3.5 *Solution Procedure*

The ARO (2.7) is difficult to solve since its multilevel structure renders an NP-hard problem. In order to solve this kind of problem, several algorithms are available in the technical literature based on extended versions of Benders decomposition [4, 16, 17] and on constraint-and-column generation methods [17, 29, 32]. In Benders-based methods, dual information from the so-called subproblem is used to build the objective function of the so-called master problem gradually. On the other hand, the constraint-and-column generation methods use cutting-plane strategies based solely on primal cuts that involve only primal decision variables. These methods generally perform computationally better than Benders' methods.

The three-level optimization problem (2.7) is decomposed into a master problem and a subproblem that exchange information on primal decision variables and that are iteratively solved until convergence to an optimal solution is achieved. Figure 2.7 schematically represents the interactions between these two problems.

The sections below provide detailed formulations of the master problem and the subproblem.

2.3.5.1 Master Problem

Considering a Benders framework, the master problem associated with problem (2.7) is formulated below:

$$\min_{x_\ell^{\mathrm{L}}, p^{\mathrm{G}}_{g,v'}, p^{\mathrm{D}}_{d,v'}, p^{\mathrm{L}}_{\ell,v'}, \theta_{n,v'}, \eta} \quad \sum_{\ell \in \Omega^{\mathrm{L}+}} \tilde{I}_\ell^{\mathrm{L}} x_\ell^{\mathrm{L}} + \eta \tag{2.8a}$$

subject to

$$\sum_{\ell \in \Omega^{\mathrm{L}+}} I_\ell^{\mathrm{L}} x_\ell^{\mathrm{L}} \leq I^{\mathrm{L,max}} \tag{2.8b}$$

$$x_\ell^{\mathrm{L}} = \{0, 1\} \quad \forall \ell \in \Omega^{\mathrm{L}+} \tag{2.8c}$$

$$\sum_{g \in \Omega_n^{\mathrm{E}}} p_{g,\nu'}^{\mathrm{E}} - \sum_{\ell | s(\ell)=n} p_{\ell,\nu'}^{\mathrm{L}} + \sum_{\ell | r(\ell)=n} p_{\ell,\nu'}^{\mathrm{L}} = \sum_{d \in \Omega_n^{\mathrm{D}}} \left(P_{d,\nu'}^{\mathrm{D^{max},*}} - p_{d,\nu'}^{\mathrm{LS}} \right) \quad \forall n, \forall \nu' \le \nu \tag{2.8d}$$

$$p_{\ell,\nu'}^{\mathrm{L}} = B_\ell \left(\theta_{s(\ell),\nu'} - \theta_{r(\ell),\nu'} \right) \quad \forall \ell \setminus \ell \in \Omega^{\mathrm{L+}}, \forall \nu' \le \nu \tag{2.8e}$$

$$p_{\ell,\nu'}^{\mathrm{L}} = x_\ell^{\mathrm{L}} B_\ell \left(\theta_{s(\ell),\nu'} - \theta_{r(\ell),\nu'} \right) \quad \forall \ell \in \Omega^{\mathrm{L+}}, \forall \nu' \le \nu \tag{2.8f}$$

$$-F_\ell^{\max} \le p_{\ell,\nu'}^{\mathrm{L}} \le F_\ell^{\max} \quad \forall \ell, \forall \nu' \le \nu \tag{2.8g}$$

$$0 \le p_{g,\nu'}^{\mathrm{E}} \le P_{g,\nu'}^{\mathrm{E^{max},*}} \quad \forall g, \forall \nu' \le \nu \tag{2.8h}$$

$$0 \le p_{d,\nu'}^{\mathrm{LS}} \le P_{d,\nu'}^{\mathrm{D^{max},*}} \quad \forall d, \forall \nu' \le \nu \tag{2.8i}$$

$$-\pi \le \theta_{n,\nu'} \le \pi \quad \forall n, \forall \nu' \le \nu \tag{2.8j}$$

$$\theta_{n,\nu'} = 0 \quad n\text{: ref.} \quad \forall \nu' \le \nu \tag{2.8k}$$

$$\eta \ge \sigma \left[\sum_g C_g^{\mathrm{G}} p_{g,\nu'}^{\mathrm{G}} + \sum_d C_d^{\mathrm{LS}} p_{d,\nu'}^{\mathrm{LS}} \right] \quad \forall \nu' \le \nu, \tag{2.8l}$$

where ν is the iteration index and $\nu' = 1, \ldots, \nu$.

The optimization variables of this master problem are the expansion decision variables x_ℓ^{L}, the operating decision variables $p_{g,\nu'}^{\mathrm{E}}, p_{d,\nu'}^{\mathrm{D}}, p_{\ell,\nu'}^{\mathrm{L}}, \theta_{n,\nu'}$ (one per iteration of the algorithm), and auxiliary variable η, which is used to reconstruct objective function (2.7a) gradually. Uncertain parameters $P_g^{\mathrm{E^{max},*}}$ and $P_d^{\mathrm{D^{max},*}}$ are fixed to their optimal values obtained from the subproblem solution at each iteration and used as input data of the master problem.

The size of master problem (2.8) increases with the iteration counter ν since a new set of constraints (2.8d)–(2.8l) is incorporated at each iteration of the algorithm. Note that if $\nu = 0$, then constraints (2.8d)–(2.8l) are not included in the master problem.

Master problem (2.8) is an MINLP problem since it includes binary variables x_ℓ^{L} and nonlinearities in constraints (2.8f). However, these nonlinear constraints can be replaced by equivalent mixed-integer linear Eqs. (2.2), as explained in Sect. 2.2.3. Therefore, the master problem is recast as an MILP problem that can be solved using conventional branch-and-cut solvers [6, 15].

2.3.5.2 Subproblem

The subproblem associated with (2.7) is given below:

$$\max_{P_d^{\mathrm{D^{max}}}, P_g^{\mathrm{E^{max}}} \in \Xi} \quad \min_{\Delta \setminus x_\ell^{\mathrm{L}} \in \Omega\left(x_\ell^{\mathrm{L},*}, P_d^{\mathrm{D^{max}}}, P_g^{\mathrm{E^{max}}}\right)} \quad \sigma \left[\sum_g C_g^{\mathrm{E}} p_g^{\mathrm{E}} + \sum_d C_d^{\mathrm{LS}} p_d^{\mathrm{LS}} \right], \tag{2.9a}$$

where the expansion decisions are considered to be fixed to $x_\ell^{\mathrm{L},*}$. These expansion decisions are obtained from the solution of master problem (2.8) at each iteration of the algorithm.

Subproblem (2.9) is a bilevel problem that can be converted into an equivalent single-level problem as explained below. The lower-level problem in (2.9) is continuous and linear (and thus convex) in its decision variables $\Delta \setminus x_\ell^{\mathrm{L}} \in \Omega \left(x_\ell^{\mathrm{L},*}, P_d^{\mathrm{D^{max}}}, P_g^{\mathrm{E^{max}}} \right)$. Therefore, it can be replaced by its Karush–Kuhn–Tucker (KKT) conditions, which are necessary and sufficient conditions for optimality. These KKT conditions are included as constraints of the upper-level problem, rendering a single-level problem as follows:

$$\max_{\Delta^{\mathrm{SUB}}} \quad \sigma \left[\sum_g C_g^{\mathrm{E}} p_g^{\mathrm{E}} + \sum_d C_d^{\mathrm{LS}} p_d^{\mathrm{LS}} \right] \tag{2.10aa}$$

subject to

$$P_g^{\mathrm{E^{max}}} \in \left[0, \overline{P}_g^{\mathrm{E^{max}}} \right] \quad \forall g \tag{2.10ab}$$

$$\frac{\sum_g \left(\overline{P}_g^{\mathrm{E^{max}}} - P_g^{\mathrm{E^{max}}} \right)}{\sum_g \overline{P}_g^{\mathrm{E^{max}}}} \leq \Gamma^{\mathrm{G}} \tag{2.10ac}$$

$$P_d^{\mathrm{D^{max}}} \in \left[\underline{P}_d^{\mathrm{D^{max}}}, \overline{P}_d^{\mathrm{D^{max}}} \right] \quad \forall d \tag{2.10ad}$$

$$\frac{\sum_d \left(P_d^{\mathrm{D^{max}}} - \underline{P}_d^{\mathrm{D^{max}}} \right)}{\sum_d \overline{P}_d^{\mathrm{D^{max}}} - \underline{P}_d^{\mathrm{D^{max}}}} \leq \Gamma^{\mathrm{D}} \tag{2.10ae}$$

$$\sum_{g \in \Omega_n^{\mathrm{E}}} p_g^{\mathrm{E}} - \sum_{\ell | s(\ell) = n} p_\ell^{\mathrm{L}} + \sum_{\ell | r(\ell) = n} p_\ell^{\mathrm{L}} = \sum_{d \in \Omega_n^{\mathrm{D}}} \left(P_d^{\mathrm{D^{max}}} - p_d^{\mathrm{LS}} \right) \quad \forall n \tag{2.10af}$$

$$p_\ell^{\mathrm{L}} = B_\ell \left(\theta_{s(\ell)} - \theta_{r(\ell)} \right) \quad \forall \ell \setminus \ell \in \Omega^{\mathrm{L+}} \tag{2.10ag}$$

$$p_\ell^{\mathrm{L}} = x_\ell^{\mathrm{L},*} B_\ell \left(\theta_{s(\ell)} - \theta_{r(\ell)} \right) \quad \forall \ell \in \Omega^{\mathrm{L+}} \tag{2.10ah}$$

$$-F_\ell^{\max} \leq p_\ell^{\mathrm{L}} \leq F_\ell^{\max} \quad \forall \ell \tag{2.10ai}$$

$$0 \leq p_g^{\mathrm{G}} \leq P_g^{\mathrm{E^{max}}} \quad \forall g \tag{2.10aj}$$

$$0 \leq p_d^{\mathrm{LS}} \leq P_d^{\mathrm{D^{max}}} \quad \forall d \tag{2.10ak}$$

$$-\pi \leq \theta_n \leq \pi \quad \forall n \tag{2.10al}$$

$$\theta_n = 0 \quad n\text{: ref.} \tag{2.10am}$$

$$\sigma C_g^{\mathrm{G}} - \lambda_{n(g)} + \phi_g^{\mathrm{E,max}} - \phi_g^{\mathrm{E,min}} = 0 \quad \forall g \tag{2.10an}$$

$$\sigma C_d^{\mathrm{D}} - \lambda_{n(d)} + \phi_d^{\mathrm{D,max}} - \phi_d^{\mathrm{D,min}} = 0 \quad \forall d \tag{2.10ao}$$

$$\lambda_{s(\ell)} - \lambda_{r(\ell)} - \phi_\ell^{\mathrm{L}} + \phi_\ell^{\mathrm{L,max}} - \phi_\ell^{\mathrm{L,min}} = 0 \quad \forall \ell \setminus \ell \in \Omega^{\mathrm{L+}} \tag{2.10ap}$$

$$\lambda_{s(\ell)} - \lambda_{r(\ell)} - \phi_\ell^{\mathrm{L+}} + \phi_\ell^{\mathrm{L,max}} - \phi_\ell^{\mathrm{L,min}} = 0 \quad \forall \ell \in \Omega^{\mathrm{L+}} \tag{2.10aq}$$

$$\sum_{\ell \setminus \ell \in \Omega^{\mathrm{L+}} | s(\ell) = n} B_\ell \phi_\ell^{\mathrm{L}} + \sum_{\ell \in \Omega^{\mathrm{L+}} | s(\ell) = n} x_\ell^{\mathrm{L},*} B_\ell \phi_\ell^{\mathrm{L+}} - \sum_{\ell \setminus \ell \in \Omega^{\mathrm{L+}} | r(\ell) = n} B_\ell \phi_\ell^{\mathrm{L}}$$
$$- \sum_{\ell \in \Omega^{\mathrm{L+}} | r(\ell) = n} x_\ell^{\mathrm{L},*} B_\ell \phi_\ell^{\mathrm{L+}} + \phi_n^{\mathrm{N,max}} - \phi_n^{\mathrm{N,min}} = 0 \quad \forall n \setminus n\text{: ref.} \tag{2.10ar}$$

$$\sum_{\ell \backslash \ell \in \Omega^{\mathrm{L+}} | s(\ell)=n} B_\ell \phi_\ell^{\mathrm{L}} + \sum_{\ell \in \Omega^{\mathrm{L+}} | s(\ell)=n} x_\ell^{\mathrm{L},*} B_\ell \phi_\ell^{\mathrm{L+}} - \sum_{\ell \backslash \ell \in \Omega^{\mathrm{L+}} | r(\ell)=n} B_\ell \phi_\ell^{\mathrm{L}}$$
$$- \sum_{\ell \in \Omega^{\mathrm{L+}} | r(\ell)=n} x_\ell^{\mathrm{L},*} B_\ell \phi_\ell^{\mathrm{L+}} + \phi_n^{\mathrm{N,max}} - \phi_n^{\mathrm{N,min}} - \chi^{\mathrm{ref}} = 0 \quad n\text{: ref.} \quad (2.10\mathrm{as})$$

$$0 \le \phi_\ell^{\mathrm{L,max}} \perp F_\ell^{\max} - p_\ell^{\mathrm{L}} \ge 0 \quad \forall \ell \quad (2.10\mathrm{at})$$
$$0 \le \phi_\ell^{\mathrm{L,min}} \perp p_\ell^{\mathrm{L}} + F_\ell^{\max} \ge 0 \quad \forall \ell \quad (2.10\mathrm{au})$$
$$0 \le \phi_g^{\mathrm{E,max}} \perp P_g^{\mathrm{E}^{\max}} - p_g^{\mathrm{E}} \ge 0 \quad \forall g \quad (2.10\mathrm{av})$$
$$0 \le \phi_g^{\mathrm{E,min}} \perp p_g^{\mathrm{E}} \ge 0 \quad \forall g \quad (2.10\mathrm{aw})$$
$$0 \le \phi_d^{\mathrm{D,max}} \perp P_d^{\mathrm{D}^{\max}} - p_d^{\mathrm{LS}} \ge 0 \quad \forall d \quad (2.10\mathrm{ax})$$
$$0 \le \phi_d^{\mathrm{D,min}} \perp p_d^{\mathrm{LS}} \ge 0 \quad \forall d \quad (2.10\mathrm{ay})$$
$$0 \le \phi_n^{\mathrm{N,max}} \perp \pi - \theta_n \ge 0 \quad \forall n \quad (2.10\mathrm{az})$$
$$0 \le \phi_n^{\mathrm{N,min}} \perp \theta_n + \pi \ge 0 \quad \forall n, \quad (2.10\mathrm{ba})$$

where variables in set $\Delta^{\mathrm{SUB}} = \{P_d^{\mathrm{D}^{\max}}, P_g^{\mathrm{E}^{\max}}, p_g^{\mathrm{E}}, p_d^{\mathrm{D}}, p_\ell^{\mathrm{L}}, \theta_n, \lambda_n, \phi_\ell^{\mathrm{L}}, \phi_\ell^{\mathrm{L+}}, \phi_\ell^{\mathrm{L,max}}, \phi_\ell^{\mathrm{L,min}}, \phi_g^{\mathrm{E,max}}, \phi_g^{\mathrm{E,min}}, \phi_d^{\mathrm{D,max}}, \phi_d^{\mathrm{D,min}}, \phi_n^{\mathrm{N,max}}, \phi_n^{\mathrm{N,min}}, \chi^{\mathrm{ref}}\}$ are the optimization variables of subproblem (2.10). These optimization variables are the worst realizations of the uncertain parameters $P_d^{\mathrm{D}^{\max}}$ and $P_g^{\mathrm{E}^{\max}}$, the operating decision variables p_g^{E}, p_d^{D}, p_ℓ^{L}, and θ_n, as well as the dual variables λ_n, ϕ_ℓ^{L}, $\phi_\ell^{\mathrm{L,max}}$, $\phi_\ell^{\mathrm{L,min}}$, $\phi_g^{\mathrm{E,max}}$, $\phi_g^{\mathrm{E,min}}$, $\phi_d^{\mathrm{D,max}}$, $\phi_d^{\mathrm{D,min}}$, $\phi_n^{\mathrm{N,max}}$, $\phi_n^{\mathrm{N,min}}$, and χ^{ref}. Note that expansion decision variables $x_\ell^{\mathrm{L},*}$ are considered to be given parameters of subproblem (2.10) with values fixed to their optimal values obtained from the solution of the master problem (2.8) at the corresponding iteration.

Constraints (2.10ab)–(2.10ae) represent the uncertainty sets, constraints (2.10af)–(2.10am) are the primal constraints of the lower-level problem in (2.9), constraints (2.10an)–(2.10as) result from differentiating the Lagrangian of the lower-level problem in (2.9) with respect to lower-level variables, and constraints (2.10at)–(2.10ba) are the complementarity conditions.

Note that dual variables λ_n in constraints (2.10an)–(2.10ap) include different subscripts, namely $n(g), n(d), s(\ell), r(\ell)$, which indicate the node in which generating unit g is located, the node in which demand d is located, the sending-end node of transmission line ℓ, and the receiving-end node of transmission line ℓ, respectively.

Subproblem (2.10) is nonlinear, due to the complementarity constraints (2.10at)–(2.10ba). Complementarity constraints have the form $0 \le a \perp b \ge 0$, which is equivalent to nonlinear constraints $a \ge 0$, $b \ge 0$, and $ab = 0$. However, these constraints can be replaced by the following exact equivalent mixed-integer linear expressions, as explained in [10]:

$$a \geq 0 \tag{2.11a}$$
$$b \geq 0 \tag{2.11b}$$
$$a \leq Mu \tag{2.11c}$$
$$b \leq M\left(1-u\right), \tag{2.11d}$$

where u is an auxiliary binary variable and M is a large enough positive constant.

The working of Eqs. (2.11) is explained below. The complementarity constraints impose that either a or b must be equal to zero. On one hand, if binary variable u is equal to 0, then $a = 0$ by Eqs. (2.11a) and (2.11c), and $0 \leq b \leq M$ by Eqs. (2.11b) and (2.11d). On the other hand, if binary variable u is equal to 1, then $b = 0$ by Eqs. (2.11b) and (2.11d) and $0 \leq a \leq M$ by Eqs. (2.11a) and (2.11c). Using Eqs. (2.11), we impose that either a or b is equal to zero and that the other one can take any value in a large enough interval defined by the disjunctive parameter M. Reference [10] discusses how to select the values of this parameter.

2.3.5.3 Algorithm

The master problem and the subproblem defined in the previous sections are solved iteratively. The optimal solution of the master problem at each iteration is used to solve the subproblem and vice versa. The iterative algorithm continues until convergence is attained. The detailed steps of this iterative algorithm are provided below:

Step 1 Set lower (LB) and upper (UB) bounds to $-\infty$ and ∞, respectively.

Step 2 Set the iteration counter to $\nu = 0$.

Step 3 Solve master problem (2.8). Obtain the optimal solution of variables $x_{\ell}^{\mathrm{L},*}$, $p_{g,\nu'}^{\mathrm{E},*}$, $p_{d,\nu'}^{\mathrm{D},*}$, $p_{\ell,\nu'}^{\mathrm{L},*}$, $\theta_{n,\nu'}^{*}$, and η^{*}. Note that if $\nu = 0$, then constraints (2.8d)–(2.8l) are not considered in the master problem.

Step 4 Update the lower bound using Eq. (2.12) below:

$$LB = \sum_{\ell \in \Omega^{\mathrm{L}+}} \tilde{I}_{\ell}^{\mathrm{L}} x_{\ell}^{\mathrm{L},*} + \eta^{*}. \tag{2.12}$$

Note that the master problem is a relaxed version of the original problem in which variable η is used to reconstruct the original problem progressively at each iteration. Therefore, the value of the lower bound increases with the iteration counter as the master problem approximates the original problem.

Step 5 Solve subproblem (2.10) by considering the optimal values of variables $x_{\ell}^{\mathrm{L},*}$ obtained in **Step 3** to be given parameters. Obtain the optimal solution of variables in the set $\Delta^{\mathrm{SUB},*}$.

Step 6 Update the upper bound using Eq. (2.13) below:

$$UB = \min\left\{UB, \sum_{\ell \in \Omega^{\mathrm{L}+}} \tilde{I}^{\mathrm{L}}_{\ell} x^{\mathrm{L},*}_{\ell} + \sigma \left[\sum_{g} C^{\mathrm{G}}_{g} p^{\mathrm{G},*}_{g} + \sum_{d} C^{\mathrm{LS}}_{d} p^{\mathrm{LS},*}_{d}\right]\right\}. \tag{2.13}$$

Note that the subproblem is a more constrained version of the original problem since variables x^{L}_{ℓ} are fixed to given values. Thus, the upper bound decreases with the iteration counter as variables x^{L}_{ℓ} change and approximate their optimal values.

Step 7 If $UB - LB$ is lower than a predefined tolerance ϵ, the algorithm terminates. The optimal solution is $x^{\mathrm{L},*}_{\ell}$. If not, then continue with the following step.

Step 8 Update the iteration counter, $\nu \leftarrow \nu + 1$, and set $P^{\mathrm{D^{max}},*}_{d,\nu} = P^{\mathrm{D^{max}},*}_{d}$ and $P^{\mathrm{E^{max}},*}_{g,\nu} = P^{\mathrm{E^{max}},*}_{g}$, where $P^{\mathrm{D^{max}},*}_{d}$ and $P^{\mathrm{E^{max}},*}_{g}$ are the optimal values obtained from the solution of the subproblem in **Step 5**.

Step 9 Continue with **Step 3**.

For the sake of clarity, the algorithm flowchart is depicted in Fig. 2.8.

Illustrative Example 2.6 *Two-stage ARO TEP*

We consider the six-node system analyzed in Illustrative Example 2.3. The technical data of generating units, demands, and both existing and prospective transmission

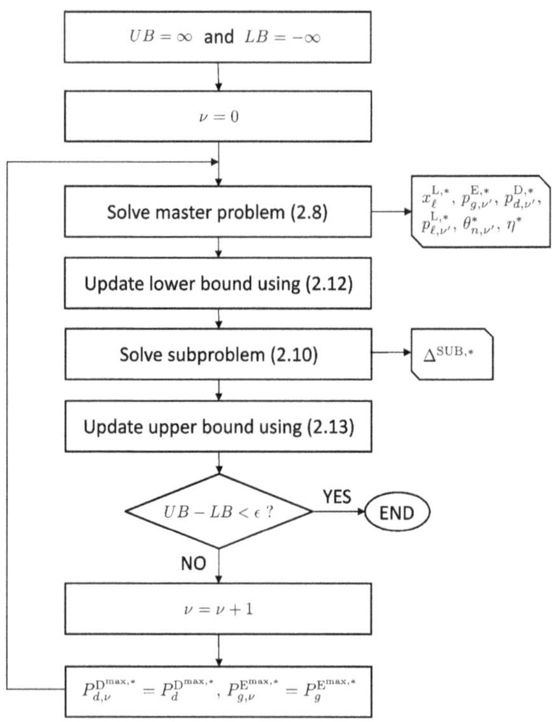

Fig. 2.8 TEP flowchart of the ARO algorithm

Table 2.7 Illustrative Example 2.6: data for uncertainty sets of demands

Demand	$\underline{P}_d^{\text{D}^{\max}}$ [MW]	$\overline{P}_d^{\text{D}^{\max}}$ [MW]
d_1	180	220
d_2	135	165
d_3	90	110
d_4	180	220

lines, as well as the available investment budget are obtained from Illustrative Example 2.3.

Here, we consider an ARO approach for the TEP problem, and therefore, we consider uncertainty in the available capacity of generating units and the maximum demand at each node of the system, as explained below:

1. We consider that $P_g^{\text{E}^{\max}}$ is uncertain and can take values between 0 and the generating capacity values provided in the third column of Table 2.1.
2. We consider that $P_d^{\text{D}^{\max}}$ is uncertain and can take values between $\underline{P}_d^{\text{D}^{\max}}$ and $\overline{P}_d^{\text{D}^{\max}}$, which are provided in the second and third columns of Table 2.7, respectively.

We consider that the uncertainty budgets for demands and generating units are $\Gamma^{\text{D}} = 0.5$ and $\Gamma^{\text{G}} = 0.2$, respectively. This means that the level of uncertainty is higher in the demand than in the capacity of generating units. For example, with the considered uncertainty budgets, up to 20 % of the generating capacity may be unavailable.

With these data, we solve the TEP problem with an ARO approach using the algorithm described in Sect. 2.3.5.3. For the sake of clarity, the steps and results of each iteration of the algorithm are provided below:

Step 1 We set the lower bound to $LB = -\infty$ and the upper bound to $UB = \infty$.

Step 2 We set the iteration counter to $\nu = 0$.

Step 3 We solve master problem (2.8). We obtain the optimal solution of variables $x_\ell^{\text{L},*} = 0, \ell = 4, \ldots, 9$; and $\eta^* = -\infty$. Note that since $\nu = 0$, constraints (2.8d)–(2.8l) are not included in the master problem at this iteration.

Step 4 We update the lower bound using Eq. (2.12), $LB = -\infty$.

Step 5 We solve subproblem (2.10) by considering the optimal value of variables $x_\ell^{\text{L},*}$ obtained in **Step 3** as given parameters. We obtain the optimal solution of variables $P_g^{\text{E}^{\max},*}$ and $P_d^{\text{D}^{\max},*}$, which are provided in Table 2.8. The first and second columns respectively give the generating unit and the corresponding $P_g^{\text{E}^{\max},*}$, while the third and fourth columns provide the demand and the corresponding $P_d^{\text{D}^{\max},*}$, respectively.

Subproblem (2.10) corresponds to the worst realization (in terms of generation and load-shedding costs) of uncertain parameters once the expansion decision variables, i.e., variables $x_\ell^{\text{L},*}$, are fixed. As observed from the results provided in Table 2.8, the worst realization corresponds to the case in which a peak load occurs for demands

Table 2.8 Illustrative Example 2.6: solution of subproblem (2.10) for iteration $\nu = 0$

Generating unit	$P_g^{\mathrm{E^{max}},*}$ [MW]	Demand	$P_d^{\mathrm{D^{max}},*}$ [MW]
g_1	300	d_1	180
g_2	250	d_2	160
g_3	400	d_3	90
g_4	170	d_4	220
g_5	60		

d_2 and d_4 (located at nodes 4 and 6, respectively, both in Region B) and when the generation capacity of generating units g_4 and g_5 (located at nodes 5 and 6, respectively, both in Region B) is limited. If we look at the data for generating units and demands provided in Tables 2.1 and 2.2, then we observe that most of the demand is located in Region B. Moreover, Region B is initially not connected to Region A, and node 6 is isolated. Therefore, a critical situation would be the case of a peak demand in Region B (and mainly at node 6) and the failure of generating units in the same region. This is consistent with the results achieved for the subproblem.

In this step, we also obtain solutions for operating decision variables $p_g^{\mathrm{E},*}$, $p_d^{\mathrm{D},*}$, $p_\ell^{\mathrm{L},*}$, and θ_n^*. However, these variables are not used in the following steps of the algorithm.

Step 6 We update the upper bound using Eq. (2.13), $UB = 2.351 \times 10^8$.

Step 7 We compute $UB - LB = \infty$. Since this difference is not small enough, we continue with the following step.

Step 8 We update the iteration counter, $\nu = 1$, and set $P_{g,\nu_1}^{\mathrm{E^{max}},*} = P_g^{\mathrm{E^{max}},*}$ and $P_{d,\nu_1}^{\mathrm{D^{max}},*} = P_d^{\mathrm{D^{max}},*}$, where $P_g^{\mathrm{E^{max}},*}$ and $P_d^{\mathrm{D^{max}},*}$ are obtained from the solution of subproblem in **Step 5**.

Step 9 Continue with **Step 3**.

Step 3 We solve master problem (2.8). We obtain the optimal solution of variables $x_\ell^{\mathrm{L},*} = 1,\ \ell = 5, 7$; $x_\ell^{\mathrm{L},*} = 0,\ \ell = 4, 6, 8, 9$; and $\eta^* = 1.169 \times 10^8$. Master problem (2.8) corresponds to the response of the TSO to the worst situation, i.e., given the worst realization of uncertain parameters obtained in subproblem in **Step 5** above, the TSO decides the optimal transmission expansion plans to minimize generation and load-shedding costs, as well as its investment costs.

Step 4 We update the lower bound using Eq. (2.12), $LB = 1.199 \cdot 10^8$.

Step 5 We solve subproblem (2.10) by considering the optimal value of variables $x_\ell^{\mathrm{L},*}$ obtained in **Step 3** as given parameters. We obtain the optimal solution of variables $P_g^{\mathrm{E^{max}},*}$ and $P_d^{\mathrm{D^{max}},*}$, which are provided in Table 2.9. The first and second columns respectively give the generating unit and the corresponding $P_g^{\mathrm{E^{max}},*}$, while the third and fourth columns provide the demand and the corresponding $P_d^{\mathrm{D^{max}},*}$, respectively.

Table 2.9 Illustrative Example 2.6: solution of subproblem (2.10) for iteration $\nu = 1$

Generating unit	$P_g^{\text{E}^{\max,*}}$ [MW]	Demand	$P_d^{\text{D}^{\max,*}}$ [MW]
g_1	300	d_1	195
g_2	250	d_2	165
g_3	400	d_3	110
g_4	20	d_4	180
g_5	150		

As previously explained, subproblem (2.10) corresponds to the worst realization (in terms of generation and load-shedding costs) of uncertain parameters once the expansion decision variables (i.e., $x_\ell^{\text{L},*}$) are fixed. However, now we consider the updated values of expansion decision variables obtained in the master problem for iteration $\nu = 1$. As a consequence, the worst realization of uncertain parameters is different from that obtained in the previous iteration. For example, considering the updated expansion decisions, initially isolated node 6 is now connected to node 3. Therefore, the worst realization that in the previous iteration was a peak load and the failure of the generating unit at this node is not that harmful in this case.

Step 6 We update the upper bound using Eq. (2.13), $UB = 1.334 \times 10^8$.

Step 7 We compute $UB - LB = 0.135 \times 10^8$. Since this difference is not small enough, we continue with the following step.

Step 8 We update the iteration counter, $\nu = 2$, and set $P_{g,\nu_1}^{\text{E}^{\max,*}} = P_g^{\text{E}^{\max,*}}$ and $P_{d,\nu_1}^{\text{D}^{\max,*}} = P_d^{\text{D}^{\max,*}}$.

Step 9 Continue with **Step 3**.

Step 3 We solve master problem (2.8). We obtain the optimal solution of variables $x_\ell^{\text{L},*} = 1, \ell = 5, 7; x_\ell^{\text{L},*} = 0, \ell = 4, 6, 8, 9;$ and $\eta^* = 1.304 \times 10^8$.

Step 4 We update the lower bound using Eq. (2.12), $LB = 1.334 \times 10^8$.

Step 5 We solve subproblem (2.10) by considering the optimal value of variables $x_\ell^{\text{L},*}$ obtained in **Step 3** as given parameters. Since the expansion decisions are the same as those obtained for $\nu = 1$, the optimal solution of subproblem (2.10) corresponds with that provided in Table 2.9.

Step 6 We update the upper bound using Eq. (2.13), $UB = 1.334 \times 10^8$.

Step 7 We compute $UB - LB = 0$. This means that the algorithm has converged, and so it terminates.

The optimal solution of the TEP problem using an ARO approach that takes into account the uncertainty in demands and the availability of generating units consists in building transmission lines ℓ_5 and ℓ_7, i.e., transmission lines connecting nodes 2 and 4 and nodes 2 and 6, respectively. □

Table 2.10 Illustrative Example 2.7: investment decisions for different uncertainty budgets[a]

		Γ^{G}				
		0	0.25	0.5	0.75	1
Γ^{D}	0	ℓ_5, ℓ_7	ℓ_5, ℓ_7	ℓ_4, ℓ_7, ℓ_9	ℓ_4, ℓ_5, ℓ_9	-
	0.25	ℓ_5, ℓ_7	ℓ_5, ℓ_7	ℓ_5, ℓ_7	ℓ_4, ℓ_7, ℓ_9	-
	0.5	ℓ_5, ℓ_7	ℓ_5, ℓ_7	ℓ_5, ℓ_7	ℓ_4, ℓ_7, ℓ_9	-
	0.75	ℓ_5, ℓ_7	ℓ_5, ℓ_7	ℓ_5, ℓ_7	ℓ_4, ℓ_7, ℓ_9	-
	1	ℓ_5, ℓ_7	ℓ_5, ℓ_7	ℓ_5, ℓ_7	ℓ_4, ℓ_7, ℓ_9	-

[a] ℓ_4: 2–3, ℓ_5: 2–4, ℓ_6: 3–4, ℓ_7: 3–6, ℓ_8: 4–6, ℓ_9: 5–6

Illustrative Example 2.7 *Two-stage ARO TEP: Impact of uncertainty budgets*

In Illustrative Example 2.6 we consider that uncertainty budgets Γ^{D} and Γ^{G} are equal to 0.5 and 0.2, respectively. Now we analyze its influence on the expansion decisions. Table 2.10 provides the optimal transmission expansion decisions for different values of these uncertainty budgets.

We obtain different expansion decisions depending on the considered uncertainty budgets. These expansion decisions are mainly conditioned by the value of Γ^{G}, as explained below:

1. For low values of Γ^{G}, it is optimal to build ℓ_5 and ℓ_7. These values of the uncertainty budget represent the case in which the uncertainty in the available capacity of generating units is not very high.
2. For values of Γ^{G} larger than or equal to 0.5, it becomes optimal to build ℓ_4, ℓ_5, ℓ_7, or ℓ_9, depending on the considered values of Γ^{D}. These values of Γ^{G} may represent the case of a system with high uncertainty in the available capacity of generating units, e.g., in a system in which building new generating units is highly uncertain.
3. The value of $\Gamma^{\mathrm{G}} = 1$ represents the case in which $P_g^{\mathrm{G,max}}$ can take any value between 0 and $\overline{P}_g^{\mathrm{G,max}}$, i.e., all units may be unavailable. Therefore, this case is not realistic.

The optimal expansion plan if no uncertainty is considered (i.e., if $\Gamma^{\mathrm{G}} = \Gamma^{\mathrm{D}} = 0$) consists in building transmission lines ℓ_5 and ℓ_7. This solution corresponds to solving the deterministic TEP problem (2.3) with $P_d^{\mathrm{D^{max}}} = \underline{P}_d^{\mathrm{D^{max}}}, \forall d$, and $P_g^{\mathrm{E^{max}}} = \overline{P}_g^{\mathrm{E^{max}}}, \forall g$. This expansion plan is significantly different from that obtained for a high level of uncertainty (e.g., if $\Gamma^{\mathrm{G}} = 0.75$ and $\Gamma^{\mathrm{D}} = 1$), which consists in building transmission lines ℓ_4, ℓ_7, and ℓ_9. This highlights the importance of modeling the uncertainty when the TEP is decided. If the deterministic solution is implemented, and then one of the worst situations considered in the ARO model occurs, the system may experience significant costs. □

2.4 Summary

This chapter analyzes the TEP problem. TEP consists in deciding on the optimal reinforcement of the transmission capacity of an existing electric energy network by appropriately selecting the type and number of transmission lines to be built. TEP is a relevant problem in electric energy systems that require a secure, economic, and reliable supply of demand.

We adopt the perspective of a TSO that decides a transmission expansion plan with the aim of facilitating energy trading among producers and consumers, by reducing the generation and load-shedding costs. To do so, two different approaches are developed:

1. A deterministic approach in which optimal transmission plans are obtained by considering the largest expected demand in the planning horizon.
2. An ARO approach in which optimal transmission expansion plans are obtained by taking into account the uncertainty in the demand and the capacity of generating units.

Different illustrative examples are provided to show the working and applicability of the two models described. From these examples, as well as from the theoretical framework described in this chapter, we obtain the following conclusions:

1. Investment costs in transmission lines are comparatively lower than the generation and load-shedding costs. Therefore, it is possible to reduce generation and load-shedding costs by employing limited resources in building new transmission lines.
2. TEP is carried out within an uncertain environment. RO is a practical tool that allows us to represent uncertain parameters by robust sets at a reduced computational cost.
3. TEP is a computationally complex problem. Therefore, it requires implementing some simplifications, e.g., a static approach or a dc power flow, especially if the system under study is very large.
4. The TEP problem is analyzed in this chapter by considering that the generation capacity of the system under study is fixed. The extension of the TEP problem to considering the joint expansion of transmission and generation capacity is analyzed in Chap. 4.

2.5 End-of-Chapter Exercises

2.1 Why is TEP needed? Who decides about it? What is its main purpose?

2.2 Describe the advantages and disadvantages of using a static approach such as that used in this chapter (not a dynamic one) for the formulation of the TEP problem.

2.3 Determine the optimal transmission expansion plan in the modified Garver's system using the deterministic model (2.3). This system is depicted in Fig. 2.9, and

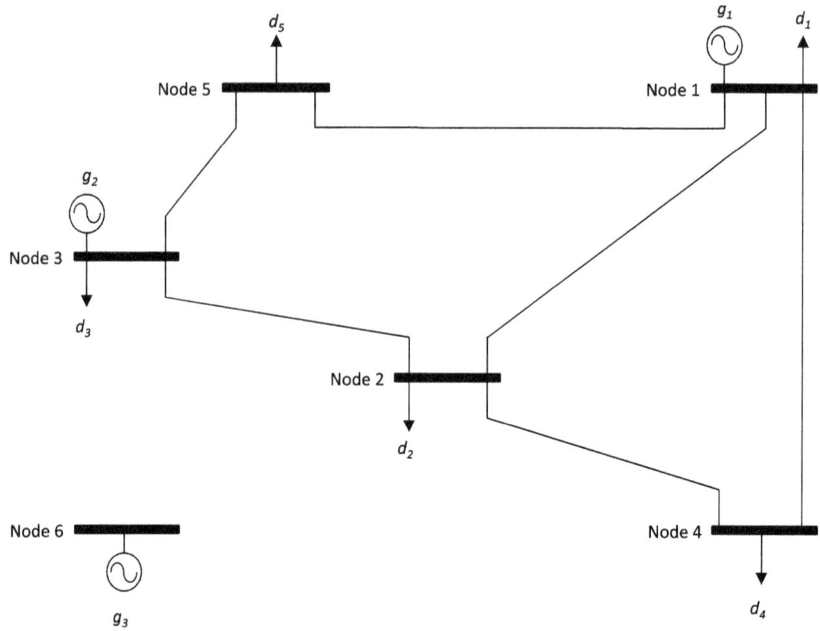

Fig. 2.9 Exercise 2.3: modified Garver's system

Table 2.11 Exercise 2.3: data for generating units of the modified Garver's system

Generating unit	Node	$P_g^{G^{max}}$ [MW]	C_g^G [$/MWh]
g_1	n_1	200	24
g_2	n_3	200	28
g_3	n_6	300	16

Table 2.12 Exercise 2.3: data for demands of the modified Garver's system

Demand	Node	$P_d^{D^{max}}$ [MW]	C_d^{LS} [$/MWh]
d_1	n_1	110	49
d_2	n_2	132	51
d_3	n_3	88	80
d_4	n_4	132	65
d_5	n_5	88	39

its data are provided in Tables 2.11, 2.12, 2.13, and 2.14. Consider an annualized investment budget equal to $12 million, a base power of 1 MW, and a base voltage of 1 kV.

Table 2.13 Exercise 2.3: data for existing transmission lines of the modified Garver's system

Line	From node	To node	B_ℓ [S]	$F_\ell^{\max}$ [MW]
ℓ_1	n_1	n_2	250	100
ℓ_2	n_1	n_4	133	80
ℓ_3	n_1	n_5	500	100
ℓ_4	n_2	n_3	500	100
ℓ_5	n_2	n_4	250	100
ℓ_6	n_3	n_5	250	100

Table 2.14 Exercise 2.3: data for prospective transmission lines of the modified Garver's system

Line	From node	To node	B_ℓ [S]	$F_\ell^{\max}$ [MW]	I_ℓ^{L} [$]
ℓ_7	n_2	n_5	323	100	3,491,000
ℓ_8	n_2	n_6	333	100	3,379,000
ℓ_9	n_3	n_6	500	100	5,406,000
ℓ_{10}	n_4	n_6	333	100	3,379,000

2.4 Expand the deterministic TEP model (2.3) to consider a dynamic approach whereby expansion decisions can be made at different points in time. Then apply this dynamic problem to obtain the optimal TEP decisions in Illustrative Example 2.3, considering three time periods and assuming a constant demand growth of 5 % at each time period.

2.5 Expand the deterministic TEP model (2.3) to consider losses through transmission lines, as done in [1].

2.6 Expand the deterministic TEP model (2.3) to consider the uncertainty in the maximum demand through a set of scenarios, i.e., obtain a stochastic programming model from the deterministic TEP model (2.3). Is this stochastic model more efficient than the ARO model provided in Sect. 2.3? Why or why not?

2.7 Write the equivalent mixed-integer linear expressions corresponding to complementarity constraints (2.10at)–(2.10ba) using the Fortuny–Amat transformation described by Eqs. (2.11).

2.8 Determine the optimal transmission expansion plan in the modified Garver's system using the ARO approach described in Sect. 2.3. This system is depicted in Fig. 2.9, and its data are provided in Tables 2.11, 2.13, 2.14, and 2.15. Consider an annualized investment budget equal to $12 million and uncertainty budgets for demands and generating units equal to $\Gamma^{\mathrm{D}} = 0.5$ and $\Gamma^{\mathrm{G}} = 0.2$, respectively.

2.9 In the ARO approach described in Sect. 2.3, the so-called uncertainty budgets, Γ^{G} and Γ^{D}, are used to model the level of uncertainty in generating units and

Table 2.15 Exercise 2.8: data for uncertainty sets of demands of the modified Garver's system

Demand	$\underline{P}_d^{D^{\max}}$ [MW]	$\overline{P}_d^{D^{\max}}$ [MW]
d_1	100	120
d_2	120	144
d_3	80	96
d_4	120	144
d_4	80	96

demands, respectively. However, uncertainty in generating units and demands may be different in different regions of a system. It is possible to define uncertainty budgets for different regions, i.e., Γ_r^{G} and Γ_r^{D}, where r indicates the region of the system, as explained in [29]. Solve Illustrative Example 2.6, considering that the uncertainties in Regions A and B are different. Region A has significant uncertainty on the available capacity of generating units ($\Gamma_{\mathrm{A}}^{\mathrm{G}} = 0.5$), but no uncertainty on demand levels ($\Gamma_{\mathrm{B}}^{\mathrm{D}} = 0$), while Region B has no uncertainty on the available capacity of generating units ($\Gamma_{\mathrm{B}}^{\mathrm{G}} = 0$), but high uncertainty on demand levels ($\Gamma_{\mathrm{B}}^{\mathrm{D}} = 0.5$). How different are the results from those obtained in Illustrative Example 2.6? Why?

2.6 GAMS Code

A GAMS code for solving problem Illustrative Example 2.3 is provided below:

```
SETS
n                  /n1*n6/
g                  /g1*g5/
d                  /d1*d4/
l                  /l1*l9/
pros(l)            /l4*l9/
ex(l)              /l1*l3/
mapG(g,n)          /g1.n1,g2.n2,g3.n3,g4.n5,g5.n6/
mapD(d,n)          /d1.n1,d2.n4,d3.n5,d4.n6/
ref(n)             /n1/
mapSL(l,n)         /l1.n1,l2.n1,l3.n4,l4.n2,l5.n2,l6.
    n3,l7.n3,l8.n4,l9.n5/
mapRL(l,n)         /l1.n2,l2.n3,l3.n5,l4.n3,l5.n4,l6.
    n4,l7.n6,l8.n6,l9.n6/;


TABLE LDATA(L,*)
          B         FLmax     IC
l1        500       150       0
l2        500       150       0
l3        500       150       0
l4        500       150       700000
l5        500       150       1400000
```

```
16          500       200       1800000
17          500       200       1600000
18          500       150       800000
19          500       150       700000;

SCALAR ILmax
/3000000/;

TABLE DDATA(d,*)
            PDmax     LScost
d1          200       40
d2          150       52
d3          100       55
d4          200       65;

TABLE GDATA(g,*)
            PEmax     Gcost
g1          300       18
g2          250       25
g3          400       16
g4          300       32
g5          150       35;

SCALAR SIGMA
/8760/;

SCALAR M
/5000/;

VARIABLES
Z
PL(l)
THETA(n);

POSITIVE VARIABLES
PG(g)
PLS(d);

BINARY VARIABLES
x(l);

EQUATIONS EQ3A, EQ3B, EQ3D, EQ3E, EQ3Fa, EQ3Fb,
    EQ3Ga, EQ3Gb, EQ3Ha, EQ3Hb, EQ3I, EQ3J, EQ3Ka,
    EQ3Kb, EQ3L;

EQ3A..                           Z=E=SUM(l$pros(l),LDATA(l,
    'IC')*x(l))+SIGMA*(SUM(g,GDATA(g,'Gcost')*PG(G)
    )+SUM(d,DDATA(d,'LSCOST')*PLS(d)));
EQ3B..                           SUM(l$pros(l),LDATA(L,'IC'
    )*x(l))=L=ILmax;
*EQUATIONS 3C ARE DEFINITION OF BINARY VARIABLES
```

```
EQ3D(n)..                    SUM(g$mapG(g,n),PG(G))-SUM
    (l$mapSL(l,n),PL(l))+SUM(l$mapRL(l,n),PL(l))=E=
    SUM(d$mapD(d,n),DDATA(d,'PDMAX')-PLS(d));
EQ3E(l)$EX(l)..              PL(l)=E=LDATA(l,'B')*(SUM(
    n$mapSL(l,n),THETA(n))-SUM(n$mapRL(l,n),THETA(n
    )));
EQ3Fa(l)$EX(l)..             -LDATA(l,'FLmax')=L=PL(l);
EQ3Fb(l)$EX(l)..             PL(l)=L=LDATA(l,'FLmax');
EQ3Ga(l)$PROS(l)..           -x(l)*LDATA(l,'FLmax')=L=
    PL(l);
EQ3Gb(l)$PROS(l)..           PL(l)=L=x(l)*LDATA(l,'
    FLmax');
EQ3Ha(l)$PROS(l)..           -(1-x(l))*M=L=PL(l)-LDATA(
    l,'B')*(SUM(n$mapSL(l,n),THETA(n))-SUM(n$mapRL(
    l,n),THETA(n)));
EQ3Hb(l)$PROS(l)..           PL(l)-LDATA(l,'B')*(SUM(
    n$mapSL(l,n),THETA(n))-SUM(N$mapRL(l,n),THETA(n
    )))=l=(1-X(l))*M;
EQ3I(g)..                    PG(g)=L=GDATA(g,'PEmax');
EQ3J(d)..                    PLS(d)=L=DDATA(d,'PDmax');
EQ3Ka(n)..                   -3.14=L=THETA(n);
EQ3Kb(n)..                   THETA(n)=L=3.14;
EQ3L(n)$REF(n)..             THETA(n)=L=0;

MODEL TEP_DET /ALL/;

SOLVE TEP_DET USING MIP MINIMIZING Z;
```

References

1. Alguacil, N., Motto, A.L., Conejo, A.J.: Transmission expansion planning: a mixed-integer LP approach. IEEE Trans. Power Syst. **18**(3), 1070–1078 (2003)
2. Bertsimas, D., Sim, M.: Robust discrete optimization and network flows. Math. Program. Ser. B **98**(1–3), 49–71 (2002)
3. Bertsimas, D., Sim, M.: The price of robustness. Oper. Res. **52**(1), 35–53 (2004)
4. Bertsimas, D., Litvinov, E., Sun, X.A., Zhao, J., Zheng, T.: Adaptive robust optimization for the security constrained unit commitment problem. IEEE Trans. Power Syst. **28**(1), 52–63 (2013)
5. Binato, S., Pereira, M.V.F., Granville, S.: A new Benders decomposition approach to solve power transmission network design problems. IEEE Trans. Power Syst. **16**(2), 235–240 (2001)
6. Castillo, E., Conejo, A.J., Pedregal, P., Alguacil, N., Garcia, R.: Building and Solving Mathematical Programming Models in Engineering and Science. Wiley, New York (2001)
7. Chen, B., Wang, J., Wang, L., He, Y., Wang, Z.: Robust optimization for transmission expansion planning: minimax cost versus minimax regret. IEEE Trans. Power Syst. **29**(6), 3069–3077 (2014)
8. Conejo, A.J., Carrión, M., Morales, J.M.: Decision Making Under Uncertainty in Electricity Markets. Springer, New York (2010)
9. Contreras, J., Wu, F.F.: A kernel-oriented algorithm for transmission expansion planning. IEEE Trans. Power Syst. **15**(4), 1434–1440 (2000)
10. Fortuny-Amat, J., McCarl, B.: A representation and economic interpretation of a two-level programming problem. J. Oper. Res. Soc. **32**(9), 783–792 (1981)

11. Garcés, L.P., Conejo, A.J., García-Bertrand, R., Romero, R.: A bilevel approach to the transmission expansion planning within a market environment. IEEE Trans. Power Syst. **24**(3), 1513–1522 (2009)
12. Garver, L.L.: Transmission network estimation using linear programming. IEEE Trans. Power Apparat. Syst. **89**(7), 1688–1697 (1970)
13. Gómez-Expósito, A., Conejo, A.J., Cañizares, C.: Electric Energy Systems: Analysis and Operation. CRC, Boca Raton, Florida (2008)
14. European Network of Transmission System Operators for Electricity, ENTSO-E (2014). http://www.entsoe.eu
15. The ILOG CPLEX Website (2014). http://www.ilog.com/products/cplex/
16. Jabr, R.A.: Robust transmission network expansion planning with uncertain renewable generation and loads. IEEE Trans. Power Syst. **28**(4), 4558–4567 (2013)
17. Jiang, R., Wang, J., Zhang, M., Guan, Y.: Two-stage minimax regret robust unit commitment. IEEE Trans. Power Syst. **28**(3), 2271–2282 (2013)
18. Latorre-Bayona, G., Pérez-Arriaga, I.J.: Chopin: a heuristic model for long term transmission expansion planning. IEEE Trans. Power Syst. **9**(4), 1886–1894 (1994)
19. Midwest Independent System Operator (2014). https://www.misoenergy.org/Pages/Home
20. MIT Energy Initiative. The future of the electric grid. Available at http://web.mit.edu/mitei/research/studies/the-electric-grid-2011 (2011)
21. Monticelli, A., Santos, A., Pereira, M.V.F., Cunha, S.H., Parker, B.J., Praca, J.C.G.: Interactive transmission network planning using a least-effort criterion. IEEE Trans. Power Apparat. Syst. **101**(10), 3919–3925 (1982)
22. Oliveira, G.C., Costa, A.P.C., Binato, S.: Large scale transmission network planning using optimization and heuristic techniques. IEEE Trans. Power Syst. **10**(4), 18283–1833 (1995)
23. Pereira, M.V.F., Pinto, L.M.V.G.: Application of sensitivity analysis of load supplying capability to interactive transmission expansion planning. IEEE Trans. Power Apparat. Syst. **104**(2), 381–389 (1985)
24. PJM Interconnection (2014). http://www.pjm.com
25. Red Eléctrica de España S.A., REE. Network provider for the electricity market of mainland Spain (2014). http://www.ree.es
26. Romero, R., Monticelli, A.: A hierarchical decomposition approach for transmission network expansion planning. IEEE Trans. Power Syst. **9**(1), 373–380 (1994)
27. Romero, R., Monticelli, A.: A zero-one implicit enumeration for optimizing the investments in transmission expansion planning. IEEE Trans. Power Syst. **9**(3), 1385–1391 (1994)
28. Romero, R., Gallego, R.A., Monticelli, A.: Transmission system expansion planning by simulated annealing. IEEE Trans. Power Syst. **11**(1), 364–369 (1996)
29. Ruiz, C., Conejo, A.J.: Robust transmission expansion planning. Eur. J. Oper. Res. **242**(2), 390–401 (2015)
30. de la Torre, S., Conejo, A.J., Contreras, J.: Transmission expansion planning in electricity markets. IEEE Trans. Power Syst. **23**(1), 238–248 (2008)
31. Tsamasphyrou, P., Renaud, A., Carpentier, P.: Transmission network planning: an efficient Benders decomposition scheme. In: Proceedings of 13th Power Systems Computer Conference, Trondheim, Norway, June/July 28–2 (1999)
32. Zeng, B., Zhao, L.: Solving two-stage robust optimization problems using a column-and-constraint generation method. Oper. Res. Lett. **41**(5), 457–461 (2013)

Chapter 3
Generation Expansion Planning

This chapter describes the generation expansion planning (GEP) problem in a given electric energy system. Here, we take the perspective of a central planner that determines the generation expansion plan that is most beneficial for the operation of the electric energy system as a whole. The central planner does not actually build the generating units; however, it encourages private investors to build the electricity production facilities, e.g., by using different types of incentives. Considering such a central view, we provide and describe different models for the GEP problem: from a very basic model based on a deterministic, static, and single-node approach to more complex models that consider the impact of investment timing, transmission constraints, and uncertainty on the GEP problem.

3.1 Introduction

One of the most relevant problems in the planning of electric energy systems is guaranteeing that demands are efficiently and reliably supplied. On the one hand, demands should be supplied by economic, flexible, and efficient generating units. On the other hand, it is necessary to guarantee that demands will be supplied even in the worst situations, e.g., if there is an unexpected peak demand or if an important generating unit fails. Note that there is a twin economic and engineering objective.

In order to tackle such a complex problem, two important issues arise. One is analyzing the transmission network, which allows the energy flows between producers and consumers. The capacity of the transmission network should be enough for demands to be met in an efficient and reliable manner. The problem of analyzing whether it is necessary to reinforce the transmission network in a given electric energy system, known as the transmission expansion planning (TEP) problem, is described in Chap. 2. The second issue is the adequacy of the generating units that are used to supply demands, i.e., analyzing whether it is necessary to invest in building new electricity production facilities. This problem, known as the generation expansion planning (GEP) problem, is the topic of this chapter.

A.J. Conejo et al., *Investment in Electricity Generation and Transmission*,
DOI 10.1007/978-3-319-29501-5_3

The GEP problem is motivated by the aging of available electricity production facilities, as well as by growth in demand with the passing of time. These two issues make it essential to analyze whether it is necessary to expand the capacity of the existing generating units and/or to build new generating units.

The GEP problem is generally tackled in two different ways: (i) considering a market framework [15, 16] and (ii) adopting a centralized approach [7, 20]. On the one hand, a market framework assumes that profit-oriented agents determine their own generation expansion plans with the aim of maximizing their expected profits. To do so, once the production facilities are built, they recover their investment costs by selling their production in a market. The GEP problem considering a market approach is the topic of Chap. 5. On the other hand, considering a centralized approach, we assume that a central planner [e.g., the independent system operator (ISO)] determines the generation expansion plan that is most efficient for the system as a whole. This is the approach considered in this chapter.

The central planner determines the generation expansion plan that results in an optimal operation of the electric energy system and in an efficient supply of demands. With this purpose, different objective functions can be considered, e.g., minimizing network bottlenecks, maximizing social welfare, minimizing generation costs. Among these different objectives, we select the maximization of social welfare. Moreover, we also consider the investment costs incurred in building new electricity production facilities since they are relevant for the agents in charge of building the new generating units. This way of tackling the GEP problem is also known in the technical literature as a command-and-control approach.

The expansion planner considered in this chapter determines the optimal generation expansion plan to be carried out. However, it does not usually build the new electricity production facilities. It encourages private investors to build them, e.g., by using different types of incentives [24]. These private investors will, in turn, recover their investment costs by selling their production in the market and, if applicable, by subsidies.

The generation expansion decisions have several aspects, as explained below.

The most important one is deciding about the type and sizing of the candidate generating units to be built in the system. Given the existing generation portfolio of an electric energy system as well as its future needs (e.g., its future demand and future changes in the system topology), we determine the optimal type and capacity of the generating units to build. These decisions are also affected by the decommissioning of old generating units as well as by the investment and production costs of the candidate generating units.

The capacities of the candidate units to build in the system constitute the basic and essential generation expansion decisions. However, it is also possible to decide on the best location to build the new generating units. Many systems have most of the demand concentrated in a given region. Moreover, with the increasing penetration of renewable energy generating units, the transmission network is usually congested [6], so that it is generally important to decide where to build the new generating units in order to alleviate system congestion. To do so, it is necessary to represent the network in the GEP problem [14, 16].

Besides deciding the optimal sizing and location of the new generating units, it is generally important to determine the optimal timing to build these electricity production facilities. In this sense, it is common to consider that the generation expansion plans are made at a single point in time, i.e., to consider a static approach [9, 23]. This allows us to formulate a relatively simple problem. However, it is sometimes required to make the generation expansion plans at different points in time, i.e., to consider a dynamic approach [3, 7]. These dynamic generation expansion decisions are generally more accurate; however, we pay the cost of formulating and solving a more complex and possibly intractable problem.

Finally, another important issue in the GEP decision-making problem is that the expansion decisions are usually made within an uncertain environment, which further complicates the problem.

Within this framework, we provide and analyze different GEP models that include the main characteristics described above.

The remainder of this chapter is organized as follows. Section 3.2 describes the main characteristics of the models described in this chapter. Section 3.3 provides a basic model for the GEP problem based on a deterministic single-node static approach. This model is extended in the following sections in order to represent the impact of different aspects on the GEP problem. Section 3.4 describes a dynamic approach in which the generation expansion plans can be made at different points in time. Section 3.5 provides a network-constrained approach in which network constraints are modeled in the GEP decision-making problem. Section 3.6 describes a stochastic approach, in which we model the impact of uncertain parameters on the GEP problem. Sections 3.3–3.6 include clarifying examples that illustrate the working and characteristics of the reported models. Section 3.7 summarizes the chapter and discusses the main conclusions stemming from the models and results. Section 3.8 provides some exercises to enable a deeper understanding of the models and concepts presented in the chapter. Finally, Sect. 3.9 provides the GAMS codes for solving some of the illustrative examples.

3.2 Problem Description

In this chapter, we describe and formulate the GEP problem, considering different approaches. However, all of them have some common characteristics, which are described in this section for the sake of clarity.

3.2.1 Notation

The main notation used in this chapter is provided below for quick reference. Other symbols are defined as needed throughout the chapter. The observations below are in order:

1. A subscript o in the symbols below indicate their values in the oth operating condition.
2. A subscript t in the symbols below indicate their values in the tth time period.
3. A subscript ω in the symbols below indicate their values in the ωth scenario.

Indices:

c Candidate generating units.
d Demands.
g Existing generating units.
ℓ Transmission lines.
n Nodes.
o Operating conditions.
t Time periods.
ω Scenarios.

Sets:

$r(\ell)$ Receiving-end node of transmission line ℓ.
$s(\ell)$ Sending-end node of transmission line ℓ.
Ω_n^{C} Candidate generating units located at node n.
Ω_n^{D} Demands located at node n.
Ω_n^{E} Existing generating units located at node n.

Parameters:

A Amortization rate [%].
B_ℓ Susceptance of transmission line ℓ [S].
C_c^{C} Production cost of candidate generating unit c [\$/MWh].
C_g^{E} Production cost of existing generating unit g [\$/MWh].
C_d^{LS} Load-shedding cost of demand d [\$/MWh].
$F_\ell^{\max}$ Capacity of transmission line ℓ [MW].
I_c^{C} Investment cost of candidate generating unit c [\$/MW].
$\tilde{I}_c^{\mathrm{C}}$ Annualized investment cost of candidate generating unit c [\$/MW].
$\overline{P}_c^{\mathrm{C}\max}$ Maximum production capacity investment of candidate generating unit c [MW].
P_{cq}^{Option} Production capacity of investment option q of candidate generating unit c [MW].
P_d^{D} Load of demand d [MW].
$P_g^{\mathrm{E}\max}$ Production capacity of existing generating unit g [MW].
φ_ω Probability of scenario ω [p.u.].
ρ_o Weight of operating condition o [h].

Binary Variables:

u_{cq}^{Option} Binary variable that is equal to 1 if option q determines the capacity to be built of candidate generating unit c and 0 otherwise.

Continuous Variables:

p_c^{C} Power produced by candidate generating unit c [MW].
$p_c^{\mathrm{C^{max}}}$ Capacity of candidate generating unit c [MW].
p_g^{E} Power produced by existing generating unit g [MW].
p_ℓ^{L} Power flow through transmission line ℓ [MW].
θ_n Voltage angle at node n [rad].

3.2.2 Aim and Assumptions

We consider a central planner (e.g., the ISO) that aims to determine the generation expansion plan that is most beneficial for the electric energy system as a whole. Subsequently, this central planner incentivizes private investors to build these electricity production facilities, e.g., by using different types of incentives.

In order to determine the generation expansion plan that is most beneficial for the system as a whole, it is possible to consider different objective functions. In this chapter, we consider that the expansion planner determines the generation expansion plan that maximizes the overall social welfare. Since building new electricity production facilities has a cost, we also include the investment costs in the objective function.

For the sake of simplicity in the formulation and description of the models, in the following we consider that demands are inelastic, and thus the social welfare is equivalent to minus the generation cost.

Finally, we assume that during the considered planning horizon, there are no changes in the system topology, i.e., the existing transmission lines remain the same. The joint generation and transmission expansion planning (G&TEP) problem is the topic of Chap. 4. Moreover, we do not consider the commissioning of old units for the sake of simplicity.

3.2.3 Time Framework

The generation expansion plan is determined for a long-term planning horizon, e.g., 20 years. In this sense, there are two different ways of tackling the GEP problem depending on when the electricity production facilities are built:

1. A *static* model in which the generation expansion decisions are made only at the beginning of the planning horizon, i.e., at a single point in time.
2. A *dynamic* model in which the generation expansion decisions are made at different points in time.

The modeling of the planning horizon depends on whether a static or a dynamic model is considered.

Considering a static model, we represent the whole planning horizon by a single year, which is considered the reference or target year. The system needs (e.g., the demand in the system) are represented for this target year, and the generation expansion decisions are made for this reference year. Since the generation expansion plan is mainly conditioned by the demand in the system, which usually increases over time, the reference year is usually selected as the last year of the planning horizon. This is so because the generation expansion plans must be made for the whole planning horizon, and thus we should consider the largest expected demand.

If a dynamic approach is considered, then the planning horizon is divided into different time periods, each one comprising a specific number of years. In turn, each time period is represented by a single year (generally, the last year of each time period), which is considered the reference or target year of the whole time period. In this case, we assume that the generation expansion decisions can be made at the beginning of each time period.

Figure 3.1 illustrates the differences between the static and dynamic approaches in the GEP problem.

The advantage of using a static approach for the GEP problem is that the resulting model is relatively simple. However, it has some disadvantages. One is that the generation expansion plan is made for the last year of the planning horizon (the reference year). Since the GEP problem is solved for a long-term planning horizon, the demand in the system in this reference year will be probably much higher than the demand in the system at the present or in the short term. Therefore, the generation expansion plan will probably result in an overcapacity that is not needed until the last years of the planning horizon. Another disadvantage of using such a static model

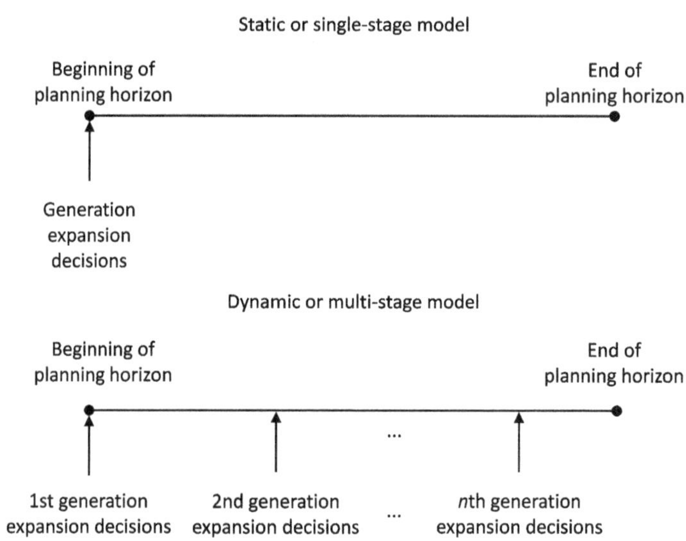

Fig. 3.1 Static and dynamic GEP models

is that for a long-term planning horizon, it is difficult to obtain an accurate forecast of the system conditions (e.g., the demand) in the future. Thus, if an unexpected change in the system occurs (e.g., if the future demand is lower than expected when the generation expansion plan is decided), then it is not possible to adapt to these changes since the generating units have already been built.

The above problems are solved by considering a dynamic approach, which allows us to increase the capacity as needed instead of building all the capacity at the beginning of the planning horizon. This also allows us to adapt to possible changes in the system conditions throughout the planning horizon.

3.2.4 Operating Conditions

We consider that the generation expansion planner determines the expansion plan that maximizes the overall social welfare and minimizes the investment costs. Thus, in order to compute the social welfare, it is necessary to model the different operating conditions that will take place in the system under study in the considered planning horizon. These operating conditions can represent different demand realizations, different renewable production conditions, or different system conditions (e.g., the failure of a transmission line).

For the sake of clarity, in the following we assume that only the demand affects the different operating conditions.

To model demand operating conditions, one alternative is to consider historical data in the system under study to predict the operating conditions in the considered planning horizon. For example, let us consider historical data of the demand in the system. If we divide these demand values by the peak demand in the system, then we obtain a set of historical values of demand levels. While the demand in the system will probably increase in the future, we can assume that the demand levels will remain approximately the same. Thus, we can use historical demand levels multiplied by the future expected peak demand to predict future operating conditions.

However, note that historical data of demand usually comprise thousands of hourly demand values. Thus, it is impractical to work with such a large data set. Nevertheless, many of these historical hourly values are similar. Thus, it is possible to obtain a reduced data set from the historical data that groups similar historical data. For this purpose, there are several techniques available in the technical literature, e.g., methods based on the load–duration curve and clustering methods [4].

These methods use as input historical data of demand levels, e.g., hourly demand levels throughout a year (i.e., 8760 demand level values). Then, the clustering methods check the similarities among the historical demand levels and group those demand levels that are similar. The output of these methods is a reduced set of operating conditions, each one defined by a demand level and the relative weight of this operating condition in the historical data analyzed. This reduced data set is used to represent the operating conditions in the considered planning horizon.

Note that these methods allow us to maintain the information of the historical data as well as possible correlations in the historical data, e.g., correlation among demands in different locations.

Illustrative Example 3.1 *Operating conditions*

Historical data of demand in an electric energy system are used as input for the K-means clustering technique [4]. The output of this method are two operating conditions. The first is characterized by a demand level of 0.4833 p.u. and a relative weight of 0.6849 p.u. The second is characterized by a demand level of 0.9167 p.u. and a relative weight of 0.3151 p.u.

Let us consider that the GEP problem involves a 20-year planning horizon. This planning horizon is represented by a single target year, whose expected peak demand is 600 MW. The two operating conditions in this target year have the following characteristics:

1. Operating condition o_1 is characterized by a demand of 290 MW (600 MW $\times$ 0.4833) and a weight of 6000 h (8760 h $\times$ 0.6849).
2. Operating condition o_2 is characterized by a demand of 550 MW (600 MW $\times$ 0.9167) and a weight of 2760 h (8760 h $\times$ 0.3151).

□

An important issue is selecting the number of operating conditions that are used in the GEP problem. This number should be large enough to represent the different operating conditions throughout the year accurately. However, if the number of operating conditions considered is very large, then the GEP problem may become computationally intractable. Thus, we should select a number of operating conditions that constitutes an appropriate tradeoff between modeling accuracy and computational tractability.

Note that in this chapter, we use the output of the clustering methods as an input to the GEP problem. We do not describe in detail the working of these clustering methods. The interested reader is referred to [4] for further details on clustering techniques to derive operating conditions.

3.2.5 Uncertainty Characterization

Under the uncertainty characterization point of view, it is possible to formulate two different GEP models:

1. A *deterministic* model: We consider that the generation expansion planner has perfect information at the time it determines the generation expansion plan, e.g., the planner knows the future demand in the system perfectly.
2. A *stochastic* model: We consider that the generation expansion decisions are made within an uncertain environment, and thus this uncertainty should be considered in the GEP decision-making problem in order to obtain informed decisions. Among

the different techniques available to deal with uncertainty, in this chapter we consider a stochastic programming framework [10] in which uncertainties are modeled through a set of scenarios indexed by ω.

Although considering a deterministic approach allows us to formulate a relatively simple problem, in the real world the generation expansion plans are decided for a long-term planning horizon, and thus it is hard to obtain a good forecast of the system conditions for the whole planning horizon. This lack of perfect information generally conditions the generation expansion plans.

For the sake of simplicity, in the following, we assume that uncertainty affects only the demand in the system. However, additional sources of uncertainty may be considered through additional scenarios.

3.2.6 Modeling of the Transmission Network

Representing the transmission network leads to two different GEP models:

1. A *single-node* model: In this case, we do not model the network constraints, i.e., we solve the GEP problem considering that all demands and generating units are connected to a single virtual node. Therefore, the solution of this GEP problem is the optimal sizing of the generating units to be built in the system, but not where to build them.
2. A *network-constrained* model: In this case, we explicitly model the network constraints. Therefore, the solution of the GEP problem in this case is the optimal sizing as well as the optimal location of the generating units to be built in the system.

Considering a single-node approach for the GEP problem allows us to formulate a relatively simple problem. However, it is important to consider the network constraints, especially in systems with congested transmission lines.

3.2.7 Complementarity Model

As previously explained, the generation expansion planner aims to determine the generation expansion plan that maximizes the overall social welfare and that minimizes the investment costs. On the one hand, the social welfare can be computed from the result of market clearing. On the other hand, the clearing of the market is affected by the generation expansion plan decided by the generation expansion planner. Thus, it is necessary to represent explicitly the clearing of the market in the GEP decision-making problem. Moreover, we should represent the market clearing for different operating conditions, time periods, and scenarios.

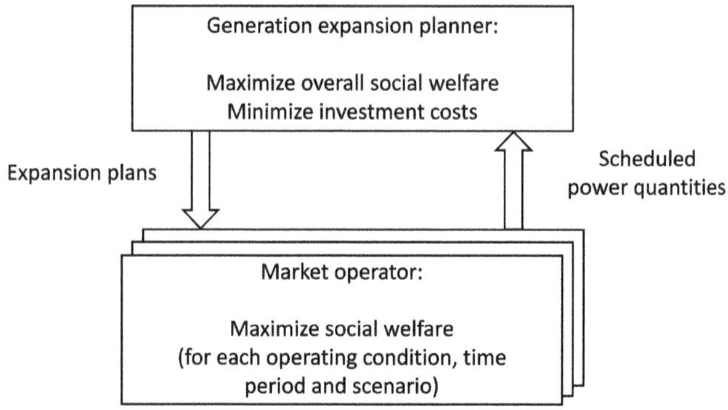

Fig. 3.2 Complementarity GEP model

However, the clearing of the market is itself an optimization problem in which the market operator (MO) receives offers and bids from producers and consumers, respectively, and determines the scheduled power quantities to be supplied by producers and consumed by demands that maximize the overall social welfare. As a result, the GEP problem becomes an optimization problem subject to other optimization problems (the market-clearing problem for each operating condition, time period, and scenario). This kind of problem is usually known in the technical literature as a bilevel, hierarchical, or complementarity model [11].

The structure of this complementarity model is schematically depicted in Fig. 3.2. The generation expansion planner decides the optimal generation expansion plan to be carried out, i.e., it determines the optimal generating units to be built in the system. On the one hand, the information about the generation expansion plan is used in market-clearing problems. On the other hand, the output of these market-clearing problems is the scheduled power quantities to be produced by the existing and candidate generating units, which are used in turn by the generation expansion planner to compute the social welfare. Additional details of these two problems are provided in the sections below.

Among the different market settings, we represent only the clearing of the day-ahead market since it is generally the market with the largest volume of energy trading.

3.3 Deterministic Single-Node Static GEP

In this section we describe the GEP problem, considering a deterministic single-node static approach that has the following characteristics:

1. We consider a deterministic approach: We assume that there is no uncertainty, i.e., the generation expansion planner has perfect information on all the parameters that affect the GEP problem.

2. We consider a single-node approach: We do not represent the network constraints. As a result, we determine the sizing of the generating units to be built but not where to build them.
3. We consider a static approach: We consider that the generation expansion plan can be made only at the beginning of the planning horizon.

The above modeling framework allows us to formulate a simple problem that will be used to illustrate the main characteristics of the GEP problem. This simple problem will be progressively enhanced in the following sections.

The next section provides the formulation of the complementarity model used for the GEP problem considering a deterministic single-node static approach. This complementarity model can be reformulated as a nonlinear programming (NLP) model, which can be in turn recast as an equivalent mixed-integer linear programming (MILP) model.

3.3.1 Complementarity Model

The GEP problem considering a deterministic single-node static approach can be formulated using the following complementarity model:

$$\min_{p_c^{\mathrm{C^{max}}}, \Delta_o^{\mathrm{D1SM}}} \quad \sum_o \rho_o \left[\sum_g C_g^{\mathrm{E}} p_{go}^{\mathrm{E}} + \sum_c C_c^{\mathrm{C}} p_{co}^{\mathrm{C}} \right] + \sum_c \tilde{I}_c^{\mathrm{C}} p_c^{\mathrm{C^{max}}} \tag{3.1a}$$

subject to

$$0 \leq p_c^{\mathrm{C^{max}}} \leq \overline{P}_c^{\mathrm{C^{max}}} \quad \forall c \tag{3.1b}$$

$$p_{go}^{\mathrm{E}}, p_{co}^{\mathrm{C}} \in \Omega_o \left(P_{do}^{\mathrm{D}}, P_g^{\mathrm{E^{max}}}, p_c^{\mathrm{C^{max}}} \right) \quad \forall g, c, o, \tag{3.1c}$$

where $p_c^{\mathrm{C^{max}}}$ and variables in set $\Delta_o^{\mathrm{D1SM}} = \{p_{go}^{\mathrm{E}},\, p_{co}^{\mathrm{C}}\}$ are the optimization variables of problem (3.1).

Objective function (3.1a) comprises two terms:

1. $\sum_g C_g^{\mathrm{E}} p_{go}^{\mathrm{E}} + \sum_c C_c^{\mathrm{C}} p_{co}^{\mathrm{C}}, \forall o$, are the generation costs (of both the existing and the candidate generating units) for each operating condition o.
2. $\sum_c \tilde{I}_c^{\mathrm{C}} p_c^{\mathrm{C^{max}}}$ is the annualized investment cost incurred in building new generating units.

The terms in 1 above are multiplied by the weight of the corresponding operating condition in the target year in order to make the generation costs and the annualized investment cost comparable.

Note that we consider as objective function of the GEP problem the overall social welfare (the generation cost is equivalent to minus the social welfare since demands are considered inelastic) and the investment cost.

Constraints of the GEP problem (3.1) include constraints (3.1b), which impose bounds on the capacity of the candidate generating units to be built, and constraints (3.1c), which state that the scheduled production of the existing and the candidate generating units is obtained from the clearing of the market, represented by sets $\Omega_o \left(P^{\mathrm{D}}_{do}, P^{\mathrm{E}^{\max}}_g, p^{\mathrm{C}^{\max}}_c\right), \forall o$.

The clearing of the market is itself another optimization problem, whose detailed formulation is provided below:

$$\Omega_o \left(P^{\mathrm{D}}_{do}, P^{\mathrm{E}^{\max}}_g, p^{\mathrm{C}^{\max}}_c\right) = \Bigg\{$$

$$\min_{\Delta_o^{\mathrm{D1SM}}} \quad \sum_g C^{\mathrm{E}}_g p^{\mathrm{E}}_{go} + \sum_c C^{\mathrm{C}}_c p^{\mathrm{C}}_{co} \tag{3.2a}$$

subject to

$$\sum_g p^{\mathrm{E}}_{go} + \sum_c p^{\mathrm{C}}_{co} = \sum_d P^{\mathrm{D}}_{do} \ : \ \lambda_o \tag{3.2b}$$

$$0 \leq p^{\mathrm{E}}_{go} \leq P^{\mathrm{E}^{\max}}_g \ : \ \mu^{\mathrm{E}}_{go} \quad \forall g \tag{3.2c}$$

$$0 \leq p^{\mathrm{C}}_{co} \leq p^{\mathrm{C}^{\max}}_c \ : \ \mu^{\mathrm{C}}_{co} \quad \forall c \tag{3.2d}$$

$$\Bigg\}, \forall o,$$

where the variables in sets $\Delta_o^{\mathrm{D1SM}} = \left\{p^{\mathrm{E}}_{go}, p^{\mathrm{C}}_{co}\right\}, \forall o$, are the optimization variables of problems (3.2), $\forall o$. The dual variables associated to constraints of problem (3.2) are provided following a colon.

Objective function (3.2a) represents the generation cost, which is equivalent in this case to minus the social welfare. Constraints (3.2b) define the generation–demand balance. Finally, constraints (3.2c) and (3.2d) impose bounds on the power quantities to be supplied by the existing and the candidate generating units, respectively. Note that the upper bounds for the existing generating units are the already built capacities, while the upper bounds for the candidate generating units are the capacities to be built. These capacities are decision variables of the GEP problem (3.1).

We consider a competitive market in which producers offer their capacities at their marginal costs. The strategic behavior of producers is analyzed in Chap. 5.

Illustrative Example 3.2 *Deterministic single-node static GEP problem: Complementarity model formulation*

Let us consider an electric energy system with the following characteristics:

1. There is one generating unit $\tilde{g}$ with capacity of 400 MW and production cost equal to \$35/MWh.
2. It is possible to build a new generating unit $\tilde{c}$ with capacity up to 500 MW and production cost equal to \$25/MWh. The annualized investment cost is \$70,000 per MW.
3. Demand conditions in the system are represented through two operating conditions. The first one, o_1, is defined by a demand of 290 MW and a weight of 6000 h, while the second one, o_2, is defined by a demand of 550 MW and a weight of 2760 h.

Given the above data, the GEP problem (3.1) results in the following complementarity model:

$$\min_{p^{\mathrm{E}}_{\tilde{g}o_1}, p^{\mathrm{C}}_{\tilde{c}o_1}, p^{\mathrm{E}}_{\tilde{g}o_2}, p^{\mathrm{C}}_{\tilde{c}o_2}, p^{\mathrm{C}^{\max}}_{\tilde{c}}} \quad 6000\left[35p^{\mathrm{E}}_{\tilde{g}o_1} + 25p^{\mathrm{C}}_{\tilde{c}o_1}\right] + 2760\left[35p^{\mathrm{E}}_{\tilde{g}o_2} + 25p^{\mathrm{C}}_{\tilde{c}o_2}\right] + 70000p^{\mathrm{C}^{\max}}_{\tilde{c}}$$

subject to

$$0 \le p^{\mathrm{C}^{\max}}_{\tilde{c}} \le 500$$

$$\left\{\begin{array}{l} \min_{p^{\mathrm{E}}_{\tilde{g}o_1}, p^{\mathrm{C}}_{\tilde{c}o_1}} \quad 35p^{\mathrm{E}}_{\tilde{g}o_1} + 25p^{\mathrm{C}}_{\tilde{c}o_1} \\ \text{s.t.} \\ \quad p^{\mathrm{E}}_{\tilde{g}o_1} + p^{\mathrm{C}}_{\tilde{c}o_1} = 290 \\ \quad 0 \le p^{\mathrm{E}}_{\tilde{g}o_1} \le 400 \\ \quad 0 \le p^{\mathrm{C}}_{\tilde{c}o_1} \le p^{\mathrm{C}^{\max}}_{\tilde{c}} \end{array}\right.$$

$$\left\{\begin{array}{l} \min_{p^{\mathrm{E}}_{\tilde{g}o_2}, p^{\mathrm{C}}_{\tilde{c}o_2}} \quad 35p^{\mathrm{E}}_{\tilde{g}o_2} + 25p^{\mathrm{C}}_{\tilde{c}o_2} \\ \text{s.t.} \\ \quad p^{\mathrm{E}}_{\tilde{g}o_2} + p^{\mathrm{C}}_{\tilde{c}o_2} = 550 \\ \quad 0 \le p^{\mathrm{E}}_{\tilde{g}o_2} \le 400 \\ \quad 0 \le p^{\mathrm{C}}_{\tilde{c}o_2} \le p^{\mathrm{C}^{\max}}_{\tilde{c}}. \end{array}\right.$$

□

Note that the complementarity model comprises problems (3.1) and (3.2), $\forall o$, which are jointly solved following the procedure described in the section below.

3.3.2 Equivalent NLP Formulation

Each market-clearing problem (3.2) (one for each operating condition o) is a linear programming (LP) problem. Thus, it is possible to replace each of these problems by its first-order optimality conditions, which are in turn used to replace sets $\Omega_o\left(P^{\mathrm{D}}_{do}, P^{\mathrm{E^{max}}}_{g}, p^{\mathrm{C^{max}}}_{c}\right)$, $\forall o$, in problem (3.1). Thus, complementarity model (3.1) results in a single-level problem, generally known in the technical literature as a mathematical program with equilibrium constraints (MPEC) [11].

The first-order optimality conditions can be formulated using one of the two approaches below:

1. Karush–Kuhn–Tucker (KKT) formulation: In this case, each market-clearing problem (3.2) is replaced by its KKT conditions [11, 17].
2. Primal–dual formulation: In this case, each market-clearing problem (3.2) is replaced by its primal constraints, its dual constraints, and its strong duality equality [11, 18].

The formulations of the MPEC based on the above two approaches are equivalent. However, using the KKT conditions requires handling complementarity constraints, which are nonlinear. Thus, it is generally easier to solve the MPEC that results from the primal–dual formulation, which does not include such complementarity conditions. Thus, the primal–dual formulation is the approach used in this chapter.

The dual problem of the market-clearing problem (3.2) is provided below:

$$\Omega_o\left(P^{\mathrm{D}}_{do}, P^{\mathrm{E^{max}}}_{g}, p^{\mathrm{C^{max}}}_{c}\right) = \Bigg\{$$

$$\max_{\Delta_o^{\mathrm{D1SM,D}}} \quad \lambda_o \sum_d P^{\mathrm{D}}_{do} - \sum_g \mu^{\mathrm{E^{max}}}_{go} P^{\mathrm{E^{max}}}_{g} - \sum_c \mu^{\mathrm{C^{max}}}_{co} p^{\mathrm{C^{max}}}_{c} \tag{3.3a}$$

subject to

$$C^{\mathrm{E}}_{g} - \lambda_o + \mu^{\mathrm{E^{max}}}_{go} \geq 0 \quad \forall g \tag{3.3b}$$

$$C^{\mathrm{C}}_{c} - \lambda_o + \mu^{\mathrm{C^{max}}}_{co} \geq 0 \quad \forall c \tag{3.3c}$$

$$\mu^{\mathrm{E^{max}}}_{go} \geq 0 \quad \forall g \tag{3.3d}$$

$$\mu^{\mathrm{C^{max}}}_{co} \geq 0 \quad \forall c \tag{3.3e}$$

$$\Bigg\}, \forall o,$$

where variables in sets $\Delta_o^{\mathrm{D1SM,D}} = \{\lambda_o, \mu^{\mathrm{E^{max}}}_{go}, \mu^{\mathrm{C^{max}}}_{co}\}$, $\forall o$, are the optimization variables of problems (3.3), $\forall o$.

Then we replace the market-clearing problem (3.2), $\forall o$, by its primal constraints, its dual constraints, and its strong duality equality, rendering an MPEC whose formulation is as follows:

$$\min_{\Delta^{\mathrm{D1S,MPEC}}} \quad \sum_{o} \rho_o \left[\sum_{g} C_g^{\mathrm{E}} p_{go}^{\mathrm{E}} + \sum_{c} C_c^{\mathrm{C}} p_{co}^{\mathrm{C}} \right] + \sum_{c} \tilde{I}_c^{\mathrm{C}} p_c^{\mathrm{C^{max}}} \tag{3.4a}$$

subject to

$$0 \leq p_c^{\mathrm{C^{max}}} \leq \overline{P}_c^{\mathrm{C^{max}}} \quad \forall c \tag{3.4b}$$

$$\left\{ \sum_{g} p_{go}^{\mathrm{E}} + \sum_{c} p_{co}^{\mathrm{C}} = \sum_{d} P_{do}^{\mathrm{D}} \right. \tag{3.4c}$$

$$0 \leq p_{go}^{\mathrm{E}} \leq P_g^{\mathrm{E^{max}}} \quad \forall g \tag{3.4d}$$

$$0 \leq p_{co}^{\mathrm{C}} \leq p_c^{\mathrm{C^{max}}} \quad \forall c \tag{3.4e}$$

$$C_g^{\mathrm{E}} - \lambda_o + \mu_{go}^{\mathrm{E^{max}}} \geq 0 \quad \forall g \tag{3.4f}$$

$$C_c^{\mathrm{C}} - \lambda_o + \mu_{co}^{\mathrm{C^{max}}} \geq 0 \quad \forall c \tag{3.4g}$$

$$\mu_{go}^{\mathrm{E^{max}}} \geq 0 \quad \forall g \tag{3.4h}$$

$$\mu_{co}^{\mathrm{C^{max}}} \geq 0 \quad \forall c \tag{3.4i}$$

$$\sum_{g} C_g^{\mathrm{E}} p_{go}^{\mathrm{E}} + \sum_{c} C_c^{\mathrm{C}} p_{co}^{\mathrm{C}} = \lambda_o \sum_{d} P_{do}^{\mathrm{D}} - \sum_{g} \mu_{go}^{\mathrm{E^{max}}} P_g^{\mathrm{E^{max}}} - \sum_{c} \mu_{co}^{\mathrm{C^{max}}} p_c^{\mathrm{C^{max}}} \tag{3.4j}$$

$$\left. \right\}, \forall o,$$

where variables in set $\Delta^{\mathrm{D1S,MPEC}} = \left\{ p_c^{\mathrm{C^{max}}}, \Delta_o^{\mathrm{D1SM}}, \Delta_o^{\mathrm{D1SM,D}} \right\}$ are the optimization variables of problem (3.4).

Constraints (3.4b) are the constraints of problem (3.1), constraints (3.4c)–(3.4e) are the primal constraints of the market-clearing problems (3.2), $\forall o$, constraints (3.4f)–(3.4h) are the dual constraints of the market-clearing problems (3.2), $\forall o$, and constraints (3.4i) are the strong duality equalities, which impose that the primal and the dual objective functions of the market-clearing problems (3.2), $\forall o$, have the same value at the optimum [8].

Illustrative Example 3.3 *Deterministic single-node static GEP problem: MPEC formulation*

Considering Illustrative Example 3.2, the corresponding MPEC formulation is provided below:

$$\min_{\Delta} \quad 6000 \left[35 p_{\tilde{g}o_1}^{\mathrm{E}} + 25 p_{\tilde{c}o_1}^{\mathrm{C}} \right] + 2760 \left[35 p_{\tilde{g}o_2}^{\mathrm{E}} + 25 p_{\tilde{c}o_2}^{\mathrm{C}} \right] + 70000 p_{\tilde{c}}^{\mathrm{C^{max}}}$$

subject to

$$0 \leq p_{\tilde{c}}^{\mathrm{C^{max}}} \leq 500$$

$$\begin{cases} p_{\tilde{g}o_1}^{\mathrm{E}} + p_{\tilde{c}o_1}^{\mathrm{C}} = 290 \\ 0 \leq p_{\tilde{g}o_1}^{\mathrm{E}} \leq 400 \\ 0 \leq p_{\tilde{c}o_1}^{\mathrm{C}} \leq p_{\tilde{c}}^{\mathrm{C^{max}}} \\ 35 - \lambda_{o_1} + \mu_{\tilde{g}o_1}^{\mathrm{E^{max}}} \geq 0 \\ 25 - \lambda_{o_1} + \mu_{\tilde{c}o_1}^{\mathrm{C^{max}}} \geq 0 \\ \mu_{\tilde{g}o_1}^{\mathrm{E^{max}}} \geq 0 \\ \mu_{\tilde{c}o_1}^{\mathrm{C^{max}}} \geq 0 \\ 35p_{\tilde{g}o_1}^{\mathrm{E}} + 25p_{\tilde{c}o_1}^{\mathrm{C}} = 290\lambda_{o_1} - 400\mu_{\tilde{g}o_1}^{\mathrm{E^{max}}} - p_{\tilde{c}}^{\mathrm{C^{max}}} \mu_{\tilde{c}o_1}^{\mathrm{C^{max}}} \end{cases}$$

$$\begin{cases} p_{\tilde{g}o_2}^{\mathrm{E}} + p_{\tilde{c}o_2}^{\mathrm{C}} = 550 \\ 0 \leq p_{\tilde{g}o_2}^{\mathrm{E}} \leq 400 \\ 0 \leq p_{\tilde{c}o_2}^{\mathrm{C}} \leq p_{\tilde{c}}^{\mathrm{C^{max}}} \\ 35 - \lambda_{o_2} + \mu_{\tilde{g}o_2}^{\mathrm{E^{max}}} \geq 0 \\ 25 - \lambda_{o_2} + \mu_{\tilde{c}o_2}^{\mathrm{C^{max}}} \geq 0 \\ \mu_{\tilde{g}o_2}^{\mathrm{E^{max}}} \geq 0 \\ \mu_{\tilde{c}o_2}^{\mathrm{C^{max}}} \geq 0 \\ 35p_{\tilde{g}o_2}^{\mathrm{E}} + 25p_{\tilde{c}o_2}^{\mathrm{C}} = 550\lambda_{o_2} - 400\mu_{\tilde{g}o_2}^{\mathrm{E^{max}}} - p_{\tilde{c}}^{\mathrm{C^{max}}} \mu_{\tilde{c}o_2}^{\mathrm{C^{max}}}, \end{cases}$$

where $\Delta = \left\{p_{\tilde{c}}^{\mathrm{C^{max}}}, p_{\tilde{g}o_1}^{\mathrm{E}}, p_{\tilde{c}o_1}^{\mathrm{C}}, p_{\tilde{g}o_2}^{\mathrm{E}}, p_{\tilde{c}o_2}^{\mathrm{C}}, \lambda_{o_1}, \mu_{\tilde{g}o_1}^{\mathrm{E^{max}}}, \mu_{\tilde{c}o_1}^{\mathrm{C^{max}}}, \lambda_{o_1}, \mu_{\tilde{g}o_1}^{\mathrm{E^{max}}}, \mu_{\tilde{c}o_1}^{\mathrm{C^{max}}}\right\}$. □

Note that problem (3.4) is an NLP problem since it includes nonlinear terms $\mu_{co}^{\mathrm{C^{max}}} p_c^{\mathrm{C^{max}}}, \forall c, o$, in constraints (3.4j), $\forall o$. NLP problems are generally hard to solve, and convergence to the optimum is not guaranteed [8]. However, it is possible to transform problem (3.4) into an equivalent MILP problem, as explained in the following section.

3.3.3 Equivalent MILP Formulation

In the GEP problem (3.4), the capacity of the candidate generating units to build in the system is defined by variables $p_c^{\mathrm{C^{max}}}, \forall c$, which can take any value between zero and $\overline{P}_c^{\mathrm{C^{max}}}$, i.e., we consider that $p_c^{\mathrm{C^{max}}}$ is a nonnegative continuous variable. However, this capacity is generally a discrete decision variable since generating units are usually built in blocks of a predetermined size. Therefore, it is necessary to reformulate constraints (3.4b), $\forall c$, in order to take into account such a fact. These constraints can be reformulated as follows:

$$p_c^{\text{C}^{\max}} = \sum_q u_{cq}^{\text{Option}} P_{cq}^{\text{Option}} \quad \forall c \tag{3.5a}$$

$$\sum_q u_{cq}^{\text{Option}} = 1 \quad \forall c \tag{3.5b}$$

$$u_{cq}^{\text{Option}} \in \{0, 1\} \quad \forall c, q. \tag{3.5c}$$

The working of constraints (3.5) is illustrated with the following example. Let us consider that it is possible to build up to 300 MW of new capacity in 100 MW blocks for a candidate generating unit $\hat{c}$. Thus, we define parameters $P_{\hat{c}q}^{\text{Option}}$, $\forall q$, as $P_{\hat{c}q_1}^{\text{Option}} = 0$, $P_{\hat{c}q_2}^{\text{Option}} = 100$, $P_{\hat{c}q_3}^{\text{Option}} = 200$, and $P_{\hat{c}q_4}^{\text{Option}} = 300$. If the optimal capacity to be built in the system is, for example, 200 MW, then binary variables $u_{\hat{c}q}^{\text{Option}}$, $\forall q$, are $u_{\hat{c}q_3}^{\text{Option}} = 1$ and $u_{\hat{c}q}^{\text{Option}} = 0$, $q = q_1, q_2, q_4$, so that $p_{\hat{c}}^{\text{C}^{\max}} = 200$ by constraints (3.5).

Using Eqs. (3.5), it is possible to reformulate nonlinear terms $\mu_{co}^{\text{C}^{\max}} p_c^{\text{C}^{\max}}$, $\forall c, o$, as $\mu_{co}^{\text{C}^{\max}} \sum_q u_{cq}^{\text{Option}} P_{cq}^{\text{Option}}$, $\forall c, o$. That is, using this reformulation we have products of a continuous variable and a binary variable, which can be transformed into exact equivalent mixed-integer linear expressions as explained below [1, 5, 12].

Let us consider $\mu_{co}^{\text{C}^{\max}} \sum_q u_{cq}^{\text{Option}} P_{cq}^{\text{Option}} = \sum_q \mu_{co}^{\text{C}^{\max}} u_{cq}^{\text{Option}} P_{cq}^{\text{Option}} = \sum_q z_{cqo}^{\text{AUX}}$, where $z_{cqo}^{\text{AUX}} = \mu_{co}^{\text{C}^{\max}} u_{cq}^{\text{Option}} P_{cq}^{\text{Option}}$, $\forall c, q, o$, are auxiliary variables equal to the products of a continuous variable, a binary variable, and a constant. These auxiliary variables can be rewritten as follows:

$$z_{cqo}^{\text{AUX}} = \mu_{co}^{\text{C}^{\max}} P_{cq}^{\text{Option}} - \hat{z}_{cqo}^{\text{AUX}} \quad \forall c, q, o \tag{3.6a}$$

$$0 \le z_{cqo}^{\text{AUX}} \le u_{cq}^{\text{Option}} M \quad \forall c, q, o \tag{3.6b}$$

$$0 \le \hat{z}_{cqo}^{\text{AUX}} \le \left(1 - u_{cq}^{\text{Option}}\right) M \quad \forall c, q, o, \tag{3.6c}$$

where $\hat{z}_{cqo}^{\text{AUX}}$, $\forall c, q, o$, are auxiliary variables and M is a large enough positive constant [5, 22].

The working of Eqs. (3.6) is explained next. On the one hand, let us consider that option $\hat{q}$ does not determine the optimal capacity of candidate generating unit $\hat{c}$, i.e., variable $u_{\hat{c}\hat{q}}^{\text{Option}} = 0$. In such a case, we have the following values of variables: $z_{\hat{c}\hat{q}o}^{\text{AUX}} = \mu_{\hat{c}o}^{\text{C}^{\max}} u_{\hat{c}\hat{q}}^{\text{Option}} P_{\hat{c}\hat{q}}^{\text{Option}} = 0$, $\forall o$. This is guaranteed by Eq. (3.6b), which impose $0 \le z_{\hat{c}\hat{q}o}^{\text{AUX}} \le 0$, $\forall o$, i.e., $z_{\hat{c}\hat{q}o}^{\text{AUX}} = 0$, $\forall o$, if $u_{\hat{c}\hat{q}}^{\text{Option}} = 0$. On the other hand, if option $\tilde{q}$ does determine the optimal capacity of candidate generating unit $\tilde{c}$, i.e., if variable $u_{\tilde{c}\tilde{q}}^{\text{Option}}$ is equal to 1, then variables $z_{\tilde{c}\tilde{q}o}^{\text{AUX}} = \mu_{\tilde{c}o}^{\text{C}^{\max}} P_{\tilde{c}\tilde{q}}^{\text{Option}}$, $\forall o$. This is guaranteed by Eq. (3.6a), which impose $z_{\tilde{c}\tilde{q}o}^{\text{AUX}} = \mu_{co}^{\text{C}^{\max}} P_{cq}^{\text{Option}} - \hat{z}_{cqo}^{\text{AUX}}$, $\forall o$, and by Eq. (3.6c), which impose $\hat{z}_{\tilde{c}\tilde{q}o}^{\mathrm{AUX}} = 0, \forall o$, if $u_{\tilde{c}\tilde{q}}^{\mathrm{Option}} = 1$. Note that Eq. (3.6b) impose in this case that $0 \leq z_{\tilde{c}\tilde{q}o}^{\mathrm{AUX}} \leq M, \forall o$, i.e., that nonnegative variables $z_{\tilde{c}\tilde{q}o}^{\mathrm{AUX}}, \forall o$, are below large enough bounds.

Constant M impose bounds for auxiliary variables z_{cqo}^{AUX} and $\hat{z}_{cqo}^{\mathrm{AUX}}$. These variables are used to compute non-linear terms $\mu_{co}^{\mathrm{C^{max}}} u_{cq}^{\mathrm{Option}} P_{cq}^{\mathrm{Option}}$ so that positive constant M must have a value larger than $\mu_{co}^{\mathrm{C^{max}}} P_{cq}^{\mathrm{Option}}$. However, note that the values of variables $\mu_{co}^{\mathrm{C^{max}}}$ are not known in advance. Additional details on how to select positive constant M are provided in [5, 22].

Note that using Eqs. (3.6) we get rid of non-linear terms $\mu_{co}^{\mathrm{C^{max}}} p_{c}^{\mathrm{C^{max}}}, \forall c, o$, and the GEP problem considering a deterministic single-node static approach can be finally formulated using the MILP model below:

$$\min_{\Delta^{\mathrm{DIS,MILP}}} \quad \sum_{o} \rho_o \left[\sum_{g} C_g^{\mathrm{E}} p_{go}^{\mathrm{E}} + \sum_{c} C_c^{\mathrm{C}} p_{co}^{\mathrm{C}} \right] + \sum_{c} \tilde{I}_c^{\mathrm{C}} p_c^{\mathrm{C^{max}}} \tag{3.7a}$$

subject to

$$p_c^{\mathrm{C^{max}}} = \sum_{q} u_{cq}^{\mathrm{Option}} P_{cq}^{\mathrm{Option}} \quad \forall c \tag{3.7b}$$

$$\sum_{q} u_{cq}^{\mathrm{Option}} = 1 \quad \forall c \tag{3.7c}$$

$$u_{cq}^{\mathrm{Option}} \in \{0, 1\} \quad \forall c, q \tag{3.7d}$$

$$\left\{ \sum_{g} p_{go}^{\mathrm{E}} + \sum_{c} p_{co}^{\mathrm{C}} = \sum_{d} P_{do}^{\mathrm{D}} \right. \tag{3.7e}$$

$$0 \leq p_{go}^{\mathrm{E}} \leq P_g^{\mathrm{E^{max}}} \quad \forall g \tag{3.7f}$$

$$0 \leq p_{co}^{\mathrm{C}} \leq p_c^{\mathrm{C^{max}}} \quad \forall c \tag{3.7g}$$

$$C_{go}^{\mathrm{E}} - \lambda_o + \mu_{go}^{\mathrm{E^{max}}} \geq 0 \quad \forall g \tag{3.7h}$$

$$C_{co}^{\mathrm{C}} - \lambda_o + \mu_{co}^{\mathrm{C^{max}}} \geq 0 \quad \forall c \tag{3.7i}$$

$$\mu_{go}^{\mathrm{E^{max}}} \geq 0 \quad \forall g \tag{3.7j}$$

$$\mu_{co}^{\mathrm{C^{max}}} \geq 0 \quad \forall c \tag{3.7k}$$

$$\sum_{g} C_g^{\mathrm{E}} p_{go}^{\mathrm{E}} + \sum_{c} C_c^{\mathrm{C}} p_{co}^{\mathrm{C}} = \lambda_o \sum_{d} P_{do}^{\mathrm{D}} - \sum_{g} \mu_{go}^{\mathrm{E^{max}}} P_g^{\mathrm{E^{max}}} - \sum_{c} \sum_{q} z_{cqo}^{\mathrm{AUX}} \tag{3.7l}$$

$$z_{cqo}^{\mathrm{AUX}} = \mu_{co}^{\mathrm{C^{max}}} P_{cq}^{\mathrm{Option}} - \hat{z}_{cqo}^{\mathrm{AUX}} \quad \forall c, q \tag{3.7m}$$

$$0 \leq z_{cqo}^{\mathrm{AUX}} \leq u_{cq}^{\mathrm{Option}} M \quad \forall c, q \tag{3.7n}$$

$$0 \leq \hat{z}_{cqo}^{\mathrm{AUX}} \leq \left(1 - u_{cq}^{\mathrm{Option}}\right) M \quad \forall c, q \tag{3.7o}$$

$$\Bigg\}, \forall o,$$

where variables in set $\Delta^{\text{D1S,MILP}} = \{p_c^{\text{C}^{\max}}, u_{cq}^{\text{Option}}, \Delta_o^{\text{D1SM}}, \Delta_o^{\text{D1SM,D}}, \Delta_o^{\text{1SM,DUAL}}, \Delta_o^{\text{1SM,AUX}}\}$ are the optimization variables of problem (3.7), $\Delta^{\text{D1S,AUX}} = \left\{z_{cqo}^{\text{AUX}}, \hat{z}_{cqo}^{\text{AUX}}\right\}$, and M is a large enough positive constant.

Illustrative Example 3.4 *Deterministic single-node static GEP problem: Solution*

Let us consider Illustrative Example 3.2. The capacity of the candidate unit is available in 100 MW blocks, up to 5 blocks.

By solving problem (3.7) for these data, we obtain that the optimal generation expansion plan consists of building 300 MW of the candidate generating unit, i.e., variable $u_{\tilde{c}q_4}^{\text{Option}} = 1$ and variables $u_{\tilde{c}q}^{\text{Option}} = 0, \forall q \neq q_4$.

The power produced by the existing generating unit is 0 and 250 MW for operating conditions o_1 and o_2, respectively. As the generation costs of the candidate generating unit built are cheaper than those of the existing one, the power produced by the candidate generating unit is 290 and 300 MW for operating conditions o_1 and o_2, respectively. □

3.3.4 Meaning of Dual Variables λ_o

Variables λ_o are the dual variables associated to the generation-demand balance Eqs. (3.2b), $\forall o$. Since the market-clearing problems (3.2), $\forall o$, represent the minimization of the generation cost (or the maximization of social welfare since demands are considered inelastic), these dual variables represent the generation cost increment in the market that results from a marginal increment of the demand in the system. Thus, these dual variables are usually referred to as the marginal market prices, i.e., the prices that demands pay for their load consumptions and the price that generating units receive for their productions.

Illustrative Example 3.5 *Deterministic single-node static GEP problem: Market prices*

The values of variables $\lambda_o, \forall o$, in Illustrative Example 3.4 are $\lambda_{o_1} = \$25/\text{MWh}$ and $\lambda_{o_2} = \$35/\text{MWh}$. The meaning of these variables is explained next.

Operating condition 1 is characterized by a demand of 290 MW. As 300 MW of capacity of the candidate generating unit $\tilde{c}$ are built and the production cost of this unit is cheaper than that of the existing generating unit $\tilde{g}$, only the newly built generating unit is used to supply the demand in the first operating condition. Moreover, the newly built generating unit is able to provide an additional 10 MW. Thus, a marginal increment in the demand would result in a marginal increment of the generation cost of \$25/MWh (i.e., λ_{o_1}). On the other hand, operating condition 2 is characterized by a demand of 550 MW. As only 300 MW of the candidate generating unit $\tilde{c}$ are built, the remaining 250 MW have to be provided by the existing generating unit $\tilde{g}$ in this case. Here, the newly built unit is used at full capacity, and thus, a marginal increment in the demand should be satisfied by the existing generating unit, which would result in a marginal increment of the generation cost of \$35/MWh (i.e., λ_{o_2}). □

3.4 Deterministic Single-Node Dynamic GEP

The GEP problem described in the previous section considers a static approach, i.e., the generation expansion decisions for a given planning horizon are made at a single point in time (usually at the beginning of the planning horizon). In this section, we describe the GEP problem considering a dynamic approach, in which the generation expansion decisions can be made at different points in time.

For the sake of simplicity, we also consider a deterministic single-node approach.

Considering this framework, the GEP problem using a deterministic single-node dynamic approach can be formulated using the following complementarity model:

$$\min_{\Delta^{\text{D1D}}, \Delta^{\text{D1DM}}_{ot}} \quad \sum_{t} \left\{ \sum_{o} \rho_o \left[\sum_{g} C^{\text{E}}_{gt} p^{\text{E}}_{got} + \sum_{c} C^{\text{C}}_{ct} p^{\text{C}}_{cot} \right] + a_t \sum_{c} I^{\text{C}}_{ct} p^{\text{C}^{\max}}_{ct} \right\} \tag{3.8a}$$

subject to

$$p^{\text{C}^{\max}}_{ct} = \sum_{q} u^{\text{Option}}_{cqt} P^{\text{Option}}_{cqt} \quad \forall c, t \tag{3.8b}$$

$$\sum_{q} u^{\text{Option}}_{cqt} = 1 \quad \forall c, t \tag{3.8c}$$

$$u^{\text{Option}}_{cqt} \in \{0, 1\} \quad \forall c, q, t \tag{3.8d}$$

$$p^{\text{E}}_{got}, p^{\text{C}}_{cot} \in \Omega_{ot} \left(P^{\text{D}}_{dt}, P^{\text{E}^{\max}}_{g}, p^{\text{C}^{\max}}_{ct} \right) \quad \forall g, c, o, t, \tag{3.8e}$$

where $\Omega_{ot} \left(P^{\text{D}}_{dt}, P^{\text{E}^{\max}}_{g}, p^{\text{C}^{\max}}_{ct} \right) = \Bigg\{$

$$\min_{\Delta^{\text{D1DM}}_{ot}} \quad \sum_{g} C^{\text{E}}_{gt} p^{\text{E}}_{got} + \sum_{c} C^{\text{C}}_{ct} p^{\text{C}}_{cot} \tag{3.9a}$$

subject to

$$\sum_{g} p^{\text{E}}_{got} + \sum_{c} p^{\text{C}}_{cot} = \sum_{d} P^{\text{D}}_{dot} \tag{3.9b}$$

$$0 \le p^{\text{E}}_{got} \le P^{\text{E}^{\max}}_{g} \quad \forall g \tag{3.9c}$$

$$0 \leq p^{\mathrm{C}}_{cot} \leq \sum_{\tau \leq t} p^{\mathrm{C}^{\max}}_{c\tau} \quad \forall c \tag{3.9d}$$

$$\Bigg\}, \forall o, t,$$

variables in sets $\Delta^{\mathrm{D1S}} = \left\{u^{\mathrm{Option}}_{cqt}, p^{\mathrm{C}^{\max}}_{ct}\right\}$ and $\Delta^{\mathrm{D1DM}}_{ot}$, $\forall o$, t, are the optimization variables of problem (3.8), and variables in sets $\Delta^{\mathrm{D1DM}}_{ot} = \{p^{\mathrm{E}}_{got}, p^{\mathrm{C}}_{cot}\}$,$\forall o$, t, are the optimization variables of problems (3.9), $\forall o$, t.

Problem (3.8) is similar to the GEP problem considering a deterministic single-node static approach (3.1) described in Sect. 3.3. The main differences are summarized below:

1. The capacities to build of each candidate generating unit (i.e., variables $p^{\mathrm{C}^{\max}}_{ct}$, $\forall c$, $\forall t$) can take different values at different time periods indexed by t.
2. The investment costs at each time period in the objective function (3.8a) are multiplied by the corresponding amortization rates, which represent the equivalent amount of money to be paid for the investments at each time period.
3. Constraints (3.9d) impose that the available capacity of the candidate generating units at time period t is equal to the capacity built at that time period plus the capacities built in the previous ones.
4. The market-clearing problems (3.9) are formulated for each operating condition o and time period t.

For the sake of simplicity, we assume that all monetary values are referred to the same point in time, and thus, it is not necessary to multiply by discount rates.

In order to solve complementarity model (3.8), we re-formulate it as a MILP problem as explained in Sect. 3.3.3.

Illustrative Example 3.6 *Deterministic single-node dynamic GEP problem*

Let us consider Illustrative Example 3.4. Here, we consider that the planning horizon is divided in two time periods, each one represented by two operating conditions.

In the first time period, operating conditions o_1 and o_2 are characterized by demands of 246.5 and 467.5 MW, and weights of 6000 h and 2760 h, respectively. In the second time period, operating conditions o_1 and o_2 are characterized by demands of 290 and 550 MW, and weights of 6000 h and 2760 h, respectively.

The investment cost is equal to \$700,000 per MW, and the amortization rates are equal to 0.2 and 0.1 for the first and second time periods, respectively.

Considering the above data, the GEP problem considering a deterministic dynamic approach results in the complementarity model below:

$$\begin{aligned}
\min_{\Delta} \quad & 6000\left[35p^{\mathrm{E}}_{\tilde{g}o_1t_1} + 25p^{\mathrm{C}}_{\tilde{c}o_1t_1}\right] + 2760\left[35p^{\mathrm{E}}_{\tilde{g}o_2t_1} + 25p^{\mathrm{C}}_{\tilde{c}o_2t_1}\right] \\
& + 6000\left[35p^{\mathrm{E}}_{\tilde{g}o_1t_2} + 25p^{\mathrm{C}}_{\tilde{c}o_1t_2}\right] + 2760\left[35p^{\mathrm{E}}_{\tilde{g}o_2t_2} + 25p^{\mathrm{C}}_{\tilde{c}o_2t_2}\right] \\
& + 140000p^{\mathrm{C}^{\max}}_{\tilde{c}t_1} + 70000p^{\mathrm{C}^{\max}}_{\tilde{c}t_2}
\end{aligned}$$

subject to

$$
\begin{aligned}
p^{\mathrm{C^{max}}}_{\tilde{c}t_1} &= 0u^{\mathrm{Option}}_{\tilde{c}q_1t_1} + 100u^{\mathrm{Option}}_{\tilde{c}q_2t_1} + 200u^{\mathrm{Option}}_{\tilde{c}q_3t_1} + 300u^{\mathrm{Option}}_{\tilde{c}q_4t_1} \\
&\quad + 400u^{\mathrm{Option}}_{\tilde{c}q_5t_1} + 500u^{\mathrm{Option}}_{\tilde{c}q_6t_1} \\
p^{\mathrm{C^{max}}}_{\tilde{c}t_2} &= 0u^{\mathrm{Option}}_{\tilde{c}q_1t_2} + 100u^{\mathrm{Option}}_{\tilde{c}q_2t_2} + 200u^{\mathrm{Option}}_{\tilde{c}q_3t_2} + 300u^{\mathrm{Option}}_{\tilde{c}q_4t_2} \\
&\quad + 400u^{\mathrm{Option}}_{\tilde{c}q_5t_2} + 500u^{\mathrm{Option}}_{\tilde{c}q_6t_2} \\
& u^{\mathrm{Option}}_{\tilde{c}q_1t_1} + u^{\mathrm{Option}}_{\tilde{c}q_2t_1} + u^{\mathrm{Option}}_{\tilde{c}q_3t_1} + u^{\mathrm{Option}}_{\tilde{c}q_4t_1} + u^{\mathrm{Option}}_{\tilde{c}q_5t_1} + u^{\mathrm{Option}}_{\tilde{c}q_6t_1} = 1 \\
& u^{\mathrm{Option}}_{\tilde{c}q_1t_2} + u^{\mathrm{Option}}_{\tilde{c}q_2t_2} + u^{\mathrm{Option}}_{\tilde{c}q_3t_2} + u^{\mathrm{Option}}_{\tilde{c}q_4t_2} + u^{\mathrm{Option}}_{\tilde{c}q_5t_2} + u^{\mathrm{Option}}_{\tilde{c}q_6t_2} = 1 \\
& u^{\mathrm{Option}}_{\tilde{c}q_1t_1}, u^{\mathrm{Option}}_{\tilde{c}q_2t_1}, u^{\mathrm{Option}}_{\tilde{c}q_3t_1}, u^{\mathrm{Option}}_{\tilde{c}q_4t_1}, u^{\mathrm{Option}}_{\tilde{c}q_5t_1}, u^{\mathrm{Option}}_{\tilde{c}q_6t_1}, u^{\mathrm{Option}}_{\tilde{c}q_1t_2}, u^{\mathrm{Option}}_{\tilde{c}q_2t_2}, u^{\mathrm{Option}}_{\tilde{c}q_3t_2}, \\
&\quad u^{\mathrm{Option}}_{\tilde{c}q_4t_2}, u^{\mathrm{Option}}_{\tilde{c}q_5t_2}, u^{\mathrm{Option}}_{\tilde{c}q_6t_2} \in \{0, 1\}
\end{aligned}
$$

$$
\left\{
\begin{aligned}
&\min_{p^{\mathrm{E}}_{\tilde{g}o_1t_1}, p^{\mathrm{C}}_{\tilde{c}o_1t_1}} \quad 35p^{\mathrm{E}}_{\tilde{g}o_1t_1} + 25p^{\mathrm{C}}_{\tilde{c}o_1t_1} \\
&\text{s.t.} \\
&\quad p^{\mathrm{E}}_{\tilde{g}o_1t_1} + p^{\mathrm{C}}_{\tilde{c}o_1t_1} = 246.5 \\
&\quad 0 \le p^{\mathrm{E}}_{\tilde{g}o_1t_1} \le 400 \\
&\quad 0 \le p^{\mathrm{C}}_{\tilde{c}o_1t_1} \le p^{\mathrm{C^{max}}}_{\tilde{c}t_1}
\end{aligned}
\right.
$$

$$
\left\{
\begin{aligned}
&\min_{p^{\mathrm{E}}_{\tilde{g}o_2t_1}, p^{\mathrm{C}}_{\tilde{c}o_2t_1}} \quad 35p^{\mathrm{E}}_{\tilde{g}o_2t_1} + 25p^{\mathrm{C}}_{\tilde{c}o_2t_1} \\
&\text{s.t.} \\
&\quad p^{\mathrm{E}}_{\tilde{g}o_2t_1} + p^{\mathrm{C}}_{\tilde{c}o_2t_1} = 467.5 \\
&\quad 0 \le p^{\mathrm{E}}_{\tilde{g}o_2t_1} \le 400 \\
&\quad 0 \le p^{\mathrm{C}}_{\tilde{c}o_2t_1} \le p^{\mathrm{C^{max}}}_{\tilde{c}t_1}
\end{aligned}
\right.
$$

$$
\left\{
\begin{aligned}
&\min_{p^{\mathrm{E}}_{\tilde{g}o_1t_2}, p^{\mathrm{C}}_{\tilde{c}o_1t_2}} \quad 35p^{\mathrm{E}}_{\tilde{g}o_1t_2} + 25p^{\mathrm{C}}_{\tilde{c}o_1t_2} \\
&\text{s.t.} \\
&\quad p^{\mathrm{E}}_{\tilde{g}o_1t_2} + p^{\mathrm{C}}_{\tilde{c}o_1t_2} = 290 \\
&\quad 0 \le p^{\mathrm{E}}_{\tilde{g}o_1t_2} \le 400 \\
&\quad 0 \le p^{\mathrm{C}}_{\tilde{c}o_1t_2} \le p^{\mathrm{C^{max}}}_{\tilde{c}t_1} + p^{\mathrm{C^{max}}}_{\tilde{c}t_2}
\end{aligned}
\right.
$$

$$
\left\{
\begin{aligned}
&\min_{p^{\mathrm{E}}_{\tilde{g}o_2t_2}, p^{\mathrm{C}}_{\tilde{c}o_2t_2}} \quad 35p^{\mathrm{E}}_{\tilde{g}o_2t_2} + 25p^{\mathrm{C}}_{\tilde{c}o_2t_2} \\
&\text{s.t.} \\
&\quad p^{\mathrm{E}}_{\tilde{g}o_2t_2} + p^{\mathrm{C}}_{\tilde{c}o_2t_2} = 550 \\
&\quad 0 \le p^{\mathrm{E}}_{\tilde{g}o_2t_2} \le 400 \\
&\quad 0 \le p^{\mathrm{C}}_{\tilde{c}o_2t_2} \le p^{\mathrm{C^{max}}}_{\tilde{c}t_1} + p^{\mathrm{C^{max}}}_{\tilde{c}t_2},
\end{aligned}
\right.
$$

where $\Delta = \Big\{p^{\mathrm{E}}_{\tilde{g}o_1t_1}, p^{\mathrm{C}}_{\tilde{c}o_1t_1}, p^{\mathrm{E}}_{\tilde{g}o_2t_1}, p^{\mathrm{C}}_{\tilde{c}o_2t_1}, p^{\mathrm{C^{max}}}_{\tilde{c}t_1}, p^{\mathrm{E}}_{\tilde{g}o_1t_2}, p^{\mathrm{C}}_{\tilde{c}o_1t_2}, p^{\mathrm{E}}_{\tilde{g}o_2t_2}, p^{\mathrm{C}}_{\tilde{c}o_2t_2}, p^{\mathrm{C^{max}}}_{\tilde{c}t_2}, u^{\mathrm{Option}}_{\tilde{c}q_1t_1}, u^{\mathrm{Option}}_{\tilde{c}q_2t_1}, u^{\mathrm{Option}}_{\tilde{c}q_3t_1}, u^{\mathrm{Option}}_{\tilde{c}q_4t_1}, u^{\mathrm{Option}}_{\tilde{c}q_5t_1}, u^{\mathrm{Option}}_{\tilde{c}q_6t_1}, u^{\mathrm{Option}}_{\tilde{c}q_1t_2}, u^{\mathrm{Option}}_{\tilde{c}q_2t_2}, u^{\mathrm{Option}}_{\tilde{c}q_3t_2}, u^{\mathrm{Option}}_{\tilde{c}q_4t_2}, u^{\mathrm{Option}}_{\tilde{c}q_5t_2}, u^{\mathrm{Option}}_{\tilde{c}q_6t_2}\Big\}$.

We obtain that it is optimal to build 200 MW of new capacity at the beginning of the planning horizon (i.e., at the beginning of the first time period) and an additional 100 MW at the beginning of the second time period.

In the first time period, the 200 MW of the newly built generating unit are used in both operating conditions since it is cheaper than the existing generating unit, from which 46.5 and 267.5 MW are used in operating conditions o_1 and o_2, respectively, to satisfy the demand. In the second time period, additional 100 MW of the candidate generating unit are built and, thus, it is possible to satisfy the demand in operating condition o_1 with this unit, while 250 MW of the existing generating unit are needed to satisfy the demand in operating condition o_2.

Note that the operating conditions in the second time period are equal to those considered in Illustrative Example 3.4. This is so because if a static approach is considered, then the generation expansion plan is determined considering the whole time period, and thus, the largest expected demand should be considered.

If we compare the solution of this example and the solution of Illustrative Example 3.4, we observe that the total capacity installed in the system in the whole planning horizon is the same (300 MW). However, considering that investment decisions can be made at two points in time, i.e., considering a dynamic approach, has the advantage that, if the operating conditions in the second time period are finally different than those considered, then it is possible to adapt by changing the generation expansion decisions. For example, let us consider that the demand in the system does not increase as expected and the operating conditions in the second time period remain the same as in the first time period. If a static approach is considered, the 300 MW of the candidate generating unit are already built at the beginning of the planning horizon, so it is not possible to adapt. However, if a dynamic approach is considered, then only 200 MW of the candidate generating unit are built at the beginning of the planning horizon, and it is possible not to build additional capacity in the second time period if these unexpected changes occur. □

3.5 Deterministic Network-Constrained Static GEP

The GEP problems described in the previous sections consider a single-node approach, i.e., an approach that does not take into account the network constraints. Thus, the generation expansion planner determines the optimal capacity of the candidate units to build in the system, i.e., their optimal sizes, but not where to build these generating units, i.e., their optimal locations. Using such a single-node approach allows us to formulate a relatively simple problem since real-world electric energy systems usually have thousands of nodes and transmission lines, whose modeling may lead to very complex problems.

Although the single-node approach is generally valid for electric energy systems whose transmission network is not congested, in those systems whose transmission lines usually become congested, it is necessary to consider the effect of network constraints on the GEP problem since generation expansion decisions may be significantly different under network congestion. Even in uncongested systems in the present, the transmission network may become an issue in the future, e.g., if the demand in the system increases or a significant amount of renewable capacity is installed, so modeling the transmission network constraints in the GEP problem is generally important. This is analyzed in this section.

For the sake of simplicity, a deterministic static approach is considered.

3.5.1 Complementarity Model

The GEP problem considering a deterministic network-constrained static approach can be formulated using the complementarity model below:

$$\min_{\Delta^{\mathrm{DNS}}, \Delta_o^{\mathrm{DNSM}}, \Delta_o^{\mathrm{DNSM,D}}} \quad \sum_o \rho_o \left[\sum_g C_g^{\mathrm{E}} p_{go}^{\mathrm{E}} + \sum_c C_c^{\mathrm{C}} p_{co}^{\mathrm{C}} \right] + \sum_c \tilde{I}_c^{\mathrm{C}} p_c^{\mathrm{C^{max}}} \tag{3.10a}$$

subject to

$$p_c^{\mathrm{C^{max}}} = \sum_q u_{cq}^{\mathrm{Option}} P_{cq}^{\mathrm{Option}} \quad \forall c \tag{3.10b}$$

$$\sum_q u_{cq}^{\mathrm{Option}} = 1 \quad \forall c \tag{3.10c}$$

$$u_{cq}^{\mathrm{Option}} \in \{0, 1\} \quad \forall c, q \tag{3.10d}$$

$$p_{go}^{\mathrm{E}}, p_{co}^{\mathrm{C}} \in \Omega_o \left(P_{do}^{\mathrm{D}}, P_g^{\mathrm{E^{max}}}, p_c^{\mathrm{C^{max}}} \right) \quad \forall g, c, o, \tag{3.10e}$$

where $\Omega_o \left(P_{do}^{\mathrm{D}}, P_g^{\mathrm{E^{max}}}, p_c^{\mathrm{C^{max}}} \right) = \Big\{$

$$\min_{\Delta_o^{\mathrm{DNSM}}} \quad \sum_g C_g^{\mathrm{E}} p_{go}^{\mathrm{E}} + \sum_c C_c^{\mathrm{C}} p_{co}^{\mathrm{C}} \tag{3.11a}$$

subject to

$$\sum_{g \in \Omega_n^{\mathrm{E}}} p_{go}^{\mathrm{E}} + \sum_{c \in \Omega_n^{\mathrm{C}}} p_{co}^{\mathrm{C}} - \sum_{\ell | s(\ell) = n} p_{\ell o}^{\mathrm{L}} + \sum_{\ell | r(\ell) = n} p_{\ell o}^{\mathrm{L}} = \sum_{d \in \Omega_n^{\mathrm{D}}} P_{do}^{\mathrm{D}} \; : \; \lambda_{no} \quad \forall n \tag{3.11b}$$

$$p_{\ell o}^{\mathrm{L}} = B_\ell \left(\theta_{s(\ell) o} - \theta_{r(\ell) o} \right) \; : \; \mu_{\ell o}^{\mathrm{L}} \quad \forall \ell \tag{3.11c}$$

$$-F_\ell^{\max} \le p_{\ell o}^{\mathrm{L}} \le F_\ell^{\max} \; : \; \mu_{\ell o}^{\mathrm{L^{min}}}, \mu_{\ell o}^{\mathrm{L^{max}}} \quad \forall \ell \tag{3.11d}$$

$$0 \le p_{go}^{\mathrm{E}} \le P_g^{\mathrm{E^{max}}} \ : \ \mu_{go}^{\mathrm{E^{max}}} \quad \forall g \tag{3.11e}$$

$$0 \le p_{co}^{\mathrm{C}} \le p_c^{\mathrm{C^{max}}} \ : \ \mu_{co}^{\mathrm{C^{max}}} \quad \forall c \tag{3.11f}$$

$$-\pi \le \theta_{no} \le \pi \ : \ \mu_{no}^{\mathrm{A^{min}}}, \mu_{no}^{\mathrm{A^{max}}} \quad \forall n \tag{3.11g}$$

$$\theta_{no} = 0 \ : \ \mu_{no}^{\mathrm{A,ref.}} \quad n\text{: ref.} \tag{3.11h}$$

$$\Bigg\}, \forall o,$$

variables in sets $\Delta^{\mathrm{DNS}} = \left\{u_{cq}^{\mathrm{Option}}, p_c^{\mathrm{C^{max}}}\right\}$, Δ_o^{DNSM}, $\forall o$, and $\Delta_o^{\mathrm{DNSM,D}}$, $\forall o$, are the optimization variables of problem (3.10), variables in sets $\Delta_o^{\mathrm{DNSM}} = \left\{p_{go}^{\mathrm{E}}, p_{co}^{\mathrm{C}}, p_{\ell o}^{\mathrm{L}}, \theta_{no}\right\}$, $\forall o$, are the primal optimization variables of problems (3.11), $\forall o$, and variables in sets $\Delta_o^{\mathrm{DNSM,D}} = \left\{\lambda_{no}, \mu_{go}^{\mathrm{E^{max}}}, \mu_{co}^{\mathrm{C^{max}}}, \mu_{lo}^{\mathrm{L^{max}}}, \mu_{lo}^{\mathrm{L^{min}}}, \mu_{no}^{\mathrm{A^{max}}}, \mu_{no}^{\mathrm{A^{min}}}, \mu_{no}^{\mathrm{A,ref.}}\right\}$, $\forall o$, are the dual optimization variables of problems (3.11), $\forall o$. The dual variables associated to constraints of problems (3.11), $\forall o$, are provided following a colon.

The differences between model (3.10) and the equivalent one considering a single-node approach (3.1) are summarized below:

1. Constraints (3.2b) are replaced by constraints (3.11b) that impose the generation-demand balance at each node of the system.
2. Constraints (3.11c) that define the power flows through transmission lines are included.
3. Constraints (3.11d) that impose bounds on the power flows through transmission lines are included.
4. Constraints (3.11g) that impose bounds on voltage angles are included.
5. Constraints (3.11h) that define the voltage angle at the reference node are included.

Note that the transmission network constraints are explicitly modeled in the market-clearing problems (3.11), $\forall o$. In doing so, we use a dc power flow model without losses for the sake of simplicity. This allows us to formulate the market-clearing problems (3.11), $\forall o$, as LP problems [13].

Illustrative Example 3.7 *Deterministic network-constrained GEP problem: Complementarity model formulation*

Let us consider Illustrative Example 3.4. We consider that the existing generating unit $\tilde{g}$ and the demand $\tilde{d}$ are located in a two-node electric energy system. Generating unit $\tilde{g}$ is located at node 1, while demand $\tilde{d}$ is located at node 2. Nodes 1 and 2 are connected through a transmission line $\tilde{\ell}$ with a susceptance equal to 500 p.u. and a transmission capacity of 500 MW. Moreover, the candidate generating unit $\tilde{c}$ can be built only at node 2. This is schematically depicted in Fig. 3.3.

Given the above data and considering a base power of 1 MW and a base voltage of 1 kV, the GEP problem (3.10) results in the following complementarity model:

Fig. 3.3 Illustrative Example 3.7: two-node electric energy system

$$\begin{aligned}\min_{\Delta} \quad & 6000\left[35p^{\mathrm{E}}_{\tilde{g}o_1} + 25p^{\mathrm{C}}_{\tilde{c}o_1}\right] + 2760\left[35p^{\mathrm{E}}_{\tilde{g}o_2} + 25p^{\mathrm{C}}_{\tilde{c}o_2}\right] \\ & + 70000p^{\mathrm{C}^{\max}}_{\tilde{c}}\end{aligned}$$

subject to

$$\begin{aligned}
& p^{\mathrm{C}^{\max}}_{\tilde{c}} = 0u^{\mathrm{Option}}_{\tilde{c}q_1} + 100u^{\mathrm{Option}}_{\tilde{c}q_2} + 200u^{\mathrm{Option}}_{\tilde{c}q_3} + 300u^{\mathrm{Option}}_{\tilde{c}q_4} + 400u^{\mathrm{Option}}_{\tilde{c}q_5} \\
& \quad + 500u^{\mathrm{Option}}_{\tilde{c}q_6} \\
& u^{\mathrm{Option}}_{\tilde{c}q_1} + u^{\mathrm{Option}}_{\tilde{c}q_2} + u^{\mathrm{Option}}_{\tilde{c}q_3} + u^{\mathrm{Option}}_{\tilde{c}q_4} + u^{\mathrm{Option}}_{\tilde{c}q_5} + u^{\mathrm{Option}}_{\tilde{c}q_6} = 1 \\
& u^{\mathrm{Option}}_{\tilde{c}q_1}, u^{\mathrm{Option}}_{\tilde{c}q_2}, u^{\mathrm{Option}}_{\tilde{c}q_3}, u^{\mathrm{Option}}_{\tilde{c}q_4}, u^{\mathrm{Option}}_{\tilde{c}q_5}, u^{\mathrm{Option}}_{\tilde{c}q_6} \in \{0, 1\}
\end{aligned}$$

$$\left\{\begin{aligned}
& \min_{p^{\mathrm{E}}_{\tilde{g}o_1}, p^{\mathrm{C}}_{\tilde{c}o_1}} \quad 35p^{\mathrm{E}}_{\tilde{g}o_1} + 25p^{\mathrm{C}}_{\tilde{c}o_1} \\
& \text{s.t.} \\
& \quad p^{\mathrm{E}}_{\tilde{g}o_1} - p^{\mathrm{L}}_{\tilde{\ell}o_1} = 0 \\
& \quad p^{\mathrm{C}}_{\tilde{c}o_1} + p^{\mathrm{L}}_{\tilde{\ell}o_1} = 290 \\
& \quad p^{\mathrm{L}}_{\tilde{\ell}o_1} = 500\left(\theta_{n_1o_1} - \theta_{n_2o_1}\right) \\
& \quad -500 \le p^{\mathrm{L}}_{\tilde{\ell}o_1} \le 500 \\
& \quad 0 \le p^{\mathrm{E}}_{\tilde{g}o_1} \le 400 \\
& \quad 0 \le p^{\mathrm{C}}_{\tilde{c}o_1} \le p^{\mathrm{C}^{\max}}_{\tilde{c}} \\
& \quad -\pi \le \theta_{n_2o_1} \le \pi \\
& \quad \theta_{n_1o_1} = 0
\end{aligned}\right.$$

$$\left\{\begin{aligned}
& \min_{p^{\mathrm{E}}_{\tilde{g}o_1}, p^{\mathrm{C}}_{\tilde{c}o_1}} \quad 35p^{\mathrm{E}}_{\tilde{g}o_1} + 25p^{\mathrm{C}}_{\tilde{c}o_1} \\
& \text{s.t.} \\
& \quad p^{\mathrm{E}}_{\tilde{g}o_2} - p^{\mathrm{L}}_{\tilde{\ell}o_2} = 0 \\
& \quad p^{\mathrm{C}}_{\tilde{c}o_2} + p^{\mathrm{L}}_{\tilde{\ell}o_2} = 550 \\
& \quad p^{\mathrm{L}}_{\tilde{\ell}o_2} = 500\left(\theta_{n_1o_2} - \theta_{n_2o_2}\right) \\
& \quad -500 \le p^{\mathrm{L}}_{\tilde{\ell}o_2} \le 500 \\
& \quad p^{\mathrm{E}}_{\tilde{g}o_2} + p^{\mathrm{C}}_{\tilde{c}o_2} = 550 \\
& \quad 0 \le p^{\mathrm{E}}_{\tilde{g}o_2} \le 400 \\
& \quad 0 \le p^{\mathrm{C}}_{\tilde{c}o_2} \le p^{\mathrm{C}^{\max}}_{\tilde{c}} \\
& \quad -\pi \le \theta_{n_2o_2} \le \pi \\
& \quad \theta_{n_1o_2} = 0,
\end{aligned}\right.$$

where $\Delta = \left\{p^{\mathrm{E}}_{\tilde{g}o_1}, p^{\mathrm{C}}_{\tilde{c}o_1}, p^{\mathrm{E}}_{\tilde{g}o_2}, p^{\mathrm{C}}_{\tilde{c}o_2}, p^{\mathrm{L}}_{\tilde{\ell}o_1}, p^{\mathrm{L}}_{\tilde{\ell}o_2}, p^{\mathrm{C^{max}}}_{\tilde{c}}, u^{\mathrm{Option}}_{\tilde{c}q_1}, u^{\mathrm{Option}}_{\tilde{c}q_2}, u^{\mathrm{Option}}_{\tilde{c}q_3}, u^{\mathrm{Option}}_{\tilde{c}q_4}, u^{\mathrm{Option}}_{\tilde{c}q_5}, u^{\mathrm{Option}}_{\tilde{c}q_6}\right\}$. □

3.5.2 Equivalent MILP Formulation

Complementarity model (3.10) is recast as the MILP problem (3.12) following the procedure explained in Sect. 3.3:

$$\min_{\Delta^{\mathrm{DNS,MILP}}} \quad \sum_o \rho_o \left[\sum_g C^{\mathrm{E}}_g p^{\mathrm{E}}_{go} + \sum_c C^{\mathrm{C}}_c p^{\mathrm{C}}_{co}\right] + \sum_c \tilde{I}^{\mathrm{C}}_c p^{\mathrm{C^{max}}}_c \tag{3.12a}$$

subject to

$$p^{\mathrm{C^{max}}}_c = \sum_q u^{\mathrm{Option}}_{cq} P^{\mathrm{Option}}_{cq} \quad \forall c \tag{3.12b}$$

$$\sum_q u^{\mathrm{Option}}_{cq} = 1 \quad \forall c \tag{3.12c}$$

$$u^{\mathrm{Option}}_{cq} \in \{0,1\} \quad \forall c, q \tag{3.12d}$$

$$\left\{ \sum_{g\in\Omega^{\mathrm{E}}_n} p^{\mathrm{E}}_{go} + \sum_{c\in\Omega^{\mathrm{C}}_n} p^{\mathrm{C}}_{co} - \sum_{\ell|s(\ell)=n} p^{\mathrm{L}}_{\ell o} + \sum_{\ell|r(\ell)=n} p^{\mathrm{L}}_{\ell o} = \sum_{d\in\Omega^{\mathrm{D}}_n} P^{\mathrm{D}}_{do} \quad \forall n \right. \tag{3.12e}$$

$$p^{\mathrm{L}}_{\ell o} = B_\ell \left(\theta_{s(\ell)o} - \theta_{r(\ell)o}\right) \quad \forall \ell \tag{3.12f}$$

$$-F^{\max}_\ell \le p^{\mathrm{L}}_{\ell o} \le F^{\max}_\ell \quad \forall \ell \tag{3.12g}$$

$$0 \le p^{\mathrm{E}}_{go} \le P^{\mathrm{E^{max}}}_g \quad \forall g \tag{3.12h}$$

$$0 \le p^{\mathrm{C}}_{co} \le P^{\mathrm{C^{max}}}_c \quad \forall c \tag{3.12i}$$

$$-\pi \le \theta_{no} \le \pi \quad \forall n \tag{3.12j}$$

$$\theta_{no} = 0 \quad n\text{: ref.} \tag{3.12k}$$

$$C^{\mathrm{E}}_g - \lambda_{n(g)o} + \mu^{\mathrm{E^{max}}}_{go} \ge 0 \quad \forall g \tag{3.12l}$$

$$C^{\mathrm{C}}_c - \lambda_{n(g)o} + \mu^{\mathrm{C^{max}}}_{co} \ge 0 \quad \forall c \tag{3.12m}$$

$$\lambda_{s(\ell)o} - \lambda_{r(\ell)o} - \mu^{\mathrm{L}}_{\ell o} + \mu^{\mathrm{L^{max}}}_{\ell o} - \mu^{\mathrm{L^{min}}}_{\ell o} = 0 \quad \forall \ell \tag{3.12n}$$

$$\sum_{\ell|s(\ell)=n} B_\ell \mu^{\mathrm{L}}_{\ell o} - \sum_{\ell|r(\ell)=n} B_\ell \mu^{\mathrm{L}}_{\ell o} + \mu^{\mathrm{A^{max}}}_{no} - \mu^{\mathrm{A^{min}}}_{no} = 0 \quad \forall n \setminus n\text{: ref.} \tag{3.12o}$$

$$\sum_{\ell|s(\ell)=n} B_\ell \mu^{\mathrm{L}}_{\ell o} - \sum_{\ell|r(\ell)=n} B_\ell \mu^{\mathrm{L}}_{\ell o} + \mu^{\mathrm{A,ref.}}_{no} = 0 \quad n\text{: ref.} \tag{3.12p}$$

$$\mu^{\mathrm{E^{max}}}_{go} \ge 0 \quad \forall g \tag{3.12q}$$

$$\mu^{\mathrm{C^{max}}}_{co} \ge 0 \quad \forall c \tag{3.12r}$$

$$\mu^{\text{L}^{\max}}_{\ell o}, \mu^{\text{L}^{\min}}_{\ell o} \geq 0 \quad \forall \ell \tag{3.12s}$$

$$\mu^{\text{A}^{\max}}_{no}, \mu^{\text{A}^{\min}}_{no} \geq 0 \quad \forall n \setminus n\text{: ref.} \tag{3.12t}$$

$$\sum_{g} C^{\text{E}}_{g} p^{\text{E}}_{go} + \sum_{c} C^{\text{C}}_{c} p^{\text{C}}_{co} = \sum_{n} \lambda_{no} \sum_{d \in \Omega^{\text{D}}_{n}} P^{\text{D}}_{do} - \sum_{g} \mu^{\text{E}^{\max}}_{go} P^{\text{E}^{\max}}_{g} - \sum_{c} \sum_{q} z^{\text{AUX}}_{cqo}$$
$$- \sum_{\ell} \left(\mu^{\text{L}^{\max}}_{\ell o} + \mu^{\text{L}^{\min}}_{\ell o} \right) F^{\max}_{\ell} - \sum_{n \setminus n:\text{ref.}} \left(\mu^{\text{A}^{\max}}_{no} + \mu^{\text{A}^{\min}}_{no} \right) \pi \tag{3.12u}$$

$$z^{\text{AUX}}_{cqo} = \mu^{\text{C}^{\max}}_{co} P^{\text{Option}}_{cq} - \hat{z}^{\text{AUX}}_{cqo} \quad \forall c, q \tag{3.12v}$$

$$0 \leq z^{\text{AUX}}_{cqo} \leq u^{\text{Option}}_{cq} M \quad \forall c, q \tag{3.12w}$$

$$0 \leq \hat{z}^{\text{AUX}}_{cqo} \leq \left(1 - u^{\text{Option}}_{cq}\right) M \quad \forall c, q \tag{3.12x}$$

$$\Bigg\}, \forall o,$$

where variables in set $\Delta^{\text{DNS,MILP}} = \left\{ u^{\text{Option}}_{cq}, p^{\text{C}^{\max}}_{c}, \Delta^{\text{DNSM}}_{o}, \Delta^{\text{DNSM,D}}_{o}, \Delta^{\text{DNSM,AUX}}_{o} \right\}$ are the optimization variables of problem (3.12) and variables in sets $\Delta^{\text{DNSM,AUX}}_{o} = \left\{ z^{\text{AUX}}_{cqo}, \hat{z}^{\text{AUX}}_{cqo} \right\}$, $\forall o$, are auxiliary variables.

Note that the differences between MILP problem (3.12) and that used for the single-node case (3.7) are:

1. New primal constraints (3.12e)–(3.12g) and (3.12j)–(3.12k) are included.
2. New dual constraints (3.12n)–(3.12p) are included.
3. As a result of 1 and 2 above, the strong duality equalities (3.12u) are different.

Illustrative Example 3.8 *Deterministic network-constrained GEP problem: MILP formulation and solution*

Let us consider the data of Illustrative Example 3.7.

Given the above data, the GEP problem (3.12) results in the following MILP model:

$$\begin{aligned} \min_{\Delta} \quad & 6000\left[35 p^{\text{E}}_{\tilde{g}o_1} + 25 p^{\text{C}}_{\tilde{c}o_1}\right] + 2760\left[35 p^{\text{E}}_{\tilde{g}o_2} + 25 p^{\text{C}}_{\tilde{c}o_2}\right] \\ & + 70000 p^{\text{C}^{\max}}_{\tilde{c}} \end{aligned}$$

subject to

$$\begin{aligned} & p^{\text{C}^{\max}}_{\tilde{c}} = 0 u^{\text{Option}}_{\tilde{c}q1} + 100 u^{\text{Option}}_{\tilde{c}q2} + 200 u^{\text{Option}}_{\tilde{c}q3} + 300 u^{\text{Option}}_{\tilde{c}q4} + 400 u^{\text{Option}}_{\tilde{c}q5} \\ & \quad + 500 u^{\text{Option}}_{\tilde{c}q6} \\ & u^{\text{Option}}_{\tilde{c}q1} + u^{\text{Option}}_{\tilde{c}q2} + u^{\text{Option}}_{\tilde{c}q3} + u^{\text{Option}}_{\tilde{c}q4} + u^{\text{Option}}_{\tilde{c}q5} + u^{\text{Option}}_{\tilde{c}q6} = 1 \\ & u^{\text{Option}}_{\tilde{c}q1}, u^{\text{Option}}_{\tilde{c}q2}, u^{\text{Option}}_{\tilde{c}q3}, u^{\text{Option}}_{\tilde{c}q4}, u^{\text{Option}}_{\tilde{c}q5}, u^{\text{Option}}_{\tilde{c}q6} \in \{0, 1\} \end{aligned}$$

$$
\left\{
\begin{array}{l}
p^{\mathrm{E}}_{\tilde{g}o_1} - p^{\mathrm{L}}_{\tilde{\ell}o_1} = 0 \\
p^{\mathrm{C}}_{\tilde{c}o_1} + p^{\mathrm{L}}_{\tilde{\ell}o_1} = 290 \\
p^{\mathrm{L}}_{\tilde{\ell}o_1} = 500\left(\theta_{n_1o_1} - \theta_{n_2o_1}\right) \\
-500 \leq p^{\mathrm{L}}_{\tilde{\ell}o_1} \leq 500 \\
0 \leq p^{\mathrm{E}}_{\tilde{g}o_1} \leq 400 \\
0 \leq p^{\mathrm{C}}_{\tilde{c}o_1} \leq p^{\mathrm{C^{max}}}_{\tilde{c}} \\
-\pi \leq \theta_{n_2o_1} \leq \pi \\
\theta_{n_1o_1} = 0 \\
35 - \lambda_{n_1o_1} + \mu^{\mathrm{E^{max}}}_{\tilde{g}o_1} \geq 0 \\
25 - \lambda_{n_2o_1} + \mu^{\mathrm{C^{max}}}_{\tilde{c}o_1} \geq 0 \\
\lambda_{n_1o_1} - \lambda_{n_2o_1} - \mu^{\mathrm{L}}_{\tilde{\ell}o_1} + \mu^{\mathrm{L^{max}}}_{\tilde{\ell}o_1} - \mu^{\mathrm{L^{min}}}_{\tilde{\ell}o_1} = 0 \\
-500\mu^{\mathrm{L}}_{\tilde{\ell}o_1} + \mu^{\mathrm{A^{max}}}_{n_2o_1} - \mu^{\mathrm{A^{min}}}_{n_2o_1} = 0 \\
500\mu^{\mathrm{L}}_{\tilde{\ell}o_1} + \mu^{\mathrm{A,ref.}}_{n_1o_1} = 0 \\
\mu^{\mathrm{E^{max}}}_{\tilde{g}o_1}, \mu^{\mathrm{C^{max}}}_{\tilde{c}o_1}, \mu^{\mathrm{L^{max}}}_{\tilde{\ell}o_1}, \mu^{\mathrm{L^{min}}}_{\tilde{\ell}o_1}, \mu^{\mathrm{A^{max}}}_{n_1o_1}, \mu^{\mathrm{A^{min}}}_{n_1o_1}, \mu^{\mathrm{A^{max}}}_{n_2o_1}, \mu^{\mathrm{A^{min}}}_{n_2o_1} \geq 0 \\
35p^{\mathrm{E}}_{\tilde{g}o_1} + 25p^{\mathrm{C}}_{\tilde{c}o_1} = 290\lambda_{n_2o_1} - 400\mu^{\mathrm{E^{max}}}_{\tilde{g}o_1} - \left(z^{\mathrm{AUX}}_{\tilde{c}q_1o_1} + z^{\mathrm{AUX}}_{\tilde{c}q_2o_1} + z^{\mathrm{AUX}}_{\tilde{c}q_3o_1} + z^{\mathrm{AUX}}_{\tilde{c}q_4o_1}\right. \\
\qquad\qquad \left. + z^{\mathrm{AUX}}_{\tilde{c}q_5o_1} + z^{\mathrm{AUX}}_{\tilde{c}q_6o_1}\right) - \left(\mu^{\mathrm{L^{max}}}_{\tilde{\ell}o_1} + \mu^{\mathrm{L^{min}}}_{\tilde{\ell}o_1}\right)500 - \left(\mu^{\mathrm{A^{max}}}_{n_2o_1} + \mu^{\mathrm{A^{min}}}_{n_2o_1}\right)\pi \\
z^{\mathrm{AUX}}_{\tilde{c}q_1o_1} = 0\mu^{\mathrm{C^{max}}}_{\tilde{c}o_1} - \hat{z}^{\mathrm{AUX}}_{\tilde{c}q_1o_1} \\
z^{\mathrm{AUX}}_{\tilde{c}q_2o_1} = 100\mu^{\mathrm{C^{max}}}_{\tilde{c}o_1} - \hat{z}^{\mathrm{AUX}}_{\tilde{c}q_2o_1} \\
z^{\mathrm{AUX}}_{\tilde{c}q_3o_1} = 200\mu^{\mathrm{C^{max}}}_{\tilde{c}o_1} - \hat{z}^{\mathrm{AUX}}_{\tilde{c}q_3o_1} \\
z^{\mathrm{AUX}}_{\tilde{c}q_4o_1} = 300\mu^{\mathrm{C^{max}}}_{\tilde{c}o_1} - \hat{z}^{\mathrm{AUX}}_{\tilde{c}q_4o_1} \\
z^{\mathrm{AUX}}_{\tilde{c}q_5o_1} = 400\mu^{\mathrm{C^{max}}}_{\tilde{c}o_1} - \hat{z}^{\mathrm{AUX}}_{\tilde{c}q_5o_1} \\
z^{\mathrm{AUX}}_{\tilde{c}q_6o_1} = 500\mu^{\mathrm{C^{max}}}_{\tilde{c}o_1} - \hat{z}^{\mathrm{AUX}}_{\tilde{c}q_6o_1} \\
0 \leq z^{\mathrm{AUX}}_{\tilde{c}q_1o_1} \leq u^{\mathrm{Option}}_{\tilde{c}q_1}M \\
0 \leq z^{\mathrm{AUX}}_{\tilde{c}q_2o_1} \leq u^{\mathrm{Option}}_{\tilde{c}q_2}M \\
0 \leq z^{\mathrm{AUX}}_{\tilde{c}q_3o_1} \leq u^{\mathrm{Option}}_{\tilde{c}q_3}M \\
0 \leq z^{\mathrm{AUX}}_{\tilde{c}q_4o_1} \leq u^{\mathrm{Option}}_{\tilde{c}q_4}M \\
0 \leq z^{\mathrm{AUX}}_{\tilde{c}q_5o_1} \leq u^{\mathrm{Option}}_{\tilde{c}q_5}M \\
0 \leq z^{\mathrm{AUX}}_{\tilde{c}q_6o_1} \leq u^{\mathrm{Option}}_{\tilde{c}q_6}M \\
0 \leq \hat{z}^{\mathrm{AUX}}_{\tilde{c}q_1o_1} \leq \left(1 - u^{\mathrm{Option}}_{\tilde{c}q_1}\right)M \\
0 \leq \hat{z}^{\mathrm{AUX}}_{\tilde{c}q_2o_1} \leq \left(1 - u^{\mathrm{Option}}_{\tilde{c}q_2}\right)M \\
0 \leq \hat{z}^{\mathrm{AUX}}_{\tilde{c}q_3o_1} \leq \left(1 - u^{\mathrm{Option}}_{\tilde{c}q_3}\right)M \\
0 \leq \hat{z}^{\mathrm{AUX}}_{\tilde{c}q_4o_1} \leq \left(1 - u^{\mathrm{Option}}_{\tilde{c}q_4}\right)M \\
0 \leq \hat{z}^{\mathrm{AUX}}_{\tilde{c}q_5o_1} \leq \left(1 - u^{\mathrm{Option}}_{\tilde{c}q_5}\right)M \\
0 \leq \hat{z}^{\mathrm{AUX}}_{\tilde{c}q_6o_1} \leq \left(1 - u^{\mathrm{Option}}_{\tilde{c}q_6}\right)M
\end{array}
\right.
$$

$$
\left\{
\begin{array}{l}
p^{\mathrm{E}}_{\tilde{g}o_2} - p^{\mathrm{L}}_{\tilde{\ell}o_2} = 0 \\
p^{\mathrm{C}}_{\tilde{c}o_2} + p^{\mathrm{L}}_{\tilde{\ell}o_2} = 550 \\
p^{\mathrm{L}}_{\tilde{\ell}o_2} = 500\left(\theta_{n_1o_2} - \theta_{n_2o_2}\right) \\
-500 \leq p^{\mathrm{L}}_{\tilde{\ell}o_2} \leq 500 \\
p^{\mathrm{E}}_{\tilde{g}o_2} + p^{\mathrm{C}}_{\tilde{c}o_2} = 550 \\
0 \leq p^{\mathrm{E}}_{\tilde{g}o_2} \leq 400 \\
0 \leq p^{\mathrm{C}}_{\tilde{c}o_2} \leq p^{\mathrm{C^{max}}}_{\tilde{c}} \\
-\pi \leq \theta_{n_2o_2} \leq \pi \\
\theta_{n_1o_2} = 0 \\
35 - \lambda_{n_1o_2} + \mu^{\mathrm{E^{max}}}_{\tilde{g}o_2} \geq 0 \\
25 - \lambda_{n_2o_2} + \mu^{\mathrm{C^{max}}}_{\tilde{c}o_2} \geq 0 \\
\lambda_{n_1o_2} - \lambda_{n_2o_2} - \mu^{\mathrm{L}}_{\tilde{\ell}o_2} + \mu^{\mathrm{L^{max}}}_{\tilde{\ell}o_2} - \mu^{\mathrm{L^{min}}}_{\tilde{\ell}o_2} = 0 \\
-500\mu^{\mathrm{L}}_{\tilde{\ell}o_2} + \mu^{\mathrm{A^{max}}}_{n_2o_2} - \mu^{\mathrm{A^{min}}}_{n_2o_2} = 0 \\
500\mu^{\mathrm{L}}_{\tilde{\ell}o_2} + \mu^{\mathrm{A,ref.}}_{n_1o_2} = 0 \\
\mu^{\mathrm{E^{max}}}_{\tilde{g}o_2}, \mu^{\mathrm{C^{max}}}_{\tilde{c}o_2}, \mu^{\mathrm{L^{max}}}_{\tilde{\ell}o_2}, \mu^{\mathrm{L^{min}}}_{\tilde{\ell}o_2}, \mu^{\mathrm{A^{max}}}_{n_1o_2}, \mu^{\mathrm{A^{min}}}_{n_1o_2}, \mu^{\mathrm{A^{max}}}_{n_2o_2}, \mu^{\mathrm{A^{min}}}_{n_2o_2} \geq 0 \\
35p^{\mathrm{E}}_{\tilde{g}o_2} + 25p^{\mathrm{C}}_{\tilde{c}o_2} = 550\lambda_{n_2o_2} - 400\mu^{\mathrm{E^{max}}}_{\tilde{g}o_2} - \left(z^{\mathrm{AUX}}_{\tilde{c}q_1o_2} + z^{\mathrm{AUX}}_{\tilde{c}q_2o_2} + z^{\mathrm{AUX}}_{\tilde{c}q_3o_2} + z^{\mathrm{AUX}}_{\tilde{c}q_4o_2}\right. \\
\qquad \left. + z^{\mathrm{AUX}}_{\tilde{c}q_5o_2} + z^{\mathrm{AUX}}_{\tilde{c}q_6o_2}\right) - \left(\mu^{\mathrm{L^{max}}}_{\tilde{\ell}o_2} + \mu^{\mathrm{L^{min}}}_{\tilde{\ell}o_2}\right)500 - \left(\mu^{\mathrm{A^{max}}}_{n_2o_2} + \mu^{\mathrm{A^{min}}}_{n_2o_2}\right)\pi \\
z^{\mathrm{AUX}}_{\tilde{c}q_1o_2} = 0\mu^{\mathrm{C^{max}}}_{\tilde{c}o_2} - \hat{z}^{\mathrm{AUX}}_{\tilde{c}q_1o_2} \\
z^{\mathrm{AUX}}_{\tilde{c}q_2o_2} = 100\mu^{\mathrm{C^{max}}}_{\tilde{c}o_2} - \hat{z}^{\mathrm{AUX}}_{\tilde{c}q_2o_2} \\
z^{\mathrm{AUX}}_{\tilde{c}q_3o_2} = 200\mu^{\mathrm{C^{max}}}_{\tilde{c}o_2} - \hat{z}^{\mathrm{AUX}}_{\tilde{c}q_3o_2} \\
z^{\mathrm{AUX}}_{\tilde{c}q_4o_2} = 300\mu^{\mathrm{C^{max}}}_{\tilde{c}o_2} - \hat{z}^{\mathrm{AUX}}_{\tilde{c}q_4o_2} \\
z^{\mathrm{AUX}}_{\tilde{c}q_5o_2} = 400\mu^{\mathrm{C^{max}}}_{\tilde{c}o_2} - \hat{z}^{\mathrm{AUX}}_{\tilde{c}q_5o_2} \\
z^{\mathrm{AUX}}_{\tilde{c}q_6o_2} = 500\mu^{\mathrm{C^{max}}}_{\tilde{c}o_2} - \hat{z}^{\mathrm{AUX}}_{\tilde{c}q_6o_2} \\
0 \leq z^{\mathrm{AUX}}_{\tilde{c}q_1o_2} \leq u^{\mathrm{Option}}_{\tilde{c}q_1}M \\
0 \leq z^{\mathrm{AUX}}_{\tilde{c}q_2o_2} \leq u^{\mathrm{Option}}_{\tilde{c}q_2}M \\
0 \leq z^{\mathrm{AUX}}_{\tilde{c}q_3o_2} \leq u^{\mathrm{Option}}_{\tilde{c}q_3}M \\
0 \leq z^{\mathrm{AUX}}_{\tilde{c}q_4o_2} \leq u^{\mathrm{Option}}_{\tilde{c}q_4}M \\
0 \leq z^{\mathrm{AUX}}_{\tilde{c}q_5o_2} \leq u^{\mathrm{Option}}_{\tilde{c}q_5}M \\
0 \leq z^{\mathrm{AUX}}_{\tilde{c}q_6o_2} \leq u^{\mathrm{Option}}_{\tilde{c}q_6}M \\
0 \leq \hat{z}^{\mathrm{AUX}}_{\tilde{c}q_1o_2} \leq \left(1 - u^{\mathrm{Option}}_{\tilde{c}q_1}\right)M \\
0 \leq \hat{z}^{\mathrm{AUX}}_{\tilde{c}q_2o_2} \leq \left(1 - u^{\mathrm{Option}}_{\tilde{c}q_2}\right)M \\
0 \leq \hat{z}^{\mathrm{AUX}}_{\tilde{c}q_3o_2} \leq \left(1 - u^{\mathrm{Option}}_{\tilde{c}q_3}\right)M \\
0 \leq \hat{z}^{\mathrm{AUX}}_{\tilde{c}q_4o_2} \leq \left(1 - u^{\mathrm{Option}}_{\tilde{c}q_4}\right)M \\
0 \leq \hat{z}^{\mathrm{AUX}}_{\tilde{c}q_5o_2} \leq \left(1 - u^{\mathrm{Option}}_{\tilde{c}q_5}\right)M \\
0 \leq \hat{z}^{\mathrm{AUX}}_{\tilde{c}q_6o_2} \leq \left(1 - u^{\mathrm{Option}}_{\tilde{c}q_6}\right)M,
\end{array}
\right.
$$

where $\Delta = \Big\{ p^{\mathrm{E}}_{\tilde{g}o_1}, p^{\mathrm{C}}_{\tilde{c}o_1}, p^{\mathrm{L}}_{\tilde{\ell}o_1}, p^{\mathrm{E}}_{\tilde{g}o_2}, p^{\mathrm{C}}_{\tilde{c}o_2}, p^{\mathrm{L}}_{\tilde{\ell}o_2}, p^{\mathrm{C}^{\max}}_{\tilde{c}}, u^{\mathrm{Option}}_{\tilde{c}q_1}, u^{\mathrm{Option}}_{\tilde{c}q_2}, u^{\mathrm{Option}}_{\tilde{c}q_3}, u^{\mathrm{Option}}_{\tilde{c}q_4}, u^{\mathrm{Option}}_{\tilde{c}q_5}, u^{\mathrm{Option}}_{\tilde{c}q_6}, \lambda_{n_1o_1}, \mu^{\mathrm{L}}_{\tilde{\ell}o_1}, \mu^{\mathrm{E}^{\max}}_{\tilde{g}o_1}, \mu^{\mathrm{C}^{\max}}_{\tilde{c}o_1}, \mu^{\mathrm{L}^{\max}}_{\tilde{\ell}o_1}, \mu^{\mathrm{L}^{\min}}_{\tilde{\ell}o_1}, \mu^{\mathrm{A}^{\max}}_{n_1o_1}, \mu^{\mathrm{A}^{\min}}_{n_2o_1}, \mu^{\mathrm{A,ref.}}_{n_1o_1}, z^{\mathrm{AUX}}_{\tilde{c}q_1o_1}, z^{\mathrm{AUX}}_{\tilde{c}q_2o_1}, z^{\mathrm{AUX}}_{\tilde{c}q_3o_1}, z^{\mathrm{AUX}}_{\tilde{c}q_4o_1}, z^{\mathrm{AUX}}_{\tilde{c}q_5o_1}, z^{\mathrm{AUX}}_{\tilde{c}q_6o_1}, \hat{z}^{\mathrm{AUX}}_{\tilde{c}q_1o_1}, \hat{z}^{\mathrm{AUX}}_{\tilde{c}q_2o_1}, \hat{z}^{\mathrm{AUX}}_{\tilde{c}q_3o_1}, \hat{z}^{\mathrm{AUX}}_{\tilde{c}q_4o_1}, \hat{z}^{\mathrm{AUX}}_{\tilde{c}q_5o_1}, \hat{z}^{\mathrm{AUX}}_{\tilde{c}q_6o_1}, \lambda_{n_1o_2}, \mu^{\mathrm{L}}_{\tilde{\ell}o_2}, \mu^{\mathrm{E}^{\max}}_{\tilde{g}o_2}, \mu^{\mathrm{C}^{\max}}_{\tilde{c}o_2}, \mu^{\mathrm{L}^{\max}}_{\tilde{\ell}o_2}, \mu^{\mathrm{L}^{\min}}_{\tilde{\ell}o_2}, \mu^{\mathrm{A}^{\max}}_{n_1o_2}, \mu^{\mathrm{A}^{\min}}_{n_2o_2}, \mu^{\mathrm{A,ref.}}_{n_1o_2}, z^{\mathrm{AUX}}_{\tilde{c}q_1o_2}, z^{\mathrm{AUX}}_{\tilde{c}q_2o_2}, z^{\mathrm{AUX}}_{\tilde{c}q_3o_2}, z^{\mathrm{AUX}}_{\tilde{c}q_4o_2}, z^{\mathrm{AUX}}_{\tilde{c}q_5o_2}, z^{\mathrm{AUX}}_{\tilde{c}q_6o_2}, \hat{z}^{\mathrm{AUX}}_{\tilde{c}q_1o_2}, \hat{z}^{\mathrm{AUX}}_{\tilde{c}q_2o_2}, \hat{z}^{\mathrm{AUX}}_{\tilde{c}q_3o_2}, \hat{z}^{\mathrm{AUX}}_{\tilde{c}q_4o_2}, \hat{z}^{\mathrm{AUX}}_{\tilde{c}q_5o_2}, \hat{z}^{\mathrm{AUX}}_{\tilde{c}q_6o_2} \Big\}$.

We obtain that the optimal solution consists of building 300 MW of capacity of the candidate generating unit, i.e., we obtain exactly the same solution as in Illustrative Example 3.4. This is so because the transmission line connecting the two nodes is not congested, and thus, it is immaterial if the transmission constraints are considered or not.

However, let us now consider that the capacity of the transmission line connecting nodes 1 and 2 is equal to 200 MW. If we solve the GEP problem (3.12) again, then we obtain that it is optimal to build 400 MW of capacity of the candidate generating unit $\tilde{c}$, i.e., the solution differs from that obtained in Illustrative Example 3.4. In Illustrative Example 3.4, the existing generating unit $\tilde{g}$, located at node 1, produces 250 MW in the second operating condition that are used to supply the demand $\tilde{d}$, located at node 2. This means that the power flow through the transmission line is 250 MW in this case. However, as the capacity of the transmission line is only 200 MW, it is possible to use only 200 MW of the existing generating unit $\tilde{g}$ to supply the demand $\tilde{d}$, and thus, it is needed to build 100 MW of additional capacity of the candidate generating unit $\tilde{c}$, which is located at the same node as the demand $\tilde{d}$. □

3.5.3 *Meaning of Dual Variables* λ_{no}

In this case, variables λ_{no}, $\forall n, o$, are the dual variables associated to the nodal generation-demand balance Eq. (3.11b), $\forall n, o$. As explained in Sect. 3.3.4, these variables are generally used as the market prices. However, note that in this case these variables may take different values at different nodes. Therefore, they are usually referred to as locational marginal prices (LMPs).

3.6 Stochastic Single-Node GEP

The models described in the previous sections consider a deterministic approach, i.e., these models assume that the expansion planner has perfect information at the time it determines the optimal generation expansion plan. Although considering such a deterministic approach allows us to formulate a relatively simple model, and comprehend its assumptions, its formulation and its outcomes, in reality generation expansion decisions are made within an uncertain environment that, if not adequately represented, may lead to a sub-optimal generation expansion plan.

This issue is analyzed in this section. We describe the GEP problem considering a stochastic approach in which uncertainties are modeled through a set of scenarios indexed by ω. For the sake of simplicity, we assume that uncertainty only affects the future demand conditions, e.g., the demand growth. However, additional sources of uncertainty may be considered through additional scenarios [3].

Considering such a stochastic framework, we describe the GEP problem using both a static approach and a dynamic one. In both cases, we consider a single-node approach for the sake of simplicity.

3.6.1 Static Model Formulation

Considering a static approach, the generation expansion plan is made at the beginning of the planning horizon. At this point in time, the generation expansion planner does not know the future scenario that will materialize. Thus, these optimal generation expansion decisions are *here-and-now* decisions since they do not depend on the scenario realizations. Figure 3.4 depicts the corresponding scenario tree. The generation expansion planner determines the expansion plans at the beginning of the planning horizon and, then, one of the scenarios used to represent the uncertainty in the decision-making problem is realized. The *wait-and-see* decisions are in this case those corresponding to the power produced by existing and candidate generating units, which do depend on the scenario realization.

Considering such a stochastic single-node stochastic framework, the GEP problem can be formulated using the following complementarity model:

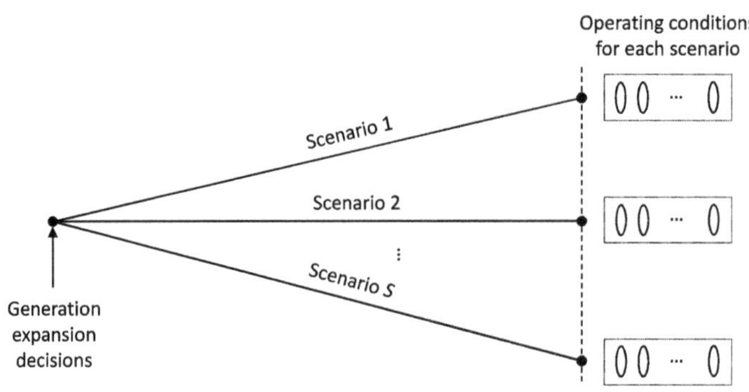

Fig. 3.4 Scenario tree for the stochastic single-node static GEP problem

$$\min_{\Delta^{\text{S1S}}, \Delta^{\text{1SSM}}_{o\omega}} \quad \sum_{\omega} \varphi_{\omega} \left\{ \sum_{o} \rho_{o} \left[\sum_{g} C^{\text{E}}_{g} p^{\text{E}}_{go\omega} + \sum_{c} C^{\text{C}}_{c} p^{\text{C}}_{co\omega} \right] \right\} + \sum_{c} \tilde{I}^{\text{C}}_{c} p^{\text{C}^{\max}}_{c} \tag{3.13a}$$

subject to

$$p^{\text{C}^{\max}}_{c} = \sum_{q} u^{\text{Option}}_{cq} P^{\text{Option}}_{cq} \quad \forall c \tag{3.13b}$$

$$\sum_{q} u^{\text{Option}}_{cq} = 1 \quad \forall c \tag{3.13c}$$

$$u^{\text{Option}}_{cq} \in \{0, 1\} \quad \forall c, q \tag{3.13d}$$

$$p^{\text{E}}_{go\omega}, p^{\text{C}}_{co\omega} \in \Omega_{o\omega} \left(P^{\text{D}}_{do\omega}, P^{\text{E}^{\max}}_{g}, p^{\text{C}^{\max}}_{c} \right) \quad \forall g, c, o, \omega, \tag{3.13e}$$

where $\Omega_{o\omega} \left(P^{\text{D}}_{do\omega}, P^{\text{E}^{\max}}_{g}, p^{\text{C}^{\max}}_{c} \right) = \Bigg\{$

$$\min_{\Delta^{\text{1SSM}}_{o\omega}} \quad \sum_{g} C^{\text{E}}_{g} p^{\text{E}}_{go\omega} + \sum_{c} C^{\text{C}}_{c} p^{\text{C}}_{co\omega} \tag{3.14a}$$

subject to

$$\sum_{g} p^{\text{E}}_{go\omega} + \sum_{c} p^{\text{C}}_{co\omega} = \sum_{d} P^{\text{D}}_{do\omega} \tag{3.14b}$$

$$0 \leq p^{\text{E}}_{go\omega} \leq P^{\text{E}^{\max}}_{g} \quad \forall g \tag{3.14c}$$

$$0 \leq p^{\text{C}}_{co\omega} \leq p^{\text{C}^{\max}}_{c} \quad \forall c \tag{3.14d}$$

$$\Bigg\}, \forall o, \omega,$$

variables in sets $\Delta^{\text{S1S}} = \left\{ p^{\text{C}^{\max}}_{c}, u^{\text{Option}}_{cq} \right\}$ and $\Delta^{\text{S1SM}}_{o\omega}$, $\forall o, \omega$, are the optimization variables of problem (3.13), and variables in sets $\Delta^{\text{S1SM}}_{o\omega} = \left\{ p^{\text{E}}_{go\omega}, p^{\text{C}}_{co\omega} \right\}$ are the optimization variables of problems (3.14), $\forall o, \omega$.

The differences between GEP problem (3.13) and the equivalent (3.1) considering a deterministic approach are summarized below:

1. In the objective function (3.13a), we compute the expected generation cost. To do so, the generation cost for each scenario, $\sum_{o} \rho_{o} \left[\sum_{g} C^{\text{E}}_{g} p^{\text{E}}_{go\omega} + \sum_{c} C^{\text{C}}_{c} p^{\text{C}}_{co\omega} \right]$, is multiplied by the probability of the corresponding scenario, φ_{ω}.
2. The market-clearing problems (3.14), $\forall o, \omega$, are formulated for each operating condition o and for each scenario ω.

Note that nonanticipativity conditions are implicitly imposed since the generation expansion decision variables, i.e., variables $p_c^{\mathrm{C^{max}}}$, do not depend on the scenario realization ω.

Finally, complementarity model (3.13) is recast as a MILP problem using the procedure explained in Sect. 3.3.3.

Illustrative Example 3.9 *Stochastic GEP problem using a static approach*

Considering the data of Illustrative Example 3.4, we assume that demand $\tilde{d}$ is subject to uncertainty as explained below.

The demand for each operating condition can be 30 % lower than and 30 % higher than the demands considered in Illustrative Example 3.4 with equal probability (0.5), i.e., we consider two scenarios. Scenario 1 is characterized by demand conditions of 203 and 385 MW for operating conditions o_1 and o_2, respectively, while scenario 2 is characterized by demand conditions of 377 and 715 MW for operating conditions o_1 and o_2, respectively.

Figure 3.5 depicts the scenario tree for this illustrative example.

Considering the above data, the GEP problem considering a stochastic static approach (3.13) results in the following complementarity model:

$$\begin{aligned}
\min_{\Delta} \quad & 0.5\left\{6000\left[35p^{\mathrm{E}}_{\tilde{g}o_1\omega_1}+25p^{\mathrm{C}}_{\tilde{c}o_1\omega_1}\right]+2760\left[35p^{\mathrm{E}}_{\tilde{g}o_2\omega_1}+25p^{\mathrm{C}}_{\tilde{c}o_2\omega_1}\right]\right\}\\
&+0.5\left\{6000\left[35p^{\mathrm{E}}_{\tilde{g}o_1\omega_2}+25p^{\mathrm{C}}_{\tilde{c}o_1\omega_2}\right]+2760\left[35p^{\mathrm{E}}_{\tilde{g}o_2\omega_2}+25p^{\mathrm{C}}_{\tilde{c}o_2\omega_2}\right]\right\}\\
&+70000p^{\mathrm{C^{max}}}_{\tilde{c}}
\end{aligned}$$

subject to

$$\begin{aligned}
&p^{\mathrm{C^{max}}}_{\tilde{c}}=0u^{\mathrm{Option}}_{\tilde{c}q_1}+100u^{\mathrm{Option}}_{\tilde{c}q_2}+200u^{\mathrm{Option}}_{\tilde{c}q_3}+300u^{\mathrm{Option}}_{\tilde{c}q_4}+400u^{\mathrm{Option}}_{\tilde{c}q_5}\\
&\quad+500u^{\mathrm{Option}}_{\tilde{c}q_6}\\
&u^{\mathrm{Option}}_{\tilde{c}q_1}+u^{\mathrm{Option}}_{\tilde{c}q_2}+u^{\mathrm{Option}}_{\tilde{c}q_3}+u^{\mathrm{Option}}_{\tilde{c}q_4}+u^{\mathrm{Option}}_{\tilde{c}q_5}+u^{\mathrm{Option}}_{\tilde{c}q_6}=1\\
&u^{\mathrm{Option}}_{\tilde{c}q_1},u^{\mathrm{Option}}_{\tilde{c}q_2},u^{\mathrm{Option}}_{\tilde{c}q_3},u^{\mathrm{Option}}_{\tilde{c}q_4},u^{\mathrm{Option}}_{\tilde{c}q_5},u^{\mathrm{Option}}_{\tilde{c}q_6}\in\{0,1\}
\end{aligned}$$

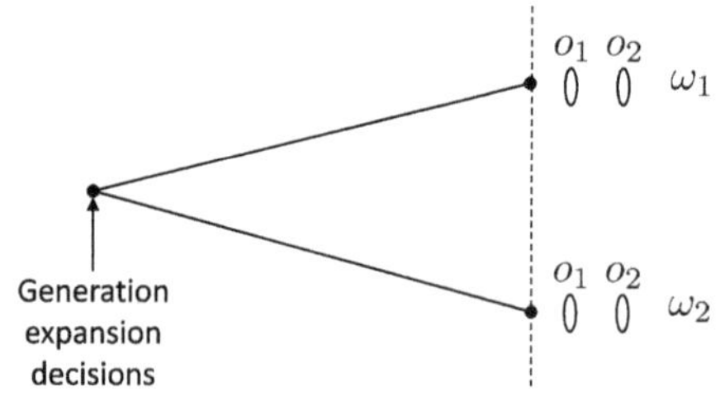

Fig. 3.5 Scenario tree for Illustrative Example 3.9

$$\left\{\begin{array}{l}\min_{p^{\mathrm{E}}_{\tilde{g}o_1\omega_1},p^{\mathrm{C}}_{\tilde{c}o_1\omega_1}} \quad 35p^{\mathrm{E}}_{\tilde{g}o_1\omega_1}+25p^{\mathrm{C}}_{\tilde{c}o_1\omega_1}\\ \text{s.t.}\\ \quad p^{\mathrm{E}}_{\tilde{g}o_1\omega_1}+p^{\mathrm{C}}_{\tilde{c}o_1\omega_1}=203\\ \quad 0\le p^{\mathrm{E}}_{\tilde{g}o_1\omega_1}\le 400\\ \quad 0\le p^{\mathrm{C}}_{\tilde{c}o_1\omega_1}\le p^{\mathrm{C}^{\max}}_{\tilde{c}}\end{array}\right.$$

$$\left\{\begin{array}{l}\min_{p^{\mathrm{E}}_{\tilde{g}o_2\omega_1},p^{\mathrm{C}}_{\tilde{c}o_2\omega_1}} \quad 35p^{\mathrm{E}}_{\tilde{g}o_2\omega_1}+25p^{\mathrm{C}}_{\tilde{c}o_2\omega_1}\\ \text{s.t.}\\ \quad p^{\mathrm{E}}_{\tilde{g}o_2\omega_1}+p^{\mathrm{C}}_{\tilde{c}o_2\omega_1}=385\\ \quad 0\le p^{\mathrm{E}}_{\tilde{g}o_2\omega_1}\le 400\\ \quad 0\le p^{\mathrm{C}}_{\tilde{c}o_2\omega_1}\le p^{\mathrm{C}^{\max}}_{\tilde{c}}\end{array}\right.$$

$$\left\{\begin{array}{l}\min_{p^{\mathrm{E}}_{\tilde{g}o_1\omega_2},p^{\mathrm{C}}_{\tilde{c}o_1\omega_2}} \quad 35p^{\mathrm{E}}_{\tilde{g}o_1\omega_2}+25p^{\mathrm{C}}_{\tilde{c}o_1\omega_2}\\ \text{s.t.}\\ \quad p^{\mathrm{E}}_{\tilde{g}o_1\omega_2}+p^{\mathrm{C}}_{\tilde{c}o_1\omega_2}=377\\ \quad 0\le p^{\mathrm{E}}_{\tilde{g}o_1\omega_2}\le 400\\ \quad 0\le p^{\mathrm{C}}_{\tilde{c}o_1\omega_2}\le p^{\mathrm{C}^{\max}}_{\tilde{c}}\end{array}\right.$$

$$\left\{\begin{array}{l}\min_{p^{\mathrm{E}}_{\tilde{g}o_2\omega_2},p^{\mathrm{C}}_{\tilde{c}o_2\omega_2}} \quad 35p^{\mathrm{E}}_{\tilde{g}o_2\omega_2}+25p^{\mathrm{C}}_{\tilde{c}o_2\omega_2}\\ \text{s.t.}\\ \quad p^{\mathrm{E}}_{\tilde{g}o_2\omega_2}+p^{\mathrm{C}}_{\tilde{c}o_2\omega_2}=715\\ \quad 0\le p^{\mathrm{E}}_{\tilde{g}o_2\omega_2}\le 400\\ \quad 0\le p^{\mathrm{C}}_{\tilde{c}o_2\omega_2}\le p^{\mathrm{C}^{\max}}_{\tilde{c}},\end{array}\right.$$

where $\Delta=\left\{p^{\mathrm{E}}_{\tilde{g}o_1\omega_1},p^{\mathrm{C}}_{\tilde{c}o_1\omega_1},p^{\mathrm{E}}_{\tilde{g}o_2\omega_1},p^{\mathrm{C}}_{\tilde{c}o_2\omega_1},p^{\mathrm{E}}_{\tilde{g}o_1\omega_2},p^{\mathrm{C}}_{\tilde{c}o_1\omega_2},p^{\mathrm{E}}_{\tilde{g}o_2\omega_2},p^{\mathrm{C}}_{\tilde{c}o_2\omega_2},p^{\mathrm{C}^{\max}}_{\tilde{c}},u^{\mathrm{Option}}_{\tilde{c}q_1},\right.$ $\left.u^{\mathrm{Option}}_{\tilde{c}q_2},u^{\mathrm{Option}}_{\tilde{c}q_3},u^{\mathrm{Option}}_{\tilde{c}q_4},u^{\mathrm{Option}}_{\tilde{c}q_5},u^{\mathrm{Option}}_{\tilde{c}q_6}\right\}$.

We obtain that the optimal GEP plan consists of building 400 MW of the candidate generating unit.

Note that in both this illustrative example and in Illustrative Example 3.4, the average demand for each operating condition is the same. However, the generation expansion decisions are different. This means that using the average value of the uncertain parameters to formulate a deterministic problem does not always result in the optimal solution. For example, in this case, building 300 MW of the candidate generating unit, i.e., the solution of the deterministic static problem in Illustrative Example 3.4, would not be enough to supply the demand in the case that scenario 2 is finally realized. Therefore, it is important to represent the uncertainties on the GEP decision-making problem in order to obtain an informed generation expansion plan.

The disadvantage of using such a stochastic framework is that the size of the problem increases. For example, in this illustrative example, the number of variables considered in the market-clearing problems is twice the number of variables needed in Illustrative Example 3.4. Moreover, in order to represent the uncertainties accurately, it is usually necessary to consider a large enough number of scenarios, which may lead to very large problems. It is, therefore, a tradeoff between modeling accuracy and tractability. □

3.6.2 Dynamic Model Formulation

The stochastic single-node static GEP model described in the previous section is extended here to consider a dynamic approach.

As explained in Sect. 3.2.3, we consider that the planning horizon is divided into a set of time periods and that the generation expansion decisions can be made at the beginning of each time period.

Uncertainty in the demand conditions at each time period is represented by a set of discrete scenarios. We assume that the generation expansion planner knows the demand scenario realization of each time period when it concludes so that it can adapt the generation expansion plan for the future time periods according to this.

Given the above framework, we consider the following decision-making sequence and corresponding scenario tree depicted in Fig. 3.6:

1. At the beginning of the planning horizon, i.e., at the beginning of the first time period, the generation expansion planner determines the optimal generation expansion plan to be carried out at this point in time. These expansion decisions are *here-and-know* decisions since they do not depend on the future scenario realizations as the generation expansion planner does not know at this point in time the future scenario realizations.
2. The first time period concludes, and the generation expansion planner knows the actual scenario realization for the first time period.
3. The generation expansion planner determines the generation expansion plan to be made at the beginning of the second time period. These expansion decisions are *wait-and-see* decisions with respect to the first time period since they do depend on the scenario realization of the first time period. However, they are *here-and-now* decisions with respect to the second and future time periods since they do not depend on the future scenario realizations.

Then, steps 2–3 above are repeated until the last time period of the planning horizon.

Considering the above framework, the GEP problem considering a stochastic single-node dynamic approach can be formulated using the complementarity model below:

$$\min_{\Delta^{\mathrm{S1D}}, \Delta^{\mathrm{S1DM}}_{ot\omega}} \quad \sum_{\omega} \varphi_{\omega} \sum_{t} \left\{ \left[\sum_{o} \rho_o \left(\sum_{g} C^{\mathrm{E}}_{g} p^{\mathrm{E}}_{got\omega} + \sum_{c} C^{\mathrm{C}}_{c} p^{\mathrm{C}}_{cot\omega} \right) + a_t \sum_{c} I^{\mathrm{C}}_{c} p^{\mathrm{C}^{\max}}_{ct\omega} \right] \right\} \tag{3.15a}$$

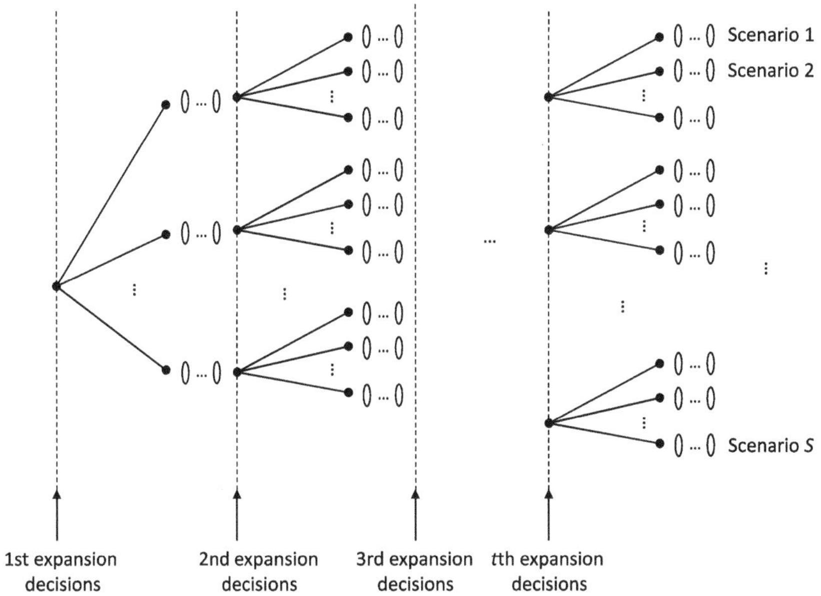

Fig. 3.6 Scenario tree for the stochastic single-node dynamic GEP problem

subject to

$$p^{\mathrm{C^{max}}}_{ct\omega} = \sum_q u^{\mathrm{Option}}_{cqt\omega} P^{\mathrm{Option}}_{cqt} \quad \forall c, t, \omega \tag{3.15b}$$

$$\sum_q u^{\mathrm{Option}}_{cqt\omega} = 1 \quad \forall c, t, \omega \tag{3.15c}$$

$$u^{\mathrm{Option}}_{cqt\omega} \in \{0, 1\} \quad \forall c, q, t, \omega \tag{3.15d}$$

$$p^{\mathrm{C^{max}}}_{ct\omega} = p^{\mathrm{C^{max}}}_{ct\tilde{\omega}} \quad \forall c, t, \omega, \tilde{\omega} | P^{\mathrm{D}}_{do\tau\omega} = P^{\mathrm{D}}_{do\tau\tilde{\omega}} \quad \forall d, o, \tau < t \tag{3.15e}$$

$$p^{\mathrm{E}}_{got\omega}, p^{\mathrm{C}}_{cot\omega} \in \Omega_{ot\omega}\left(P^{\mathrm{D}}_{dot\omega}, P^{\mathrm{E^{max}}}_{g}, p^{\mathrm{C^{max}}}_{ct\omega}\right) \quad \forall g, c, o, t, \omega, \tag{3.15f}$$

where $\Omega_{ot\omega}\left(P^{\mathrm{D}}_{dot\omega}, P^{\mathrm{E^{max}}}_{g}, p^{\mathrm{C^{max}}}_{ct\omega}\right) = \Bigg\{$

$$\min_{\Delta^{\mathrm{S1DM}}_{ot\omega}} \quad \sum_g C^{\mathrm{E}}_g p^{\mathrm{E}}_{got\omega} + \sum_c C^{\mathrm{C}}_c p^{\mathrm{C}}_{cot\omega} \tag{3.16a}$$

subject to

$$\sum_g p^{\mathrm{E}}_{got\omega} + \sum_c p^{\mathrm{C}}_{cot\omega} = \sum_d P^{\mathrm{D}}_{dot\omega} \tag{3.16b}$$

$$0 \le p^{\mathrm{E}}_{got\omega} \le P^{\mathrm{E^{max}}}_{g} \quad \forall g \tag{3.16c}$$

$$0 \le p^{\mathrm{C}}_{cot\omega} \le \sum_{\tau \le t} p^{\mathrm{C^{max}}}_{c\tau\omega} \quad \forall c \tag{3.16d}$$

$$\Bigg\}, \forall o, t, \omega,$$

variables in sets $\Delta^{\mathrm{S1D}} = \left\{p^{\mathrm{C^{max}}}_{ct\omega}, u^{\mathrm{Option}}_{cqt\omega}\right\}$ and $\Delta^{\mathrm{S1DM}}_{ot\omega}$, $\forall o, t, \omega$, are the optimization variables of problem (3.15), and variables in sets $\Delta^{\mathrm{S1DM}}_{ot\omega} = \left\{p^{\mathrm{E}}_{got\omega}, p^{\mathrm{C}}_{cot\omega}\right\}$, $\forall o, t, \omega$, are the optimization variables of problems (3.16), $\forall o, t, \omega$.

The main differences between model (3.15) and the equivalent one using a stochastic static approach (3.13) are summarized below:

1. In the objective function (3.15a), the investment costs at each time period are multiplied by the corresponding amortization rates.
2. We include nonanticipativity constraints (3.15e), i.e., constraints that avoid anticipating information. These constraints impose that for a given time period t, if the characteristics of two scenarios ω and $\tilde{\omega}$ in the previous time periods $\tau < t$ are the same, i.e., if $P^{\mathrm{D}}_{do\tau\omega} = P^{\mathrm{D}}_{do\tau\tilde{\omega}}$, $\forall d, o, \tau < t$, then the generation expansion plans of these two scenarios for time period t are also the same.
3. Constraints (3.16d) impose that the available capacity of the candidate generating units at time period t is equal to the capacity built at that time period and in the previous ones.
4. The market-clearing problems (3.16), $\forall o, t, \omega$, are formulated for each operating condition o, time period t, and scenario ω.

Finally, the GEP problem considering a stochastic dynamic approach is recast as an MILP problem by following the procedure described in Sect. 3.3.3.

Illustrative Example 3.10 *Stochastic GEP problem using a dynamic approach*

Let us consider the data of Illustrative Example 3.4. The planning horizon is divided in this case into two time periods, so that generation expansion decisions can be made at the beginning of both time periods.

There are two possible scenario realizations in the first time period, a and b. Realizations a and b consider that operating conditions o_1 and o_2 in the first time period are characterized by a low and high demand, respectively. Specifically, scenario a considers that operating conditions o_1 and o_2 are defined by demands equal to 212 MW and 402 MW, respectively, while scenario b considers that operating conditions o_1 and o_2 are defined by demands equal to 281 MW and 533 MW, respectively, The probability of each of scenario realizations a and b is equal to 0.5.

On the other hand, there are also two possible scenario realizations in the second time period, c and d. Realizations c and d consider that operating conditions o_1 and o_2 in the second time period are characterized by a low and high demand, respectively. The operating conditions for these two scenarios in the second time period depend on the scenario realization in the first time period, as explained next and illustrated using the corresponding scenario tree depicted in Fig. 3.7:

1. If scenario a is realized in the first time period, then (i) scenario c considers that operating conditions o_1 and o_2 in the second time period are defined by demands equal to 214 MW and 407 MW, respectively, while (ii) scenario d considers that operating conditions o_1 and o_2 in the second time period are defined by demands equal to 284 MW and 539 MW, respectively.
2. If scenario b is realized in the first time period, then (i) scenario c considers that operating conditions o_1 and o_2 in the second time period are defined by demands equal to 284 MW and 539 MW, respectively, while (ii) scenario d considers that operating conditions o_1 and o_2 in the second time period are defined by demands equal to 377 MW and 715 MW, respectively.

The probability of each of realizations c and d is equal to 0.5.

The weights of operating conditions o_1 and o_2 in both time periods are 6000 h and 2760 h, respectively.

This results in four scenarios for all time periods (combinations of possible realizations in both time periods), whose data are summarized in Table 3.1. Note that the data of scenarios and operating conditions have been selected so that the results of this illustrative example can be compared with the results of the previous ones.

Considering the above data, the GEP problem considering a stochastic dynamic approach (3.15) results in the following complementarity model:

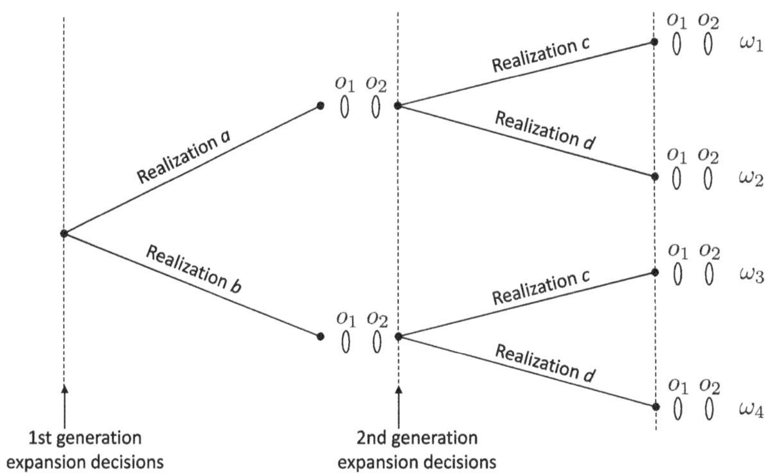

Fig. 3.7 Scenario tree for Illustrative Example 3.10

Table 3.1 Illustrative Example 3.10: data for scenarios

Scenario	Period 1	Period 2	Probability
Scenario 1 (*a* and *c*)	$P^{\mathrm{D}}_{o_1t_1\omega_1}$ = 212 MW	$P^{\mathrm{D}}_{o_1t_2\omega_1}$ = 214 MW	0.5 × 0.5 = 0.25
	$P^{\mathrm{D}}_{o_2t_1\omega_1}$ = 402 MW	$P^{\mathrm{D}}_{o_2t_2\omega_1}$ = 407 MW	
Scenario 2 (*a* and *d*)	$P^{\mathrm{D}}_{o_1t_1\omega_2}$ = 212 MW	$P^{\mathrm{D}}_{o_1t_2\omega_2}$ = 284 MW	0.5 × 0.5 = 0.25
	$P^{\mathrm{D}}_{o_2t_1\omega_2}$ = 402 MW	$P^{\mathrm{D}}_{o_2t_2\omega_2}$ = 539 MW	
Scenario 3 (*b* and *c*)	$P^{\mathrm{D}}_{o_1t_1\omega_3}$ = 281 MW	$P^{\mathrm{D}}_{o_1t_2\omega_3}$ = 284 MW	0.5 × 0.5 = 0.25
	$P^{\mathrm{D}}_{o_2t_1\omega_3}$ = 533 MW	$P^{\mathrm{D}}_{o_2t_2\omega_3}$ = 539 MW	
Scenario 4 (*b* and *d*)	$P^{\mathrm{D}}_{o_1t_1\omega_4}$ = 281 MW	$P^{\mathrm{D}}_{o_1t_2\omega_4}$ = 377 MW	0.5 × 0.5 = 0.25
	$P^{\mathrm{D}}_{o_2t_1\omega_4}$ = 533 MW	$P^{\mathrm{D}}_{o_2t_2\omega_4}$ = 715 MW	

$$\begin{aligned}
\min_{\Delta} \quad & 0.25\left\{6000\left[35p^{\mathrm{E}}_{\tilde{g}o_1t_1\omega_1}+25p^{\mathrm{C}}_{\tilde{c}o_1t_1\omega_1}\right]+2760\left[35p^{\mathrm{E}}_{\tilde{g}o_2t_1\omega_1}+25p^{\mathrm{C}}_{\tilde{c}o_2t_1\omega_1}\right]\right.\\
&\quad+6000\left[35p^{\mathrm{E}}_{\tilde{g}o_1t_2\omega_1}+25p^{\mathrm{C}}_{\tilde{c}o_1t_2\omega_1}\right]+2760\left[35p^{\mathrm{E}}_{\tilde{g}o_2t_2\omega_1}+25p^{\mathrm{C}}_{\tilde{c}o_2t_2\omega_1}\right]\\
&\quad\left.+140000p^{\mathrm{C^{max}}}_{\tilde{c}t_1\omega_1}+70000p^{\mathrm{C^{max}}}_{\tilde{c}t_2\omega_1}\right\}\\
&+0.25\left\{6000\left[35p^{\mathrm{E}}_{\tilde{g}o_1t_1\omega_2}+25p^{\mathrm{C}}_{\tilde{c}o_1t_1\omega_2}\right]+2760\left[35p^{\mathrm{E}}_{\tilde{g}o_2t_1\omega_2}+25p^{\mathrm{C}}_{\tilde{c}o_2t_1\omega_2}\right]\right.\\
&\quad+6000\left[35p^{\mathrm{E}}_{\tilde{g}o_1t_2\omega_2}+25p^{\mathrm{C}}_{\tilde{c}o_1t_2\omega_2}\right]+2760\left[35p^{\mathrm{E}}_{\tilde{g}o_2t_2\omega_2}+25p^{\mathrm{C}}_{\tilde{c}o_2t_2\omega_2}\right]\\
&\quad\left.+140000p^{\mathrm{C^{max}}}_{\tilde{c}t_1\omega_2}+70000p^{\mathrm{C^{max}}}_{\tilde{c}t_2\omega_2}\right\}\\
&+0.25\left\{6000\left[35p^{\mathrm{E}}_{\tilde{g}o_1t_1\omega_3}+25p^{\mathrm{C}}_{\tilde{c}o_1t_1\omega_3}\right]+2760\left[35p^{\mathrm{E}}_{\tilde{g}o_2t_1\omega_3}+25p^{\mathrm{C}}_{\tilde{c}o_2t_1\omega_3}\right]\right.\\
&\quad+6000\left[35p^{\mathrm{E}}_{\tilde{g}o_1t_2\omega_3}+25p^{\mathrm{C}}_{\tilde{c}o_1t_2\omega_3}\right]+2760\left[35p^{\mathrm{E}}_{\tilde{g}o_2t_2\omega_3}+25p^{\mathrm{C}}_{\tilde{c}o_2t_2\omega_3}\right]\\
&\quad\left.+140000p^{\mathrm{C^{max}}}_{\tilde{c}t_1\omega_3}+70000p^{\mathrm{C^{max}}}_{\tilde{c}t_2\omega_3}\right\}\\
&+0.25\left\{6000\left[35p^{\mathrm{E}}_{\tilde{g}o_1t_1\omega_4}+25p^{\mathrm{C}}_{\tilde{c}o_1t_1\omega_4}\right]+2760\left[35p^{\mathrm{E}}_{\tilde{g}o_2t_1\omega_4}+25p^{\mathrm{C}}_{\tilde{c}o_2t_1\omega_4}\right]\right.\\
&\quad+6000\left[35p^{\mathrm{E}}_{\tilde{g}o_1t_2\omega_4}+25p^{\mathrm{C}}_{\tilde{c}o_1t_2\omega_4}\right]+2760\left[35p^{\mathrm{E}}_{\tilde{g}o_2t_2\omega_4}+25p^{\mathrm{C}}_{\tilde{c}o_2t_2\omega_4}\right]\\
&\quad\left.+140000p^{\mathrm{C^{max}}}_{\tilde{c}t_1\omega_4}+70000p^{\mathrm{C^{max}}}_{\tilde{c}t_2\omega_4}\right\}
\end{aligned}$$

subject to

$$\begin{aligned}
p^{\mathrm{C^{max}}}_{\tilde{c}t_1\omega_1} &= 0u^{\mathrm{Option}}_{\tilde{c}q_1t_1\omega_1}+100u^{\mathrm{Option}}_{\tilde{c}q_2t_1\omega_1}+200u^{\mathrm{Option}}_{\tilde{c}q_3t_1\omega_1}+300u^{\mathrm{Option}}_{\tilde{c}q_4t_1\omega_1}+400u^{\mathrm{Option}}_{\tilde{c}q_5t_1\omega_1}\\
&\quad+500u^{\mathrm{Option}}_{\tilde{c}q_6t_1\omega_1}\\
p^{\mathrm{C^{max}}}_{\tilde{c}t_2\omega_1} &= 0u^{\mathrm{Option}}_{\tilde{c}q_1t_2\omega_1}+100u^{\mathrm{Option}}_{\tilde{c}q_2t_2\omega_1}+200u^{\mathrm{Option}}_{\tilde{c}q_3t_2\omega_1}+300u^{\mathrm{Option}}_{\tilde{c}q_4t_2\omega_1}+400u^{\mathrm{Option}}_{\tilde{c}q_5t_2\omega_1}\\
&\quad+500u^{\mathrm{Option}}_{\tilde{c}q_6t_2\omega_1}
\end{aligned}$$

$$
\begin{aligned}
&p^{C^{\max}}_{\tilde{c}t_1\omega_2} = 0u^{\text{Option}}_{\tilde{c}q_1t_1\omega_2} + 100u^{\text{Option}}_{\tilde{c}q_2t_1\omega_2} + 200u^{\text{Option}}_{\tilde{c}q_3t_1\omega_2} + 300u^{\text{Option}}_{\tilde{c}q_4t_1\omega_2} + 400u^{\text{Option}}_{\tilde{c}q_5t_1\omega_2} \\
&\quad + 500u^{\text{Option}}_{\tilde{c}q_6t_1\omega_2} \\
&p^{C^{\max}}_{\tilde{c}t_2\omega_2} = 0u^{\text{Option}}_{\tilde{c}q_1t_2\omega_2} + 100u^{\text{Option}}_{\tilde{c}q_2t_2\omega_2} + 200u^{\text{Option}}_{\tilde{c}q_3t_2\omega_2} + 300u^{\text{Option}}_{\tilde{c}q_4t_2\omega_2} + 400u^{\text{Option}}_{\tilde{c}q_5t_2\omega_2} \\
&\quad + 500u^{\text{Option}}_{\tilde{c}q_6t_2\omega_2} \\
&p^{C^{\max}}_{\tilde{c}t_1\omega_3} = 0u^{\text{Option}}_{\tilde{c}q_1t_1\omega_3} + 100u^{\text{Option}}_{\tilde{c}q_2t_1\omega_3} + 200u^{\text{Option}}_{\tilde{c}q_3t_1\omega_3} + 300u^{\text{Option}}_{\tilde{c}q_4t_1\omega_3} + 400u^{\text{Option}}_{\tilde{c}q_5t_1\omega_3} \\
&\quad + 500u^{\text{Option}}_{\tilde{c}q_6t_1\omega_3} \\
&p^{C^{\max}}_{\tilde{c}t_2\omega_3} = 0u^{\text{Option}}_{\tilde{c}q_1t_2\omega_3} + 100u^{\text{Option}}_{\tilde{c}q_2t_2\omega_3} + 200u^{\text{Option}}_{\tilde{c}q_3t_2\omega_3} + 300u^{\text{Option}}_{\tilde{c}q_4t_2\omega_3} + 400u^{\text{Option}}_{\tilde{c}q_5t_2\omega_3} \\
&\quad + 500u^{\text{Option}}_{\tilde{c}q_6t_3\omega_1} \\
&p^{C^{\max}}_{\tilde{c}t_1\omega_4} = 0u^{\text{Option}}_{\tilde{c}q_1t_1\omega_4} + 100u^{\text{Option}}_{\tilde{c}q_2t_1\omega_4} + 200u^{\text{Option}}_{\tilde{c}q_3t_1\omega_4} + 300u^{\text{Option}}_{\tilde{c}q_4t_1\omega_4} + 400u^{\text{Option}}_{\tilde{c}q_5t_1\omega_4} \\
&\quad + 500u^{\text{Option}}_{\tilde{c}q_6t_1\omega_4} \\
&p^{C^{\max}}_{\tilde{c}t_2\omega_4} = 0u^{\text{Option}}_{\tilde{c}q_1t_2\omega_4} + 100u^{\text{Option}}_{\tilde{c}q_2t_2\omega_4} + 200u^{\text{Option}}_{\tilde{c}q_3t_2\omega_4} + 300u^{\text{Option}}_{\tilde{c}q_4t_2\omega_4} + 400u^{\text{Option}}_{\tilde{c}q_5t_2\omega_4} \\
&\quad + 500u^{\text{Option}}_{\tilde{c}q_6t_2\omega_4} \\
&u^{\text{Option}}_{\tilde{c}q_1t_1\omega_1} + u^{\text{Option}}_{\tilde{c}q_2t_1\omega_1} + u^{\text{Option}}_{\tilde{c}q_3t_1\omega_1} + u^{\text{Option}}_{\tilde{c}q_4t_1\omega_1} + u^{\text{Option}}_{\tilde{c}q_5t_1\omega_1} + u^{\text{Option}}_{\tilde{c}q_6t_1\omega_1} = 1 \\
&u^{\text{Option}}_{\tilde{c}q_1t_2\omega_1} + u^{\text{Option}}_{\tilde{c}q_2t_2\omega_1} + u^{\text{Option}}_{\tilde{c}q_3t_2\omega_1} + u^{\text{Option}}_{\tilde{c}q_4t_2\omega_1} + u^{\text{Option}}_{\tilde{c}q_5t_2\omega_1} + u^{\text{Option}}_{\tilde{c}q_6t_2\omega_1} = 1 \\
&u^{\text{Option}}_{\tilde{c}q_1t_1\omega_2} + u^{\text{Option}}_{\tilde{c}q_2t_1\omega_2} + u^{\text{Option}}_{\tilde{c}q_3t_1\omega_2} + u^{\text{Option}}_{\tilde{c}q_4t_1\omega_2} + u^{\text{Option}}_{\tilde{c}q_5t_1\omega_2} + u^{\text{Option}}_{\tilde{c}q_6t_1\omega_2} = 1 \\
&u^{\text{Option}}_{\tilde{c}q_1t_2\omega_2} + u^{\text{Option}}_{\tilde{c}q_2t_2\omega_2} + u^{\text{Option}}_{\tilde{c}q_3t_2\omega_2} + u^{\text{Option}}_{\tilde{c}q_4t_2\omega_2} + u^{\text{Option}}_{\tilde{c}q_5t_2\omega_2} + u^{\text{Option}}_{\tilde{c}q_6t_2\omega_2} = 1 \\
&u^{\text{Option}}_{\tilde{c}q_1t_1\omega_3} + u^{\text{Option}}_{\tilde{c}q_2t_1\omega_3} + u^{\text{Option}}_{\tilde{c}q_3t_1\omega_3} + u^{\text{Option}}_{\tilde{c}q_4t_1\omega_3} + u^{\text{Option}}_{\tilde{c}q_5t_1\omega_3} + u^{\text{Option}}_{\tilde{c}q_6t_1\omega_3} = 1 \\
&u^{\text{Option}}_{\tilde{c}q_1t_2\omega_3} + u^{\text{Option}}_{\tilde{c}q_2t_2\omega_3} + u^{\text{Option}}_{\tilde{c}q_3t_2\omega_3} + u^{\text{Option}}_{\tilde{c}q_4t_2\omega_3} + u^{\text{Option}}_{\tilde{c}q_5t_2\omega_3} + u^{\text{Option}}_{\tilde{c}q_6t_2\omega_3} = 1 \\
&u^{\text{Option}}_{\tilde{c}q_1t_1\omega_4} + u^{\text{Option}}_{\tilde{c}q_2t_1\omega_4} + u^{\text{Option}}_{\tilde{c}q_3t_1\omega_4} + u^{\text{Option}}_{\tilde{c}q_4t_1\omega_4} + u^{\text{Option}}_{\tilde{c}q_5t_1\omega_4} + u^{\text{Option}}_{\tilde{c}q_6t_1\omega_4} = 1 \\
&u^{\text{Option}}_{\tilde{c}q_1t_2\omega_4} + u^{\text{Option}}_{\tilde{c}q_2t_2\omega_4} + u^{\text{Option}}_{\tilde{c}q_3t_2\omega_4} + u^{\text{Option}}_{\tilde{c}q_4t_2\omega_4} + u^{\text{Option}}_{\tilde{c}q_5t_2\omega_4} + u^{\text{Option}}_{\tilde{c}q_6t_2\omega_4} = 1 \\
&u^{\text{Option}}_{\tilde{c}q_1t_1\omega_1}, u^{\text{Option}}_{\tilde{c}q_2t_1\omega_1}, u^{\text{Option}}_{\tilde{c}q_3t_1\omega_1}, u^{\text{Option}}_{\tilde{c}q_4t_1\omega_1}, u^{\text{Option}}_{\tilde{c}q_5t_1\omega_1}, u^{\text{Option}}_{\tilde{c}q_6t_1\omega_1} \in \{0, 1\} \\
&p^{C^{\max}}_{\tilde{c}t_1\omega_1} = p^{C^{\max}}_{\tilde{c}t_1\omega_2} = p^{C^{\max}}_{\tilde{c}t_1\omega_3} = p^{C^{\max}}_{\tilde{c}t_1\omega_4} \\
&p^{C^{\max}}_{\tilde{c}t_2\omega_1} = p^{C^{\max}}_{\tilde{c}t_2\omega_2} \\
&p^{C^{\max}}_{\tilde{c}t_2\omega_3} = p^{C^{\max}}_{\tilde{c}t_2\omega_3}
\end{aligned}
$$

$$\begin{cases} \min_{p^{\mathrm{E}}_{\tilde{g}o_1t_1\omega_1},p^{\mathrm{C}}_{\tilde{c}o_1t_1\omega_1}} & 35p^{\mathrm{E}}_{\tilde{g}o_1t_1\omega_1}+25p^{\mathrm{C}}_{\tilde{c}o_1t_1\omega_1} \\ \text{s.t.} & \\ \quad p^{\mathrm{E}}_{\tilde{g}o_1t_1\omega_1}+p^{\mathrm{C}}_{\tilde{c}o_1t_1\omega_1}=212 & \\ \quad 0\le p^{\mathrm{E}}_{\tilde{g}o_1t_1\omega_1}\le 400 & \\ \quad 0\le p^{\mathrm{C}}_{\tilde{c}o_1t_1\omega_1}\le p^{\mathrm{C^{max}}}_{\tilde{c}t_1\omega_1} & \end{cases}$$

$$\begin{cases} \min_{p^{\mathrm{E}}_{\tilde{g}o_2t_1\omega_1},p^{\mathrm{C}}_{\tilde{c}o_2t_1\omega_1}} & 35p^{\mathrm{E}}_{\tilde{g}o_2t_1\omega_1}+25p^{\mathrm{C}}_{\tilde{c}o_2t_1\omega_1} \\ \text{s.t.} & \\ \quad p^{\mathrm{E}}_{\tilde{g}o_2t_1\omega_1}+p^{\mathrm{C}}_{\tilde{c}o_2t_1\omega_1}=402 & \\ \quad 0\le p^{\mathrm{E}}_{\tilde{g}o_2t_1\omega_1}\le 400 & \\ \quad 0\le p^{\mathrm{C}}_{\tilde{c}o_2t_1\omega_1}\le p^{\mathrm{C^{max}}}_{\tilde{c}t_1\omega_1} & \end{cases}$$

$$\begin{cases} \min_{p^{\mathrm{E}}_{\tilde{g}o_1t_2\omega_1},p^{\mathrm{C}}_{\tilde{c}o_1t_2\omega_1}} & 35p^{\mathrm{E}}_{\tilde{g}o_1t_2\omega_1}+25p^{\mathrm{C}}_{\tilde{c}o_1t_2\omega_1} \\ \text{s.t.} & \\ \quad p^{\mathrm{E}}_{\tilde{g}o_1t_2\omega_1}+p^{\mathrm{C}}_{\tilde{c}o_1t_2\omega_1}=214 & \\ \quad 0\le p^{\mathrm{E}}_{\tilde{g}o_1t_2\omega_1}\le 400 & \\ \quad 0\le p^{\mathrm{C}}_{\tilde{c}o_1t_2\omega_1}\le p^{\mathrm{C^{max}}}_{\tilde{c}t_1\omega_1}+p^{\mathrm{C^{max}}}_{\tilde{c}t_2\omega_1} & \end{cases}$$

$$\begin{cases} \min_{p^{\mathrm{E}}_{\tilde{g}o_2t_2\omega_1},p^{\mathrm{C}}_{\tilde{c}o_2t_2\omega_1}} & 35p^{\mathrm{E}}_{\tilde{g}o_2t_2\omega_1}+25p^{\mathrm{C}}_{\tilde{c}o_2t_2\omega_1} \\ \text{s.t.} & \\ \quad p^{\mathrm{E}}_{\tilde{g}o_2t_2\omega_1}+p^{\mathrm{C}}_{\tilde{c}o_2t_2\omega_1}=407 & \\ \quad 0\le p^{\mathrm{E}}_{\tilde{g}o_2t_2\omega_1}\le 400 & \\ \quad 0\le p^{\mathrm{C}}_{\tilde{c}o_2t_2\omega_1}\le p^{\mathrm{C^{max}}}_{\tilde{c}t_1\omega_1}+p^{\mathrm{C^{max}}}_{\tilde{c}t_2\omega_1} & \end{cases}$$

$$\begin{cases} \min_{p^{\mathrm{E}}_{\tilde{g}o_1t_1\omega_2},p^{\mathrm{C}}_{\tilde{c}o_1t_1\omega_2}} & 35p^{\mathrm{E}}_{\tilde{g}o_1t_1\omega_2}+25p^{\mathrm{C}}_{\tilde{c}o_1t_1\omega_2} \\ \text{s.t.} & \\ \quad p^{\mathrm{E}}_{\tilde{g}o_1t_1\omega_2}+p^{\mathrm{C}}_{\tilde{c}o_1t_1\omega_2}=212 & \\ \quad 0\le p^{\mathrm{E}}_{\tilde{g}o_1t_1\omega_2}\le 400 & \\ \quad 0\le p^{\mathrm{C}}_{\tilde{c}o_1t_1\omega_2}\le p^{\mathrm{C^{max}}}_{\tilde{c}t_1\omega_2} & \end{cases}$$

$$\begin{cases} \min_{p^{\mathrm{E}}_{\tilde{g}o_2t_1\omega_2},p^{\mathrm{C}}_{\tilde{c}o_2t_1\omega_2}} & 35p^{\mathrm{E}}_{\tilde{g}o_2t_1\omega_2}+25p^{\mathrm{C}}_{\tilde{c}o_2t_1\omega_2} \\ \text{s.t.} & \\ \quad p^{\mathrm{E}}_{\tilde{g}o_2t_1\omega_2}+p^{\mathrm{C}}_{\tilde{c}o_2t_1\omega_2}=402 & \\ \quad 0\le p^{\mathrm{E}}_{\tilde{g}o_2t_1\omega_2}\le 400 & \\ \quad 0\le p^{\mathrm{C}}_{\tilde{c}o_2t_1\omega_2}\le p^{\mathrm{C^{max}}}_{\tilde{c}t_1\omega_2} & \end{cases}$$

$$\left\{\begin{array}{l}\min_{p^{\mathrm{E}}_{\tilde{g}o_1t_2\omega_2},p^{\mathrm{C}}_{\tilde{c}o_1t_2\omega_2}} \quad 35p^{\mathrm{E}}_{\tilde{g}o_1t_2\omega_2}+25p^{\mathrm{C}}_{\tilde{c}o_1t_2\omega_2}\\ \text{s.t.}\\ \quad p^{\mathrm{E}}_{\tilde{g}o_1t_2\omega_2}+p^{\mathrm{C}}_{\tilde{c}o_1t_2\omega_2}=284\\ \quad 0\le p^{\mathrm{E}}_{\tilde{g}o_1t_2\omega_2}\le 400\\ \quad 0\le p^{\mathrm{C}}_{\tilde{c}o_1t_2\omega_2}\le p^{\mathrm{C^{max}}}_{\tilde{c}t_1\omega_2}+p^{\mathrm{C^{max}}}_{\tilde{c}t_2\omega_2}\end{array}\right.$$

$$\left\{\begin{array}{l}\min_{p^{\mathrm{E}}_{\tilde{g}o_2t_2\omega_2},p^{\mathrm{C}}_{\tilde{c}o_2t_2\omega_2}} \quad 35p^{\mathrm{E}}_{\tilde{g}o_2t_2\omega_2}+25p^{\mathrm{C}}_{\tilde{c}o_2t_2\omega_2}\\ \text{s.t.}\\ \quad p^{\mathrm{E}}_{\tilde{g}o_2t_2\omega_2}+p^{\mathrm{C}}_{\tilde{c}o_2t_2\omega_2}=539\\ \quad 0\le p^{\mathrm{E}}_{\tilde{g}o_2t_2\omega_2}\le 400\\ \quad 0\le p^{\mathrm{C}}_{\tilde{c}o_2t_2\omega_2}\le p^{\mathrm{C^{max}}}_{\tilde{c}t_1\omega_2}+p^{\mathrm{C^{max}}}_{\tilde{c}t_2\omega_2}\end{array}\right.$$

$$\left\{\begin{array}{l}\min_{p^{\mathrm{E}}_{\tilde{g}o_1t_1\omega_3},p^{\mathrm{C}}_{\tilde{c}o_1t_1\omega_3}} \quad 35p^{\mathrm{E}}_{\tilde{g}o_1t_1\omega_3}+25p^{\mathrm{C}}_{\tilde{c}o_1t_1\omega_3}\\ \text{s.t.}\\ \quad p^{\mathrm{E}}_{\tilde{g}o_1t_1\omega_3}+p^{\mathrm{C}}_{\tilde{c}o_1t_1\omega_3}=281\\ \quad 0\le p^{\mathrm{E}}_{\tilde{g}o_1t_1\omega_3}\le 400\\ \quad 0\le p^{\mathrm{C}}_{\tilde{c}o_1t_1\omega_3}\le p^{\mathrm{C^{max}}}_{\tilde{c}t_1\omega_3}\end{array}\right.$$

$$\left\{\begin{array}{l}\min_{p^{\mathrm{E}}_{\tilde{g}o_2t_1\omega_3},p^{\mathrm{C}}_{\tilde{c}o_2t_1\omega_3}} \quad 35p^{\mathrm{E}}_{\tilde{g}o_2t_1\omega_3}+25p^{\mathrm{C}}_{\tilde{c}o_2t_1\omega_3}\\ \text{s.t.}\\ \quad p^{\mathrm{E}}_{\tilde{g}o_2t_1\omega_3}+p^{\mathrm{C}}_{\tilde{c}o_2t_1\omega_3}=533\\ \quad 0\le p^{\mathrm{E}}_{\tilde{g}o_2t_1\omega_3}\le 400\\ \quad 0\le p^{\mathrm{C}}_{\tilde{c}o_2t_1\omega_3}\le p^{\mathrm{C^{max}}}_{\tilde{c}t_1\omega_3}\end{array}\right.$$

$$\left\{\begin{array}{l}\min_{p^{\mathrm{E}}_{\tilde{g}o_1t_2\omega_3},p^{\mathrm{C}}_{\tilde{c}o_1t_2\omega_3}} \quad 35p^{\mathrm{E}}_{\tilde{g}o_1t_2\omega_3}+25p^{\mathrm{C}}_{\tilde{c}o_1t_2\omega_3}\\ \text{s.t.}\\ \quad p^{\mathrm{E}}_{\tilde{g}o_1t_2\omega_3}+p^{\mathrm{C}}_{\tilde{c}o_1t_2\omega_3}=284\\ \quad 0\le p^{\mathrm{E}}_{\tilde{g}o_1t_2\omega_3}\le 400\\ \quad 0\le p^{\mathrm{C}}_{\tilde{c}o_1t_2\omega_3}\le p^{\mathrm{C^{max}}}_{\tilde{c}t_1\omega_3}+p^{\mathrm{C^{max}}}_{\tilde{c}t_2\omega_3}\end{array}\right.$$

$$\left\{\begin{array}{l}\min_{p^{\mathrm{E}}_{\tilde{g}o_2t_2\omega_3},p^{\mathrm{C}}_{\tilde{c}o_2t_2\omega_3}} \quad 35p^{\mathrm{E}}_{\tilde{g}o_2t_2\omega_3}+25p^{\mathrm{C}}_{\tilde{c}o_2t_2\omega_3}\\ \text{s.t.}\\ \quad p^{\mathrm{E}}_{\tilde{g}o_2t_2\omega_3}+p^{\mathrm{C}}_{\tilde{c}o_2t_2\omega_3}=593\\ \quad 0\le p^{\mathrm{E}}_{\tilde{g}o_2t_2\omega_3}\le 400\\ \quad 0\le p^{\mathrm{C}}_{\tilde{c}o_2t_2\omega_3}\le p^{\mathrm{C^{max}}}_{\tilde{c}t_1\omega_3}+p^{\mathrm{C^{max}}}_{\tilde{c}t_2\omega_3}\end{array}\right.$$

$$\begin{cases} \min_{p^{\mathrm{E}}_{\tilde{g}o_1t_1\omega_4}, p^{\mathrm{C}}_{\tilde{c}o_1t_1\omega_4}} & 35p^{\mathrm{E}}_{\tilde{g}o_1t_1\omega_4} + 25p^{\mathrm{C}}_{\tilde{c}o_1t_1\omega_4} \\ \text{s.t.} & \\ \quad p^{\mathrm{E}}_{\tilde{g}o_1t_1\omega_4} + p^{\mathrm{C}}_{\tilde{c}o_1t_1\omega_4} = 281 & \\ \quad 0 \le p^{\mathrm{E}}_{\tilde{g}o_1t_1\omega_4} \le 400 & \\ \quad 0 \le p^{\mathrm{C}}_{\tilde{c}o_1t_1\omega_4} \le p^{\mathrm{C^{max}}}_{\tilde{c}t_1\omega_4} & \end{cases}$$

$$\begin{cases} \min_{p^{\mathrm{E}}_{\tilde{g}o_2t_1\omega_4}, p^{\mathrm{C}}_{\tilde{c}o_2t_1\omega_4}} & 35p^{\mathrm{E}}_{\tilde{g}o_2t_1\omega_4} + 25p^{\mathrm{C}}_{\tilde{c}o_2t_1\omega_4} \\ \text{s.t.} & \\ \quad p^{\mathrm{E}}_{\tilde{g}o_2t_1\omega_4} + p^{\mathrm{C}}_{\tilde{c}o_2t_1\omega_4} = 533 & \\ \quad 0 \le p^{\mathrm{E}}_{\tilde{g}o_2t_1\omega_4} \le 400 & \\ \quad 0 \le p^{\mathrm{C}}_{\tilde{c}o_2t_1\omega_4} \le p^{\mathrm{C^{max}}}_{\tilde{c}t_1\omega_4} & \end{cases}$$

$$\begin{cases} \min_{p^{\mathrm{E}}_{\tilde{g}o_1t_2\omega_4}, p^{\mathrm{C}}_{\tilde{c}o_1t_2\omega_4}} & 35p^{\mathrm{E}}_{\tilde{g}o_1t_2\omega_4} + 25p^{\mathrm{C}}_{\tilde{c}o_1t_2\omega_4} \\ \text{s.t.} & \\ \quad p^{\mathrm{E}}_{\tilde{g}o_1t_2\omega_4} + p^{\mathrm{C}}_{\tilde{c}o_1t_2\omega_4} = 377 & \\ \quad 0 \le p^{\mathrm{E}}_{\tilde{g}o_1t_2\omega_4} \le 400 & \\ \quad 0 \le p^{\mathrm{C}}_{\tilde{c}o_1t_2\omega_4} \le p^{\mathrm{C^{max}}}_{\tilde{c}t_1\omega_4} + p^{\mathrm{C^{max}}}_{\tilde{c}t_2\omega_4} & \end{cases}$$

$$\begin{cases} \min_{p^{\mathrm{E}}_{\tilde{g}o_2t_2\omega_4}, p^{\mathrm{C}}_{\tilde{c}o_2t_2\omega_4}} & 35p^{\mathrm{E}}_{\tilde{g}o_2t_2\omega_4} + 25p^{\mathrm{C}}_{\tilde{c}o_2t_2\omega_4} \\ \text{s.t.} & \\ \quad p^{\mathrm{E}}_{\tilde{g}o_2t_2\omega_4} + p^{\mathrm{C}}_{\tilde{c}o_2t_2\omega_4} = 715 & \\ \quad 0 \le p^{\mathrm{E}}_{\tilde{g}o_2t_2\omega_4} \le 400 & \\ \quad 0 \le p^{\mathrm{C}}_{\tilde{c}o_2t_2\omega_4} \le p^{\mathrm{C^{max}}}_{\tilde{c}t_1\omega_4} + p^{\mathrm{C^{max}}}_{\tilde{c}t_2\omega_4}, & \end{cases}$$

where $\Delta = \Big\{p^{\mathrm{E}}_{\tilde{g}o_1t_1\omega_1}, p^{\mathrm{C}}_{\tilde{c}o_1t_1\omega_1}, p^{\mathrm{E}}_{\tilde{g}o_2t_1\omega_1}, p^{\mathrm{C}}_{\tilde{c}o_2t_1\omega_1}, p^{\mathrm{E}}_{\tilde{g}o_1t_2\omega_1}, p^{\mathrm{C}}_{\tilde{c}o_1t_2\omega_1}, p^{\mathrm{E}}_{\tilde{g}o_2t_2\omega_1}, p^{\mathrm{C}}_{\tilde{c}o_2t_2\omega_1},$ $p^{\mathrm{E}}_{\tilde{g}o_1t_1\omega_2}, p^{\mathrm{C}}_{\tilde{c}o_1t_1\omega_2}, p^{\mathrm{E}}_{\tilde{g}o_2t_1\omega_2}, p^{\mathrm{C}}_{\tilde{c}o_2t_1\omega_2}, p^{\mathrm{E}}_{\tilde{g}o_1t_2\omega_2}, p^{\mathrm{C}}_{\tilde{c}o_1t_2\omega_2}, p^{\mathrm{E}}_{\tilde{g}o_2t_2\omega_2}, p^{\mathrm{C}}_{\tilde{c}o_2t_2\omega_2}, p^{\mathrm{E}}_{\tilde{g}o_1t_1\omega_3}, p^{\mathrm{C}}_{\tilde{c}o_1t_1\omega_3},$ $p^{\mathrm{E}}_{\tilde{g}o_2t_1\omega_3}, p^{\mathrm{C}}_{\tilde{c}o_2t_1\omega_3}, p^{\mathrm{E}}_{\tilde{g}o_1t_2\omega_3}, p^{\mathrm{C}}_{\tilde{c}o_1t_2\omega_3}, p^{\mathrm{E}}_{\tilde{g}o_2t_2\omega_3}, p^{\mathrm{C}}_{\tilde{c}o_2t_2\omega_3}, p^{\mathrm{E}}_{\tilde{g}o_1t_1\omega_4}, p^{\mathrm{C}}_{\tilde{c}o_1t_1\omega_4}, p^{\mathrm{E}}_{\tilde{g}o_2t_1\omega_4}, p^{\mathrm{C}}_{\tilde{c}o_2t_1\omega_4},$ $p^{\mathrm{E}}_{\tilde{g}o_1t_2\omega_4}, p^{\mathrm{C}}_{\tilde{c}o_1t_2\omega_4}, p^{\mathrm{E}}_{\tilde{g}o_2t_2\omega_4}, p^{\mathrm{C}}_{\tilde{c}o_2t_2\omega_4}, u^{\mathrm{Option}}_{\tilde{c}q_1t_1\omega_1}, u^{\mathrm{Option}}_{\tilde{c}q_2t_1\omega_1}, u^{\mathrm{Option}}_{\tilde{c}q_3t_1\omega_1}, u^{\mathrm{Option}}_{\tilde{c}q_4t_1\omega_1}, u^{\mathrm{Option}}_{\tilde{c}q_5t_1\omega_1}, u^{\mathrm{Option}}_{\tilde{c}q_6t_1\omega_1},$ $u^{\mathrm{Option}}_{\tilde{c}q_1t_2\omega_1}, u^{\mathrm{Option}}_{\tilde{c}q_2t_2\omega_1}, u^{\mathrm{Option}}_{\tilde{c}q_3t_2\omega_1}, u^{\mathrm{Option}}_{\tilde{c}q_4t_2\omega_1}, u^{\mathrm{Option}}_{\tilde{c}q_5t_2\omega_1}, u^{\mathrm{Option}}_{\tilde{c}q_6t_2\omega_1}, u^{\mathrm{Option}}_{\tilde{c}q_1t_1\omega_2}, u^{\mathrm{Option}}_{\tilde{c}q_2t_1\omega_2}, u^{\mathrm{Option}}_{\tilde{c}q_3t_1\omega_2}, u^{\mathrm{Option}}_{\tilde{c}q_4t_1\omega_2},$ $u^{\mathrm{Option}}_{\tilde{c}q_5t_1\omega_2}, u^{\mathrm{Option}}_{\tilde{c}q_6t_1\omega_2}, u^{\mathrm{Option}}_{\tilde{c}q_1t_2\omega_2}, u^{\mathrm{Option}}_{\tilde{c}q_2t_2\omega_2}, u^{\mathrm{Option}}_{\tilde{c}q_3t_2\omega_2}, u^{\mathrm{Option}}_{\tilde{c}q_4t_2\omega_2}, u^{\mathrm{Option}}_{\tilde{c}q_5t_2\omega_2}, u^{\mathrm{Option}}_{\tilde{c}q_6t_2\omega_2}, u^{\mathrm{Option}}_{\tilde{c}q_1t_1\omega_3}, u^{\mathrm{Option}}_{\tilde{c}q_2t_1\omega_3},$ $u^{\mathrm{Option}}_{\tilde{c}q_3t_1\omega_3}, u^{\mathrm{Option}}_{\tilde{c}q_4t_1\omega_3}, u^{\mathrm{Option}}_{\tilde{c}q_5t_1\omega_3}, u^{\mathrm{Option}}_{\tilde{c}q_6t_1\omega_3}, u^{\mathrm{Option}}_{\tilde{c}q_1t_2\omega_3}, u^{\mathrm{Option}}_{\tilde{c}q_2t_2\omega_3}, u^{\mathrm{Option}}_{\tilde{c}q_3t_2\omega_3}, u^{\mathrm{Option}}_{\tilde{c}q_4t_2\omega_3}, u^{\mathrm{Option}}_{\tilde{c}q_5t_2\omega_3}, u^{\mathrm{Option}}_{\tilde{c}q_6t_2\omega_3},$ $u^{\mathrm{Option}}_{\tilde{c}q_1t_1\omega_4}, u^{\mathrm{Option}}_{\tilde{c}q_2t_1\omega_4}, u^{\mathrm{Option}}_{\tilde{c}q_3t_1\omega_4}, u^{\mathrm{Option}}_{\tilde{c}q_4t_1\omega_4}, u^{\mathrm{Option}}_{\tilde{c}q_5t_1\omega_4}, u^{\mathrm{Option}}_{\tilde{c}q_6t_1\omega_4}, u^{\mathrm{Option}}_{\tilde{c}q_1t_2\omega_4}, u^{\mathrm{Option}}_{\tilde{c}q_2t_2\omega_4}, u^{\mathrm{Option}}_{\tilde{c}q_3t_2\omega_4}, u^{\mathrm{Option}}_{\tilde{c}q_4t_2\omega_4},$ $u^{\mathrm{Option}}_{\tilde{c}q_5t_2\omega_4}, u^{\mathrm{Option}}_{\tilde{c}q_6t_2\omega_4}\Big\}$.

We obtain that the optimal generation expansion plan consists in building 200 MW of capacity at the beginning of the planning horizon. This decision does not depend on the future scenario realizations. Then, if scenario a is realized in the first time period, i.e., the demand in the first time period is low, it is optimal not to build additional capacity at the beginning of the second time period. However, if scenario *b* is realized in the first time period, i.e., the demand in the first time period is high, then it is optimal to build 200 MW of additional capacity at the beginning of the second time period. Note that the generation expansion decisions for the second time period depend on the scenario realization in the first time period but are independent of the scenario realizations in the second time period.

On the one hand, if we compare these results with those obtained in Illustrative Example 3.6 (considering a deterministic single-node dynamic approach), then we observe that the optimal generation expansion plan is the same in the first time period but not in the second one. Although the average values of operating conditions in the second time period in both examples are the same, by using a stochastic approach we have the option of making different generation expansion plans for different scenario realizations in the first time period. Building 200 MW in the first time period and 100 MW in the second time period, i.e., the optimal results of Illustrative Example 3.6, would have the result that the demand in the second time period could not be satisfied if scenario 4 is realized.

On the other hand, if we compare the results with those obtained in Illustrative Example 3.9 (considering a stochastic single-node static approach), then we observe that using a static approach results in building a larger capacity at the beginning of the planning horizon, which in reality is not always needed but is needed only if some of the scenarios are realized in the future. □

Note that the stochastic dynamic GEP problem described in this section is the most complex model described in this chapter since it considers a multiperiod framework and uncertainty in the system variables. It can be further complicated by including transmission constraints, but the main problem in such a *complete* model is its size. In this chapter, the models have been solved for a very simple system, which allows us to illustrate the main characteristics of the models described. However, in reality, electric energy systems are much more complex, with thousands of nodes and transmission lines. Moreover, if a large enough number of scenarios is considered in order to represent the uncertain parameters adequately and the planning horizon is divided into a large enough number of time periods, then the GEP problem may become intractable. Therefore, in solving these kinds of problems, it is necessary to achieve a tradeoff between modeling accuracy and computational tractability.

In order to solve this kind of problem for real-world systems, the recommendations below are generally useful:

1. To use scenario-reduction methods that retain most of the information of the original scenarios but reduce significantly the number of scenarios [19, 21].
2. To use decomposition techniques, e.g., Benders decomposition, to solve the resulting MILP models [2, 3, 15].

3.7 Summary and Conclusions

In this chapter, we describe and analyze the GEP problem in a given electric energy system. We adopt a central perspective in which a central planner (e.g., the ISO) determines the generation expansion plan that is most appropriate for the system as a whole, e.g., by determining the generation expansion plan that maximizes the overall social welfare and minimizes the investment costs. To do so, we develop different models that increasingly complicate the problem by considering the impact of different aspects on the GEP problem. Once the generation expansion planner has determined the optimal expansion plan, it encourages private investors to build the generating units.

The models described in this chapter are complementarity models that can be reformulated as MILP problems, which can be solved using branch-and-cut solvers.

Considering the theoretical framework described in the chapter and the numerical results of the illustrative examples, the conclusions below are in order:

1. As different details are incorporated in the GEP problem, the size and complexity of the problem increases. This may result in intractability issues for very large systems. Therefore, it is necessary to achieve a tradeoff between modeling accuracy and computational tractability.
2. Considering a dynamic approach for the GEP problem allows the generation expansion planner to build generating units at different points in time. This results in more flexibility to adapt to possible future changes in the system under study.
3. In uncongested systems, the modeling of the transmission network in the GEP problem is immaterial. However, in systems whose transmission lines are usually (or may become in the future) congested, it is necessary to model the transmission network in order to determine the optimal generation expansion plan.
4. Generation expansion plans are made for a long-term planning horizon, which means that these decisions are usually made with uncertain data. Therefore, it is necessary to represent this uncertainty in the GEP decision-making problem in order to obtain informed generation expansion plans.

3.8 End-of-Chapter Exercises

3.1 Why is GEP needed? Who decides about it?

3.2 List the advantages and disadvantages of the different GEP models described in this chapter.

3.3 Expand the GEP problem considering a stochastic single-node dynamic approach approach [model (3.15)] to include also network constraints. Then solve the problem using the data of Illustrative Examples 3.8 and 3.10.

3.4 Determine the optimal generation expansion plan for the six-node system depicted in Fig. 3.8. This system comprises three existing generating units, four demands, six transmission lines, and it is possible to build two new generating units. The data for the existing generating units, demands, transmission lines, and candidate generating units are provided in Tables 3.2, 3.3, 3.4, and 3.5. The reference node is node 1. Apply the different models analyzed in this chapter and compare the solutions achieved by them.

3.5 Robust optimization is used in Chap. 2 to address the uncertainty in the TEP problem, while stochastic programming is used in this chapter to represent the uncertainty in the GEP problem. What are the advantages and disadvantages of both

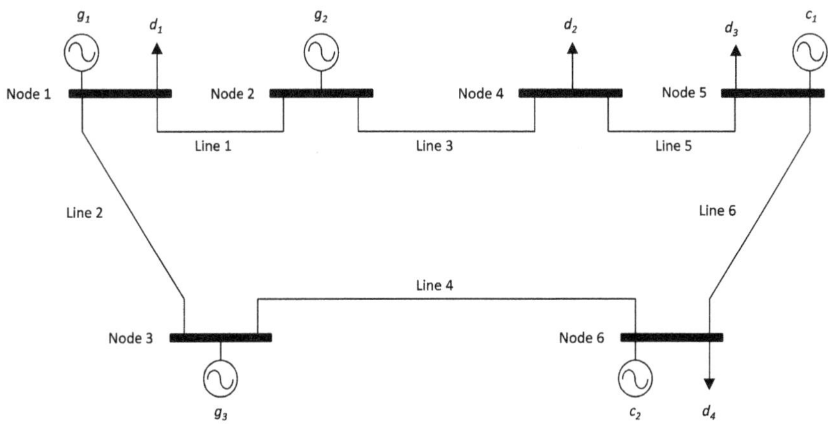

Fig. 3.8 Exercise 3.4: six-node system

Table 3.2 Exercise 3.4: data for generating units

Generating unit	Node	$P_g^{\mathrm{G}^{\max}}$ [MW]	C_g^{E} [\$/MWh]
g_1	n_1	300	18
g_2	n_2	250	25
g_3	n_3	400	16

Table 3.3 Exercise 3.4: data for demands

Demand	Node	$P_{do_1}^{\mathrm{D}}$ [MW]	$P_{do_2}^{\mathrm{D}}$ [MW]
d_1	n_1	200	300
d_2	n_4	150	250
d_3	n_5	100	200
d_4	n_6	200	300

Table 3.4 Exercise 3.4: data for transmission lines

Line	From node	To node	B_ℓ [p.u.]	$F_\ell^{\max}$ [MW]
ℓ_1	n_1	n_2	500	150
ℓ_2	n_1	n_3	500	150
ℓ_3	n_2	n_4	500	200
ℓ_4	n_3	n_6	500	200
ℓ_5	n_4	n_5	500	150
ℓ_9	n_5	n_6	500	150

Table 3.5 Exercise 3.4: data for candidate generating units

Generating unit	Node	P_{cq}^{Option} [MW]	C_c^{C} [$/MWh]
c_1	n_5	0, 100, 200, 300	32
c_2	n_6	0, 50, 100, 150	35

techniques? Based on the adaptive robust optimization approach used in Sect. 2.3 of Chap. 2, formulate the GEP problem (3.1) using an adaptive robust optimization approach and considering that uncertainty affects only P_{do}^{D}.

3.6 Consider the model developed in Exercise 3.5. Solve Illustrative Example 3.4, considering that demands can vary within ±10 % of their values.

3.9 GAMS Codes

A GAMS code for solving Illustrative Example 3.4 is provided below:

```
SETS
g                  /g1*g1/
c                  /c1*c1/
d                  /d1*d1/
o                  /o1*o2/
q                  /q1*q6/;

PARAMETER IC(c)
/C1          70000/;

TABLE PD(d,o)
             o1         o2
d1           290        550;

TABLE EDATA(g,*)
             PEmax      Ecost
g1           400        35;
```

```
TABLE CandidateData(c,q)
           q1        q2        q3        q4        q5        q6
c1         0         100       200       300       400
    500;

TABLE CDATA(c,*)
           Ccost
c1         25;

PARAMETER RHO(o)
/o1        6000
o2         2760/;

SCALAR M
/300000/;

VARIABLES
Z;

POSITIVE VARIABLES
PGE(g,o)
PGC(c,o)
PCmax(c)
LAMBDA(o);

BINARY VARIABLES
uOp(c,q);

POSITIVE VARIABLES
MU_EMAX(g,o)
MU_CMAX(c,o);

VARIABLES
ZAUX(c,q,o)
ZAUX2(c,q,o);

EQUATIONS EQ7A, EQ7B, EQ7C, EQ7E, EQ7F, EQ7G, EQ7H,
    EQ7I, EQ7L, EQ7M, EQ7Na, EQ7NB, EQ7Oa, EQ7OB;

EQ7A..             Z=E=SUM(o,RHO(o)*(SUM(g,EDATA(g,'
    Ecost')*PGE(g,o))+SUM(c,CDATA(c,'Ccost')*PGC(c,
    o))))+SUM(c,IC(c)*PCmax(c));
EQ7B(c)..          PCmax(c)=E=SUM(q,uOp(c,q)*
    CandidateData(c,q));
EQ7C(c)..          SUM(q,uOp(c,q))=E=1;
*EQ7D IS BINARY VARIABLE DECLARATION
EQ7E(o)..          SUM(g,PGE(g,o))+SUM(c,PGC(c,o))=E=
    SUM(d,PD(d,o));
EQ7F(g,o)..        PGE(g,o)=L=EDATA(G,'PEmax');
EQ7G(c,o)..        PGC(c,o)=L=PCmax(c);
EQ7H(g,o)..        EDATA(g,'Ecost')-LAMBDA(o)+MU_EMAX
    (g,o)=G=0;
```

```
EQ7I(c,o)..         CDATA(c,'Ccost')-LAMBDA(o)+MU_CMAX(c,o)=G=0;
*EQ7J AND EQ7K ARE NON-NEGATIVE VARIABLE DEFINITIONS
EQ7L(o)..           SUM(g,EDATA(g,'Ecost')*PGE(g,o))+SUM(c,CDATA(c,'Ccost')*PGC(c,o))=E=LAMBDA(o)*SUM(d,PD(d,o))-SUM(g,MU_EMAX(g,o)*EDATA(g,'PEmax'))-SUM(c,SUM(q,ZAUX(c,q,o)));
EQ7M(c,q,o)..       ZAUX(c,q,o)=E=MU_CMAX(c,o)*CandidateData(c,q)-ZAUX2(c,q,o);
EQ7Na(c,q,o)..      0=L=ZAUX(c,q,o);
EQ7Nb(c,q,o)..      ZAUX(c,q,o)=L=uOp(c,q)*M;
EQ7Oa(c,q,o)..      0=L=ZAUX2(c,q,o);
EQ7Ob(c,q,o)..      ZAUX2(c,q,o)=L=(1-uOp(c,q))*M;

MODEL GEP_DetSta1N /ALL/;

SOLVE GEP_DetSta1N USING MIP MINIMIZING Z;
```

A GAMS code for solving Illustrative Example 3.8 is provided below:

```
SETS
n               /n1*n2/
g               /g1*g1/
c               /c1*c1/
d               /d1*d1/
o               /o1*o2/
q               /q1*q6/
l               /l1*l1/
mapE(g,n)       /g1.n1/
mapC(c,n)       /c1.n2/
mapD(d,n)       /d1.n2/
ref(n)          /n1/
mapSL(l,n)      /l1.n1/
mapRL(l,n)      /l1.n2/;

TABLE LDATA(l,*)
        B       FLmax
l1      500     500;

PARAMETER IC(c)
/C1     70000/;

TABLE PD(d,o)
        o1      o2
d1      290     550;

TABLE EDATA(g,*)
        PEmax   Ecost
g1      400     35;

TABLE CandidateData(c,q)
        q1      q2      q3      q4      q5      q6
```

```
c1          0          100        200        300        400
    500;

TABLE CDATA(c,*)
            Ccost
C1          25;

PARAMETER RHO(o)
/o1         6000
o2          2760/;

SCALAR M
/300000/;

VARIABLES
Z
PL(l,o)
THETA(n,o);

POSITIVE VARIABLES
PGE(g,o)
PGC(c,o)
PCmax(c);

BINARY VARIABLES
uOp(c,q);

VARIABLE
LAMBDA(n,o)
MU_L(l,o)
MU_AREF(n,o);

POSITIVE VARIABLES
MU_EMAX(g,o)
MU_CMAX(c,o)
MU_LMAX(l,o)
MU_AMAX(n,o)
MU_EMIN(g,o)
MU_CMIN(c,o)
MU_LMIN(l,o)
MU_AMIN(n,o);

VARIABLES
ZAUX(c,q,o)
ZAUX2(c,q,o);

EQUATIONS EQ12A, EQ12B, EQ12C, EQ12E, EQ12F, EQ12Ga
    , EQ12Gb, EQ12H, EQ12I, EQ12Ja, EQ12Jb, EQ12K,
    EQ12L, EQ12M, EQ12N, EQ12O, EQ12P, EQ12U, EQ12V
    , EQ12Wa, EQ12Wb, EQ12Xa, EQ12XB;
```

```
EQ12A..          Z=E=SUM(o,RHO(o)*(SUM(g,EDATA(g,'Ecost')*PGE(g,o))+SUM(c,CDATA(c,'Ccost')*PGC(c,o))))+SUM(c,IC(c)*PCmax(c));
EQ12B(c)..       PCmax(c)=E=SUM(q,uOp(c,q)*CandidateData(c,q));
EQ12C(c)..       SUM(q,uOp(c,q))=E=1;
*EQ12D IS BINARY VARIABLE DECLARATION
EQ12E(n,o)..     SUM(g$mapE(g,n),PGE(g,o))+SUM(c$mapC(c,n),PGC(c,o))-SUM(l$mapSL(l,n),PL(l,o))+SUM(L$mapRL(l,n),PL(l,o))=E=SUM(d$mapD(d,n),PD(d,o));
EQ12F(l,o)..     PL(l,o)=E=LDATA(l,'B')*(SUM(n$mapSL(l,n),THETA(n,o))-SUM(n$mapRL(l,n),THETA(n,o)));
EQ12Ga(l,o)..    -LDATA(l,'FLmax')=L=PL(l,o);
EQ12Gb(l,o)..    PL(l,o)=L=LDATA(l,'FLmax');
EQ12H(g,o)..     PGE(g,o)=L=EDATA(g,'PEmax');
EQ12I(c,o)..     PGC(c,o)=L=PCmax(c);
EQ12Ja(n,o)..    -3.14=L=THETA(n,o);
EQ12Jb(n,o)..    THETA(n,o)=L=3.14;
EQ12K(n,o)$REF(n)..  THETA(n,o)=E=0;
EQ12L(g,o)..     EDATA(g,'Ecost')-SUM(n$mapE(g,n),LAMBDA(n,o))+MU_EMAX(g,o)=G=0;
EQ12M(c,o)..     CDATA(c,'Ccost')-SUM(n$mapC(c,n),LAMBDA(n,o))+MU_CMAX(c,o)=G=0;


EQ12N(l,o)..
SUM(n$mapSL(l,n),LAMBDA(n,o))-SUM(n$mapRL(l,n),LAMBDA(n,o))-MU_L(l,o)+MU_LMAX(l,o)-MU_LMIN(l,o)=E=0;
EQ12O(n,o)$(NOT REF(n))..
SUM(l$mapSL(l,n),LDATA(l,'B')*MU_L(l,o))-SUM(l$mapRL(l,n),LDATA(l,'B')*MU_L(l,o))+MU_AMAX(n,o)-MU_AMIN(n,o)=E=0;
EQ12P(n,o)$REF(n)..
SUM(l$mapSL(l,n),LDATA(l,'B')*MU_L(l,o))-SUM(l$mapRL(l,n),LDATA(l,'B')*MU_L(l,o))+MU_AREF(n,o)=E=0;
*EQ12Q-EQ12T ARE NON-NEGATIVE VARIABLE DEFINITIONS EQ12U(o)..
SUM(g,EDATA(g,'Ecost')*PGE(g,O))+SUM(c,CDATA(c,'Ccost')*PGC(c,o))=E=SUM(n,LAMBDA(n,o)*SUM(d$mapD(d,n),PD(d,o)))-SUM(g,MU_EMAX(g,o)*EDATA(g,'PEmax'))-SUM(c,SUM(q,ZAUX(c,q,o)))-SUM(l,(MU_LMAX(l,o)+MU_LMIN(l,o))*LDATA(l,'FLmax'))-SUM(n$(NOT
```

```
REF(n)),(MU_AMAX(n,o)-MU_AMIN(n,o))*3.14); EQ12V(c,q,o)..
ZAUX(c,q,o)=E=MU_CMAX(c,o)*CandidateData(c,q)-ZAUX2(c,q,o);
EQ12Wa(c,q,o)..                    0=L=ZAUX(c,q,o); EQ12Wb(c,q,o)..
ZAUX(c,q,o)=L=uOp(c,q)*M; EQ12Xa(c,q,o).. 0=L=ZAUX2(c,q,o);
EQ12Xb(c,q,o).. ZAUX2(c,q,o)=L=(1-uOp(c,q))*M;

MODEL GEP_DetStaNetwork /ALL/;

SOLVE GEP_DetStaNetwork USING MIP MINIMIZING Z;
```

References

1. Alguacil, N., Motto, A.L., Conejo, A.J.: Transmission expansion planning: a mixed-integer LP approach. IEEE Trans. Power Syst. **18**(3), 1070–1078 (2003)
2. Baringo, L., Conejo, A.J.: Wind power investment: a Benders' decomposition approach. IEEE Trans. Power Syst. **27**(1), 433–441 (2012)
3. Baringo, L., Conejo, A.J.: Risk-constrained multi-stage wind power investment. IEEE Trans. Power Syst. **28**(1), 401–411 (2013)
4. Baringo, L., Conejo, A.J.: Correlated wind-power production and electric load scenarios for investment decisions. Appl. Energy **101**, 475–482 (2013)
5. Binato, S., Pereira, M.V.F., Granville, S.: A new Benders decomposition approach to solve power transmission network design problems. IEEE Trans. Power Syst. **16**(2), 235–240 (2001)
6. Bird, L., Cochran, J., Wang, X.: Wind and solar energy curtailment: experience and practices in the United States. NREL (2014)
7. Botterud, A., Ilic, M.D., Wangensteen, I.: Optimal investments in power generation under centralized and decentralized decision making. IEEE Trans. Power Syst. **20**(1), 254–263 (2005)
8. Castillo, E., Conejo, A.J., Pedregal, P., Garcia, R., Alguacil, N.: Building and Solving Mathematical Programming Models in Engineering and Science. Wiley, New York (2001)
9. Chuang, A.S., Wu, F., Varaiya, P.: A game-theoretic model for generation expansion planning: problem formulation and numerical comparisons. IEEE Trans. Power Syst. **16**(4), 885–891 (2001)
10. Conejo, A.J., Carrión, M., Morales, J.M.: Decision Making Under Uncertainty in Electricity Markets. Springer, New York (2010)
11. Gabriel, S., Conejo, A.J., Hobbs, B., Fuller, D., Ruiz, C.: Complementarity Modeling in Energy Markets. Springer, New York (2012)
12. Garcés, L.P., Conejo, A.J., García-Bertrand, R., Romero, R.: A bilevel approach to the transmission expansion planning within a market environment. IEEE Trans. Power Syst. **24**(3), 1513–1522 (2009)
13. Gómez-Expósito, A., Conejo, A.J., Cañizares, C.: Electric Energy Systems: Analysis and Operation. CRC, Boca Raton (2008)
14. Kaymaz, P., Valenzuela, J., Park, C.S.: Transmission congestion and competition on power generation expansion. IEEE Trans. Power Syst. **22**(1), 156–163 (2007)
15. Kazempour, S.J., Conejo, A.J.: Strategic generation investment under uncertainty via Benders' decomposition. IEEE Trans. Power Syst. **27**(1), 424–432 (2012)
16. Kazempour, S.J., Conejo, A.J., Ruiz, C.: Strategic generation investment using a complementarity approach. IEEE Trans. Power Syst. **26**(2), 940–948 (2011)

17. Luenberger, D.G.: Introduction to Linear and Nonlinear Programming. Addison-Wesley Publishing Company, New York (1973)
18. Motto, A.L., Arroyo, J.M., Galiana, F.D.: A mixed-integer LP procedure for the analysis of electric grid security under disruptive threat. IEEE Trans. Power Syst. **20**(3), 1357–1365 (2005)
19. Morales, J.M., Pineda, S., Conejo, A.J., Carrión, M.: Scenario reduction for futures market trading in electricity markets. IEEE Trans. Power Syst. **24**(2), 878–888 (2009)
20. Murphy, F.H., Smeers, Y.: Generation capacity expansion in imperfectly competitive restructured electricity markets. Oper. Res. **53**(4), 646–661 (2005)
21. Pineda, S., Conejo, A.J.: Scenario reduction for risk-averse electricity trading. IET Gener. Transm. Distrib. **4**(6), 694–705 (2010)
22. Tsamasphyrou, P., Renaud, A., Carpentier, P.: Transmission network planning: an efficient Benders decomposition scheme. In: Proceedings of the 13th Power Systems Computation Conference. Trondheim, Norway, 28–2 June/July 1999
23. Wang, J., Shahidehpour, M., Li, Z., Botterud, A.: Strategic generation capacity expansion planning with incomplete information. IEEE Trans. Power Syst. **24**(2), 1002–1010 (2009)
24. Zhou, I., Wang, L., McCalley, J.D.: Designing effective and efficient incentive policies for renewable energy in generation expansion planning. Appl. Energy **88**(6), 2201–2209 (2011)

Chapter 4
Generation and Transmission Expansion Planning

The two previous chapters describe and analyze the transmission expansion planning problem (TEP, Chap. 2) and the generation expansion planning problem (GEP, Chap. 3). These two problems are critical for the optimal expansion planning of electric energy systems. However, these chapters analyze the GEP and TEP problems independently. In this chapter, we describe the joint generation and transmission expansion planning (G&TEP) problem in a given electric energy system. The G&TEP problem is analyzed from the perspective of a central planner that determines the generation and transmission expansion plan that is most beneficial for the system as a whole. To do so, we provide and analyze different models that progressively incorporate additional details within the G&TEP problem with a special emphasis on the modeling of the risk associated with expansion plans.

4.1 Introduction

Chapters 2 and 3 describe the transmission expansion planning problem (TEP) and the generation expansion problem (GEP), respectively. These two problems allow decision makers to determine the optimal transmission and generation-capacity expansion plans, respectively, to be carried out in a given electric energy system. In these two chapters, these two different expansion plans are analyzed independently, i.e., the transmission expansion plan is determined by considering that it is not possible to build new generating units and vice versa.

However, generation and transmission expansion plans are clearly interrelated. For example, if we solved the GEP problem by considering that the transmission capacity is fixed, then we would be restricted to building new generating units at those locations with sufficient transmission capacity. This is especially important in deciding where to build new renewable generating units since the locations with optimal characteristics for building such units, e.g., the locations with the best wind or solar conditions, are usually located far away from demand centers and poorly connected through the transmission system. Similarly, if we solved the TEP problem

A.J. Conejo et al., *Investment in Electricity Generation and Transmission*,
DOI 10.1007/978-3-319-29501-5_4

by considering that it is impossible to build new generating units, then we would decide the transmission expansion plan based mainly on current network bottlenecks and expected future demand, but we would not take into account the impact of a possible expansion of the generation capacity of the system.

Therefore, although the GEP and TEP problems are relevant for the optimal planning of electric energy systems, their outputs are usually suboptimal since generation and transmission expansion plans are derived independently. These plans are derived independently because the involved planners are different market agents: GEP pertains to profit-focused producers, while TEP pertains to a welfare-focused agent, i.e., the independent system operator (ISO). To overcome such a drawback, we describe in this chapter the generation and transmission expansion planning (G&TEP) problem, which allows us to determine generation and transmission expansion plans jointly.

To do so, we consider the perspective of a central planner, e.g., the ISO, that determines the generation and transmission expansion plan that is optimal for the operation of the electric energy system as a whole. Note that although the ISO determines the optimal expansion plans, the actual building of the facilities is generally carried out by other market agents, e.g., by private investors that build the generation-capacity facilities with the aim of maximizing profits. Therefore, after determining optimal generation and transmission expansion plans, the ISO may set incentives to stimulate the building of these facilities.

The generation and transmission expansion plans are generally made for a long-term planning horizon, e.g., 20 years. Thus, it is important to represent in the decision-making problem the system conditions throughout this planning horizon, e.g., future demand, decommissioning of old generating units and increasing penetration of renewable generating units. However, these future conditions are generally uncertain at the time the expansion plans are decided. Therefore, it is important to represent this uncertainty in the decision-making framework accurately. To do so, in this chapter we use a set of scenarios that model the future realization of the uncertain parameters in the considered planning horizon, i.e., we consider a stochastic programming approach [6].

Making expansion decisions within an uncertain environment is usually a risky endeavor. We decide the expansion plan that is optimal for all scenarios as a whole, e.g., we select the expansion plan that minimizes the expected cost. However, this expansion plan is decided before the actual scenario realization is known. Thus the expansion plan that minimizes the expected cost may have a very high cost for some scenarios. This constitutes a risk for the decision maker since such a high cost may not be admissible for the system. Thus, it is generally important to consider in the G&TEP problem the management of the risk associated with the expansion plan. This is done in this chapter using the conditional value-at-risk (CVaR) metric [14, 15].

The relevance of the G&TEP problem has recently generated research in this area [1, 3, 7, 9, 13, 16, 17]. The work [1] proposes a stochastic model with probabilistic constraints for the generation and transmission expansion plan considering uncertainties in demand, availability of generating units, and transmission capacity. Considering the G&TEP problem within a market framework, it is worthwhile to mention references [9, 16, 17]. Also, considering market issues, [13] addresses the G&TEP problem using a three-level equilibrium model. Finally, [3, 7] consider joint investment in renewable generating units and transmission facilities.

The remainder of this chapter is organized as follows. Section 4.2 describes the main features of the G&TEP problem described in this chapter. Sections 4.3–4.6 provide and analyze the formulation of the G&TEP problem considering different approaches. Section 4.3 describes the G&TEP problem considering a deterministic static approach. This basic model is extended in Sect. 4.4 to consider a dynamic framework, in Sect. 4.5 to represent the impact of uncertain parameters using a stochastic approach, and in Sect. 4.6 to incorporate risk management. Sections 4.3–4.6 include clarifying examples. Section 4.7 summarizes the chapter and discusses the main conclusions of the models and results reported. Section 4.8 proposes some exercises to gain a deeper understanding of the G&TEP problem. Finally, Sect. 4.9 includes the GAMS code for one of the illustrative examples.

4.2 Problem Description

This section describes the main characteristics of the G&TEP problem analyzed in this chapter. In addition to these characteristics, we consider the operating conditions, time framework, and uncertainty characterization described in Sect. 3.2 of Chap. 3 to develop the proposed models.

4.2.1 Notation

The main notation used in this chapter is provided below for quick reference. Other symbols are defined as needed throughout the chapter. The observations below are in order:

1. A subscript o in the symbols below indicates their values in the oth operating condition.
2. A subscript t in the symbols below indicates their values in the tth time period.
3. A subscript ω in the symbols below indicates their values in the ωth scenario.

Indices

c	Candidate generating units.
d	Demands.
g	Existing generating units.
ℓ	Transmission lines.
n	Nodes.
o	Operating conditions.
t	Time periods.

ω Scenarios.

Sets

$r(\ell)$ Receiving-end node of transmission line ℓ.
$s(\ell)$ Sending-end node of transmission line ℓ.
Ω_n^{C} Candidate generating units located at node n.
Ω_n^{D} Demands located at node n.
Ω_n^{E} Existing generating units located at node n.
$\Omega^{\mathrm{L+}}$ Prospective transmission lines n.

Parameters

A Amortization rate [%].
B_ℓ Susceptance of transmission line ℓ [S].
C_c^{C} Production cost of candidate generating unit c [\$/MWh].
C_d^{LS} Load-shedding cost of demand d [\$/MWh].
C_g^{E} Production cost of existing generating unit g [\$/MWh].
$F_\ell^{\max}$ Capacity of transmission line ℓ [MW].
I_c^{C} Investment cost of candidate generating unit c [\$/MW].
$I^{\mathrm{C,max}}$ Investment budget for building candidate generating units [\$].
I_ℓ^{L} Investment cost of prospective transmission line ℓ [\$].
$I^{\mathrm{L,max}}$ Investment budget for building prospective transmission lines [\$].
$\tilde{I}_c^{\mathrm{C}}$ Annualized investment cost of candidate generating unit c [\$/MW].
$\tilde{I}_\ell^{\mathrm{L}}$ Annualized investment cost of prospective transmission line ℓ [\$].
$\overline{P}_c^{\mathrm{C}^{\max}}$ Maximum production capacity of candidate generating unit c [MW].
$P_d^{\mathrm{D}^{\max}}$ Load of demand d [MW].
$P_g^{\mathrm{E}^{\max}}$ Production capacity of existing generating unit g [MW].
α Confidence level used to compute the CVaR.
η Weighting parameter used to model the tradeoff between expected cost and CVaR.
φ_ω Probability of scenario ω [p.u.].
ρ_o Weight of operating condition o [h].

Binary Variables

x_ℓ^{L} Binary variable that is equal to 1 if prospective transmission line ℓ is built and 0 otherwise.

Continuous Variables

p_c^{C} Power produced by candidate generating unit c [MW].
$p_c^{\mathrm{C}^{\max}}$ Capacity of candidate generating unit c [MW].
p_g^{E} Power produced by existing generating unit g [MW].
p_ℓ^{L} Power flow through transmission line ℓ [MW].
p_d^{LS} Load shed of demand d [MW].

θ_n Voltage angle at node n [rad].
ς_ω Auxiliary variable used to compute the CVaR [$].
ϑ Value-at-risk [$].

4.2.2 Approach

We consider the perspective of a central planner that aims to determine the generation and transmission expansion plan that results in the optimal operation of the electric energy system under study throughout the planning horizon. With this purpose, different objective functions may be selected, e.g., minimization of generation costs, maximization of social welfare, minimization of load-shedding costs. Among them, we select the minimization of both generation and load-shedding costs. However, the models described in this chapter are general, and alternative objective functions may be considered. We also include in the objective function of the G&TEP problem the minimization of the investment costs in both generating units and transmission facilities since they are relevant to the G&TEP decision-making problem.

In order to determine the optimal expansion plan, it is necessary to consider different constraints that influence the generation and transmission expansion plan, namely:

1. Investment budgets in building new generating units and transmission facilities.
2. Power balance at all nodes of the electric energy system under study.
3. Power flow limits through both existing and prospective transmission lines.
4. Capacity of both candidate and existing generating units.
5. Load-demand characteristics.
6. Voltage angle limits.

By considering the above framework, the G&TEP problem is formulated as a mathematical programming problem. This problem includes binary variables that represent whether a prospective transmission line is built or not. Moreover, it includes products of these binary variables and continuous variables. As a result, the G&TEP model is a mixed-integer nonlinear programming (MINLP) problem. However, it is possible to recast this MINLP problem as a mixed-integer linear programming (MILP) problem that can be solved by traditional branch-and-cut solvers.

4.2.3 Risk Management

Generation and transmission expansion decisions involve a long-term planning horizon. This implies that the expansion plan has to be derived under uncertainty. In this chapter, we model the uncertain parameters through a set of scenarios. Then we decide the generation and transmission expansion plans that minimize the expected generation, load-shedding, and investment costs.

However, selecting the expansion plan that minimizes the expected costs is not always the best alternative since this expansion plan may be harmful if some of the *extreme* scenarios are realized. This means that the expansion plan that minimizes the expected costs may be too risky for the central planner.

In order to deal with this, the risk associated with the expansion plan needs to be controlled. With this purpose, we incorporate the conditional value-at-risk (CVaR) in the G&TEP decision-making problem. Further details on this issue are provided in [14, 15], in Appendix D of this book, and in Sect. 4.6 of this chapter.

4.3 Deterministic Static G&TEP

In this section we provide and describe a very simple model to determine the optimal generation and transmission expansion plan. It is based on a deterministic static approach with the following characteristics:

1. No uncertainties are considered in the decision-making problem. We assume that the central planner has a perfect forecast of the future system demand.
2. Expansion decisions are made at a single point in time, in particular, at the beginning of the planning horizon.

This basic model allows us to illustrate the working of the G&TEP problem and will be progressively enriched in the following sections by incorporating different details in the G&TEP problem.

4.3.1 MINLP Formulation

The G&TEP problem considering a deterministic static approach can be formulated using the MINLP model below:

$$\min_{\Delta^{\text{DS}}} \quad \sum_{o} \rho_o \left[\sum_{g} C_g^{\text{E}} p_{go}^{\text{E}} + \sum_{c} C_c^{\text{C}} p_{co}^{\text{C}} + \sum_{d} C_d^{\text{LS}} p_{do}^{\text{LS}} \right] + \sum_{c} \tilde{I}_c^{\text{C}} p_c^{\text{C}^{\max}} + \sum_{\ell \in \Omega^{\text{L}+}} \tilde{I}_\ell^{\text{L}} x_\ell^{\text{L}} \tag{4.1a}$$

subject to

$$0 \leq p_c^{\text{C}^{\max}} \leq \overline{P}_c^{\text{C}^{\max}} \quad \forall c \tag{4.1b}$$

$$x_\ell^{\text{L}} \in \{0, 1\} \quad \forall \ell \in \Omega^{\text{L}+} \tag{4.1c}$$

$$\sum_{c} I_c^{\text{C}} p_c^{\text{C}^{\max}} \leq I^{\text{C},\max} \tag{4.1d}$$

$$\sum_{\ell \in \Omega^{\mathrm{L+}}} I_{\ell}^{\mathrm{L}} x_{\ell}^{\mathrm{L}} \leq I^{\mathrm{L,max}} \quad (4.1\mathrm{e})$$

$$\Bigg\{ \sum_{g \in \Omega_n^{\mathrm{E}}} p_{go}^{\mathrm{E}} + \sum_{c \in \Omega_n^{\mathrm{C}}} p_{co}^{\mathrm{C}} - \sum_{\ell | s(\ell)=n} p_{\ell o}^{\mathrm{L}} + \sum_{\ell | r(\ell)=n} p_{\ell o}^{\mathrm{L}} = \sum_{d \in \Omega_n^{\mathrm{D}}} \left(P_{do}^{\mathrm{D^{max}}} - p_{do}^{\mathrm{LS}} \right) \quad \forall n \quad (4.1\mathrm{f})$$

$$p_{\ell o}^{\mathrm{L}} = B_{\ell} \left(\theta_{s(\ell)o} - \theta_{r(\ell)o} \right) \quad \forall \ell \setminus \ell \in \Omega^{\mathrm{L+}} \quad (4.1\mathrm{g})$$

$$p_{\ell o}^{\mathrm{L}} = x_{\ell}^{\mathrm{L}} B_{\ell} \left(\theta_{s(\ell)o} - \theta_{r(\ell)o} \right) \quad \forall \ell \in \Omega^{\mathrm{L+}} \quad (4.1\mathrm{h})$$

$$- F_{\ell}^{\max} \leq p_{\ell o}^{\mathrm{L}} \leq F_{\ell}^{\max} \quad \forall \ell \quad (4.1\mathrm{i})$$

$$0 \leq p_{go}^{\mathrm{E}} \leq P_g^{\mathrm{E^{max}}} \quad \forall g \quad (4.1\mathrm{j})$$

$$0 \leq p_{co}^{\mathrm{C}} \leq p_c^{\mathrm{C^{max}}} \quad \forall c \quad (4.1\mathrm{k})$$

$$0 \leq p_{do}^{\mathrm{LS}} \leq P_{do}^{\mathrm{D^{max}}} \quad \forall d \quad (4.1\mathrm{l})$$

$$- \pi \leq \theta_{no} \leq \pi \quad \forall n \quad (4.1\mathrm{m})$$

$$\theta_{no} = 0 \quad n\text{: ref.} \quad (4.1\mathrm{n})$$

$$\Bigg\}, \forall o,$$

where variables in set $\Delta^{\mathrm{DS}} = \left\{ p_{go}^{\mathrm{E}}, p_{do}^{\mathrm{LS}}, p_{co}^{\mathrm{C}}, p_{\ell o}^{\mathrm{L}}, \theta_{no}, p_c^{\mathrm{C^{max}}}, x_{\ell}^{\mathrm{L}} \right\}$ are the optimization variables of problem (4.1).

The objective function (4.1a) comprises the four terms below:

1. $\sum_g C_g^{\mathrm{E}} p_{go}^{\mathrm{E}} + \sum_c C_c^{\mathrm{C}} p_{co}^{\mathrm{C}}, \forall o$, are the generation costs of both existing and candidate generating units.
2. $\sum_d C_d^{\mathrm{LS}} p_d^{\mathrm{LS}}, \forall o$, are the load-shedding costs.
3. $\sum_c \tilde{I}_c^{\mathrm{C}} p_c^{\mathrm{C^{max}}}$ is the annualized investment cost in new generating units.
4. $\sum_{\ell \in \Omega^{\mathrm{L+}}} \tilde{I}_{\ell}^{\mathrm{L}} x_{\ell}^{\mathrm{L}}$ is the annualized investment cost in new transmission lines.

Note that the terms in items 1 and 2 above are multiplied by the weight of the corresponding operating condition, ρ_o, in order to make the generation/load-shedding costs and the annualized investment costs comparable.

There are two groups of constraints, namely investment constraints (4.1b)–(4.1e) and operational constraints (4.1f)–(4.1n).

On the one hand, investment constraints comprise constraints (4.1b), which impose bounds on the production capacity of each candidate generating unit to be built; constraints (4.1c), which define binary variables x_{ℓ}^{L}, which indicate whether prospective transmission line $\ell \in \Omega^{\mathrm{L+}}$ is built ($x_{\ell}^{\mathrm{L}} = 1$) or not ($x_{\ell}^{\mathrm{L}} = 0$); and constraints (4.1d) and (4.1e), which impose investment budgets for building candidate generating units and prospective transmission lines, respectively.

On the other hand, operational constraints comprise constraints (4.1f), which impose the generation–demand balance at each node; constraints (4.1g) and (4.1h), which define the power flows through existing and prospective transmission lines, respectively; constraints (4.1i), which limit the power flows through transmission lines by the corresponding transmission capacities; constraints (4.1j), (4.1k), and (4.1l), which impose bounds on the power quantities of existing generating units, candidate generating units, and load shed of demands, respectively; constraints (4.1m), which impose bounds for voltage angles; and constraints (4.1n), which define the voltage angle at the reference node. Note that we define operational constraints (4.1f)–(4.1n) for each operating condition o.

Note that the network constraints are explicitly represented in the G&TEP problem using a dc model without loss for the sake of simplicity [8].

Illustrative Example 4.1 *2-node system: Deterministic static G&TEP problem (MINLP Formulation)*

Let us consider the two-node electric energy system depicted in Fig. 4.1, which has the following characteristics:

1. There is one generating unit $\tilde{g}$ located at node 1 with an installed capacity of 400 MW and a production cost equal to \$35/MWh.
2. It is possible to build a new generating unit $\tilde{c}$ at node 2 with a maximum capacity of 300 MW and a production cost equal to \$25/MWh. The investment cost is \$700,000 per MW. The annualized investment cost is 10 % of the total cost.
3. There is one demand $\tilde{d}$ located at node 2, whose demand conditions are represented by two operating conditions. The first one, o_1, is defined by a demand of 290 MW and a weight of 6000 h, while the second one, o_2, is defined by a demand of 550 MW and a weight of 2760 h. The load-shedding cost is \$80/MWh.
4. Nodes 1 and 2 are connected through a transmission line ℓ_1 with a susceptance equal to 500 S and a transmission capacity of 200 MW.
5. It is possible to build an additional transmission line ℓ_2 between nodes 1 and 2 with a susceptance equal to 500 S, a transmission capacity of 200 MW, and an investment cost of \$1 million. The annualized investment cost is 10 % of the total cost.
6. The investment budgets in building new generating units and new transmission lines are equal to \$400 million and \$2 million, respectively.
7. Node 1 is the reference node, and the base power and voltage are 1 MW and 1 kV, respectively.

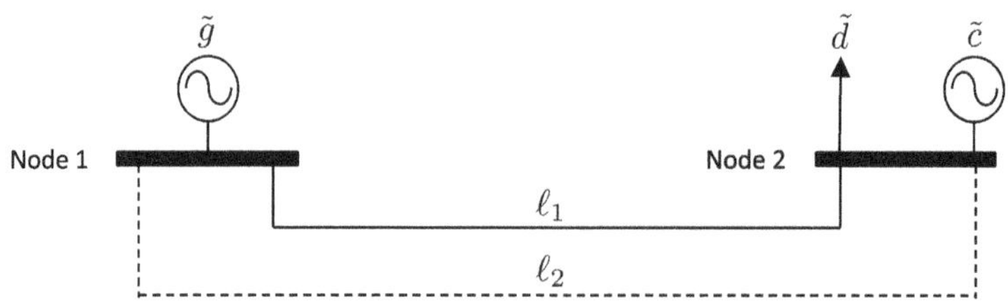

Fig. 4.1 Illustrative Example 4.1: two-node electric energy system

Given the above data, the G&TEP problem (4.1) results in the following MINLP model:

$$\min_{\tilde{\Delta}^{\mathrm{DS}}} \quad 6000\left[35p^{\mathrm{E}}_{\tilde{g}o_1} + 25p^{\mathrm{C}}_{\tilde{c}o_1} + 80p^{\mathrm{LS}}_{\tilde{d}o_1}\right] + 2760\left[35p^{\mathrm{E}}_{\tilde{g}o_2} + 25p^{\mathrm{C}}_{\tilde{c}o_2} + 80p^{\mathrm{LS}}_{\tilde{d}o_2}\right] + 70000p^{\mathrm{C^{max}}}_{\tilde{c}} + 100000x^{\mathrm{L}}_{\ell_2}$$

subject to

$$\begin{aligned}
&0 \le p^{\mathrm{C^{max}}}_{\tilde{c}} \le 300\\
&x^{\mathrm{L}}_{\ell_2} \in \{0, 1\}\\
&700000p^{\mathrm{C^{max}}}_{\tilde{c}} \le 400000000\\
&1000000x^{\mathrm{L}}_{\ell_2} \le 2000000\\
&\begin{cases}
p^{\mathrm{E}}_{\tilde{g}o_1} - p^{\mathrm{L}}_{\ell_1 o_1} - p^{\mathrm{L}}_{\ell_2 o_1} = 0\\
p^{\mathrm{C}}_{\tilde{c}o_1} + p^{\mathrm{L}}_{\ell_1 o_1} + p^{\mathrm{L}}_{\ell_2 o_1} = 290 - p^{\mathrm{LS}}_{\tilde{d}o_1}\\
p^{\mathrm{L}}_{\ell_1 o_1} = 500\left(\theta_{n_1 o_1} - \theta_{n_2 o_1}\right)\\
p^{\mathrm{L}}_{\ell_2 o_1} = 500x^{\mathrm{L}}_{\ell_2}\left(\theta_{n_1 o_1} - \theta_{n_2 o_1}\right)\\
-200 \le p^{\mathrm{L}}_{\ell_1 o_1} \le 200\\
-200 \le p^{\mathrm{L}}_{\ell_2 o_1} \le 200\\
0 \le p^{\mathrm{E}}_{\tilde{c}o_1} \le 400\\
0 \le p^{\mathrm{C}}_{\tilde{c}o_1} \le p^{\mathrm{C^{max}}}_{\tilde{c}}\\
0 \le p^{\mathrm{LS}}_{\tilde{d}o_1} \le 290\\
-\pi \le \theta_{n_2 o_1} \le \pi\\
\theta_{n_1 o_1} = 0
\end{cases}\\
&\begin{cases}
p^{\mathrm{E}}_{\tilde{g}o_2} - p^{\mathrm{L}}_{\ell_1 o_2} - p^{\mathrm{L}}_{\ell_2 o_2} = 0\\
p^{\mathrm{C}}_{\tilde{c}o_2} + p^{\mathrm{L}}_{\ell_1 o_2} + p^{\mathrm{L}}_{\ell_2 o_2} = 550 - p^{\mathrm{LS}}_{\tilde{d}o_1}\\
p^{\mathrm{L}}_{\ell_1 o_2} = 500\left(\theta_{n_1 o_2} - \theta_{n_2 o_2}\right)\\
p^{\mathrm{L}}_{\ell_2 o_2} = 500x^{\mathrm{L}}_{\ell_2}\left(\theta_{n_1 o_2} - \theta_{n_2 o_2}\right)\\
-200 \le p^{\mathrm{L}}_{\ell_1 o_2} \le 200\\
-200 \le p^{\mathrm{L}}_{\ell_2 o_2} \le 200\\
0 \le p^{\mathrm{E}}_{\tilde{c}o_2} \le 400\\
0 \le p^{\mathrm{C}}_{\tilde{c}o_2} \le p^{\mathrm{C^{max}}}_{\tilde{c}}\\
0 \le p^{\mathrm{LS}}_{\tilde{d}o_2} \le 550\\
-\pi \le \theta_{n_2 o_2} \le \pi\\
\theta_{n_1 o_2} = 0,
\end{cases}
\end{aligned}$$

where $\tilde{\Delta}^{\mathrm{DS}} = \left\{p^{\mathrm{E}}_{\tilde{g}o_1}, p^{\mathrm{E}}_{\tilde{g}o_2}, p^{\mathrm{C}}_{\tilde{c}o_1}, p^{\mathrm{C}}_{\tilde{c}o_2}, p^{\mathrm{LS}}_{\tilde{d}o_1}, p^{\mathrm{LS}}_{\tilde{d}o_2}, p^{\mathrm{L}}_{\ell_1 o_1}, p^{\mathrm{L}}_{\ell_1 o_2}, p^{\mathrm{L}}_{\ell_2 o_1}, p^{\mathrm{L}}_{\ell_2 o_2}, \theta_{n_1 o_1}, \theta_{n_1 o_2}, \theta_{n_2 o_1}, \theta_{n_2 o_2}, p^{\mathrm{C^{max}}}_{\tilde{c}}, x^{\mathrm{L}}_{\ell_2}\right\}$. □

4.3.2 MILP Formulation

The G&TEP problem (4.1) is an MINLP problem since it includes the products of binary variables x_ℓ^{L} and continuous variables θ_{no}, i.e., terms $x_\ell^{\mathrm{L}} B_\ell \left(\theta_{s(\ell)o} - \theta_{r(\ell)o}\right)$, in constraints (4.1h). Such MINLP problems are usually hard to solve, and their convergence to an optimum is not guaranteed [5]. However, it is possible to replace these nonlinear terms by the following set of exact equivalent mixed-integer linear expressions:

$$- x_\ell^{\mathrm{L}} F_\ell^{\max} \le p_{\ell o}^{\mathrm{L}} \le x_\ell^{\mathrm{L}} F_\ell^{\max} \quad \forall \ell \in \Omega^{\mathrm{L}+}, o \tag{4.2a}$$

$$- \left(1 - x_\ell^{\mathrm{L}}\right) M \le p_{\ell o}^{\mathrm{L}} - B_\ell \left(\theta_{s(\ell)o} - \theta_{r(\ell)o}\right) \le \left(1 - x_\ell^{\mathrm{L}}\right) M \quad \forall \ell \in \Omega^{\mathrm{L}+}, o, \tag{4.2b}$$

where M is a large enough positive constant [4, 18]. The working of the above Eq. (4.2) is explained in Sect. 2.2.3 of Chap. 2.

Note that if Eq. (4.2) are used to replace constraints (4.1h) in the G&TEP problem (4.1), the resulting model is an MILP problem, which can be solved using branch-and-cut solvers, and its convergence to an optimum is guaranteed [5].

Illustrative Example 4.2 *2-node system: Deterministic static G&TEP problem (MILP formulation)*

Let us consider the MINLP problem formulation from Illustrative Example 4.1. There are two nonlinear constraints, namely $p_{\ell_2 o_1}^{\mathrm{L}} = 500 x_{\ell_2}^{\mathrm{L}} \left(\theta_{n_1 o_1} - \theta_{n_2 o_1}\right)$ and $p_{\ell_2 o_2}^{\mathrm{L}} = 500 x_{\ell_2}^{\mathrm{L}} \left(\theta_{n_1 o_2} - \theta_{n_2 o_2}\right)$. These two constraints can be recast as mixed-integer linear constraints using Eq. (4.2):

$$\begin{cases} -200 x_{\ell_2}^{\mathrm{L}} \le p_{\ell_2 o_1}^{\mathrm{L}} \le 200 x_{\ell_2}^{\mathrm{L}} \\ - \left(1 - x_{\ell_2}^{\mathrm{L}}\right) M \le p_{\ell_2 o_1}^{\mathrm{L}} - 500 \left(\theta_{n_1 o_1} - \theta_{n_2 o_1}\right) \le \left(1 - x_{\ell_2}^{\mathrm{L}}\right) M \end{cases}$$

$$\begin{cases} -200 x_{\ell_2}^{\mathrm{L}} \le p_{\ell_2 o_2}^{\mathrm{L}} \le 200 x_{\ell_2}^{\mathrm{L}} \\ - \left(1 - x_{\ell_2}^{\mathrm{L}}\right) M \le p_{\ell_2 o_2}^{\mathrm{L}} - 500 \left(\theta_{n_1 o_2} - \theta_{n_2 o_2}\right) \le \left(1 - x_{\ell_2}^{\mathrm{L}}\right) M, \end{cases}$$

where M is a large enough positive constant.

If these equations are used to replace the two nonlinear constraints in the MINLP problem formulation obtained from Illustrative Example 4.1, then we obtain an MILP problem, whose solution is provided and analyzed in the following illustrative example. □

Illustrative Example 4.3 *2-node system: Deterministic static G&TEP problem (solution)*

Let us consider the data of Illustrative Example 4.1. First, we assume that the investment budget in building candidate generating units is \$400 million and that the investment budget in building prospective transmission lines is zero, i.e., we assume that it is possible to expand only the generation capacity of the system. In such a case, the optimal solution consists in building 300 MW of the candidate generating unit. Since there is a peak demand of 550 MW at node 2 and it is possible to use only 200 MW of the existing generating unit due to the capacity of the existing transmission line ℓ_1, it is necessary to build 350 MW of the candidate generating unit to satisfy the demand. However, there are only 300 MW of the candidate generating unit available, and, thus, 50 MW of the load demand cannot be supplied, and it is shed in operating condition o_2. Note that it would also be possible to shed a larger load demand in operating condition o_2 while building a lower capacity of the candidate generating unit; however, this would result in a higher cost. The total annualized cost (generation, load-shedding, and investment costs) is equal to \$115.56 million.

Second, we assume that the investment budget in building prospective transmission lines is \$2 million and that the investment budget in building candidate generating units is zero, i.e., we assume that it is possible to expand only the transmission capacity of the system. In this case, the optimal solution consists in building the additional transmission line ℓ_2. By building this transmission line, it is possible to use 400 MW of the existing generating unit to supply the demand. Note that in operating condition o_2, 150 MW of load demand cannot be supplied, and, therefore, it is shed. The total annualized cost is equal to \$132.76 million in this case.

Third, we assume that the investment budgets in building candidate generating units and prospective transmission lines are \$400 million and \$2 million, respectively, i.e., we assume that it is possible to expand both the generation capacity and the transmission capacity of the system. In such a case, the optimal solution consists in building 290 MW of the candidate generating unit and building transmission line ℓ_2. In this case, all the load demand can be supplied in both operating conditions. The total annualized cost is equal to \$109.03 million.

Note that in the three different cases analyzed, the lowest total cost corresponds to the third one, i.e., the case in which it is possible to expand both the generation and transmission capacity. □

4.4 Deterministic Dynamic G&TEP

In the G&TEP problem described in the previous section, we consider that the generation and transmission expansion decisions are made at a single point in time (generally at the beginning of the planning horizon), i.e., we consider a static approach. Note that when we decide about expansion plans, it is necessary to model the future needs of the system under study throughout the planning horizon. However, the expansion decisions are usually made for a long-term planning horizon, and thus the conditions of the system at the end of the planning horizon are generally different from those at the present. If it is possible to make expansion plans only at a single point in time, then these expansion decisions need to be made for a *worst-case* scenario, e.g., the largest expected demand in the planning horizon. This usually corresponds to the last years of the planning horizon. As a result, if a static approach is considered for the G&TEP problem, then the optimal solution generally consists in an overexpansion at the beginning of the planning horizon that is not needed until the last years of the planning horizon.

Considering such a static approach has some problems. On the one hand, it requires a large investment at the beginning of the planning horizon and thus a high investment budget, which may not be available. On the other hand, if investment changes are required throughout the planning horizon, then it might not be possible to adapt expansion decisions to these changes.

To overcome this, in this section we model the G&TEP problem considering a dynamic approach that allows the ISO to make expansion decisions regarding the generation and transmission facilities at different points in time. This gives the expansion planner the flexibility to adapt the system to changing future conditions as needed throughout the planning horizon. For the sake of simplicity, we consider a deterministic approach.

The G&TEP problem considering a deterministic dynamic approach can be formulated using the MILP model below:

$$\min_{\Delta^{\text{DS}}} \quad \sum_{t} \left\{ \sum_{o} \rho_o \left[\sum_{g} C_g^{\text{E}} p_{got}^{\text{E}} + \sum_{c} C_c^{\text{C}} p_{cot}^{\text{C}} + \sum_{d} C_d^{\text{LS}} p_{dot}^{\text{LS}} \right] + a_t \left[\sum_{c} I_c^{\text{C}} p_{ct}^{\text{C}^{\max}} + \sum_{\ell \in \Omega^{\text{L}+}} I_\ell^{\text{L}} x_{\ell t}^{\text{L}} \right] \right\} \tag{4.3a}$$

subject to

$$0 \leq \sum_{t} p_{ct}^{\text{C}^{\max}} \leq \overline{P}_c^{\text{C}^{\max}} \quad \forall c \tag{4.3b}$$

$$x_{\ell t}^{\text{L}} \in \{0, 1\} \quad \forall \ell \in \Omega^{\text{L}+}, t \tag{4.3c}$$

$$\sum_{t} x_{\ell t}^{\text{L}} \leq 1 \quad \forall \ell \in \Omega^{\text{L}+} \tag{4.3d}$$

$$\sum_{c} I_c^{\text{C}} p_{ct}^{\text{C}^{\max}} \leq I_t^{\text{C,max}} \quad \forall t \tag{4.3e}$$

$$\sum_{\ell \in \Omega^{\text{L}+}} I_\ell^{\text{L}} x_{\ell t}^{\text{L}} \leq I_t^{\text{L,max}} \quad \forall t \tag{4.3f}$$

$$\left\{ \sum_{g \in \Omega_n^{\text{E}}} p_{got}^{\text{E}} + \sum_{c \in \Omega_n^{\text{C}}} p_{cot}^{\text{C}} - \sum_{\ell | s(\ell)=n} p_{\ell ot}^{\text{L}} + \sum_{\ell | r(\ell)=n} p_{\ell ot}^{\text{L}} = \sum_{d \in \Omega_n^{\text{D}}} \left(P_{dot}^{\text{D}^{\max}} - p_{dot}^{\text{LS}}\right) \quad \forall n \right. \tag{4.3g}$$

$$p_{\ell ot}^{\text{L}} = B_\ell \left(\theta_{s(\ell)ot} - \theta_{r(\ell)ot}\right) \quad \forall \ell \setminus \ell \in \Omega^{\text{L}+} \tag{4.3h}$$

$$-\sum_{\tau \leq t} x_{\ell \tau}^{\text{L}} F_\ell^{\max} \leq p_{\ell ot}^{\text{L}} \leq \sum_{\tau \leq t} x_{\ell \tau}^{\text{L}} F_\ell^{\max} \quad \forall \ell \in \Omega^{\text{L}+} \tag{4.3i}$$

$$-\left(1 - \sum_{\tau \leq t} x_{\ell \tau}^{\text{L}}\right) M \leq p_{\ell ot}^{\text{L}} - B_\ell \left(\theta_{s(\ell)ot} - \theta_{r(\ell)ot}\right) \leq \left(1 - \sum_{\tau \leq t} x_{\ell \tau}^{\text{L}}\right) M \quad \forall \ell \in \Omega^{\text{L}+} \tag{4.3j}$$

$$- F_{\ell}^{\max} \leq p_{\ell ot}^{\mathrm{L}} \leq F_{\ell}^{\max} \quad \forall \ell \tag{4.3k}$$

$$0 \leq p_{got}^{\mathrm{E}} \leq P_{g}^{\mathrm{E}^{\max}} \quad \forall g \tag{4.3l}$$

$$0 \leq p_{cot}^{\mathrm{C}} \leq \sum_{\tau \leq t} p_{c\tau}^{\mathrm{C}^{\max}} \quad \forall c \tag{4.3m}$$

$$0 \leq p_{dot}^{\mathrm{LS}} \leq P_{dot}^{\mathrm{D}^{\max}} \quad \forall d \tag{4.3n}$$

$$- \pi \leq \theta_{not} \leq \pi \quad \forall n \tag{4.3o}$$

$$\theta_{not} = 0 \quad n\text{: ref.} \tag{4.3p}$$

$$\Big\}, \forall o, t,$$

where variables in set $\Delta^{\mathrm{DS}} = \left\{p_{got}^{\mathrm{E}}, p_{dot}^{\mathrm{LS}}, p_{cot}^{\mathrm{C}}, p_{\ell ot}^{\mathrm{L}}, \theta_{not}, p_{ct}^{\mathrm{C}^{\max}}, x_{\ell t}^{\mathrm{L}}\right\}$ are the optimization variables of problems (4.3).

The main differences between problem (4.3) and the G&TEP problem considering a deterministic static approach (4.1) are summarized below:

1. The capacities of each candidate generating unit to be built (i.e., variables $p_{ct}^{\mathrm{C}^{\max}}$, $\forall c, t$) can take different values in different time periods indexed by t.
2. Binary variables defining whether a prospective line is built or not (i.e., variables $x_{\ell t}^{\mathrm{L}}$, $\forall \ell \in \Omega^{\mathrm{L}+}, t$) can take different values in different time periods indexed by t.
3. The investment costs at each time period in the objective function (4.3a) are multiplied by the corresponding amortization rates, which represent the equivalent amount of money to be paid for the expansions made at each time period.
4. We include constraints (4.3b), which bound the capacity of each candidate generating unit built over the whole planning horizon.
5. We include constraints (4.3d), which impose that a prospective transmission line can be built only once for the whole planning horizon.
6. The investment budget constraints (4.3e)–(4.3f) are imposed for each time period t.
7. Constraints (4.3i)–(4.3j) define the power flows through prospective transmission lines and impose that a prospective transmission line must be available at time period t if it has been built at the beginning of that time period or in the previous ones.
8. Constraints (4.3m) impose that the available capacity of the candidate generating units at time period t must be equal to the capacity built at the beginning of that time period plus the capacities built in the previous ones.
9. The operational constraints (4.3g)–(4.3p) are formulated for each operating condition o and also for each time period t.

Note that for the sake of clarity, we assume that all monetary values are referred to the same point in time, so that it is not necessary to multiply by discount factors.

Illustrative Example 4.4 *2-node system: Deterministic dynamic G&TEP problem*

Let us consider the two-node electric energy system described in Illustrative Example 4.1. Now we consider that the planning horizon is split into two 10-year time periods and that it is possible to build both the candidate generating unit and the prospective transmission line at the beginning of either of these two time periods, i.e., at the beginning of the first or eleventh years.

The operating conditions in the two considered time periods have the following characteristics:

1. There are two operating conditions o_1 and o_2 in the first time period defined by load demands of 246.5 and 467.5 MW, respectively.
2. There are two operating conditions o_1 and o_2 in the second time period defined by load demands of 290 and 550 MW, respectively.

The weights of operating conditions o_1 and o_2 in both time periods are equal to 6000 and 2760 h, respectively.

The investment budgets for building additional generation capacity and additional transmission lines are \$400 million and \$2 million, respectively, in both time periods. Finally, the amortization rates for the first and second time periods are equal to 0.2 and 0.1, respectively.

Considering the above data, the G&TEP problem considering a deterministic dynamic approach results in the following MILP model:

$$
\begin{aligned}
\min_{\tilde{\Delta}^{\mathrm{DS}}} \quad & 6000\left[35p^{\mathrm{E}}_{\tilde{g}o_1t_1} + 25p^{\mathrm{C}}_{\tilde{c}o_1t_1} + 80p^{\mathrm{LS}}_{\tilde{d}o_1t_1}\right] + 2760\left[35p^{\mathrm{E}}_{\tilde{g}o_2t_1} + 25p^{\mathrm{C}}_{\tilde{c}o_2t_1}\right.\\
& \left. + 80p^{\mathrm{LS}}_{\tilde{d}o_2t_1}\right] + 6000\left[35p^{\mathrm{E}}_{\tilde{g}o_1t_2} + 25p^{\mathrm{C}}_{\tilde{c}o_1t_2} + 80p^{\mathrm{LS}}_{\tilde{d}o_1t_2}\right]\\
& + 2760\left[35p^{\mathrm{E}}_{\tilde{g}o_2t_2} + 25p^{\mathrm{C}}_{\tilde{c}o_2t_2} + 80p^{\mathrm{LS}}_{\tilde{d}o_2t_2}\right] + 140000p^{\mathrm{C^{max}}}_{\tilde{c}t_1}\\
& + 200000x^{\mathrm{L}}_{\ell_2t_1} + 70000p^{\mathrm{C^{max}}}_{\tilde{c}t_2} + 100000x^{\mathrm{L}}_{\ell_2t_2}
\end{aligned}
$$

subject to

$$
\begin{aligned}
& 0 \le p^{\mathrm{C^{max}}}_{\tilde{c}t_1} + p^{\mathrm{C^{max}}}_{\tilde{c}t_2} \le 300\\
& x^{\mathrm{L}}_{\ell_2t_1}, x^{\mathrm{L}}_{\ell_2t_2} \in \{0, 1\}\\
& x^{\mathrm{L}}_{\ell_2t_1} + x^{\mathrm{L}}_{\ell_2t_2} \le 1\\
& 700000p^{\mathrm{C^{max}}}_{\tilde{c}t_1} \le 400000000\\
& 700000p^{\mathrm{C^{max}}}_{\tilde{c}t_2} \le 400000000\\
& 1000000x^{\mathrm{L}}_{\ell_2t_1} \le 2000000\\
& 1000000x^{\mathrm{L}}_{\ell_2t_2} \le 2000000
\end{aligned}
$$

$$
\left\{
\begin{array}{l}
p^{\mathrm{E}}_{\tilde{g}o_1t_1} - p^{\mathrm{L}}_{\ell_1o_1t_1} - p^{\mathrm{L}}_{\ell_2o_1t_1} = 0 \\
p^{\mathrm{C}}_{\tilde{c}o_1t_1} + p^{\mathrm{L}}_{\ell_1o_1t_1} + p^{\mathrm{L}}_{\ell_2o_1t_1} = 246.5 - p^{\mathrm{LS}}_{\tilde{d}o_1t_1} \\
p^{\mathrm{L}}_{\ell_1o_1t_1} = 500\left(\theta_{n_1o_1t_1} - \theta_{n_2o_1t_1}\right) \\
-200x^{\mathrm{L}}_{\ell_2t_1} \le p^{\mathrm{L}}_{\ell_2o_1t_1} \le 200x^{\mathrm{L}}_{\ell_2t_1} \\
-\left(1 - x^{\mathrm{L}}_{\ell_2t_1}\right)M \le p^{\mathrm{L}}_{\ell_2o_1t_1} - 500\left(\theta_{n_1o_1t_1} - \theta_{n_2o_1t_1}\right) \le \left(1 - x^{\mathrm{L}}_{\ell_2t_1}\right)M \\
-200 \le p^{\mathrm{L}}_{\ell_1o_1t_1} \le 200 \\
-200 \le p^{\mathrm{L}}_{\ell_2o_1t_1} \le 200 \\
0 \le p^{\mathrm{E}}_{\tilde{c}o_1t_1} \le 400 \\
0 \le p^{\mathrm{C}}_{\tilde{c}o_1t_1} \le p^{\mathrm{C^{max}}}_{\tilde{c}t_1} \\
0 \le p^{\mathrm{LS}}_{\tilde{d}o_1t_1} \le 246.5 \\
-\pi \le \theta_{n_2o_1t_1} \le \pi \\
\theta_{n_1o_1t_1} = 0
\end{array}
\right.
$$

$$
\left\{
\begin{array}{l}
p^{\mathrm{E}}_{\tilde{g}o_2t_1} - p^{\mathrm{L}}_{\ell_1o_2t_1} - p^{\mathrm{L}}_{\ell_2o_2t_1} = 0 \\
p^{\mathrm{C}}_{\tilde{c}o_2t_1} + p^{\mathrm{L}}_{\ell_1o_2t_1} + p^{\mathrm{L}}_{\ell_2o_2t_1} = 467.5 - p^{\mathrm{LS}}_{\tilde{d}o_1t_1} \\
p^{\mathrm{L}}_{\ell_1o_2t_1} = 500\left(\theta_{n_1o_2t_1} - \theta_{n_2o_2t_1}\right) \\
-200x^{\mathrm{L}}_{\ell_2t_1} \le p^{\mathrm{L}}_{\ell_2o_2t_1} \le 200x^{\mathrm{L}}_{\ell_2t_1} \\
-\left(1 - x^{\mathrm{L}}_{\ell_2t_1}\right)M \le p^{\mathrm{L}}_{\ell_2o_2t_1} - 500\left(\theta_{n_1o_2t_1} - \theta_{n_2o_2t_1}\right) \le \left(1 - x^{\mathrm{L}}_{\ell_2t_1}\right)M \\
-200 \le p^{\mathrm{L}}_{\ell_1o_2t_1} \le 200 \\
-200 \le p^{\mathrm{L}}_{\ell_2o_2t_1} \le 200 \\
0 \le p^{\mathrm{E}}_{\tilde{c}o_2t_1} \le 400 \\
0 \le p^{\mathrm{C}}_{\tilde{c}o_2t_1} \le p^{\mathrm{C^{max}}}_{\tilde{c}t_1} \\
0 \le p^{\mathrm{LS}}_{\tilde{d}o_2t_1} \le 467.5 \\
-\pi \le \theta_{n_2o_2t_1} \le \pi \\
\theta_{n_1o_2t_1} = 0
\end{array}
\right.
$$

$$
\begin{cases}
p^{\mathrm{E}}_{\tilde{g}o_1t_2} - p^{\mathrm{L}}_{\ell_1o_1t_2} - p^{\mathrm{L}}_{\ell_2o_1t_2} = 0 \\
p^{\mathrm{C}}_{\tilde{c}o_1t_2} + p^{\mathrm{L}}_{\ell_1o_1t_2} + p^{\mathrm{L}}_{\ell_2o_1t_2} = 290 - p^{\mathrm{LS}}_{\tilde{d}o_1t_2} \\
p^{\mathrm{L}}_{\ell_1o_1t_2} = 500\left(\theta_{n_1o_1t_2} - \theta_{n_2o_1t_2}\right) \\
-200\left(x^{\mathrm{L}}_{\ell_2t_1} + x^{\mathrm{L}}_{\ell_2t_2}\right) \le p^{\mathrm{L}}_{\ell_2o_1t_2} \le 200\left(x^{\mathrm{L}}_{\ell_2t_1} + x^{\mathrm{L}}_{\ell_2t_2}\right) \\
-\left[1 - \left(x^{\mathrm{L}}_{\ell_2t_1} + x^{\mathrm{L}}_{\ell_2t_2}\right)\right]M \le p^{\mathrm{L}}_{\ell_2o_1t_2} - 500\left(\theta_{n_1o_1t_2} - \theta_{n_2o_1t_2}\right) \\
p^{\mathrm{L}}_{\ell_2o_1t_1} - 500\left(\theta_{n_1o_1t_2} - \theta_{n_2o_1t_2}\right) \le \left[1 - \left(x^{\mathrm{L}}_{\ell_2t_1} + x^{\mathrm{L}}_{\ell_2t_2}\right)\right]M \\
-200 \le p^{\mathrm{L}}_{\ell_1o_1t_2} \le 200 \\
-200 \le p^{\mathrm{L}}_{\ell_2o_1t_2} \le 200 \\
0 \le p^{\mathrm{E}}_{\tilde{c}o_1t_2} \le 400 \\
0 \le p^{\mathrm{C}}_{\tilde{c}o_1t_2} \le p^{\mathrm{C^{max}}}_{\tilde{c}t_1} + p^{\mathrm{C^{max}}}_{\tilde{c}t_2} \\
0 \le p^{\mathrm{LS}}_{\tilde{d}o_1t_2} \le 290 \\
-\pi \le \theta_{n_2o_1t_2} \le \pi \\
\theta_{n_1o_1t_2} = 0
\end{cases}
$$

$$
\begin{cases}
p^{\mathrm{E}}_{\tilde{g}o_2t_2} - p^{\mathrm{L}}_{\ell_1o_2t_2} - p^{\mathrm{L}}_{\ell_2o_2t_2} = 0 \\
p^{\mathrm{C}}_{\tilde{c}o_2t_2} + p^{\mathrm{L}}_{\ell_1o_2t_2} + p^{\mathrm{L}}_{\ell_2o_2t_2} = 550 - p^{\mathrm{LS}}_{\tilde{d}o_1t_2} \\
p^{\mathrm{L}}_{\ell_1o_2t_2} = 500\left(\theta_{n_1o_2t_2} - \theta_{n_2o_2t_2}\right) \\
-200\left(x^{\mathrm{L}}_{\ell_2t_1} + x^{\mathrm{L}}_{\ell_2t_2}\right) \le p^{\mathrm{L}}_{\ell_2o_2t_2} \le 200\left(x^{\mathrm{L}}_{\ell_2t_1} + x^{\mathrm{L}}_{\ell_2t_2}\right) \\
-\left[1 - \left(x^{\mathrm{L}}_{\ell_2t_1} + x^{\mathrm{L}}_{\ell_2t_2}\right)\right]M \le p^{\mathrm{L}}_{\ell_2o_2t_2} - 500\left(\theta_{n_1o_2t_2} - \theta_{n_2o_2t_2}\right) \\
p^{\mathrm{L}}_{\ell_2o_1t_1} - 500\left(\theta_{n_1o_2t_2} - \theta_{n_2o_2t_2}\right) \le \left[1 - \left(x^{\mathrm{L}}_{\ell_2t_1} + x^{\mathrm{L}}_{\ell_2t_2}\right)\right]M \\
-200 \le p^{\mathrm{L}}_{\ell_1o_2t_2} \le 200 \\
-200 \le p^{\mathrm{L}}_{\ell_2o_2t_2} \le 200 \\
0 \le p^{\mathrm{E}}_{\tilde{c}o_2t_2} \le 400 \\
0 \le p^{\mathrm{C}}_{\tilde{c}o_2t_2} \le p^{\mathrm{C^{max}}}_{\tilde{c}t_1} + p^{\mathrm{C^{max}}}_{\tilde{c}t_2} \\
0 \le p^{\mathrm{LS}}_{\tilde{d}o_2t_2} \le 550 \\
-\pi \le \theta_{n_2o_2t_2} \le \pi \\
\theta_{n_1o_2t_2} = 0,
\end{cases}
$$

where $\tilde{\Delta}^{\text{DS}} = \Big\{p^{\text{E}}_{\tilde{g}o_1t_1}, p^{\text{E}}_{\tilde{g}o_2t_1}, p^{\text{C}}_{\tilde{c}o_1t_1}, p^{\text{C}}_{\tilde{c}o_2t_1}, p^{\text{LS}}_{\tilde{d}o_1t_1}, p^{\text{LS}}_{\tilde{d}o_2t_1}, p^{\text{L}}_{\ell_1o_1t_1}, p^{\text{L}}_{\ell_1o_2t_1}, p^{\text{L}}_{\ell_2o_1t_1}, p^{\text{L}}_{\ell_2o_2t_1}, \theta_{n_1o_1t_1}, \theta_{n_1o_2t_1}, \theta_{n_2o_1t_1}, \theta_{n_2o_2t_1}, p^{\text{C}^{\max}}_{\tilde{c}t_1}, x^{\text{L}}_{l_2t_1}, p^{\text{E}}_{\tilde{g}o_1t_2}, p^{\text{E}}_{\tilde{g}o_2t_2}, p^{\text{C}}_{\tilde{c}o_1t_2}, p^{\text{C}}_{\tilde{c}o_2t_2}, p^{\text{LS}}_{\tilde{d}o_1t_2}, p^{\text{LS}}_{\tilde{d}o_2t_2}, p^{\text{L}}_{\ell_1o_1t_2}, p^{\text{L}}_{\ell_1o_2t_2}, p^{\text{L}}_{\ell_2o_1t_2}, p^{\text{L}}_{\ell_2o_2t_2}, \theta_{n_1o_1t_2}, \theta_{n_1o_2t_2}, \theta_{n_2o_1t_2}, \theta_{n_2o_2t_2}, p^{\text{C}^{\max}}_{\tilde{c}t_2}, x^{\text{L}}_{\ell_2t_1}, x^{\text{L}}_{\ell_2t_2}\Big\}$ and M is a large enough number.

We obtain that it is optimal to build ℓ_2 in the first time period. Regarding the candidate generating unit, it is optimal to build 246.5 MW at the beginning of the first time period and an additional 43.5 MW at the beginning of the second time period.

If we compare the optimal solution of this illustrative example with that of Illustrative Example 4.3, then we obtain that the optimal solutions for the whole planning horizon are the same. However, here, instead of building 290 MW of the candidate generating unit at the beginning of the planning horizon, we first build 246.5 MW and then we add capacity as required in the second time period. This has some advantages:

1. The investment budget required at the beginning of the planning horizon is lower.
2. We build additional capacity depending on the needs of the system. For example, if the demand of the system in the second time period is lower than expected, then we have the option of not building additional capacity if we are considering a dynamic approach.

Despite the above advantages of considering a dynamic approach, note that this approach involves solving a more complex problem with a larger number of variables than if a static approach is considered. For example, in this simple illustrative example with only two time periods, the number of variables is twice the number of variables needed in Illustrative Example 4.3.

4.5 Stochastic G&TEP

Sections 4.3 and 4.4 describe the G&TEP problem considering a deterministic approach, i.e., the generation and transmission expansion plans are determined by the ISO considering that it has perfect information about the future needs of the system under study. However, this perfect information is generally not available since the expansion plans are made for a long-term planning horizon. Therefore, the expansion plans obtained from the models considering a deterministic approach may be not optimal if the future conditions of the system differ from those considered.

In this section, we analyze the G&TEP problem considering a stochastic approach that enables the expansion planner to model the uncertain nature of the future needs of the system under study in the decision-making problem. Uncertainties are represented through a set of scenarios [6]. For the sake of simplicity and clarity in the development of the models, we assume that uncertainty affects only the future demand of the system. However, other sources of uncertainty (e.g., production cost or investment cost) may also be considered through additional scenarios.

The sections below describe two models, namely a stochastic static model that considers that the expansion plan can be made only at the beginning of the planning horizon and a stochastic dynamic one that considers that expansion plans can be made at different points in time throughout the planning horizon. Further details about the characteristics of these two models are provided in Sect. 3.6 of Chap. 3.

4.5.1 Static Approach

The G&TEP problem considering a stochastic static approach can be formulated using the MILP model below:

$$\min_{\Delta^{\text{SS}}} \quad \sum_{\omega} \varphi_{\omega} \left\{ \sum_{o} \rho_{o} \left[\sum_{g} C_{g}^{\text{E}} p_{go\omega}^{\text{E}} + \sum_{c} C_{c}^{\text{C}} p_{co\omega}^{\text{C}} + \sum_{d} C_{d}^{\text{LS}} p_{do\omega}^{\text{LS}} \right] \right\} + \sum_{c} \tilde{I}_{c}^{\text{C}} p_{c}^{\text{C}^{\max}} + \sum_{\ell \in \Omega^{\text{L}+}} \tilde{I}_{\ell}^{\text{L}} x_{\ell}^{\text{L}} \tag{4.4a}$$

subject to

$$0 \le p_{c}^{\text{C}^{\max}} \le \overline{P}_{c}^{\text{C}^{\max}} \quad \forall c \tag{4.4b}$$

$$x_{\ell}^{\text{L}} \in \{0, 1\} \quad \forall \ell \in \Omega^{\text{L}+} \tag{4.4c}$$

$$\sum_{c} I_{c}^{\text{C}} p_{c}^{\text{C}^{\max}} \le I^{\text{C},\max} \tag{4.4d}$$

$$\sum_{\ell \in \Omega^{\text{L}+}} I_{\ell}^{\text{L}} x_{\ell}^{\text{L}} \le I^{\text{L},\max} \tag{4.4e}$$

$$\left\{ \sum_{g \in \Omega_{n}^{\text{E}}} p_{go\omega}^{\text{E}} + \sum_{c \in \Omega_{n}^{\text{C}}} p_{co\omega}^{\text{C}} - \sum_{\ell | s(\ell)=n} p_{\ell o\omega}^{\text{L}} + \sum_{\ell | r(\ell)=n} p_{\ell o\omega}^{\text{L}} = \sum_{d \in \Omega_{n}^{\text{D}}} \left(P_{do\omega}^{\text{D}^{\max}} - p_{do\omega}^{\text{LS}} \right) \quad \forall n \tag{4.4f}$$

$$p_{\ell o\omega}^{\text{L}} = B_{\ell} \left(\theta_{s(\ell)o\omega} - \theta_{r(\ell)o\omega} \right) \quad \forall \ell \setminus \ell \in \Omega^{\text{L}+} \tag{4.4g}$$

$$- x_{\ell}^{\text{L}} F_{\ell}^{\max} \le p_{\ell o\omega}^{\text{L}} \le x_{\ell}^{\text{L}} F_{\ell}^{\max} \quad \forall \ell \in \Omega^{\text{L}+} \tag{4.4h}$$

$$- \left(1 - x_{\ell}^{\text{L}}\right) M \le p_{\ell\omega}^{\text{L}} - B_{\ell} \left(\theta_{s(\ell)o\omega} - \theta_{r(\ell)o\omega} \right) \le \left(1 - x_{\ell}^{\text{L}}\right) M \quad \forall \ell \in \Omega^{\text{L}+} \tag{4.4i}$$

$$- F_{\ell}^{\max} \le p_{\ell o\omega}^{\text{L}} \le F_{\ell}^{\max} \quad \forall \ell \tag{4.4j}$$

$$0 \le p_{go\omega}^{\text{E}} \le P_{g}^{\text{E}^{\max}} \quad \forall g \tag{4.4k}$$

$$0 \le p_{co\omega}^{\text{C}} \le p_{c}^{\text{C}^{\max}} \quad \forall c \tag{4.4l}$$

$$0 \le p_{do\omega}^{\text{LS}} \le P_{do\omega}^{\text{D}^{\max}} \quad \forall d \tag{4.4m}$$

$$-\pi \le \theta_{no\omega} \le \pi \quad \forall n \tag{4.4n}$$

$$\theta_{no\omega} = 0 \quad n\text{: ref.} \tag{4.4o}$$

$$\Bigg\}, \forall o, \omega,$$

where the variables in set $\Delta^{\mathrm{SS}} = \left\{p^{\mathrm{E}}_{go\omega}, p^{\mathrm{LS}}_{do\omega}, p^{\mathrm{C}}_{co\omega}, p^{\mathrm{L}}_{\ell o\omega}, \theta_{no\omega}, p^{\mathrm{C^{max}}}_{c}, x^{\mathrm{L}}_{\ell}\right\}$ are the optimization variables of problem (4.4).

The main differences between problem (4.4) and problem (4.1) considering a deterministic static approach are summarized below:

1. In the objective function (4.4a), we compute the expected generation and load-shedding costs. To do so, the generation and load-shedding costs for each scenario, $\sum_{o} \rho_{o} \left[\sum_{g} C^{\mathrm{E}}_{g} p^{\mathrm{E}}_{go\omega} + \sum_{c} C^{\mathrm{C}}_{c} p^{\mathrm{C}}_{co\omega} + \sum_{d} C^{\mathrm{LS}}_{d} p^{\mathrm{LS}}_{do\omega}\right]$, are multiplied by the probability of the corresponding scenario, φ_{ω}.
2. The operation constraints (4.4f)–(4.4o), $\forall o, \omega$, are formulated for each operating condition o and for each scenario ω.

Finally, note that generation and transmission expansion decisions do not depend on the scenario realization since these decisions are made before the actual scenario realization is known. Therefore, they are denoted by *here-and-now* expansion decisions. On the other hand, operation decisions, e.g., productions of candidate and existing generating units, depend on the scenario realization, and, therefore, they are called *wait-and-see* operation decisions.

Illustrative Example 4.5 *2-node system: Stochastic static G&TEP problem (formulation)*

Considering the data of Illustrative Example 4.1, we assume that demand $\tilde{d}$ is subject to uncertainty, explained as follows. The demand for each operating condition can be 30 % lower or 30 % higher than the demands considered in Illustrative Example 4.1 with equal probability (0.5), i.e., we consider two scenarios. Scenario 1 is characterized by demand conditions of 203 and 385 MW for operating conditions o_1 and o_2, respectively, while scenario 2 is characterized by demand conditions of

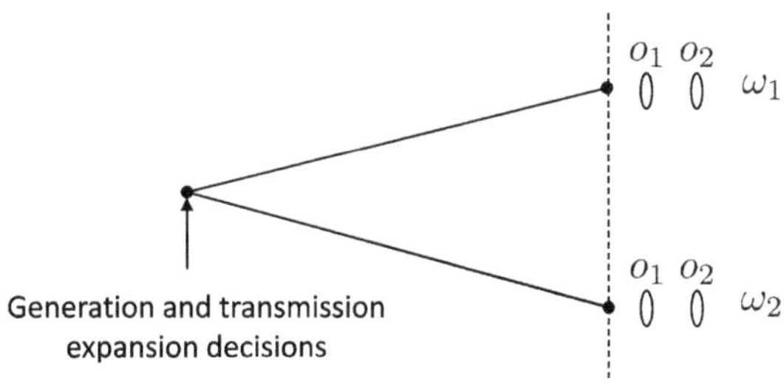

Fig. 4.2 Scenario tree for Illustrative Example 4.5

377 and 715 MW for operating conditions o_1 and o_2, respectively. Figure 4.2 depicts the scenario tree for this illustrative example.

Considering the above data, the G&TEP problem considering a stochastic static approach (4.4) results in the following MILP model:

$$\begin{aligned}\min_{\tilde{\Delta}^{\mathrm{SS}}} \quad & 0.5\left\{6000\left[35p^{\mathrm{E}}_{\tilde{g}o_1\omega_1} + 25p^{\mathrm{C}}_{\tilde{c}o_1\omega_1} + 80p^{\mathrm{LS}}_{\tilde{d}o_1\omega_1}\right] + 2760\left[35p^{\mathrm{E}}_{\tilde{g}o_2\omega_1}\right.\right.\\ & \left.\left. + 25p^{\mathrm{C}}_{\tilde{c}o_2\omega_1} + 80p^{\mathrm{LS}}_{\tilde{d}o_2\omega_1}\right]\right\} + 0.5\left\{6000\left[35p^{\mathrm{E}}_{\tilde{g}o_1\omega_2} + 25p^{\mathrm{C}}_{\tilde{c}o_1\omega_2}\right.\right.\\ & \left.\left. + 80p^{\mathrm{LS}}_{\tilde{d}o_1\omega_2}\right] + 2760\left[35p^{\mathrm{E}}_{\tilde{g}o_2\omega_2} + 25p^{\mathrm{C}}_{\tilde{c}o_2\omega_2} + 80p^{\mathrm{LS}}_{\tilde{d}o_2\omega_2}\right]\right\}\\ & + 70000p^{\mathrm{C^{max}}}_{\tilde{c}} + 100000x^{\mathrm{L}}_{\ell_2}\end{aligned}$$

subject to

$$\begin{aligned}& 0 \le p^{\mathrm{C^{max}}}_{\tilde{c}} \le 300\\ & x^{\mathrm{L}}_{\ell_2} \in \{0, 1\}\\ & 700000p^{\mathrm{C^{max}}}_{\tilde{c}} \le 400000000\\ & 1000000x^{\mathrm{L}}_{\ell_2} \le 2000000\\ & \left\{\begin{array}{l} p^{\mathrm{E}}_{\tilde{g}o_1\omega_1} - p^{\mathrm{L}}_{\ell_1 o_1\omega_1} - p^{\mathrm{L}}_{\ell_2 o_1\omega_1} = 0\\ p^{\mathrm{C}}_{\tilde{c}o_1\omega_1} + p^{\mathrm{L}}_{\ell_1 o_1\omega_1} + p^{\mathrm{L}}_{\ell_2 o_1\omega_1} = 203 - p^{\mathrm{LS}}_{\tilde{d}o_1\omega_1}\\ p^{\mathrm{L}}_{\ell_1 o_1\omega_1} = 500\left(\theta_{n_1 o_1\omega_1} - \theta_{n_2 o_1\omega_1}\right)\\ -200 \le p^{\mathrm{L}}_{\ell_1 o_1\omega_1} \le 200\\ -200x^{\mathrm{L}}_{\ell_2} \le p^{\mathrm{L}}_{\ell_2 o_1\omega_1} \le 200x^{\mathrm{L}}_{\ell_2}\\ -\left(1 - x^{\mathrm{L}}_{\ell_2}\right)M \le p^{\mathrm{L}}_{\ell_2 o_1\omega_1} - 500\left(\theta_{n_1 o_1\omega_1} - \theta_{n_2 o_1\omega_1}\right) \le \left(1 - x^{\mathrm{L}}_{\ell_2}\right)M\\ 0 \le p^{\mathrm{E}}_{\tilde{c}o_1\omega_1} \le 400\\ 0 \le p^{\mathrm{C}}_{\tilde{c}o_1\omega_1} \le p^{\mathrm{C^{max}}}_{\tilde{c}}\\ 0 \le p^{\mathrm{LS}}_{\tilde{d}o_1\omega_1} \le 203\\ -\pi \le \theta_{n_2 o_1\omega_1} \le \pi\\ \theta_{n_1 o_1\omega_1} = 0 \end{array}\right.\end{aligned}$$

$$\left\{\begin{array}{l}
p^{\mathrm{E}}_{\tilde{g}o_2\omega_1} - p^{\mathrm{L}}_{\ell_1 o_2\omega_1} - p^{\mathrm{L}}_{\ell_2 o_2\omega_1} = 0\\
p^{\mathrm{C}}_{\tilde{c}o_2\omega_1} + p^{\mathrm{L}}_{\ell_1 o_2\omega_1} + p^{\mathrm{L}}_{\ell_2 o_2\omega_1} = 385 - p^{\mathrm{LS}}_{\tilde{d}o_1\omega_1}\\
p^{\mathrm{L}}_{\ell_1 o_2\omega_1} = 500\left(\theta_{n_1 o_2\omega_1} - \theta_{n_2 o_2\omega_1}\right)\\
-200 \le p^{\mathrm{L}}_{\ell_1 o_2\omega_1} \le 200\\
-200 x^{\mathrm{L}}_{\ell_2} \le p^{\mathrm{L}}_{\ell_2 o_2\omega_1} \le 200 x^{\mathrm{L}}_{\ell_2}\\
-\left(1 - x^{\mathrm{L}}_{\ell_2}\right) M \le p^{\mathrm{L}}_{\ell_2 o_2\omega_1} - 500\left(\theta_{n_1 o_2\omega_1} - \theta_{n_2 o_2\omega_1}\right) \le \left(1 - x^{\mathrm{L}}_{\ell_2}\right) M\\
0 \le p^{\mathrm{E}}_{\tilde{c}o_2\omega_1} \le 400\\
0 \le p^{\mathrm{C}}_{\tilde{c}o_2\omega_1} \le p^{\mathrm{C^{max}}}_{\tilde{c}}\\
0 \le p^{\mathrm{LS}}_{\tilde{d}o_2\omega_1} \le 385\\
-\pi \le \theta_{n_2 o_2\omega_1} \le \pi\\
\theta_{n_1 o_2\omega_1} = 0
\end{array}\right.$$

$$\left\{\begin{array}{l}
p^{\mathrm{E}}_{\tilde{g}o_1\omega_2} - p^{\mathrm{L}}_{\ell_1 o_1\omega_2} - p^{\mathrm{L}}_{\ell_2 o_1\omega_2} = 0\\
p^{\mathrm{C}}_{\tilde{c}o_1\omega_2} + p^{\mathrm{L}}_{\ell_1 o_1\omega_2} + p^{\mathrm{L}}_{\ell_2 o_1\omega_2} = 377 - p^{\mathrm{LS}}_{\tilde{d}o_1\omega_2}\\
p^{\mathrm{L}}_{\ell_1 o_1\omega_2} = 500\left(\theta_{n_1 o_1\omega_2} - \theta_{n_2 o_1\omega_2}\right)\\
-200 \le p^{\mathrm{L}}_{\ell_1 o_1\omega_2} \le 200\\
-200 x^{\mathrm{L}}_{\ell_2} \le p^{\mathrm{L}}_{\ell_2 o_1\omega_2} \le 200 x^{\mathrm{L}}_{\ell_2}\\
-\left(1 - x^{\mathrm{L}}_{\ell_2}\right) M \le p^{\mathrm{L}}_{\ell_2 o_1\omega_2} - 500\left(\theta_{n_1 o_1\omega_2} - \theta_{n_2 o_1\omega_2}\right) \le \left(1 - x^{\mathrm{L}}_{\ell_2}\right) M\\
0 \le p^{\mathrm{E}}_{\tilde{c}o_1\omega_2} \le 400\\
0 \le p^{\mathrm{C}}_{\tilde{c}o_1\omega_2} \le p^{\mathrm{C^{max}}}_{\tilde{c}}\\
0 \le p^{\mathrm{LS}}_{\tilde{d}o_1\omega_2} \le 377\\
-\pi \le \theta_{n_2 o_1\omega_2} \le \pi\\
\theta_{n_1 o_1\omega_2} = 0
\end{array}\right.$$

$$
\begin{cases}
p^{\mathrm{E}}_{\tilde{g}o_2\omega_2} - p^{\mathrm{L}}_{\ell_1 o_2\omega_2} - p^{\mathrm{L}}_{\ell_2 o_2\omega_2} = 0 \\
p^{\mathrm{C}}_{\tilde{c}o_2\omega_2} + p^{\mathrm{L}}_{\ell_1 o_2\omega_2} + p^{\mathrm{L}}_{\ell_2 o_2\omega_2} = 715 - p^{\mathrm{LS}}_{\tilde{d}o_1\omega_2} \\
p^{\mathrm{L}}_{\ell_1 o_2\omega_2} = 500\left(\theta_{n_1 o_2\omega_2} - \theta_{n_2 o_2\omega_2}\right) \\
-200 \le p^{\mathrm{L}}_{\ell_1 o_2\omega_2} \le 200 \\
-200x^{\mathrm{L}}_{\ell_2} \le p^{\mathrm{L}}_{\ell_2 o_2\omega_2} \le 200x^{\mathrm{L}}_{\ell_2} \\
-\left(1 - x^{\mathrm{L}}_{\ell_2}\right)M \le p^{\mathrm{L}}_{\ell_2 o_2\omega_2} - 500\left(\theta_{n_1 o_2\omega_2} - \theta_{n_2 o_2\omega_2}\right) \le \left(1 - x^{\mathrm{L}}_{\ell_2}\right)M \\
0 \le p^{\mathrm{E}}_{\tilde{c}o_2\omega_2} \le 400 \\
0 \le p^{\mathrm{C}}_{\tilde{c}o_2\omega_2} \le p^{\mathrm{C^{max}}}_{\tilde{c}} \\
0 \le p^{\mathrm{LS}}_{\tilde{d}o_2\omega_2} \le 715 \\
-\pi \le \theta_{n_2 o_2\omega_2} \le \pi \\
\theta_{n_1 o_2\omega_2} = 0,
\end{cases}
$$

where $\tilde{\Delta}^{\mathrm{SS}} = \Big\{p^{\mathrm{E}}_{\tilde{g}o_1\omega_1}, p^{\mathrm{E}}_{\tilde{g}o_2\omega_1}, p^{\mathrm{C}}_{\tilde{c}o_1\omega_1}, p^{\mathrm{C}}_{\tilde{c}o_2\omega_1}, p^{\mathrm{LS}}_{\tilde{d}o_1\omega_1}, p^{\mathrm{LS}}_{\tilde{d}o_2\omega_1}, p^{\mathrm{L}}_{\ell_1 o_1\omega_1}, p^{\mathrm{L}}_{\ell_1 o_2\omega_1}, p^{\mathrm{L}}_{\ell_2 o_1\omega_1}, p^{\mathrm{L}}_{\ell_2 o_2\omega_1}, \theta_{n_1 o_1\omega_1}, \theta_{n_1 o_2\omega_1}, \theta_{n_2 o_1\omega_1}, \theta_{n_2 o_2\omega_1}, p^{\mathrm{E}}_{\tilde{g}o_1\omega_2}, p^{\mathrm{E}}_{\tilde{g}o_2\omega_2}, p^{\mathrm{C}}_{\tilde{c}o_1\omega_2}, p^{\mathrm{C}}_{\tilde{c}o_2\omega_2}, p^{\mathrm{LS}}_{\tilde{d}o_1\omega_2}, p^{\mathrm{LS}}_{\tilde{d}o_2\omega_2}, p^{\mathrm{L}}_{\ell_1 o_1\omega_2}, p^{\mathrm{L}}_{\ell_1 o_2\omega_2}, p^{\mathrm{L}}_{\ell_2 o_1\omega_2}, p^{\mathrm{L}}_{\ell_2 o_2\omega_2}, \theta_{n_1 o_1\omega_2}, \theta_{n_1 o_2\omega_2}, \theta_{n_2 o_1\omega_2}, \theta_{n_2 o_2\omega_2}, p^{\mathrm{C^{max}}}_{\tilde{c}}, x^{\mathrm{L}}_{\ell_2}\Big\}$.

We obtain that the optimal generation and transmission expansion plan consists in building transmission line ℓ_2 and 300 MW of the candidate generating unit.

For each operating condition, the average demand over the two considered scenarios is equal to the demand values considered in Illustrative Example 4.3, the equivalent problem considering a deterministic approach. However, the results of the deterministic and the stochastic approaches are different. By considering a stochastic approach, we take into account the possible future realizations of the demand, which is ignored by the deterministic approach. As a result, considering the average value of the uncertain parameters for solving a relatively simpler problem does not always provide the optimal solution.

Nevertheless, the number of variables and constraints in this illustrative example is approximately twice the number of variables and constraints needed in Illustrative Example 4.3. Moreover, here we use only two scenarios to model the uncertainty in the demand for the sake of clarity. However, in order to correctly represent the uncertain parameters, it is generally necessary to consider a large enough number of scenarios, and, thus, the size and computation burden of the stochastic G&TEP problem increase. □

4.5.2 Dynamic Approach

The G&TEP problem described in the previous section is extended here to consider a dynamic approach that enables making expansion decisions at different points in time. This problem can be formulated using the MILP model below:

$$\min_{\Delta^{\mathrm{SD}}} \quad \sum_{\omega} \varphi_{\omega} \left\{ \sum_{t} \left[\sum_{o} \rho_{o} \left(\sum_{g} C_{g}^{\mathrm{E}} p_{got\omega}^{\mathrm{E}} + \sum_{c} C_{c}^{\mathrm{C}} p_{cot\omega}^{\mathrm{C}} + \sum_{d} C_{d}^{\mathrm{LS}} p_{dot\omega}^{\mathrm{LS}} \right) + a_{t} \left(\sum_{c} I_{c}^{\mathrm{C}} p_{ct\omega}^{\mathrm{C^{max}}} + \sum_{\ell \in \Omega^{\mathrm{L+}}} I_{\ell}^{\mathrm{L}} x_{\ell t\omega}^{\mathrm{L}} \right) \right] \right\} \tag{4.5a}$$

subject to

$$0 \leq \sum_{t} p_{ct\omega}^{\mathrm{C^{max}}} \leq \overline{P}_{c}^{\mathrm{C^{max}}} \quad \forall c, \omega \tag{4.5b}$$

$$x_{\ell t\omega}^{\mathrm{L}} \in \{0, 1\} \quad \forall \ell \in \Omega^{\mathrm{L+}}, t, \omega \tag{4.5c}$$

$$\sum_{t} x_{\ell t\omega}^{\mathrm{L}} \leq 1 \quad \forall \ell \in \Omega^{\mathrm{L+}}, \omega \tag{4.5d}$$

$$\sum_{c} I_{c}^{\mathrm{C}} p_{ct\omega}^{\mathrm{C^{max}}} \leq I_{t}^{\mathrm{C,max}} \quad \forall t \tag{4.5e}$$

$$\sum_{\ell \in \Omega^{\mathrm{L+}}} I_{\ell}^{\mathrm{L}} x_{\ell t\omega}^{\mathrm{L}} \leq I_{t}^{\mathrm{L,max}} \quad \forall t \tag{4.5f}$$

$$p_{ct\omega}^{\mathrm{C^{max}}} = p_{ct\tilde{\omega}}^{\mathrm{C^{max}}} \quad \forall c, t, \omega, \tilde{\omega} | P_{do\tau\omega}^{\mathrm{D^{max}}} = P_{do\tau\tilde{\omega}}^{\mathrm{D^{max}}} \quad \forall d, o, \tau < t \tag{4.5g}$$

$$x_{\ell t\omega}^{\mathrm{L}} = x_{\ell t\tilde{\omega}}^{\mathrm{L}} \quad \forall \ell, t, \omega, \tilde{\omega} | P_{do\tau\omega}^{\mathrm{D^{max}}} = P_{do\tau\tilde{\omega}}^{\mathrm{D^{max}}} \quad \forall d, o, \tau < t \tag{4.5h}$$

$$\left\{ \sum_{g \in \Omega_{n}^{\mathrm{E}}} p_{got\omega}^{\mathrm{E}} + \sum_{c \in \Omega_{n}^{\mathrm{C}}} p_{cot\omega}^{\mathrm{C}} - \sum_{\ell | s(\ell)=n} p_{lot\omega}^{\mathrm{L}} + \sum_{\ell | r(\ell)=n} p_{\ell ot\omega}^{\mathrm{L}} = \sum_{d \in \Omega_{n}^{\mathrm{D}}} \left(P_{dot\omega}^{\mathrm{D^{max}}} - p_{dot\omega}^{\mathrm{LS}} \right) \quad \forall n \right. \tag{4.5i}$$

$$p_{\ell ot\omega}^{\mathrm{L}} = B_{\ell} \left(\theta_{s(\ell)ot\omega} - \theta_{r(\ell)ot\omega} \right) \quad \forall \ell \setminus \ell \in \Omega^{\mathrm{L+}} \tag{4.5j}$$

$$-\sum_{\tau \leq t} x_{\ell \tau\omega}^{\mathrm{L}} F_{\ell}^{\max} \leq p_{\ell ot\omega}^{\mathrm{L}} \leq \sum_{\tau \leq t} x_{\ell \tau\omega}^{\mathrm{L}} F_{\ell}^{\max} \quad \forall \ell \in \Omega^{\mathrm{L+}} \tag{4.5k}$$

$$-\left(1 - \sum_{\tau \leq t} x_{\ell \tau\omega}^{\mathrm{L}}\right) M \leq p_{\ell ot\omega}^{\mathrm{L}} - B_{\ell} \left(\theta_{s(\ell)ot\omega} - \theta_{r(\ell)ot\omega} \right) \leq \left(1 - \sum_{\tau \leq t} x_{\ell \tau\omega}^{\mathrm{L}}\right) M \quad \forall \ell \in \Omega^{\mathrm{L+}} \tag{4.5l}$$

$$-F_{\ell}^{\max} \leq p_{\ell ot\omega}^{\mathrm{L}} \leq F_{\ell}^{\max} \quad \forall \ell \tag{4.5m}$$

$$0 \leq p_{got\omega}^{\mathrm{E}} \leq P_{g}^{\mathrm{E^{max}}} \quad \forall g \tag{4.5n}$$

$$0 \le p^{\mathrm{C}}_{cot\omega} \le \sum_{\tau \le t} p^{\mathrm{C^{max}}}_{c\tau\omega} \quad \forall c \tag{4.5o}$$

$$0 \le p^{\mathrm{LS}}_{dot\omega} \le P^{\mathrm{D^{max}}}_{dot\omega} \quad \forall d \tag{4.5p}$$

$$-\pi \le \theta_{not\omega} \le \pi \quad \forall n \tag{4.5q}$$

$$\theta_{not\omega} = 0 \quad n\text{: ref.} \tag{4.5r}$$

$$\Big\}, \forall o, t, \omega,$$

where the variables in sets $\Delta^{\mathrm{SD}} = \left\{p^{\mathrm{E}}_{got\omega}, p^{\mathrm{LS}}_{dot\omega}, p^{\mathrm{C}}_{cot\omega}, p^{\mathrm{L}}_{\ell ot\omega}, \theta_{not\omega}, p^{\mathrm{C^{max}}}_{ct\omega}, x^{\mathrm{L}}_{\ell t\omega}\right\}$ are the optimization variables of problem (4.5).

The main differences between problem (4.5) and problem (4.4) considering a stochastic static approach are summarized below:

1. In the objective function (4.5a), the investment costs at each time period are multiplied by the corresponding amortization rates.
2. We include nonanticipativity constraints (4.5g) and (4.5h), i.e., constraints that avoid anticipating information. These constraints impose that for a given time period t, if the characteristics of two scenarios ω and $\tilde{\omega}$ in the previous time periods $\tau < t$ are the same, then the generation and transmission expansion plans of these two scenarios for time period t are also the same.
3. Constraints (4.5b), which impose bounds on the capacity of each candidate generating unit for the whole planning horizon, are included.
4. Constraints (4.5d), which impose that a prospective transmission line can be built only once during the whole planning horizon, are included.
5. Constraints (4.5k)–(4.5l), which impose that a prospective transmission line must be available at time period t if it has been built at the beginning of that time period or in the previous time periods, are included.
6. Constraints (4.5o), which impose that the available capacity of the candidate generating units at time period t must be equal to the capacity built at that time period plus the capacities built in the previous periods, are included.
7. The operation constraints (4.5i)–(4.5r), $\forall o, t, \omega$, are formulated for each operating condition o, time period t, and scenario ω.

For the sake of clarity, the decision-making sequence is summarized below:

1. The generation and transmission expansion planner determines the expansion plan to be made at the beginning of the planning horizon, i.e., at the beginning of the first time period. This expansion plan does not depend on the future scenario realizations since this information is not available for the planner at the time it makes these expansion plans. Therefore, these expansion decisions are referred to as *here-and-now* decisions.
2. One of the scenarios modeling the uncertainty in the first time period is realized. This information becomes available to the expansion planner.
3. The expansion planner determines the expansion plan to be made at the beginning of the second time period. This expansion plan depends on the scenario realization

in the first time period since it depends on the scenario realization in that period. However, this expansion plan does not depend on the future scenario realizations. Therefore, these expansion decisions are *wait-and-see* with respect to the scenario realizations in the first time period but *here-and-now* with respect to the scenario realizations in the second and following time periods.

4. One of the scenarios modeling the uncertainty in the second time period is realized. This information becomes available for the expansion planner.

The above steps 3 and 4 are repeated until the last period of the planning horizon.

Illustrative Example 4.6 *2-node system: Stochastic dynamic G&TEP problem*

Let us consider the data of Illustrative Example 4.1. The planning horizon is divided in this case into two time periods, so that generation and transmission expansion plans can be made at the beginning of both time periods.

There are two possible scenario realizations in the first time period, a and b. Realizations a and b consider that operating conditions o_1 and o_2 in the first time period are characterized respectively by a low and high demand. Scenario a considers that operating conditions o_1 and o_2 are defined by demands equal to 212 and 402 MW, respectively, while scenario b considers that operating conditions o_1 and o_2 are defined by demands equal to 281 and 533 MW, respectively. The probability of each scenario realization, a and b, is equal to 0.5.

On the other hand, there are also two possible scenario realizations in the second time period, c and d. Realizations c and d consider that operating conditions o_1 and o_2 in the second time period are characterized respectively by a low and high demand. The operating conditions for these two scenarios in the second time period depend on the scenario realization in the first time period, as explained next and illustrated with the corresponding scenario tree depicted in Fig. 4.3:

1. If scenario a is realized in the first time period, then (i) realization c considers that operating conditions o_1 and o_2 in the second time period are defined by demands equal to 214 and 407 MW, respectively, while (ii) realization d considers that operating conditions o_1 and o_2 in the second time period are defined by demands equal to 284 and 539 MW, respectively.
2. If scenario b is realized in the first time period, then (i) realization c considers that operating conditions o_1 and o_2 in the second time period are defined by demands equal to 284 and 539 MW, respectively, while (ii) realization d considers that operating conditions o_1 and o_2 in the second time period are defined by demands equal to 377 and 715 MW, respectively.

The probability of each realization, c and d, is equal to 0.5.

The weights of operating conditions o_1 and o_2 in both time periods and for all scenario realizations are 6000 and 2760 h, respectively. The above data result in four scenarios for the whole planning horizon (combinations of possible realizations in both time periods), whose data are summarized in Table 4.1. The second and third columns provide the demand conditions in the first and second time periods, respectively; finally, the fourth column gives the probability of each scenario. Note

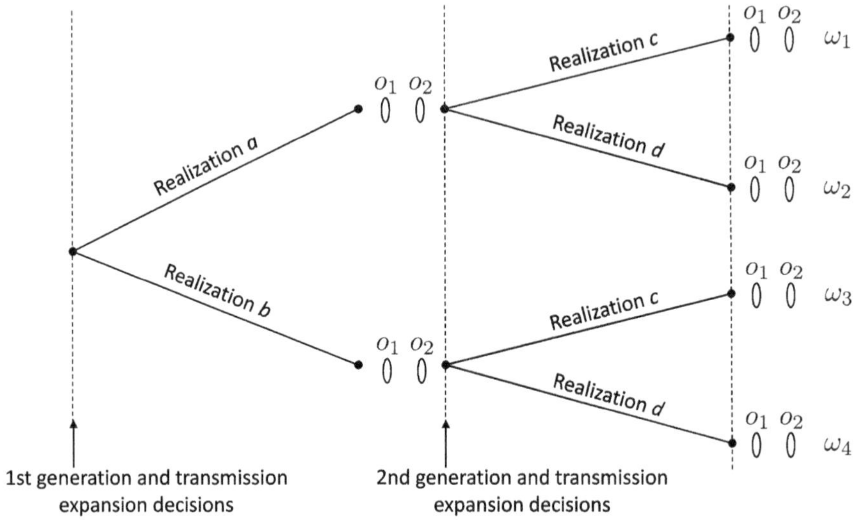

Fig. 4.3 Scenario tree for Illustrative Example 4.6

Table 4.1 Illustrative Example 4.6: data for scenarios

Scenario	Period 1	Period 2	Probability
Scenario 1 (*a* and *c*)	$P^{\mathrm{D}}_{o_1 t_1 \omega_1} = 212\,\mathrm{MW}$	$P^{\mathrm{D}}_{o_1 t_2 \omega_1} = 214\,\mathrm{MW}$	$0.5 \times 0.5 = 0.25$
	$P^{\mathrm{D}}_{o_2 t_1 \omega_1} = 402\,\mathrm{MW}$	$P^{\mathrm{D}}_{o_2 t_2 \omega_1} = 407\,\mathrm{MW}$	
Scenario 2 (*a* and *d*)	$P^{\mathrm{D}}_{o_1 t_1 \omega_2} = 212\,\mathrm{MW}$	$P^{\mathrm{D}}_{o_1 t_2 \omega_2} = 284\,\mathrm{MW}$	$0.5 \times 0.5 = 0.25$
	$P^{\mathrm{D}}_{o_2 t_1 \omega_2} = 402\,\mathrm{MW}$	$P^{\mathrm{D}}_{o_2 t_2 \omega_2} = 539\,\mathrm{MW}$	
Scenario 3 (*b* and *c*)	$P^{\mathrm{D}}_{o_1 t_1 \omega_3} = 281\,\mathrm{MW}$	$P^{\mathrm{D}}_{o_1 t_2 \omega_3} = 284\,\mathrm{MW}$	$0.5 \times 0.5 = 0.25$
	$P^{\mathrm{D}}_{o_2 t_1 \omega_3} = 533\,\mathrm{MW}$	$P^{\mathrm{D}}_{o_2 t_2 \omega_3} = 539\,\mathrm{MW}$	
Scenario 4 (*b* and *d*)	$P^{\mathrm{D}}_{o_1 t_1 \omega_4} = 281\,\mathrm{MW}$	$P^{\mathrm{D}}_{o_1 t_2 \omega_4} = 377\,\mathrm{MW}$	$0.5 \times 0.5 = 0.25$
	$P^{\mathrm{D}}_{o_2 t_1 \omega_4} = 533\,\mathrm{MW}$	$P^{\mathrm{D}}_{o_2 t_2 \omega_4} = 715\,\mathrm{MW}$	

that the data for scenarios and operating conditions have been selected so that the results of this illustrative example can be compared with those of previous ones.

Considering the above data, the G&TEP problem taking a stochastic dynamic approach (4.5) results in the following optimization model:

$$\min_{\tilde{\Delta}^{\mathrm{SD}}} \quad 0.25 \Big\{ 6000 \Big[35 p^{\mathrm{E}}_{\tilde{g} o_1 t_1 \omega_1} + 25 p^{\mathrm{C}}_{\tilde{c} o_1 t_1 \omega_1} + 80 p^{\mathrm{LS}}_{\tilde{d} o_1 t_1 \omega_1} \Big] + 2760 \Big[35 p^{\mathrm{E}}_{\tilde{g} o_2 t_1 \omega_1}$$

$$+ 25 p^{\mathrm{C}}_{\tilde{c} o_2 t_1 \omega_1} + 80 p^{\mathrm{LS}}_{\tilde{d} o_2 t_1 \omega_1} \Big] + 6000 \Big[35 p^{\mathrm{E}}_{\tilde{g} o_1 t_2 \omega_1} + 25 p^{\mathrm{C}}_{\tilde{c} o_1 t_2 \omega_1}$$

$$+ 80p^{\mathrm{LS}}_{\tilde{d}o_1t_2\omega_1}\Big] + 2760\Big[35p^{\mathrm{E}}_{\tilde{g}o_2t_2\omega_1} + 25p^{\mathrm{C}}_{\tilde{c}o_2t_2\omega_1} + 80p^{\mathrm{LS}}_{\tilde{d}o_2t_2\omega_1}\Big]$$
$$+ 140000p^{\mathrm{C^{max}}}_{\tilde{c}t_1\omega_1} + 200000x^{\mathrm{L}}_{\ell_2t_1\omega_1} + 70000p^{\mathrm{C^{max}}}_{\tilde{c}t_2\omega_1} + 100000x^{\mathrm{L}}_{\ell_2t_2\omega_1}\Big\}$$
$$0.25\Big\{6000\Big[35p^{\mathrm{E}}_{\tilde{g}o_1t_1\omega_2} + 25p^{\mathrm{C}}_{\tilde{c}o_1t_1\omega_2} + 80p^{\mathrm{LS}}_{\tilde{d}o_1t_1\omega_2}\Big] + 2760\Big[35p^{\mathrm{E}}_{\tilde{g}o_2t_1\omega_2}$$
$$+ 25p^{\mathrm{C}}_{\tilde{c}o_2t_1\omega_2} + 80p^{\mathrm{LS}}_{\tilde{d}o_2t_1\omega_2}\Big] + 6000\Big[35p^{\mathrm{E}}_{\tilde{g}o_1t_2\omega_2} + 25p^{\mathrm{C}}_{\tilde{c}o_1t_2\omega_2}$$
$$+ 80p^{\mathrm{LS}}_{\tilde{d}o_1t_2\omega_2}\Big] + 2760\Big[35p^{\mathrm{E}}_{\tilde{g}o_2t_2\omega_2} + 25p^{\mathrm{C}}_{\tilde{c}o_2t_2\omega_1} + 80p^{\mathrm{LS}}_{\tilde{d}o_2t_2\omega_2}\Big]$$
$$+ 140000p^{\mathrm{C^{max}}}_{\tilde{c}t_1\omega_2} + 200000x^{\mathrm{L}}_{\ell_2t_1\omega_2} + 70000p^{\mathrm{C^{max}}}_{\tilde{c}t_2\omega_2} + 100000x^{\mathrm{L}}_{\ell_2t_2\omega_2}\Big\}$$
$$0.25\Big\{6000\Big[35p^{\mathrm{E}}_{\tilde{g}o_1t_1\omega_3} + 25p^{\mathrm{C}}_{\tilde{c}o_1t_1\omega_3} + 80p^{\mathrm{LS}}_{\tilde{d}o_1t_1\omega_3}\Big] + 2760\Big[35p^{\mathrm{E}}_{\tilde{g}o_2t_1\omega_3}$$
$$+ 25p^{\mathrm{C}}_{\tilde{c}o_2t_1\omega_3} + 80p^{\mathrm{LS}}_{\tilde{d}o_2t_1\omega_3}\Big] + 6000\Big[35p^{\mathrm{E}}_{\tilde{g}o_1t_2\omega_3} + 25p^{\mathrm{C}}_{\tilde{c}o_1t_2\omega_3}$$
$$+ 80p^{\mathrm{LS}}_{\tilde{d}o_1t_2\omega_3}\Big] + 2760\Big[35p^{\mathrm{E}}_{\tilde{g}o_2t_2\omega_3} + 25p^{\mathrm{C}}_{\tilde{c}o_2t_2\omega_3} + 80p^{\mathrm{LS}}_{\tilde{d}o_2t_2\omega_3}\Big]$$
$$+ 140000p^{\mathrm{C^{max}}}_{\tilde{c}t_1\omega_3} + 200000x^{\mathrm{L}}_{\ell_2t_1\omega_3} + 70000p^{\mathrm{C^{max}}}_{\tilde{c}t_2\omega_3} + 100000x^{\mathrm{L}}_{\ell_2t_2\omega_3}\Big\}$$
$$0.25\Big\{6000\Big[35p^{\mathrm{E}}_{\tilde{g}o_1t_1\omega_4} + 25p^{\mathrm{C}}_{\tilde{c}o_1t_1\omega_4} + 80p^{\mathrm{LS}}_{\tilde{d}o_1t_1\omega_4}\Big] + 2760\Big[35p^{\mathrm{E}}_{\tilde{g}o_2t_1\omega_4}$$
$$+ 25p^{\mathrm{C}}_{\tilde{c}o_2t_1\omega_4} + 80p^{\mathrm{LS}}_{\tilde{d}o_2t_1\omega_4}\Big] + 6000\Big[35p^{\mathrm{E}}_{\tilde{g}o_1t_2\omega_4} + 25p^{\mathrm{C}}_{\tilde{c}o_1t_2\omega_4}$$
$$+ 80p^{\mathrm{LS}}_{\tilde{d}o_1t_2\omega_4}\Big] + 2760\Big[35p^{\mathrm{E}}_{\tilde{g}o_2t_2\omega_4} + 25p^{\mathrm{C}}_{\tilde{c}o_2t_2\omega_4} + 80p^{\mathrm{LS}}_{\tilde{d}o_2t_2\omega_4}\Big]$$
$$+ 140000p^{\mathrm{C^{max}}}_{\tilde{c}t_1\omega_4} + 200000x^{\mathrm{L}}_{\ell_2t_1\omega_4} + 70000p^{\mathrm{C^{max}}}_{\tilde{c}t_2\omega_4} + 100000x^{\mathrm{L}}_{\ell_2t_2\omega_4}\Big\}$$

subject to

$$0 \le p^{\mathrm{C^{max}}}_{\tilde{c}t_1\omega_1} + p^{\mathrm{C^{max}}}_{\tilde{c}t_2\omega_1} \le 300$$
$$0 \le p^{\mathrm{C^{max}}}_{\tilde{c}t_1\omega_2} + p^{\mathrm{C^{max}}}_{\tilde{c}t_2\omega_2} \le 300$$
$$0 \le p^{\mathrm{C^{max}}}_{\tilde{c}t_1\omega_3} + p^{\mathrm{C^{max}}}_{\tilde{c}t_2\omega_3} \le 300$$
$$0 \le p^{\mathrm{C^{max}}}_{\tilde{c}t_1\omega_4} + p^{\mathrm{C^{max}}}_{\tilde{c}t_2\omega_4} \le 300$$
$$x^{\mathrm{L}}_{\ell_2t_1\omega_1}, x^{\mathrm{L}}_{\ell_2t_1\omega_2}, x^{\mathrm{L}}_{\ell_2t_1\omega_3}, x^{\mathrm{L}}_{\ell_2t_4\omega_1} \in \{0, 1\}$$
$$x^{\mathrm{L}}_{\ell_2t_2\omega_1}, x^{\mathrm{L}}_{\ell_2t_2\omega_2}, x^{\mathrm{L}}_{\ell_2t_2\omega_3}, x^{\mathrm{L}}_{\ell_2t_2\omega_1} \in \{0, 1\}$$
$$x^{\mathrm{L}}_{\ell_2t_1\omega_1} + x^{\mathrm{L}}_{\ell_2t_2\omega_1} \le 1$$
$$x^{\mathrm{L}}_{\ell_2t_1\omega_2} + x^{\mathrm{L}}_{\ell_2t_2\omega_2} \le 1$$
$$x^{\mathrm{L}}_{\ell_2t_1\omega_3} + x^{\mathrm{L}}_{\ell_2t_2\omega_3} \le 1$$
$$x^{\mathrm{L}}_{\ell_2t_1\omega_4} + x^{\mathrm{L}}_{\ell_2t_2\omega_4} \le 1$$

$$
\begin{aligned}
&700000 p^{\mathrm{C^{max}}}_{\tilde{c}t_1\omega_1} \leq 400000000\\
&700000 p^{\mathrm{C^{max}}}_{\tilde{c}t_1\omega_2} \leq 400000000\\
&700000 p^{\mathrm{C^{max}}}_{\tilde{c}t_1\omega_3} \leq 400000000\\
&700000 p^{\mathrm{C^{max}}}_{\tilde{c}t_1\omega_4} \leq 400000000\\
&700000 p^{\mathrm{C^{max}}}_{\tilde{c}t_2\omega_1} \leq 400000000\\
&700000 p^{\mathrm{C^{max}}}_{\tilde{c}t_2\omega_2} \leq 400000000\\
&700000 p^{\mathrm{C^{max}}}_{\tilde{c}t_2\omega_3} \leq 400000000\\
&700000 p^{\mathrm{C^{max}}}_{\tilde{c}t_2\omega_4} \leq 400000000\\
&1000000 x^{\mathrm{L}}_{\ell_2 t_1\omega_1} \leq 2000000\\
&1000000 x^{\mathrm{L}}_{\ell_2 t_1\omega_2} \leq 2000000\\
&1000000 x^{\mathrm{L}}_{\ell_2 t_1\omega_3} \leq 2000000\\
&1000000 x^{\mathrm{L}}_{\ell_2 t_1\omega_4} \leq 2000000\\
&1000000 x^{\mathrm{L}}_{\ell_2 t_2\omega_1} \leq 2000000\\
&1000000 x^{\mathrm{L}}_{\ell_2 t_2\omega_2} \leq 2000000\\
&1000000 x^{\mathrm{L}}_{\ell_2 t_2\omega_3} \leq 2000000\\
&1000000 x^{\mathrm{L}}_{\ell_2 t_2\omega_4} \leq 2000000\\
&p^{\mathrm{C^{max}}}_{\tilde{c}t_1\omega_1} = p^{\mathrm{C^{max}}}_{\tilde{c}t_1\omega_2} = p^{\mathrm{C^{max}}}_{\tilde{c}t_1\omega_3} = p^{\mathrm{C^{max}}}_{\tilde{c}t_1\omega_4}\\
&p^{\mathrm{C^{max}}}_{\tilde{c}t_2\omega_1} = p^{\mathrm{C^{max}}}_{\tilde{c}t_2\omega_3}\\
&p^{\mathrm{C^{max}}}_{\tilde{c}t_2\omega_3} = p^{\mathrm{C^{max}}}_{\tilde{c}t_2\omega_4}\\
&x^{\mathrm{L}}_{\ell_2 t_1\omega_1} = x^{\mathrm{L}}_{\ell_2 t_1\omega_2} = x^{\mathrm{L}}_{\ell_2 t_1\omega_3} = x^{\mathrm{L}}_{\ell_2 t_4\omega_1}\\
&x^{\mathrm{L}}_{\ell_2 t_2\omega_1} = x^{\mathrm{L}}_{\ell_2 t_2\omega_2}\\
&x^{\mathrm{L}}_{\ell_2 t_2\omega_3} = x^{\mathrm{L}}_{\ell_2 t_2\omega_4}
\end{aligned}
$$

$$\begin{cases} p^{\mathrm{E}}_{\tilde{g}o_1t_1\omega_1} - p^{\mathrm{L}}_{\ell_1o_1t_1\omega_1} - p^{\mathrm{L}}_{\ell_2o_1t_1\omega_1} = 0 \\ p^{\mathrm{C}}_{\tilde{c}o_1t_1\omega_1} + p^{\mathrm{L}}_{\ell_1o_1t_1\omega_1} + p^{\mathrm{L}}_{\ell_2o_1t_1\omega_1} = 212 - p^{\mathrm{LS}}_{\tilde{d}o_1t_1\omega_1} \\ p^{\mathrm{L}}_{\ell_1o_1t_1\omega_1} = 500\left(\theta_{n_1o_1t_1\omega_1} - \theta_{n_2o_1t_1\omega_1}\right) \\ -200 \leq p^{\mathrm{L}}_{\ell_1o_1t_1\omega_1} \leq 200 \\ -200x^{\mathrm{L}}_{\ell_2t_1\omega_1} \leq p^{\mathrm{L}}_{\ell_2o_1t_1\omega_1} \leq 200x^{\mathrm{L}}_{\ell_2t_1\omega_2} \\ -\left(1 - x^{\mathrm{L}}_{\ell_2t_1\omega_1}\right)M \leq p^{\mathrm{L}}_{\ell_2o_1t_1\omega_1} - 500\left(\theta_{n_1o_1t_1\omega_1} - \theta_{n_2o_1t_1\omega_1}\right) \leq \left(1 - x^{\mathrm{L}}_{\ell_2t_1\omega_1}\right)M \\ 0 \leq p^{\mathrm{E}}_{\tilde{c}o_1t_1\omega_1} \leq 400 \\ 0 \leq p^{\mathrm{C}}_{\tilde{c}o_1t_1\omega_1} \leq p^{\mathrm{C^{max}}}_{\tilde{c}t_1\omega_1} \\ 0 \leq p^{\mathrm{LS}}_{\tilde{d}o_1t_1\omega_1} \leq 212 \\ -\pi \leq \theta_{n_2o_1t_1\omega_1} \leq \pi \\ \theta_{n_1o_1t_1\omega_1} = 0 \end{cases}$$

$$\begin{cases} p^{\mathrm{E}}_{\tilde{g}o_2t_1\omega_1} - p^{\mathrm{L}}_{\ell_1o_2t_1\omega_1} - p^{\mathrm{L}}_{\ell_2o_2t_1\omega_1} = 0 \\ p^{\mathrm{C}}_{\tilde{c}o_2t_1\omega_1} + p^{\mathrm{L}}_{\ell_1o_2t_1\omega_1} + p^{\mathrm{L}}_{\ell_2o_2t_1\omega_1} = 402 - p^{\mathrm{LS}}_{\tilde{d}o_1t_1\omega_1} \\ p^{\mathrm{L}}_{\ell_1o_2t_1\omega_1} = 500\left(\theta_{n_1o_2t_1\omega_1} - \theta_{n_2o_2t_1\omega_1}\right) \\ -200 \leq p^{\mathrm{L}}_{\ell_1o_2t_1\omega_1} \leq 200 \\ -200x^{\mathrm{L}}_{\ell_2t_1\omega_1} \leq p^{\mathrm{L}}_{\ell_2o_2t_1\omega_1} \leq 200x^{\mathrm{L}}_{\ell_2t_1\omega_1} \\ -\left(1 - x^{\mathrm{L}}_{\ell_2t_1\omega_1}\right)M \leq p^{\mathrm{L}}_{\ell_2o_2t_1\omega_1} - 500\left(\theta_{n_1o_2t_1\omega_1} - \theta_{n_2o_2t_1\omega_1}\right) \leq \left(1 - x^{\mathrm{L}}_{\ell_2t_1\omega_1}\right)M \\ 0 \leq p^{\mathrm{E}}_{\tilde{c}o_2t_1\omega_1} \leq 400 \\ 0 \leq p^{\mathrm{C}}_{\tilde{c}o_2t_1\omega_1} \leq p^{\mathrm{C^{max}}}_{\tilde{c}t_1\omega_1} \\ 0 \leq p^{\mathrm{LS}}_{\tilde{d}o_2t_1\omega_1} \leq 402 \\ -\pi \leq \theta_{n_2o_2t_1\omega_1} \leq \pi \\ \theta_{n_1o_2t_1\omega_1} = 0 \end{cases}$$

$$\left\{\begin{array}{l}
p^{\mathrm{E}}_{\tilde{g}o_1t_2\omega_1} - p^{\mathrm{L}}_{\ell_1o_1t_2\omega_1} - p^{\mathrm{L}}_{\ell_2o_1t_2\omega_1} = 0\\
p^{\mathrm{C}}_{\tilde{c}o_1t_2\omega_1} + p^{\mathrm{L}}_{\ell_1o_1t_2\omega_1} + p^{\mathrm{L}}_{\ell_2o_1t_2\omega_1} = 214 - p^{\mathrm{LS}}_{\tilde{d}o_1t_2\omega_1}\\
p^{\mathrm{L}}_{\ell_1o_1t_2\omega_1} = 500\left(\theta_{n_1o_1t_2\omega_1} - \theta_{n_2o_1t_2\omega_1}\right)\\
-200 \le p^{\mathrm{L}}_{\ell_1o_1t_2\omega_1} \le 200\\
-200\left(x^{\mathrm{L}}_{\ell_2t_1\omega_1} + x^{\mathrm{L}}_{\ell_2t_2\omega_1}\right) \le p^{\mathrm{L}}_{\ell_2o_1t_2\omega_1} \le 200\left(x^{\mathrm{L}}_{\ell_2t_1\omega_1} + x^{\mathrm{L}}_{\ell_2t_2\omega_1}\right)\\
-\left[1-\left(x^{\mathrm{L}}_{\ell_2t_1\omega_1} + x^{\mathrm{L}}_{\ell_2t_2\omega_1}\right)\right]M \le p^{\mathrm{L}}_{\ell_2o_1t_2\omega_1} - 500\left(\theta_{n_1o_1t_2\omega_1} - \theta_{n_2o_1t_2\omega_1}\right)\\
p^{\mathrm{L}}_{\ell_2o_1t_2\omega_1} - 500\left(\theta_{n_1o_1t_2\omega_1} - \theta_{n_2o_1t_2\omega_1}\right) \le \left[1-\left(x^{\mathrm{L}}_{\ell_2t_1\omega_1} + x^{\mathrm{L}}_{\ell_2t_2\omega_1}\right)\right]M\\
0 \le p^{\mathrm{E}}_{\tilde{c}o_1t_2\omega_1} \le 400\\
0 \le p^{\mathrm{C}}_{\tilde{c}o_1t_2\omega_1} \le p^{\mathrm{C^{max}}}_{\tilde{c}t_1\omega_1} + p^{\mathrm{C^{max}}}_{\tilde{c}t_2\omega_1}\\
0 \le p^{\mathrm{LS}}_{\tilde{d}o_1t_2\omega_1} \le 214\\
-\pi \le \theta_{n_2o_1t_2\omega_1} \le \pi\\
\theta_{n_1o_1t_2\omega_1} = 0
\end{array}\right.$$

$$\left\{\begin{array}{l}
p^{\mathrm{E}}_{\tilde{g}o_2t_2\omega_1} - p^{\mathrm{L}}_{\ell_1o_2t_2\omega_1} - p^{\mathrm{L}}_{\ell_2o_2t_2\omega_1} = 0\\
p^{\mathrm{C}}_{\tilde{c}o_2t_2\omega_1} + p^{\mathrm{L}}_{\ell_1o_2t_2\omega_1} + p^{\mathrm{L}}_{\ell_2o_2t_2\omega_1} = 407 - p^{\mathrm{LS}}_{\tilde{d}o_2t_2\omega_1}\\
p^{\mathrm{L}}_{\ell_1o_2t_2\omega_1} = 500\left(\theta_{n_1o_2t_2\omega_1} - \theta_{n_2o_2t_2\omega_1}\right)\\
-200 \le p^{\mathrm{L}}_{\ell_1o_2t_2\omega_1} \le 200\\
-200\left(x^{\mathrm{L}}_{\ell_2t_1\omega_1} + x^{\mathrm{L}}_{\ell_2t_2\omega_1}\right) \le p^{\mathrm{L}}_{\ell_2o_2t_2\omega_1} \le 200\left(x^{\mathrm{L}}_{\ell_2t_1\omega_1} + x^{\mathrm{L}}_{\ell_2t_2\omega_1}\right)\\
-\left[1-\left(x^{\mathrm{L}}_{\ell_2t_1\omega_1} + x^{\mathrm{L}}_{\ell_2t_2\omega_1}\right)\right]M \le p^{\mathrm{L}}_{\ell_2o_2t_2\omega_1} - 500\left(\theta_{n_1o_2t_2\omega_1} - \theta_{n_2o_2t_2\omega_1}\right)\\
p^{\mathrm{L}}_{\ell_2o_2t_2\omega_1} - 500\left(\theta_{n_1o_2t_2\omega_1} - \theta_{n_2o_2t_2\omega_1}\right) \le \left[1-\left(x^{\mathrm{L}}_{\ell_2t_1\omega_1} + x^{\mathrm{L}}_{\ell_2t_2\omega_1}\right)\right]M\\
0 \le p^{\mathrm{E}}_{\tilde{c}o_2t_2\omega_1} \le 400\\
0 \le p^{\mathrm{C}}_{\tilde{c}o_2t_2\omega_1} \le p^{\mathrm{C^{max}}}_{\tilde{c}t_1\omega_1} + p^{\mathrm{C^{max}}}_{\tilde{c}t_2\omega_1}\\
0 \le p^{\mathrm{LS}}_{\tilde{d}o_2t_2\omega_1} \le 407\\
-\pi \le \theta_{n_2o_2t_2\omega_1} \le \pi\\
\theta_{n_1o_2t_2\omega_1} = 0
\end{array}\right.$$

$$
\begin{cases}
p^{\mathrm{E}}_{\tilde{g}o_1t_1\omega_2} - p^{\mathrm{L}}_{\ell_1o_1t_1\omega_2} - p^{\mathrm{L}}_{\ell_2o_1t_1\omega_2} = 0 \\
p^{\mathrm{C}}_{\tilde{c}o_1t_1\omega_2} + p^{\mathrm{L}}_{\ell_1o_1t_1\omega_2} + p^{\mathrm{L}}_{\ell_2o_1t_1\omega_2} = 212 - p^{\mathrm{LS}}_{\tilde{d}o_1t_1\omega_2} \\
p^{\mathrm{L}}_{\ell_1o_1t_1\omega_2} = 500\left(\theta_{n_1o_1t_1\omega_2} - \theta_{n_2o_1t_1\omega_2}\right) \\
-200 \le p^{\mathrm{L}}_{\ell_1o_1t_1\omega_2} \le 200 \\
-200x^{\mathrm{L}}_{\ell_2t_1\omega_2} \le p^{\mathrm{L}}_{\ell_2o_1t_1\omega_2} \le 200x^{\mathrm{L}}_{\ell_2t_1\omega_2} \\
-\left(1 - x^{\mathrm{L}}_{\ell_2t_1\omega_2}\right)M \le p^{\mathrm{L}}_{\ell_2o_1t_1\omega_2} - 500\left(\theta_{n_1o_1t_1\omega_2} - \theta_{n_2o_1t_1\omega_2}\right) \le \left(1 - x^{\mathrm{L}}_{\ell_2t_1\omega_2}\right)M \\
0 \le p^{\mathrm{E}}_{\tilde{c}o_1t_1\omega_2} \le 400 \\
0 \le p^{\mathrm{C}}_{\tilde{c}o_1t_1\omega_2} \le p^{\mathrm{C^{max}}}_{\tilde{c}t_1\omega_2} \\
0 \le p^{\mathrm{LS}}_{\tilde{d}o_1t_1\omega_2} \le 212 \\
-\pi \le \theta_{n_2o_1t_1\omega_2} \le \pi \\
\theta_{n_1o_1t_1\omega_2} = 0
\end{cases}
$$

$$
\begin{cases}
p^{\mathrm{E}}_{\tilde{g}o_2t_1\omega_2} - p^{\mathrm{L}}_{\ell_1o_2t_1\omega_2} - p^{\mathrm{L}}_{\ell_2o_2t_1\omega_2} = 0 \\
p^{\mathrm{C}}_{\tilde{c}o_2t_1\omega_2} + p^{\mathrm{L}}_{\ell_1o_2t_1\omega_2} + p^{\mathrm{L}}_{\ell_2o_2t_1\omega_2} = 402 - p^{\mathrm{LS}}_{\tilde{d}o_1t_1\omega_2} \\
p^{\mathrm{L}}_{\ell_1o_2t_1\omega_2} = 500\left(\theta_{n_1o_2t_1\omega_2} - \theta_{n_2o_2t_1\omega_2}\right) \\
-200 \le p^{\mathrm{L}}_{\ell_1o_2t_1\omega_2} \le 200 \\
-200x^{\mathrm{L}}_{\ell_2t_1\omega_2} \le p^{\mathrm{L}}_{\ell_2o_2t_1\omega_2} \le 200x^{\mathrm{L}}_{\ell_2t_1\omega_2} \\
-\left(1 - x^{\mathrm{L}}_{\ell_2t_1\omega_2}\right)M \le p^{\mathrm{L}}_{\ell_2o_2t_1\omega_2} - 500\left(\theta_{n_1o_2t_1\omega_2} - \theta_{n_2o_2t_1\omega_2}\right) \le \left(1 - x^{\mathrm{L}}_{\ell_2t_1\omega_2}\right)M \\
0 \le p^{\mathrm{E}}_{\tilde{c}o_2t_1\omega_2} \le 400 \\
0 \le p^{\mathrm{C}}_{\tilde{c}o_2t_1\omega_2} \le p^{\mathrm{C^{max}}}_{\tilde{c}t_1\omega_2} \\
0 \le p^{\mathrm{LS}}_{\tilde{d}o_2t_1\omega_2} \le 402 \\
-\pi \le \theta_{n_2o_2t_1\omega_2} \le \pi \\
\theta_{n_1o_2t_1\omega_2} = 0
\end{cases}
$$

$$
\begin{cases}
p^{\mathrm{E}}_{\tilde{g}o_1t_2\omega_2} - p^{\mathrm{L}}_{\ell_1o_1t_2\omega_2} - p^{\mathrm{L}}_{\ell_2o_1t_2\omega_2} = 0 \\
p^{\mathrm{C}}_{\tilde{c}o_1t_2\omega_2} + p^{\mathrm{L}}_{\ell_1o_1t_2\omega_2} + p^{\mathrm{L}}_{\ell_2o_1t_2\omega_2} = 284 - p^{\mathrm{LS}}_{\tilde{d}o_1t_2\omega_2} \\
p^{\mathrm{L}}_{\ell_1o_1t_2\omega_2} = 500\left(\theta_{n_1o_1t_2\omega_2} - \theta_{n_2o_1t_2\omega_2}\right) \\
-200 \le p^{\mathrm{L}}_{\ell_1o_1t_2\omega_2} \le 200 \\
-200\left(x^{\mathrm{L}}_{\ell_2t_1\omega_2} + x^{\mathrm{L}}_{\ell_2t_2\omega_2}\right) \le p^{\mathrm{L}}_{\ell_2o_1t_2\omega_1} \le 200\left(x^{\mathrm{L}}_{\ell_2t_1\omega_2} + x^{\mathrm{L}}_{\ell_2t_2\omega_2}\right) \\
-\left[1 - \left(x^{\mathrm{L}}_{\ell_2t_1\omega_2} + x^{\mathrm{L}}_{\ell_2t_2\omega_2}\right)\right]M \le p^{\mathrm{L}}_{\ell_2o_1t_2\omega_2} - 500\left(\theta_{n_1o_1t_2\omega_1} - \theta_{n_2o_1t_2\omega_2}\right) \\
p^{\mathrm{L}}_{\ell_2o_1t_2\omega_2} - 500\left(\theta_{n_1o_1t_2\omega_1} - \theta_{n_2o_1t_2\omega_2}\right) \le \left[1 - \left(x^{\mathrm{L}}_{\ell_2t_1\omega_2} + x^{\mathrm{L}}_{\ell_2t_2\omega_2}\right)\right]M \\
0 \le p^{\mathrm{E}}_{\tilde{c}o_1t_2\omega_2} \le 400 \\
0 \le p^{\mathrm{C}}_{\tilde{c}o_1t_2\omega_2} \le p^{\mathrm{C^{max}}}_{\tilde{c}t_1\omega_2} + p^{\mathrm{C^{max}}}_{\tilde{c}t_2\omega_2} \\
0 \le p^{\mathrm{LS}}_{\tilde{d}o_1t_2\omega_2} \le 284 \\
-\pi \le \theta_{n_2o_1t_2\omega_2} \le \pi \\
\theta_{n_1o_1t_2\omega_2} = 0
\end{cases}
$$

$$
\begin{cases}
p^{\mathrm{E}}_{\tilde{g}o_2t_2\omega_2} - p^{\mathrm{L}}_{\ell_1o_2t_2\omega_2} - p^{\mathrm{L}}_{\ell_2o_2t_2\omega_2} = 0 \\
p^{\mathrm{C}}_{\tilde{c}o_2t_2\omega_2} + p^{\mathrm{L}}_{\ell_1o_2t_2\omega_2} + p^{\mathrm{L}}_{\ell_2o_2t_2\omega_2} = 539 - p^{\mathrm{LS}}_{\tilde{d}o_2t_2\omega_2} \\
p^{\mathrm{L}}_{\ell_1o_2t_2\omega_2} = 500\left(\theta_{n_1o_2t_2\omega_2} - \theta_{n_2o_2t_2\omega_2}\right) \\
-200 \le p^{\mathrm{L}}_{\ell_1o_2t_2\omega_2} \le 200 \\
-200\left(x^{\mathrm{L}}_{\ell_2t_1\omega_2} + x^{\mathrm{L}}_{\ell_2t_2\omega_2}\right) \le p^{\mathrm{L}}_{\ell_2o_2t_2\omega_2} \le 200\left(x^{\mathrm{L}}_{\ell_2t_1\omega_2} + x^{\mathrm{L}}_{\ell_2t_2\omega_2}\right) \\
-\left[1 - \left(x^{\mathrm{L}}_{\ell_2t_1\omega_2} + x^{\mathrm{L}}_{\ell_2t_2\omega_2}\right)\right]M \le p^{\mathrm{L}}_{\ell_2o_2t_2\omega_2} - 500\left(\theta_{n_1o_2t_2\omega_2} - \theta_{n_2o_2t_2\omega_2}\right) \\
p^{\mathrm{L}}_{\ell_2o_2t_2\omega_2} - 500\left(\theta_{n_1o_2t_2\omega_2} - \theta_{n_2o_2t_2\omega_2}\right) \le \left[1 - \left(x^{\mathrm{L}}_{\ell_2t_1\omega_2} + x^{\mathrm{L}}_{\ell_2t_2\omega_2}\right)\right]M \\
0 \le p^{\mathrm{E}}_{\tilde{c}o_2t_2\omega_2} \le 400 \\
0 \le p^{\mathrm{C}}_{\tilde{c}o_2t_2\omega_2} \le p^{\mathrm{C^{max}}}_{\tilde{c}t_1\omega_2} + p^{\mathrm{C^{max}}}_{\tilde{c}t_2\omega_2} \\
0 \le p^{\mathrm{LS}}_{\tilde{d}o_2t_2\omega_2} \le 407 \\
-\pi \le \theta_{n_2o_2t_2\omega_2} \le \pi \\
\theta_{n_1o_2t_2\omega_2} = 0
\end{cases}
$$

$$\left\{\begin{array}{l}
p^{\mathrm{E}}_{\tilde{g}o_1t_1\omega_3} - p^{\mathrm{L}}_{\ell_1o_1t_1\omega_3} - p^{\mathrm{L}}_{\ell_2o_1t_1\omega_3} = 0 \\
p^{\mathrm{C}}_{\tilde{c}o_1t_1\omega_3} + p^{\mathrm{L}}_{\ell_1o_1t_1\omega_3} + p^{\mathrm{L}}_{\ell_2o_1t_1\omega_3} = 281 - p^{\mathrm{LS}}_{\tilde{d}o_1t_1\omega_3} \\
p^{\mathrm{L}}_{\ell_1o_1t_1\omega_3} = 500\left(\theta_{n_1o_1t_1\omega_3} - \theta_{n_2o_1t_1\omega_3}\right) \\
-200 \le p^{\mathrm{L}}_{\ell_1o_1t_1\omega_3} \le 200 \\
-200x^{\mathrm{L}}_{\ell_2t_1\omega_3} \le p^{\mathrm{L}}_{\ell_2o_1t_1\omega_3} \le 200x^{\mathrm{L}}_{\ell_2t_1\omega_3} \\
-\left(1 - x^{\mathrm{L}}_{\ell_2t_1\omega_3}\right)M \le p^{\mathrm{L}}_{\ell_2o_1t_1\omega_3} - 500\left(\theta_{n_1o_1t_1\omega_3} - \theta_{n_2o_1t_1\omega_3}\right) \le \left(1 - x^{\mathrm{L}}_{\ell_2t_1\omega_3}\right)M \\
0 \le p^{\mathrm{E}}_{\tilde{c}o_1t_1\omega_3} \le 400 \\
0 \le p^{\mathrm{C}}_{\tilde{c}o_1t_1\omega_3} \le p^{\mathrm{C^{max}}}_{\tilde{c}t_1\omega_3} \\
0 \le p^{\mathrm{LS}}_{\tilde{d}o_1t_1\omega_3} \le 281 \\
-\pi \le \theta_{n_2o_1t_1\omega_3} \le \pi \\
\theta_{n_1o_1t_1\omega_3} = 0
\end{array}\right.$$

$$\left\{\begin{array}{l}
p^{\mathrm{E}}_{\tilde{g}o_2t_1\omega_3} - p^{\mathrm{L}}_{\ell_1o_2t_1\omega_3} - p^{\mathrm{L}}_{\ell_2o_2t_1\omega_3} = 0 \\
p^{\mathrm{C}}_{\tilde{c}o_2t_1\omega_3} + p^{\mathrm{L}}_{\ell_1o_2t_1\omega_3} + p^{\mathrm{L}}_{\ell_2o_2t_1\omega_3} = 533 - p^{\mathrm{LS}}_{\tilde{d}o_2t_1\omega_3} \\
p^{\mathrm{L}}_{\ell_1o_2t_1\omega_3} = 500\left(\theta_{n_1o_2t_1\omega_3} - \theta_{n_2o_2t_1\omega_3}\right) \\
-200 \le p^{\mathrm{L}}_{\ell_1o_2t_1\omega_3} \le 200 \\
-200x^{\mathrm{L}}_{\ell_2t_1\omega_3} \le p^{\mathrm{L}}_{\ell_2o_2t_1\omega_3} \le 200x^{\mathrm{L}}_{\ell_2t_1\omega_3} \\
-\left(1 - x^{\mathrm{L}}_{\ell_2t_1\omega_3}\right)M \le p^{\mathrm{L}}_{\ell_2o_2t_1\omega_3} - 500\left(\theta_{n_1o_2t_1\omega_3} - \theta_{n_2o_2t_1\omega_3}\right) \le \left(1 - x^{\mathrm{L}}_{\ell_2t_1\omega_3}\right)M \\
0 \le p^{\mathrm{E}}_{\tilde{c}o_2t_1\omega_3} \le 400 \\
0 \le p^{\mathrm{C}}_{\tilde{c}o_2t_1\omega_3} \le p^{\mathrm{C^{max}}}_{\tilde{c}t_1\omega_3} \\
0 \le p^{\mathrm{LS}}_{\tilde{d}o_2t_1\omega_1} \le 533 \\
-\pi \le \theta_{n_2o_2t_1\omega_3} \le \pi \\
\theta_{n_1o_2t_1\omega_3} = 0
\end{array}\right.$$

$$\left\{\begin{array}{l}
p^{\mathrm{E}}_{\tilde{g}o_1t_2\omega_3} - p^{\mathrm{L}}_{\ell_1o_1t_2\omega_3} - p^{\mathrm{L}}_{\ell_2o_1t_2\omega_3} = 0\\
p^{\mathrm{C}}_{\tilde{c}o_1t_2\omega_3} + p^{\mathrm{L}}_{\ell_1o_1t_2\omega_3} + p^{\mathrm{L}}_{\ell_2o_1t_2\omega_3} = 284 - p^{\mathrm{LS}}_{\tilde{d}o_1t_2\omega_3}\\
p^{\mathrm{L}}_{\ell_1o_1t_2\omega_3} = 500\left(\theta_{n_1o_1t_2\omega_3} - \theta_{n_2o_1t_2\omega_3}\right)\\
-200 \le p^{\mathrm{L}}_{\ell_1o_1t_2\omega_3} \le 200\\
-200\left(x^{\mathrm{L}}_{\ell_2t_1\omega_3} + x^{\mathrm{L}}_{\ell_2t_2\omega_3}\right) \le p^{\mathrm{L}}_{\ell_2o_1t_2\omega_3} \le 200\left(x^{\mathrm{L}}_{\ell_2t_1\omega_3} + x^{\mathrm{L}}_{\ell_2t_2\omega_3}\right)\\
-\left[1 - \left(x^{\mathrm{L}}_{\ell_2t_1\omega_3} + x^{\mathrm{L}}_{\ell_2t_2\omega_3}\right)\right]M \le p^{\mathrm{L}}_{\ell_2o_1t_2\omega_3} - 500\left(\theta_{n_1o_1t_2\omega_3} - \theta_{n_2o_1t_2\omega_3}\right)\\
p^{\mathrm{L}}_{\ell_2o_1t_2\omega_3} - 500\left(\theta_{n_1o_1t_2\omega_3} - \theta_{n_2o_1t_2\omega_3}\right) \le \left[1 - \left(x^{\mathrm{L}}_{\ell_2t_1\omega_3} + x^{\mathrm{L}}_{\ell_2t_2\omega_3}\right)\right]M\\
0 \le p^{\mathrm{E}}_{\tilde{c}o_1t_2\omega_3} \le 400\\
0 \le p^{\mathrm{C}}_{\tilde{c}o_1t_2\omega_3} \le p^{\mathrm{C^{max}}}_{\tilde{c}t_1\omega_3} + p^{\mathrm{C^{max}}}_{\tilde{c}t_2\omega_3}\\
0 \le p^{\mathrm{LS}}_{\tilde{d}o_1t_2\omega_3} \le 214\\
-\pi \le \theta_{n_2o_1t_2\omega_3} \le \pi\\
\theta_{n_1o_1t_2\omega_3} = 0
\end{array}\right.$$

$$\left\{\begin{array}{l}
p^{\mathrm{E}}_{\tilde{g}o_2t_2\omega_3} - p^{\mathrm{L}}_{\ell_1o_2t_2\omega_3} - p^{\mathrm{L}}_{\ell_2o_2t_2\omega_3} = 0\\
p^{\mathrm{C}}_{\tilde{c}o_2t_2\omega_3} + p^{\mathrm{L}}_{\ell_1o_2t_2\omega_3} + p^{\mathrm{L}}_{\ell_2o_2t_2\omega_3} = 539 - p^{\mathrm{LS}}_{\tilde{d}o_2t_2\omega_3}\\
p^{\mathrm{L}}_{\ell_1o_2t_2\omega_3} = 500\left(\theta_{n_1o_2t_2\omega_3} - \theta_{n_2o_2t_2\omega_3}\right)\\
-200 \le p^{\mathrm{L}}_{\ell_1o_2t_2\omega_3} \le 200\\
-200\left(x^{\mathrm{L}}_{\ell_2t_1\omega_3} + x^{\mathrm{L}}_{\ell_2t_2\omega_3}\right) \le p^{\mathrm{L}}_{\ell_2o_2t_2\omega_3} \le 200\left(x^{\mathrm{L}}_{\ell_2t_1\omega_3} + x^{\mathrm{L}}_{\ell_2t_2\omega_3}\right)\\
-\left[1 - \left(x^{\mathrm{L}}_{\ell_2t_1\omega_3} + x^{\mathrm{L}}_{\ell_2t_2\omega_3}\right)\right]M \le p^{\mathrm{L}}_{\ell_2o_2t_2\omega_3} - 500\left(\theta_{n_1o_2t_2\omega_3} - \theta_{n_2o_2t_2\omega_3}\right)\\
p^{\mathrm{L}}_{\ell_2o_2t_2\omega_3} - 500\left(\theta_{n_1o_2t_2\omega_3} - \theta_{n_2o_2t_2\omega_3}\right) \le \left[1 - \left(x^{\mathrm{L}}_{\ell_2t_1\omega_3} + x^{\mathrm{L}}_{\ell_2t_2\omega_3}\right)\right]M\\
0 \le p^{\mathrm{E}}_{\tilde{c}o_2t_2\omega_3} \le 400\\
0 \le p^{\mathrm{C}}_{\tilde{c}o_2t_2\omega_3} \le p^{\mathrm{C^{max}}}_{\tilde{c}t_1\omega_3} + p^{\mathrm{C^{max}}}_{\tilde{c}t_2\omega_3}\\
0 \le p^{\mathrm{LS}}_{\tilde{d}o_2t_2\omega_3} \le 539\\
-\pi \le \theta_{n_2o_2t_2\omega_3} \le \pi\\
\theta_{n_1o_2t_2\omega_3} = 0
\end{array}\right.$$

$$\begin{cases} p^{\mathrm{E}}_{\tilde{g}o_1t_1\omega_4} - p^{\mathrm{L}}_{\ell_1o_1t_1\omega_4} - p^{\mathrm{L}}_{\ell_2o_1t_1\omega_4} = 0 \\ p^{\mathrm{C}}_{\tilde{c}o_1t_1\omega_4} + p^{\mathrm{L}}_{\ell_1o_1t_1\omega_4} + p^{\mathrm{L}}_{\ell_2o_1t_1\omega_4} = 281 - p^{\mathrm{LS}}_{\tilde{d}o_1t_1\omega_4} \\ p^{\mathrm{L}}_{\ell_1o_1t_1\omega_4} = 500\left(\theta_{n_1o_1t_1\omega_4} - \theta_{n_2o_1t_1\omega_4}\right) \\ -200 \le p^{\mathrm{L}}_{\ell_1o_1t_1\omega_4} \le 200 \\ -200x^{\mathrm{L}}_{\ell_2t_1\omega_4} \le p^{\mathrm{L}}_{\ell_2o_1t_1\omega_4} \le 200x^{\mathrm{L}}_{\ell_2t_1\omega_4} \\ -\left(1 - x^{\mathrm{L}}_{\ell_2t_1\omega_4}\right)M \le p^{\mathrm{L}}_{\ell_2o_1t_1\omega_4} - 500\left(\theta_{n_1o_1t_1\omega_4} - \theta_{n_2o_1t_1\omega_4}\right) \le \left(1 - x^{\mathrm{L}}_{\ell_2t_1\omega_4}\right)M \\ 0 \le p^{\mathrm{E}}_{\tilde{c}o_1t_1\omega_4} \le 400 \\ 0 \le p^{\mathrm{C}}_{\tilde{c}o_1t_1\omega_4} \le p^{\mathrm{C}^{\max}}_{\tilde{c}t_1\omega_4} \\ 0 \le p^{\mathrm{LS}}_{\tilde{d}o_1t_1\omega_4} \le 281 \\ -\pi \le \theta_{n_2o_1t_1\omega_4} \le \pi \\ \theta_{n_1o_1t_1\omega_4} = 0 \end{cases}$$

$$\begin{cases} p^{\mathrm{E}}_{\tilde{g}o_2t_1\omega_4} - p^{\mathrm{L}}_{\ell_1o_2t_1\omega_4} - p^{\mathrm{L}}_{\ell_2o_2t_1\omega_4} = 0 \\ p^{\mathrm{C}}_{\tilde{c}o_2t_1\omega_4} + p^{\mathrm{L}}_{\ell_1o_2t_1\omega_4} + p^{\mathrm{L}}_{\ell_2o_2t_1\omega_4} = 533 - p^{\mathrm{LS}}_{\tilde{d}o_2t_1\omega_4} \\ p^{\mathrm{L}}_{\ell_1o_2t_1\omega_4} = 500\left(\theta_{n_1o_2t_1\omega_4} - \theta_{n_2o_2t_1\omega_4}\right) \\ -200 \le p^{\mathrm{L}}_{\ell_1o_2t_1\omega_4} \le 200 \\ -200x^{\mathrm{L}}_{\ell_2t_1\omega_4} \le p^{\mathrm{L}}_{\ell_2o_2t_1\omega_4} \le 200x^{\mathrm{L}}_{\ell_2t_1\omega_4} \\ -\left(1 - x^{\mathrm{L}}_{\ell_2t_1\omega_4}\right)M \le p^{\mathrm{L}}_{\ell_2o_2t_1\omega_4} - 500\left(\theta_{n_1o_2t_1\omega_4} - \theta_{n_2o_2t_1\omega_4}\right) \le \left(1 - x^{\mathrm{L}}_{\ell_2t_1\omega_4}\right)M \\ 0 \le p^{\mathrm{E}}_{\tilde{c}o_2t_1\omega_4} \le 400 \\ 0 \le p^{\mathrm{C}}_{\tilde{c}o_2t_1\omega_4} \le p^{\mathrm{C}^{\max}}_{\tilde{c}t_1\omega_4} \\ 0 \le p^{\mathrm{LS}}_{\tilde{d}o_2t_1\omega_4} \le 533 \\ -\pi \le \theta_{n_2o_2t_1\omega_4} \le \pi \\ \theta_{n_1o_2t_1\omega_4} = 0 \end{cases}$$

$$\begin{cases} p^{\mathrm{E}}_{\tilde{g}o_1t_2\omega_4} - p^{\mathrm{L}}_{\ell_1o_1t_2\omega_4} - p^{\mathrm{L}}_{\ell_2o_1t_2\omega_4} = 0 \\ p^{\mathrm{C}}_{\tilde{c}o_1t_2\omega_4} + p^{\mathrm{L}}_{\ell_1o_1t_2\omega_4} + p^{\mathrm{L}}_{\ell_2o_1t_2\omega_4} = 377 - p^{\mathrm{LS}}_{\tilde{d}o_1t_2\omega_4} \\ p^{\mathrm{L}}_{\ell_1o_1t_2\omega_4} = 500\left(\theta_{n_1o_1t_2\omega_4} - \theta_{n_2o_1t_2\omega_4}\right) \\ -200 \le p^{\mathrm{L}}_{\ell_1o_1t_2\omega_4} \le 200 \\ -200\left(x^{\mathrm{L}}_{\ell_2t_1\omega_4} + x^{\mathrm{L}}_{\ell_2t_2\omega_4}\right) \le p^{\mathrm{L}}_{\ell_2o_1t_2\omega_4} \le 200\left(x^{\mathrm{L}}_{\ell_2t_1\omega_4} + x^{\mathrm{L}}_{\ell_2t_2\omega_4}\right) \\ -\left[1 - \left(x^{\mathrm{L}}_{\ell_2t_1\omega_4} + x^{\mathrm{L}}_{\ell_2t_2\omega_4}\right)\right]M \le p^{\mathrm{L}}_{\ell_2o_1t_2\omega_4} - 500\left(\theta_{n_1o_1t_2\omega_4} - \theta_{n_2o_1t_2\omega_4}\right) \\ p^{\mathrm{L}}_{\ell_2o_1t_2\omega_4} - 500\left(\theta_{n_1o_1t_2\omega_4} - \theta_{n_2o_1t_2\omega_4}\right) \le \left[1 - \left(x^{\mathrm{L}}_{\ell_2t_1\omega_4} + x^{\mathrm{L}}_{\ell_2t_2\omega_4}\right)\right]M \\ 0 \le p^{\mathrm{E}}_{\tilde{c}o_1t_2\omega_4} \le 400 \\ 0 \le p^{\mathrm{C}}_{\tilde{c}o_1t_2\omega_4} \le p^{\mathrm{C^{max}}}_{\tilde{c}t_1\omega_4} + p^{\mathrm{C^{max}}}_{\tilde{c}t_2\omega_4} \\ 0 \le p^{\mathrm{LS}}_{\tilde{d}o_1t_2\omega_4} \le 377 \\ -\pi \le \theta_{n_2o_1t_2\omega_4} \le \pi \\ \theta_{n_1o_1t_2\omega_4} = 0 \end{cases}$$

$$\begin{cases} p^{\mathrm{E}}_{\tilde{g}o_2t_2\omega_4} - p^{\mathrm{L}}_{\ell_1o_2t_2\omega_4} - p^{\mathrm{L}}_{\ell_2o_2t_2\omega_4} = 0 \\ p^{\mathrm{C}}_{\tilde{c}o_2t_2\omega_4} + p^{\mathrm{L}}_{\ell_1o_2t_2\omega_4} + p^{\mathrm{L}}_{\ell_2o_2t_2\omega_4} = 715 p^{\mathrm{LS}}_{\tilde{d}o_2t_2\omega_4} \\ p^{\mathrm{L}}_{\ell_1o_2t_2\omega_4} = 500\left(\theta_{n_1o_2t_2\omega_4} - \theta_{n_2o_2t_2\omega_4}\right) \\ -200 \le p^{\mathrm{L}}_{\ell_1o_2t_2\omega_4} \le 200 \\ -200\left(x^{\mathrm{L}}_{\ell_2t_1\omega_4} + x^{\mathrm{L}}_{\ell_2t_2\omega_4}\right) \le p^{\mathrm{L}}_{\ell_2o_2t_2\omega_4} \le 200\left(x^{\mathrm{L}}_{\ell_2t_1\omega_4} + x^{\mathrm{L}}_{\ell_2t_2\omega_4}\right) \\ -\left[1 - \left(x^{\mathrm{L}}_{\ell_2t_1\omega_4} + x^{\mathrm{L}}_{\ell_2t_2\omega_4}\right)\right]M \le p^{\mathrm{L}}_{\ell_2o_2t_2\omega_4} - 500\left(\theta_{n_1o_2t_2\omega_4} - \theta_{n_2o_2t_2\omega_4}\right) \\ p^{\mathrm{L}}_{\ell_2o_2t_2\omega_4} - 500\left(\theta_{n_1o_2t_2\omega_4} - \theta_{n_2o_2t_2\omega_4}\right) \le \left[1 - \left(x^{\mathrm{L}}_{\ell_2t_1\omega_4} + x^{\mathrm{L}}_{\ell_2t_2\omega_4}\right)\right]M \\ 0 \le p^{\mathrm{E}}_{\tilde{c}o_2t_2\omega_4} \le 400 \\ 0 \le p^{\mathrm{C}}_{\tilde{c}o_2t_2\omega_4} \le p^{\mathrm{C^{max}}}_{\tilde{c}t_1\omega_4} + p^{\mathrm{C^{max}}}_{\tilde{c}t_2\omega_4} \\ 0 \le p^{\mathrm{LS}}_{\tilde{d}o_2t_2\omega_4} \le 715 \\ -\pi \le \theta_{n_2o_2t_2\omega_4} \le \pi \\ \theta_{n_1o_2t_2\omega_4} = 0, \end{cases}$$

where $\tilde{\Delta}^{\mathrm{SD}} = \Big\{p^{\mathrm{E}}_{\tilde{g}o_1\omega_1}, p^{\mathrm{E}}_{\tilde{g}o_2\omega_1}, p^{\mathrm{C}}_{\tilde{c}o_1\omega_1}, p^{\mathrm{C}}_{\tilde{c}o_2\omega_1}, p^{\mathrm{LS}}_{\tilde{d}o_1\omega_1}, p^{\mathrm{LS}}_{\tilde{d}o_2\omega_1}, p^{\mathrm{L}}_{\ell_1o_1\omega_1}, p^{\mathrm{L}}_{\ell_1o_2\omega_1}, p^{\mathrm{L}}_{\ell_2o_1\omega_1},$ $p^{\mathrm{L}}_{\ell_2o_2\omega_1}, \theta_{n_1o_1\omega_1}, \theta_{n_1o_2\omega_1}, \theta_{n_2o_1\omega_1}, \theta_{n_2o_2\omega_1}, p^{\mathrm{E}}_{\tilde{g}o_1\omega_2}, p^{\mathrm{E}}_{\tilde{g}o_2\omega_2}, p^{\mathrm{C}}_{\tilde{c}o_1\omega_2}, p^{\mathrm{C}}_{\tilde{c}o_2\omega_2}, p^{\mathrm{LS}}_{\tilde{d}o_1\omega_2}, p^{\mathrm{LS}}_{\tilde{d}o_2\omega_2},$ $p^{\mathrm{L}}_{\ell_1o_1\omega_2}, p^{\mathrm{L}}_{\ell_1o_2\omega_2}, p^{\mathrm{L}}_{\ell_2o_1\omega_2}, p^{\mathrm{L}}_{\ell_2o_2\omega_2}, \theta_{n_1o_1\omega_2}, \theta_{n_1o_2\omega_2}, \theta_{n_2o_1\omega_2}, \theta_{n_2o_2\omega_2}, p^{\mathrm{E}}_{\tilde{g}o_1\omega_3}, p^{\mathrm{E}}_{\tilde{g}o_2\omega_3}, p^{\mathrm{C}}_{\tilde{c}o_1\omega_3},$ $p^{\mathrm{C}}_{\tilde{c}o_2\omega_3}, p^{\mathrm{LS}}_{\tilde{d}o_1\omega_3}, p^{\mathrm{LS}}_{\tilde{d}o_2\omega_3}, p^{\mathrm{L}}_{\ell_1o_1\omega_3}, p^{\mathrm{L}}_{\ell_1o_2\omega_3}, p^{\mathrm{L}}_{\ell_2o_1\omega_3}, p^{\mathrm{L}}_{\ell_2o_2\omega_3}, \theta_{n_1o_1\omega_3}, \theta_{n_1o_2\omega_3}, \theta_{n_2o_1\omega_3}, \theta_{n_2o_2\omega_3},$ $p^{\mathrm{E}}_{\tilde{g}o_1\omega_4}, p^{\mathrm{E}}_{\tilde{g}o_2\omega_4}, p^{\mathrm{C}}_{\tilde{c}o_1\omega_4}, p^{\mathrm{C}}_{\tilde{c}o_2\omega_4}, p^{\mathrm{LS}}_{\tilde{d}o_1\omega_4}, p^{\mathrm{LS}}_{\tilde{d}o_2\omega_4}, p^{\mathrm{L}}_{\ell_1o_1\omega_4}, p^{\mathrm{L}}_{\ell_1o_2\omega_4}, p^{\mathrm{L}}_{\ell_2o_1\omega_4}, p^{\mathrm{L}}_{\ell_2o_2\omega_4}, \theta_{n_1o_1\omega_4},$

$\theta_{n_1 o_2 \omega_4}, \theta_{n_2 o_1 \omega_4}, \theta_{n_2 o_2 \omega_4}, p^{\mathrm{C}^{\max}}_{\tilde{c} t_1 \omega_1}, p^{\mathrm{C}^{\max}}_{\tilde{c} t_2 \omega_1}, p^{\mathrm{C}^{\max}}_{\tilde{c} t_1 \omega_2}, p^{\mathrm{C}^{\max}}_{\tilde{c} t_2 \omega_2}, p^{\mathrm{C}^{\max}}_{\tilde{c} t_1 \omega_3}, p^{\mathrm{C}^{\max}}_{\tilde{c} t_2 \omega_3}, p^{\mathrm{C}^{\max}}_{\tilde{c} t_1 \omega_4}, p^{\mathrm{C}^{\max}}_{\tilde{c} t_2 \omega_4}, x^{\mathrm{L}}_{\ell_2 t_1 \omega_1}, x^{\mathrm{L}}_{\ell_2 t_2 \omega_1}, x^{\mathrm{L}}_{\ell_2 t_1 \omega_2}, x^{\mathrm{L}}_{\ell_2 t_2 \omega_2}, x^{\mathrm{L}}_{\ell_2 t_1 \omega_3}, x^{\mathrm{L}}_{\ell_2 t_2 \omega_3}, x^{\mathrm{L}}_{\ell_2 t_1 \omega_4}, x^{\mathrm{L}}_{\ell_2 t_2 \omega_4}\}$ and M is a large enough constant.

The optimal solution of the stochastic dynamic G&TEP problem formulated in this illustrative example consists in building transmission line ℓ_2 in the first time period. Regarding the expansion of the generation capacity, it is optimal to build 212 MW at the beginning of the planning horizon and then building an additional 2 MW if scenario a is realized in the first time period, i.e., for scenarios 1 and 2, and building an additional 88 MW if scenario b is realized in the first time period, i.e., for scenarios 3 and 4.

The expansion decisions at the beginning of the planning horizon do not depend on the scenario realization since this information is not available to the ISO. When the first period concludes, the decision maker knows whether scenario a or b has been realized, and thus it can make two different expansion plans at the beginning of the second time period depending on this scenario realization.

Note that although the scenario data of this illustrative example have been selected so that the average value of demands for each operating condition are the same as those considered in the previous illustrative examples, the optimal results are significantly different, which highlights the importance of an accurate modeling of the uncertain parameters in the G&TEP problem. □

The stochastic dynamic G&TEP problem provided in this section has some advantages over the previous models described in this chapter:

1. It allows us to model the uncertainty in the future conditions of the system under study. This generally results in more informed expansion decisions.
2. It is a more flexible model than a static one since we have the option of building new facilities at different points in time.
3. As a result of item 2 above, we can adapt to the conditions of the system.
4. The investment budget required at the beginning of the planning horizon is lower than that needed if the expansion decisions can be made only at a single point in time.

However, in real-world world applications it is important to note the following aspects:

1. It is necessary to include additional sources of uncertainty through additional scenarios.
2. A large enough number of scenarios is needed to represent the uncertain parameters accurately.
3. It is important to consider more than two time periods in the decision-making problem.
4. Systems have thousands of nodes and transmission lines.

The above facts make the stochastic dynamic G&TEP problem a complex problem in the real world. Therefore, it is generally necessary to achieve a tradeoff between modeling accuracy and computation tractability. On the other hand, it is generally useful to use mathematical tools such as scenario reduction methods [11, 12] or decomposition techniques [2, 10] to deal with the computational burden of the G&TEP problem.

4.6 Stochastic Dynamic Risk-Constrained G&TEP

Making generation and transmission expansion decisions generally requires investing a large amount of money. Moreover, as explained in the previous section, these expansion decisions are usually made under uncertainty. This means that making expansion decisions may be risky if uncertainty is not adequately accounted for. Even if uncertainty is well represented in the decision-making problem, the expansion plan may be risky if certain of the scenarios are realized. The example below illustrates the concept of risk in the G&TEP decision-making problem.

Illustrative Example 4.7 *Risk modeling in the G&TEP problem*

The ISO is considering two possible generation and transmission expansion plans to be carried out, namely *expansion plan A* and *expansion plan B*. Uncertainty in the system conditions in the considered planning horizon is modeled using five scenarios each with the same probability (0.2). The ISO computes the cost (operation and expansion costs) of each expansion plan for each of the five scenarios. These costs are provided in Table 4.2.

With the cost data of Table 4.2, the ISO computes the expected costs of expansions plans A and B:

$$\mathbb{E}\{C_{\text{A}}\} = 0.2 \times 10 + 0.2 \times 11 + 0.2 \times 14 + 0.2 \times 15 + 0.2 \times 50 = \$20 \text{ million}$$
$$\mathbb{E}\{C_{\text{B}}\} = 0.2 \times 19 + 0.2 \times 20 + 0.2 \times 21 + 0.2 \times 22 + 0.2 \times 23 = \$21 \text{ million}$$

Following the criterion considered in the previous sections of this chapter, i.e., selecting the expansion plan that minimizes the expected cost, this would result in the ISO selecting expansion plan A. However, if we observe the costs provided in Table 4.2, then we note that the cost of expansion plan A for one of the scenarios is very high (in particular, for scenario 5). Since the expansion plans are made before the actual scenario realization is known, if the ISO chooses expansion plan A and scenario 5 is realized, then the ISO would have to face a very high cost, which it may not be willing to assume.

Table 4.2 Illustrative Example 4.7: costs for expansion plans A and B [M$]

Expansion plan\Scenario	1	2	3	4	5
A	10	11	14	15	50
B	19	20	21	22	23

On the other hand, the expected cost of expansion plan B is \$21 million. However, the largest cost is only \$23 million (for scenario 5) in this case, i.e., less than half of the largest cost for expansion plan A. This means that if expansion plan B is carried out instead of expansion plan A, then the cost will likely be higher (the costs for expansion plan B are higher than those for expansion plan A in four of the five scenarios). Nevertheless, the ISO also knows that if it implements expansion plan B, then the cost will be \$23 million in the worst case (and not \$50 million, as in the case of choosing expansion plan A).

In other words, using expansion plan A is a risky decision since the actual cost will be very high with probability 20 %. On the other hand, expansion plan B is a more conservative solution since its expected cost is higher than that of expansion plan A but with a much lower highest possible cost. □

The previous illustrative example highlights the importance of considering not only the expected cost in the G&TEP decision-making problem but also the cost risk associated with the expansion plans. This is analyzed below.

In order to model the cost risk associated with the generation and transmission expansion plans, we consider the conditional value-at-risk (CVaR). This risk metric is incorporated in the stochastic dynamic G&TEP problem (4.5) described in Sect. 4.5.2. Further details on the CVaR are provided in [14, 15] and in Appendix D of this book.

4.6.1 Formulation

The G&TEP problem considering a stochastic dynamic risk-constrained approach can be formulated using the MILP model below:

$$\begin{aligned} \min_{\Delta^{\text{Risk}}} \quad & \sum_{\omega} \varphi_{\omega} \left\{ \sum_{t} \left[\sum_{o} \rho_{o} \left(\sum_{g} C_{g}^{\text{E}} p_{got\omega}^{\text{E}} + \sum_{c} C_{c}^{\text{C}} p_{cot\omega}^{\text{C}} + \sum_{d} C_{d}^{\text{LS}} p_{dot\omega}^{\text{LS}} \right) \right. \right. \\ & \left. \left. + a_{t} \left(\sum_{c} I_{c}^{\text{C}} p_{ct\omega}^{\text{C}^{\max}} + \sum_{\ell \in \Omega^{\text{L}+}} I_{\ell}^{\text{L}} x_{\ell t\omega}^{\text{L}} \right) \right] \right\} \\ & - \eta \left(\vartheta - \frac{1}{1-\alpha} \sum_{\omega} \varphi_{\omega} \varsigma_{\omega} \right) \end{aligned} \quad (4.6a)$$

subject to

$$0 \le \sum_{t} p_{ct\omega}^{\text{C}^{\max}} \le \overline{P}_{c}^{\text{C}^{\max}} \quad \forall c, \omega \quad (4.6b)$$

$$x_{\ell t\omega}^{\text{L}} \in \{0, 1\} \quad \forall \ell \in \Omega^{\text{L}+}, t, \omega \quad (4.6c)$$

$$\sum_{t} x^{\mathrm{L}}_{\ell t\omega} = 1 \quad \forall \ell \in \Omega^{\mathrm{L+}}, \omega \tag{4.6d}$$

$$\sum_{c} I^{\mathrm{C}}_{c} p^{\mathrm{C^{max}}}_{c} \leq I^{\mathrm{C,max}}_{t} \quad \forall t \tag{4.6e}$$

$$\sum_{\ell \in \Omega^{\mathrm{L+}}} I^{\mathrm{L}}_{\ell} x^{\mathrm{L}}_{\ell t\omega} \leq I^{\mathrm{L,max}}_{t} \quad \forall t \tag{4.6f}$$

$$p^{\mathrm{C^{max}}}_{ct\omega} = p^{\mathrm{C^{max}}}_{ct\tilde{\omega}} \quad \forall c, t, \omega, \tilde{\omega} | P^{\mathrm{D^{max}}}_{do\tau\omega} = P^{\mathrm{D^{max}}}_{do\tau\tilde{\omega}} \quad \forall d, o, \tau < t \tag{4.6g}$$

$$x^{\mathrm{L}}_{\ell t\omega} = x^{\mathrm{L}}_{\ell t\tilde{\omega}} \quad \forall \ell, t, \omega, \tilde{\omega} | P^{\mathrm{D^{max}}}_{do\tau\omega} = P^{\mathrm{D^{max}}}_{do\tau\tilde{\omega}} \quad \forall d, o, \tau < t \tag{4.6h}$$

$$\vartheta + \sum_{t} \left\{ \sum_{o} \rho_o \left[\sum_{g} C^{\mathrm{E}}_{g} p^{\mathrm{E}}_{got\omega} + \sum_{c} C^{\mathrm{C}}_{c} p^{\mathrm{C}}_{cot\omega} + \sum_{d} C^{\mathrm{LS}}_{d} p^{\mathrm{LS}}_{dot\omega} \right] \right.$$
$$\left. + a_t \left[\sum_{c} I^{\mathrm{C}}_{c} p^{\mathrm{C^{max}}}_{ct\omega} + \sum_{\ell \in \Omega^{\mathrm{L+}}} I^{\mathrm{L}}_{\ell} x^{\mathrm{L}}_{\ell t\omega} \right] \right\} \leq \varsigma_\omega \quad \forall \omega \tag{4.6i}$$

$$\varsigma_\omega \geq 0 \quad \forall \omega \tag{4.6j}$$

$$\left\{ \sum_{g \in \Omega^{\mathrm{G}}_{n}} p^{\mathrm{E}}_{got\omega} + \sum_{c \in \Omega^{\mathrm{C}}_{n}} p^{\mathrm{C}}_{cot\omega} - \sum_{\ell | s(\ell) = n} p^{\mathrm{L}}_{\ell ot\omega} \right.$$
$$+ \sum_{\ell | r(\ell) = n} p^{\mathrm{L}}_{\ell ot\omega} = \sum_{d \in \Omega^{\mathrm{D}}_{n}} \left(P^{\mathrm{D^{max}}}_{dot\omega} - p^{\mathrm{LS}}_{dot\omega} \right) \quad \forall n \tag{4.6k}$$

$$p^{\mathrm{L}}_{\ell ot\omega} = B_\ell \left(\theta_{s(\ell)ot\omega} - \theta_{r(\ell)ot\omega} \right) \quad \forall \ell \setminus \ell \in \Omega^{\mathrm{L+}} \tag{4.6l}$$

$$-\sum_{\tau \leq t} x^{\mathrm{L}}_{\ell\tau\omega} F^{\max}_{\ell} \leq p^{\mathrm{L}}_{\ell ot\omega} \leq \sum_{\tau \leq t} x^{\mathrm{L}}_{\ell\tau\omega} F^{\max}_{\ell} \quad \forall \ell \in \Omega^{\mathrm{L+}} \tag{4.6m}$$

$$-\left(1 - \sum_{\tau \leq t} x^{\mathrm{L}}_{\ell\tau\omega}\right) M \leq p^{\mathrm{L}}_{\ell ot\omega} - B_\ell \left(\theta_{s(\ell)ot\omega} - \theta_{r(\ell)ot\omega} \right) \leq \left(1 - \sum_{\tau \leq t} x^{\mathrm{L}}_{\ell\tau\omega}\right) M$$
$$\forall \ell \in \Omega^{\mathrm{L+}} \tag{4.6n}$$

$$-F^{\max}_{\ell} \leq p^{\mathrm{L}}_{\ell ot\omega} \leq F^{\max}_{\ell} \quad \forall \ell \tag{4.6o}$$

$$0 \leq p^{\mathrm{E}}_{got\omega} \leq P^{\mathrm{E^{max}}}_{g} \quad \forall g \tag{4.6p}$$

$$0 \leq p^{\mathrm{C}}_{cot\omega} \leq \sum_{\tau \leq t} p^{\mathrm{C^{max}}}_{c\tau\omega} \quad \forall c \tag{4.6q}$$

$$0 \leq p^{\mathrm{LS}}_{dot\omega} \leq P^{\mathrm{D^{max}}}_{dot\omega} \quad \forall d \tag{4.6r}$$

$$-\pi \leq \theta_{not\omega} \leq \pi \quad \forall n \tag{4.6s}$$

$$\theta_{not\omega} = 0 \quad n\text{: ref.} \tag{4.6t}$$

$$\Bigg\}, \forall o, t, \omega,$$

where the variables in set $\Delta^{\mathrm{Risk}} = \left\{ p^{\mathrm{E}}_{got\omega}, p^{\mathrm{LS}}_{dot\omega}, p^{\mathrm{C}}_{cot\omega}, p^{\mathrm{L}}_{\ell ot\omega}, \theta_{not\omega}, p^{\mathrm{C^{max}}}_{ct\omega}, x^{\mathrm{L}}_{\ell t\omega}, \vartheta, \varsigma_\omega \right\}$ are the optimization variables of problem (4.6).

The main differences between problem (4.6) and problem (4.5) are summarized below:

1. In the objective function (4.6a) we include an additional term (third line), which is the CVaR.
2. The CVaR is multiplied in the objective function (4.6a) by the weighting parameter η, which enables us to model the tradeoff between expected cost and CVaR, and therefore to represent different expansion strategies. If η is equal to zero, we minimize only the expected cost, i.e., we represent a risk-neutral expansion planner. This case is equivalent to problem (4.5), i.e., a problem where the expected cost is minimized but the cost risk is neglected. On the other hand, increasing values of η represent increasing risk-averse expansion strategies, i.e., strategies that consider the minimization of the expected profit but also the CVaR.
3. We include constraints (4.6i) and (4.6j), which are used to compute the CVaR.

Illustrative Example 4.8 *6-node system: Stochastic dynamic risk-constrained G&TEP problem*

The stochastic dynamic risk-constrained G&TEP problem (4.6) is applied to the six-node system depicted in Fig. 4.4. This system comprises six nodes, five generating units, four demands, and three transmission lines. It is possible to build up to six additional transmission lines and two candidate generating units.

The system is divided in two zones: region A (nodes 1–3) and region B (nodes 4–6), which are initially not interconnected. Node six is initially isolated, and, thus, demand at this node can be supplied only by generating unit g_5.

Table 4.3 provides data for the existing generating units. The second column identifies the node location, while the third and fourth columns provide the capacity and the production cost of each existing generating unit, respectively.

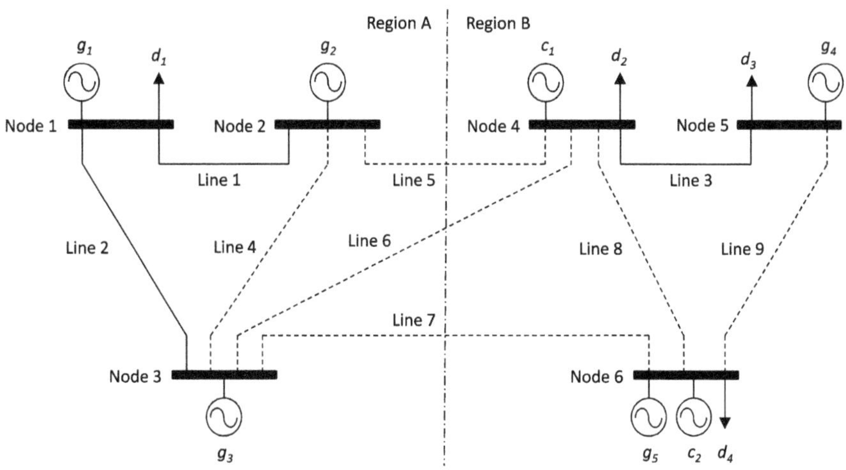

Fig. 4.4 Illustrative Example 4.8: six-node system

Table 4.3 Illustrative Example 4.8: data for existing generating units

Generating unit	Node	$P_g^{E^{max}}$ [MW]	C_g^E [$/MWh]
g_1	n_1	300	18
g_2	n_2	250	25
g_3	n_3	400	16
g_4	n_5	300	32
g_5	n_6	150	35

Table 4.4 Illustrative Example 4.8: data for candidate generating units

Generating unit	Node	$P_c^{C^{max}}$ [MW]	C_c^C [$/MWh]	I_c^C [$/MW]
c_1	n_4	300	10	700,000
c_2	n_6	250	15	600,000

Table 4.5 Illustrative Example 4.8: data for demands

Demand	Node	$P_{do_1}^{D^{max}}$ [MW]	$P_{do_2}^{D^{max}}$ [MW]	C_d^{LS} [$/MWh]
d_1	n_1	150	300	70
d_2	n_4	120	240	72
d_3	n_5	80	160	75
d_4	n_6	150	300	85

It is possible to build two additional generating units, whose data are provided in Table 4.4. The second column identifies the node location, while the third, fourth, and fifth columns provide the capacity, production cost, and investment cost of each candidate generating unit, respectively. The investment budget per time period is considered equal to $600 million.

Table 4.5 provides data for the demands. The second column identifies the node location, the third and fourth columns provide the demand values for operating conditions o_1 and o_2, respectively, and the fifth column gives the load-shedding cost. The weights of operating conditions o_1 and o_2 are 5000 and 3760 h, respectively. Note that these demand values are the expected values at the beginning of the planning horizon. The evolution throughout the planning horizon and the modeling of the uncertainty of these demand conditions are explained below.

Table 4.6 provides data for the existing transmission lines. The second and third columns identify the sending-end and receiving-end nodes, respectively, while the fourth and fifth columns provide the susceptance and capacity of each existing transmission line, respectively.

We consider that it is possible to build up to six additional transmission lines, whose data are provided in Table 4.7. The second and third columns identify the sending-end and receiving-end nodes, respectively, while the fourth and fifth columns provide the susceptance and capacity of each prospective transmission line,

Table 4.6 Illustrative Example 4.8: data for existing transmission lines

Line	From node	To node	B_ℓ [S]	$F_\ell^{\max}$ [MW]
ℓ_1	n_1	n_2	500	150
ℓ_2	n_1	n_3	500	150
ℓ_3	n_4	n_5	500	150

Table 4.7 Illustrative Example 4.8: data for prospective transmission lines

Line	From node	To node	B_ℓ [S]	$F_\ell^{\max}$ [MW]	I_ℓ^{L} [\$]
ℓ_4	n_2	n_3	500	150	700,000
ℓ_5	n_2	n_4	500	200	1,400,000
ℓ_6	n_3	n_4	500	200	1,800,000
ℓ_7	n_3	n_6	500	200	1,600,000
ℓ_8	n_4	n_6	500	150	800,000
ℓ_9	n_5	n_6	500	150	700,000

respectively. The sixth column gives the investment cost. The investment budget per time period is considered equal to \$30 million.

The reference node is node 1, and the base power and voltage are 1 MW and 1 kV, respectively.

The planning horizon comprises two 10-years time periods, so that the expansion plans can be made at the beginning of each time period, i.e., at the beginning of the first and eleventh time periods. The amortization rates are considered equal to 0.2 and 0.1 in the first and second time periods, respectively.

The uncertainties in the demand conditions are modeled through a set of scenarios as explained below. The demand values of the two operating conditions in the first time period can be 10 % lower than (realization LL1), 5 % lower than (realization L1), equal to (realization M1), 10 % higher than (realization H1), and 20 % higher than (realization HH1) the values provided in Table 4.5, with probabilities equal to 0.1, 0.2, 0.2, 0.3, and 0.2, respectively. On the other hand, the demand values of the two operating conditions in the second time period can be 10 % lower than (realization LL2), 5 % lower than (realization L2), equal to (realization M2), 10 % higher than (realization H2), and 20 % higher than (realization HH2) the values in the first time period, with probabilities equal to 0.05, 0.1, 0.15, 0.4, and 0.3, respectively. Figure 4.5 and Table 4.8 summarize these scenario data. For example, the demand values of scenario 1 (realizations LL1 and LL2) for operating condition o_1 are obtained as $150 \times 0.9 = 135$ MW and $150 \times 0.9 \times 0.9 = 121.5$ MW for the first and second time periods, respectively. On the other hand, the weight of scenario 1 is computed as $0.1 \times 0.05 = 0.005$.

The confidence level α for computing the CVaR is fixed to 0.95. Then problem (4.6) is solved for different values of the weighting parameter η. As previously

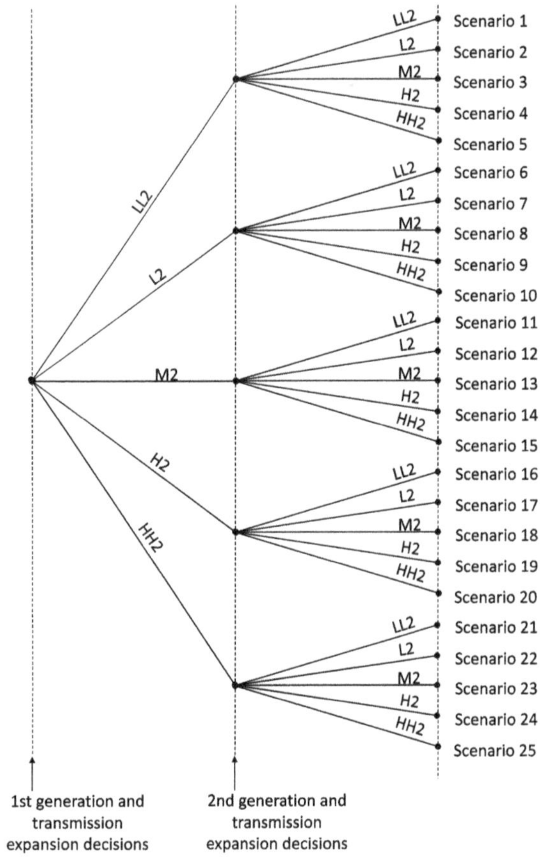

Fig. 4.5 Illustrative Example 4.8: scenario tree

explained, η allows us to model the tradeoff between expected costs and CVaR. As a result of varying parameter η, the transmission and expansion plans are different. Figure 4.6 depicts the so-called efficient frontier, which shows how the expected costs and the CVaR vary as we vary the value of parameter η.

A value of η equal to 0 represents a risk-neutral expansion strategy, i.e., the expansion strategy that minimizes the expected costs but neglects the risk associated with it. Therefore, as noted in Fig. 4.6, this strategy results in the lowest expected cost but in the largest CVaR, i.e., $\eta = 0$ is the riskiest strategy. On the other hand, increasing values of η represent increasing risk-averse expansion strategies that result in higher expected costs but also in a reduction of the CVaR, i.e., by considering increasing values of η we reduce the cost risk associated with the expansion decisions.

Table 4.9 provides the transmission and expansion plans for different values of the weighting parameter η. The second column indicates the scenarios, while the third/fourth and fifth/sixth columns provide the transmission/generation expansion decisions in the first and second time periods, respectively.

Regarding the results provided in Table 4.9, the observations below are in order:

Table 4.8 Illustrative Example 4.8: scenario data

	Scenario	First time period	Second time period	Weight
1	(LL1+LL2)	−10 %	−10 %	0.005
2	(LL1+L2)	−10 %	−5 %	0.010
3	(LL1+M2)	−10 %	+0 %	0.015
4	(LL1+H2)	−10 %	+10 %	0.040
5	(LL1+HH2)	−10 %	+20 %	0.030
6	(L1+LL2)	−5 %	−10 %	0.010
7	(L1+L2)	−5 %	−5 %	0.020
8	(L1+M2)	−5 %	+0 %	0.030
9	(L1+H2)	−5 %	+10 %	0.080
10	(L1+HH2)	−5 %	+20 %	0.060
11	(M1+LL2)	+0 %	−10 %	0.010
12	(M1+L2)	+0 %	−5 %	0.020
13	(M1+M2)	+0 %	+0 %	0.030
14	(M1+H2)	+0 %	+10 %	0.080
15	(M1+HH2)	+0 %	+20 %	0.060
16	(H1+LL2)	+10 %	−10 %	0.015
17	(H1+L2)	+10 %	−5 %	0.030
18	(H1+M2)	+10 %	+0 %	0.045
19	(H1+H2)	+10 %	+10 %	0.120
20	(H1+HH2)	+10 %	+20 %	0.090
21	(HH1+LL2)	+20 %	−10 %	0.010
22	(HH1+L2)	+20 %	−5 %	0.020
23	(HH1+M2)	+20 %	+0 %	0.030
24	(HH1+H2)	+20 %	+10 %	0.080
25	(HH1+HH2)	+20 %	+20 %	0.060

1. The expansion decisions at the beginning of the planning horizon do not depend on the scenario realizations.
2. There are five possible expansion decisions for the second time period, one for each scenario realization in the first one (LL1, L1, M1, H1, HH1).
3. A risk-neutral expansion strategy considers a comparatively lower expansion of the generation capacity at the beginning of the planning horizon. In this way, the expected cost throughout the planning horizon is minimized since making a larger investment in generation capacity may result in an overinvestment that is not needed if some of the scenarios are realized.
4. Increasing risk-averse expansion strategies consider comparatively larger expansions of the generation capacity at the beginning of the planning horizon. In this way, the variability of the operation cost throughout the planning horizon is reduced and also the CVaR.

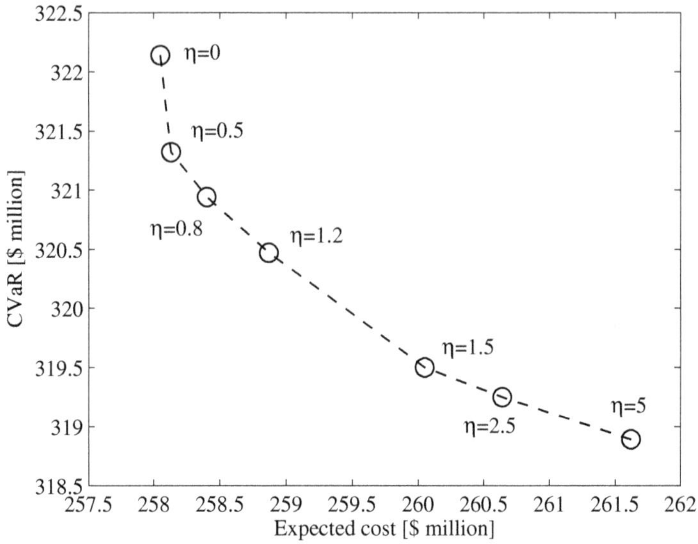

Fig. 4.6 Illustrative Example 4.8: efficient frontier

Table 4.9 Illustrative Example 4.8: generation and transmission expansion plans for different risk-aversion strategies[a]

η	Scenarios	First time period		Second time period	
		Lines built	Capacity built [MW]	Lines built	Capacity built [MW]
0	1–5	ℓ_6, ℓ_7, ℓ_8	300 (c_1), 40.8 (c_2)	–	–
	6–10			–	23.1 (c_2)
	11–15			–	46.2 (c_2)
	16–20			–	92.4 (c_2)
	21–15			ℓ_5	89.4 (c_2)
1.5	1–5	ℓ_6, ℓ_7	298 (c_1), 130 (c_2)	–	–
	6–10			–	–
	11–15			–	–
	16–20			–	2 (c_1), 33 (c_2)
	21–15			–	2 (c_1), 102 (c_2)
5	1–5	ℓ_6, ℓ_7	298 (c_1), 160 (c_2)	–	–
	6–10			–	–
	11–15			–	–
	16–20			–	2 (c_1), 3 (c_2)
	21–15			–	2 (c_1), 72 (c_2)

[a] ℓ_4: 2–3, ℓ_5: 2–4, ℓ_6: 3–4, ℓ_7: 3–6, ℓ_8: 4–6, ℓ_9: 5–6

5. Regarding the transmission expansion decisions, it is optimal to build transmission lines ℓ_6, ℓ_7, and ℓ_8, i.e., those prospective transmission lines that connect Region A and node 6 with the rest of the system.
6. Regarding the generation-capacity expansion decisions, it is optimal to build first candidate generating unit c_1 and then, if needed, additional capacity of generating unit c_2.

□

4.7 Summary and Conclusions

In this chapter, we describe the joint expansion of the generation capacity and the transmission network of a given electric energy system. This is done considering the perspective of a central planner, the ISO, whose aim is maximizing the overall social welfare. In particular, we determine the generation and transmission expansion plan that minimizes the generation and load-shedding costs as well as the investment costs. To do so, we provide different models that incorporate an increasing level of detail on the G&TEP decision-making problem, namely a dynamic framework, stochastic parameters, and risk management.

Considering the theoretical framework and the results of the examples reported in this chapter, the conclusions below are in order:

1. A deterministic static approach allows us to formulate a simple model for the G&TEP problem. However, the modeling accuracy of this model is modest.
2. Considering a dynamic approach enables us to make expansion decisions at different points in time, which increases the flexibility of the decision-maker and reduces the investment budget required at the beginning of the planning horizon.
3. It is important to model the impact of uncertain parameters on the G&TEP problem. Using the expected values of these uncertain parameters to formulate an equivalent deterministic problem usually results in suboptimal expansion plans.
4. Making generation and transmission expansion plans is a risky task. Therefore, it is important to manage the cost risk associated with these expansion plans.
5. Different risk strategies result in different expansion plans.
6. It is important to model accurately different details in the G&TEP problem. However, this generally results in a complex problem for real-world systems. Therefore, it is necessary to establish a tradeoff between modeling accuracy and computational complexity.

4.8 End-of-Chapter Exercises

4.1 Why is G&TEP needed? Who decides about it?

4.2 Enumerate the advantages and disadvantages of the different approaches described in this chapter for the G&TEP problem.

4.3 Determine the optimal generation and transmission expansion plan in the modified Garver's system depicted in Fig. 4.7, whose data are provided in Tables 4.10, 4.11, 4.12, 4.13, and 4.14. Apply the deterministic static model (4.1).

4.4 Robust optimization is used in Chap. 2 to address the uncertainty in the TEP problem, while stochastic programming is used in this chapter to represent the uncertainty in the G&TEP problem. What are the advantages and disadvantages of both techniques? Based on the adaptive robust optimization approach used in Sect. 2.3 of

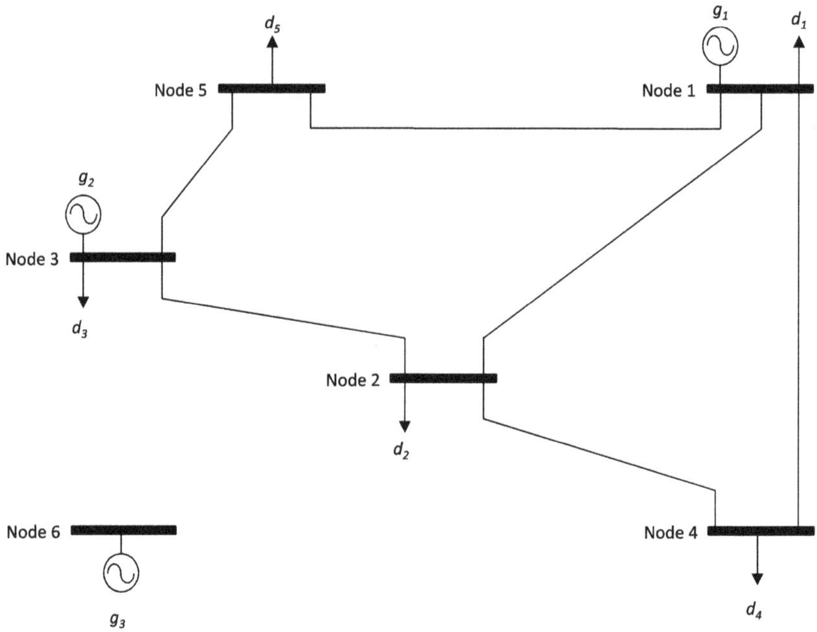

Fig. 4.7 Exercise 4.3: modified Garver's system

Table 4.10 Exercise 4.3: data for existing generating units of the modified Garver's system

Generating unit	Node	$P_g^{\mathrm{E}^{\max}}$ [MW]	C_g^{E} [\$/MWh]
g_1	n_1	200	24
g_2	n_3	200	28
g_3	n_6	300	16

Table 4.11 Exercise 4.3: data for candidate generating units of the modified Garver's system

Generating unit	Node	$P_c^{\mathrm{C^{max}}}$ [MW]	C_c^{C} [$/MWh]	I_c^{C} [$/MW]
c_1	n_1	300	10	600,000
c_2	n_3	300	15	400,000

Table 4.12 Exercise 4.3: data for demands of the modified Garver's system

Demand	Node	$P_{do_1}^{\mathrm{D^{max}}}$ [MW]	$P_{do_2}^{\mathrm{D^{max}}}$ [MW]	C_d^{LS} [$/MWh]
d_1	n_1	77	110	49
d_2	n_2	92	132	51
d_3	n_3	62	88	80
d_4	n_4	92	132	65
d_5	n_5	62	88	39

Table 4.13 Exercise 4.3: data for existing transmission lines of the modified Garver's system

Line	From node	To node	B_ℓ [S]	$F_\ell^{\max}$ [MW]
ℓ_1	n_1	n_2	250	100
ℓ_2	n_1	n_4	133	80
ℓ_3	n_1	n_5	500	100
ℓ_4	n_2	n_3	500	100
ℓ_5	n_2	n_4	250	100
ℓ_6	n_3	n_5	250	100

Table 4.14 Exercise 4.3: data for prospective transmission lines of the modified Garver's system

Line	From node	To node	B_ℓ [S]	$F_\ell^{\max}$ [MW]	I_ℓ^{L} [$]
ℓ_7	n_2	n_5	323	100	3,491,000
ℓ_8	n_2	n_6	333	100	3,379,000
ℓ_9	n_3	n_6	500	100	5,406,000
ℓ_{10}	n_4	n_6	333	100	3,379,000

Chap. 2, formulate the G&TEP problem (4.1) using an adaptive robust optimization approach and considering that uncertainty affects only $P_{do}^{\mathrm{D^{max}}}$.

4.5 Solve Illustrative Example 4.6 considering the robust model formulated in Exercise 4.4. Consider that demands vary within $\pm 10\,\%$ of their expected values. Compare the results with those obtained considering a stochastic approach.

4.6 In this chapter, the CVaR is incorporated in the G&TEP problem to manage the cost risk associated with the expansion plans. However, there are other risk metrics that may be used with the same purpose, e.g., the value-at-risk (VaR) and the variance. Incorporate these two risk metrics in the stochastic dynamic G&TEP problem (4.5).

4.7 Solve Illustrative Example 4.8 considering the models formulated in Exercise 4.6 and compare their solutions with those of Illustrative Example 4.8.

4.9 GAMS Code

A GAMS code for solving Illustrative Example 4.3 is provided below:

```
SETS
n                   /n1*n2/
g                   /g1*g1/
c                   /c1*c1/
d                   /d1*d1/
o                   /o1*o2/
l                   /l1*l2/
pros(l)             /l2*l2/
ex(l)               /l1*l1/
mapE(g,n)           /g1.n1/
mapC(c,n)           /c1.n2/
mapD(d,n)           /d1.n2/
ref(n)              /n1/
mapSL(l,n)          /l1.n1,l2.n1/
mapRL(l,n)          /l1.n2,l2.n2/;

TABLE LDATA(l,*)
          B         FLmax
l1        500       200
l2        500       200;

PARAMETER IC(c)
/c1       70000/;

PARAMETER IC_L(l)
/l2       100000/;

SCALAR IB
/40000000/;

SCALAR IB_L
/200000/;

TABLE PDmax(d,o)
          o1        o2
d1        290       550;

TABLE EDATA(g,*)
          PEmax     Ecost
g1        400       35;
```

```
TABLE DDATA(d,*)
          LSCost
d1        80;

TABLE CDATA(c,*)
          Ccost    PCmax
c1        25       300;

PARAMETER RHO(o)
/o1       6000
o2        2760/;

SCALAR M
/300000/;

VARIABLES
Z
PL(l,o)
THETA(n,o);

POSITIVE VARIABLES
PGE(g,o)
PGC(c,o)
PCmax(c)
PLS(d,o);

BINARY VARIABLES
XL(l);

EQUATIONS EQ1A, EQ1B, EQ1D, EQ1E, EQ1F, EQ1G, EQ2Aa,
   EQ2Ab, EQ2Ba, EQ2Bb, EQ1Ia, EQ1Ib, EQ1J, EQ1K, EQ1L
   , EQ1Ma, EQ1Mb, EQ1N;

EQ1A..                     Z=E=SUM(o,RHO(o)*(SUM(g,EDATA
   (g,'Ecost')*PGE(g,o))+SUM(c,CDATA(c,'Ccost')*PGC(c,
   o))+SUM(d,DDATA(d,'LSCOST')*PLS(d,o))))+SUM(c,IC(c)
   *PCmax(c))+SUM(l$pros(l),IC_L(l)*xL(l));
EQ1B(c)..                  PCmax(c)=L=CDATA(C,'PCmax');
*EQ1C IS BINARY VARIABLE DECLARATION
EQ1D..                     SUM(c,IC(c)*PCmax(c))=L=IB;
EQ1E..                     SUM(l$pros(l),IC_L(L)*xL(l))=
   L=IB_L;
EQ1F(n,o)..                SUM(g$mapE(g,n),PGE(g,o))+SUM
   (c$mapC(c,n),PGC(c,o))-SUM(l$mapSL(l,n),PL(l,o))+
   SUM(l$mapRL(l,n),PL(l,o))=E=SUM(d$mapD(d,n),(PDmax(
   d,o)-PLS(d,o)));
EQ1G(l,o)$EX(l)..          PL(l,o)=E=LDATA(l,'B')*(SUM(
   n$mapSL(l,n),THETA(n,o))-SUM(n$mapRL(l,n),THETA(n,o
   )));
EQ2Aa(l,o)$pros(l)..       -XL(l)*LDATA(l,'FLmax')=L=PL(
   l,o);
```

```
EQ2Ab(l,o)$pros(l)..      PL(l,o)=L=XL(l)*LDATA(l,'
    FLmax');
EQ2Ba(l,o)$pros(l)..      -(1-XL(l))*M=L=PL(l,o)-LDATA(
    l,'B')*(SUM(n$mapSL(l,n),THETA(n,o))-SUM(n$mapRL(l,
    n),THETA(n,o)));
EQ2Bb(l,o)$pros(l)..      PL(l,o)-LDATA(l,'B')*(SUM(
    n$mapSL(l,n),THETA(n,o))-SUM(n$mapRL(l,n),THETA(n,o
    )))=l=(1-XL(l))*M;
EQ1Ia(l,o)..              -LDATA(L,'FLmax')=L=PL(l,o);
EQ1Ib(l,o)..              PL(l,o)=L=LDATA(l,'FLmax');
EQ1J(g,o)..               PGE(g,O)=L=EDATA(g,'PEmax');
EQ1K(c,o)..               PGC(c,o)=L=PCmax(c);
EQ1L(d,o)..               PLS(d,o)=L=PDmax(d,o);
EQ1Ma(n,o)..              -3.14=L=THETA(n,o);
EQ1Mb(n,o)..              THETA(n,o)=L=3.14;
EQ1N(n,o)$REF(n)..        THETA(n,o)=E=0;

MODEL GaTEP_DetSta /ALL/;

option OPTCR=0;

option OPTCA=0;

SOLVE GaTEP_DetSta USING MIP MINIMIZING Z;
```

References

1. Alvarez, J., Ponnambalam, K., Quintana, V.H.: Generation and transmission expansion under risk using stochastic programming. IEEE Trans. Power Syst. **22**(3), 1369–1377 (2007)
2. Baringo, L., Conejo, A.J.: Wind power investment: a Benders' decomposition approach. IEEE Trans. Power Syst. **27**(1), 433–441 (2012)
3. Baringo, L., Conejo, A.J.: Transmission and wind power investment. IEEE Trans. Power Syst. **27**(2), 885–893 (2012)
4. Binato, S., Pereira, M.V.F., Granville, S.: A new Benders decomposition approach to solve power transmission network design problems. IEEE Trans. Power Syst. **16**(2), 235–240 (2001)
5. Castillo, E., Conejo, A.J., Pedregal, P., García, R., Alguacil, N.: Building and Solving Mathematical Programming Models in Engineering and Science. Wiley, New York (2001)
6. Conejo, A.J., Carrión, M., Morales, J.M.: Decision Making Under Uncertainty in Electricity Markets. Springer, New York (2010)
7. Domínguez, R., Conejo, A.J., Carrión, M.: Toward fully renewable electric energy systems. IEEE Trans. Power Syst. **30**(1), 316–326 (2015)
8. Gómez-Expósito, A., Conejo, A.J., Cañizares, C.: Electric Energy Systems: Analysis and Operation. CRC, Boca Raton (2008)
9. Jenabi, M., Ghomi, S.M.T.F., Smeers, Y.: Bi-level game approaches for coordination of generation and transmission expansion planning within a market environment. IEEE Trans. Power Syst. **28**(3), 2639–2650 (2013)
10. Kazempour, S.J., Conejo, A.J.: Strategic generation investment under uncertainty via Benders' decomposition. IEEE Trans. Power Syst. **27**(1), 424–432 (2012)
11. Morales, J.M., Pineda, S., Conejo, A.J., Carrión, M.: Scenario reduction for futures market trading in electricity markets. IEEE Trans. Power Syst. **24**(2), 878–888 (2009)

12. Pineda, S., Conejo, A.J.: Scenario reduction for risk-averse electricity trading. IET Gener. Trans. Distrib. **4**(6), 694–705 (2010)
13. Pozo, D., Sauma, E.E., Contreras, J.: A three-level static MILP model for generation and transmission expansion planning. IEEE Trans. Power Syst. **28**(1), 202–210 (2013)
14. Rockafellar, R.T., Uryasev, S.: Optimization of conditional value-at-risk. J. Risk. **2**, 21–41 (2000)
15. Rockafellar, R.T., Uryasev, S.: Conditional value-at-risk for general loss distributions. J. Bank. Finance **26**(7), 1443–1471 (2002)
16. Roh, J.H., Shahidehpour, M., Wu, L.: Market-based coordination of transmission and generation capacity planning. IEEE Trans. Power Syst. **22**(4), 1406–1419 (2007)
17. Roh, J.H., Shahidehpour, M., Wu, L.: Market-based generation and transmission planning with uncertainties. IEEE Trans. Power Syst. **24**(3), 1587–1598 (2009)
18. Tsamasphyrou, P., Renaud, A., Carpentier, P.: Transmission network planning: An efficient Benders decomposition scheme. In: Proceedings of 13th Power Systems Computer Conference Trondheim, Norway, June/July 28–2 (1999)

Chapter 5
Investment in Production Capacity

The two previous chapters describe and analyze the generation expansion planning (GEP) problem as well as the joint generation and transmission expansion planning (G&TEP) problem, both from the perspective of a central planner that determines the expansion plan that is most beneficial for the system as a whole. Unlike the problems in those two chapters, referred to as *centralized* expansion problems, this chapter addresses a *market-oriented* generation investment problem, in which a single power producer competing in an electricity pool seeks to maximize its own profit through making the best generation investment decisions. It is important to note that a market-oriented decision-making problem is generally more complex than a centralized one since the former is subject to additional uncertainty related to the behavior of other producers (rivals). In this chapter, we provide and analyze different investment models that progressively incorporate additional details with emphasis on solution techniques.

5.1 Introduction

This chapter develops models for making production capacity investment decisions by a producer that competes with other producers (rivals) in an electricity market. We consider that this producer is *strategic*, i.e., it owns a significant share of the production capacity in the industry, and therefore is able to exert market power. The aim of this strategic producer is to maximize its own profit. To this end, it intends to alter the market-clearing outcomes to its own benefit. Specifically, this producer makes:

1. *Strategic investment decisions*, i.e., long-term investment decisions to build new production units strategically including conventional (e.g., gas-based) and stochastic (e.g., wind-based) facilities.
2. *Strategic offering decisions*, i.e., short-term operation decisions to offer the production of conventional units at strategic prices. For simplicity, the offer price of stochastic production units is assumed to be zero.

A.J. Conejo et al., *Investment in Electricity Generation and Transmission*,
DOI 10.1007/978-3-319-29501-5_5

We assume that the investment and offering decisions of the rival producers are given as exogenous data. An investment model considering multiple strategic producers in which each producer makes its own decisions seeking its own profit is addressed in Chap. 6, which leads to an investment equilibrium problem.

The investment decision-making problem considered in this chapter allows the strategic producer to select the best production technologies, to find the best investment years throughout the planning horizon, and to locate the new production units optimally throughout the network. This is embodied in a bilevel model, which is recast as a mixed-integer linear programming (MILP) problem, solvable using available branch-and-cut solvers.

The following sections explain in detail features of the bilevel models considered in this chapter.

5.1.1 Electricity Pool

We consider a pool-based electricity market in which an independent system operator (ISO) clears the pool once a day, one day ahead, and on an hourly basis. The market operator seeks to maximize the social welfare of the pool considering the offering curves submitted by producers and the bidding curves submitted by consumers. The market-clearing results are hourly productions, consumptions, and clearing prices. For the sake of simplicity, a single trading floor, i.e., the day-ahead market, is considered since it is generally the market with the largest volume of energy trading. A more general formulation to include different trading floors (e.g., long-run futures market, short-run real-time market, and bilateral contracting) is an extension to the proposed model that does not change its nature.

5.1.2 Network Representation

A dc representation of the transmission network is embedded within the investment model provided in this chapter since such a representation is linear, simple, and appropriate for planning models. In this way, the effect of locating new units at different nodes and the impact of transmission congestion are adequately represented. For simplicity, active power losses are neglected; however, they can be easily incorporated using piecewise linear approximations [29, 31]. Further details on network modeling are available in [20]. We also assume that the network does not change throughout the investment horizon.

5.1.3 Static and Dynamic Investment Models

Two different models are generally considered in investment studies:

1. Static models [1, 6, 14, 22–26, 32, 33].
2. Dynamic (multistage) models [3, 11, 17–19, 35].

The characteristics of these two models are briefly explained next.

In the static investment model, the producer considers a *single* future year and decides its optimal production portfolio for that target year. Note that the optimal timing for building new units from the current year to the target year cannot be obtained explicitly using this static model. However, it can be derived through an ex-post analysis once the optimal portfolio of the strategic producer for the target year has been obtained through a profit-maximization problem in which the generation capacity portfolios in the current and target years are given (exogenous parameters), but the optimal investment years are decision variables. This ex-post analysis is outside the scope of this chapter. Existing production units that will be decommissioned over the planning horizon should not be included in the static analysis.

In contrast to the static model, in the dynamic model, the investment decisions are endogenously made at several points in time throughout the planning horizon. For example, let us consider three ten-year time periods within a thirty-year planning horizon. Then the investment decisions are made at the beginning of each time period, i.e., at the beginning of years 1, 11, and 21. In this way, the time schedule for building new units from the current year to the target year (year 30) is optimally derived. Note that within each time period, e.g., the first time period spanning from the beginning of year 1 to the end of year 10, a static investment model needs to be considered, i.e., year 10 is the target year for that time period. In general, the dynamic model results in more accurate investment decisions with respect to the static model, but at the cost of high computational burden and potential intractability. In this chapter, both static and dynamic models are considered.

5.1.4 Operating Conditions: Demand Level and Stochastic Production

Since the demand level and the production level in the case of stochastic units (e.g., wind power producers) vary over the target year of the static model or during each time period of the dynamic model, a number of operating conditions needs to be considered to capture such variability. Each operating condition contains the following two factors expressed in per unit (p.u.):

1. A demand factor for each consumer,
2. A power capacity factor for each stochastic production unit.

The demand factor of a consumer refers to its consumption level divided by its peak demand. In addition, the power capacity factor of a stochastic unit corresponds to its production level divided by its installed capacity. For example, the following items represent two different operating conditions for a system including a single consumer and a single stochastic production unit:

- Condition 1: demand factor equal to 1.0 p.u. and power capacity factor equal to 0.3 p.u.
- Condition 2: demand factor equal to 0.7 p.u. and power capacity factor equal to 0.6 p.u.

Condition 1 above represents the peak demand since the consumer's demand factor is 1.0 p.u., whereas the stochastic unit produces at 30 % of its installed capacity. In Condition 2, the stochastic production is comparatively higher than that in Condition 1, i.e., it generates at 60 % of its installed capacity. However, the consumption level of the consumer is comparatively lower than that in Condition 1, i.e., the consumer's demand in Condition 2 is 70 % of its peak demand.

It is important to note that a *weighting factor* is associated with each operating condition, indicating the number of hours during the target year that are represented by that condition. Clearly, the summation of weighting factors corresponding to all operating conditions is 8760, i.e., the number of hours in a year. For example, the weighting factors of Conditions 1 and 2 above might be 1200 and 7560 h, respectively.

In practice, the operating conditions of a given electric system can be appropriately derived from historical data using a clustering technique, e.g., the K-means clustering method [4]. In this way, the hours during the target year with similar demand and stochastic production levels can be grouped in the same cluster. Then, these operating conditions are projected into the future by appropriate growth factors. This technique allows the representation of the possible temporal and/or spatial correlation between demand and stochastic production.

5.1.5 Uncertainty

The production capacity investment problem of a producer is subject to diverse sources of uncertainty throughout the planning horizon, such as:

- Behavior of rival producers regarding investment decisions.
- Demand growth.
- Investment and operating costs of different production technologies.
- Regulatory changes.

For the sake of simplicity, among the uncertainties above, we consider the investment decisions of rival producers and the investment costs of different production technologies (in the case of dynamic investment model) as uncertain parameters. Other parameters, e.g., demand growth, operation costs, and regulatory changes, are assumed to be known. However, it is not complex, but computationally costly, to consider all potential uncertainties.

In the literature, different approaches are available to model uncertainty within an optimization problem, e.g., stochastic programming [10, 13], robust optimization [9, 30], and interval optimization [21]. A comparison of those approaches applied to a problem related to power systems is available in [30]. In this chapter, we use the stochastic programming approach so that the uncertain parameters are represented through a set of plausible scenarios. For example, the investment decisions of a rival producer are represented by two scenarios: no investment (scenario 1) and investment in a 100 MW coal unit (scenario 2). Each scenario embodies a probability, e.g., 0.6 for scenario 1 and 0.4 for scenario 2. Note that an appropriate set of scenarios corresponding to an uncertain parameter can be generated using historical data [4]. However, the scenario generation and/or reduction techniques are outside the scope of this chapter.

5.1.6 Bilevel Model

In the context of an electricity market, the strategic producer sells its production at the market-clearing price. However, the market-clearing price is affected by the offering and investment decisions of the strategic producer. Therefore, the strategic producer's decisions are constrained by market-clearing problems. Note that each market-clearing problem (one per operating condition and scenario) is itself an optimization problem. Thus, the strategic producer needs to solve an optimization problem constrained by other optimization problems. These type of models are known in the literature as *bilevel models* [16].

In this chapter, we use a bilevel model to represent the strategic behavior of a producer competing with its rival producers in an electricity pool. This bilevel model consists of an upper-level problem and a set of lower-level problems. The general structure of bilevel models is explained in Chap. 1. Additionally, mathematical details of bilevel models are provided in Appendix C.

Specifically, the upper-level problem of the bilevel model maximizes the expected profit of the strategic producer and determines its strategic decisions, i.e., strategic investment actions and strategic offer prices. Note that the offer prices of a strategic producer are generally different from its actual production costs. The upper-level problem is constrained by two sets of constraints as follows:

1. Upper-level constraints.
2. A collection of lower-level problems, one per operating condition and scenario. Each lower-level problem represents the clearing of the market by the Independent System Operator (ISO) maximizing social welfare.

5.1.7 *Alternative Solution Approaches*

To solve investment bilevel models, two alternative solution approaches are provided in this chapter, namely:

1. Direct solution [1, 22, 23],
2. Benders decomposition [2, 26].

In the direct solution approach, all operating conditions and scenarios are simultaneously taken into account, and as such, the bilevel model is solved. In this solution approach, a large number of operating conditions and scenarios may result in high computational burden and eventual intractability, especially if a dynamic investment model is considered.

In the second approach, the operating conditions and scenarios are considered separately using a Benders decomposition technique [12]. The objective of this solution approach is to make the investment models computationally tractable even if many operating conditions and scenarios are considered. Both solution approaches are presented in this chapter.

The remainder of this chapter is organized as follows. Section 5.2 provides a production capacity investment model using the static investment model, while Sect. 5.3 refers to a similar problem but using the dynamic model. Section 5.4 provides the resulting mathematical program with equilibrium constraints (MPEC) and the MILP problem corresponding to the investment models presented in Sects. 5.2 and 5.3. Section 5.5 provides a computationally efficient solution approach based on Benders decomposition to be applied to both static and dynamic investment models. Section 5.6 summarizes the chapter and discusses the main conclusions of the models and results reported in the chapter. Section 5.7 proposes some exercises to enable a deeper understanding of the models and concepts described in the chapter. Finally, Sect. 5.8 includes the GAMS code for an illustrative example.

5.2 Static Production Capacity Investment Model

Figure 5.1 shows the bilevel structure of a static production capacity investment model. The upper-level problem represents the expected profit maximization (or minus the expected profit minimization) of the strategic producer subject to upper-level constraints and to the lower-level problems. The upper-level constraints pertain to investment options, available investment budget, and nonnegativity of strategic offer prices.

Each lower-level problem, one per operating condition and scenario, represents the market clearing with the target of maximizing the social welfare (or minimizing minus the social welfare) and is subject to power balance at every node, power limits for production and consumption levels, transmission line capacity limits, voltage angle bounds, and reference node identification.

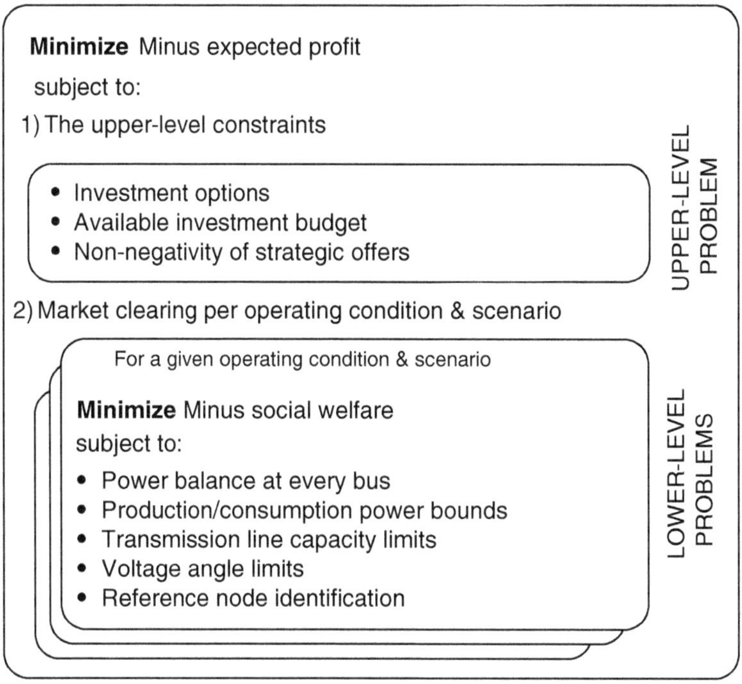

Fig. 5.1 Bilevel structure of the static production capacity investment model

Note that the upper-level and lower-level problems are interrelated. On the one hand, the lower-level problems determine the market-clearing prices and the power production quantities, which directly influence the strategic producer's expected profit in the upper-level problem. On the other hand, the offering and investment decisions made by the strategic producer in the upper-level problem affect the market-clearing outcomes in the lower-level problems.

The notation used in this chapter is defined below:

Indices

h Production capacity investment options (conventional technologies).
n, m Nodes.
o Operating conditions.
ω Scenarios.

Sets

Ω_n Set of nodes connected to node n.

Parameters

B_{nm} Susceptance of the transmission line connecting node n to node m [S].
C_n^{C} Production cost of the candidate conventional unit of the strategic producer located at node n [\$/MWh].
C_n^{E} Production cost of the existing conventional unit of the strategic producer located at node n [\$/MWh].
C_{no}^{R} Offer price of the rival unit located at node n in operating condition o [\$/MWh].
$F_{nm}^{\max}$ Transmission capacity of the line connecting node n to node m [MW].
K_n^{C} Annualized investment cost of the candidate conventional unit located at node n [\$/MW].
K_n^{S} Annualized investment cost of the candidate stochastic unit located at node n [\$/MW].
$K^{\max}$ Available annualized investment budget of the strategic producer [\$].
$P_n^{\mathrm{E}^{\max}}$ Capacity of the existing conventional unit of the strategic producer located at node n [MW].
$P_n^{\mathrm{D}^{\max}}$ Maximum load of the consumer located at node n [MW].
$P_{n\omega}^{\mathrm{R}^{\max}}$ Capacity of the rival unit located at node n under scenario ω [MW].
Q_{no}^{S} Power capacity factor of the candidate stochastic unit of the strategic producer located at node n in operating condition o [p.u.].
Q_{no}^{D} Demand factor of the consumer located at node n in operating condition o [p.u.].
U_{no}^{D} Bid price of the consumer located at node n in operating condition o [\$/MWh].
X_{nh}^{C} Option h for production capacity investment of the candidate conventional unit located at node n [MW].
$X_n^{\mathrm{S}^{\max}}$ Maximum production capacity investment of the candidate stochastic unit located at node n [MW].
φ_ω Probability associated with scenario ω [p.u.].
ρ_o Weighting factor associated with operating condition o [h].

Binary Variables

u_{nh}^{C} Binary variable that is equal to 1 if the conventional production investment option h is selected to be built at node n.

Continuous Variables

$p_{no\omega}^{\mathrm{C}}$ Power produced by the candidate conventional unit of the strategic producer located at node n in operating condition o under scenario ω [MW].
$p_{no\omega}^{\mathrm{D}}$ Power consumed by the consumer located at node n in operating condition o under scenario ω [MW].
$p_{no\omega}^{\mathrm{E}}$ Power produced by the existing conventional unit of the strategic producer located at node n in operating condition o under scenario ω [MW].

$p^{\mathrm{R}}_{no\omega}$ Power produced by the rival unit located at node n in operating condition o under scenario ω [MW].

$p^{\mathrm{S}}_{no\omega}$ Power produced by the candidate stochastic unit of the strategic producer located at node n in operating condition o under scenario ω [MW].

x^{C}_{n} Capacity investment of the candidate conventional unit of the strategic producer located at node n [MW].

x^{S}_{n} Capacity investment of the candidate stochastic unit of the strategic producer located at node n [MW].

$\alpha^{\mathrm{C}}_{no\omega}$ Offer price by the candidate conventional unit of the strategic producer located at node n in operating condition o under scenario ω [\$/MWh].

$\alpha^{\mathrm{E}}_{no\omega}$ Offer price by the existing conventional unit of the strategic producer located at node n in operating condition o under scenario ω [\$/MWh].

$\theta_{no\omega}$ Voltage angle of node n in operating condition o under scenario ω [rad].

$\lambda_{no\omega}$ Market-clearing price at node n in operating condition o under scenario ω [\$/MWh].

The formulation of the static investment bilevel model is given by (5.1). Note that (5.1a)–(5.1h) is the upper-level problem, and (5.1j)–(5.1s) included in (5.1i) pertains to the lower-level problems, one per operating condition o and scenario ω. Note also that lower-level problems (5.1j)–(5.1s) are actually constraints of the upper-level problem. This bilevel problem is as follows:

$$\min_{\Xi^{\mathrm{UL}} \cup \Xi^{\mathrm{Primal}}_{o\omega} \cup \Xi^{\mathrm{Dual}}_{o\omega}} \quad \sum_{n} \left(K^{\mathrm{C}}_{n}\, x^{\mathrm{C}}_{n} + K^{\mathrm{S}}_{n}\, x^{\mathrm{S}}_{n}\right)$$
$$- \sum_{\omega} \varphi_{\omega} \sum_{o} \rho_{o} \sum_{n} \Big[\lambda_{no\omega} \left(p^{\mathrm{C}}_{no\omega} + p^{\mathrm{E}}_{no\omega} + p^{\mathrm{S}}_{no\omega}\right)$$
$$- p^{\mathrm{C}}_{no\omega}\, C^{\mathrm{C}}_{n} - p^{\mathrm{E}}_{no\omega}\, C^{\mathrm{E}}_{n} \Big] \tag{5.1a}$$

subject to

$$x^{\mathrm{C}}_{n} = \sum_{h} u^{\mathrm{C}}_{nh}\, X^{\mathrm{C}}_{nh} \quad \forall n \tag{5.1b}$$

$$\sum_{h} u^{\mathrm{C}}_{nh} = 1 \quad \forall n \tag{5.1c}$$

$$u^{\mathrm{C}}_{nh} \in \{0, 1\} \quad \forall n, \forall h \tag{5.1d}$$

$$0 \le x^{\mathrm{S}}_{n} \le X^{\mathrm{S}^{\max}}_{n} \quad \forall n \tag{5.1e}$$

$$\sum_{n} \left(K^{\mathrm{C}}_{n}\, x^{\mathrm{C}}_{n} + K^{\mathrm{S}}_{n}\, x^{\mathrm{S}}_{n}\right) \le K^{\max} \tag{5.1f}$$

$$\alpha^{\mathrm{C}}_{no\omega} \ge 0 \quad \forall o, \forall n, \forall \omega \tag{5.1g}$$

$$\alpha^{\mathrm{E}}_{no\omega} \ge 0 \quad \forall o, \forall n, \forall \omega \tag{5.1h}$$

$$\text{Lower-level problems (5.1j)–(5.1s)} \quad \forall o, \forall \omega. \tag{5.1i}$$

The primal variables of the upper-level problem (5.1a)–(5.1h) are those in set $\Xi^{\mathrm{UL}} = \{\alpha^{\mathrm{C}}_{no\omega}, \alpha^{\mathrm{E}}_{no\omega}, x^{\mathrm{C}}_{n}, u_{nh}, x^{\mathrm{S}}_{n}\}$ plus all primal and dual variables of lower-level problems (5.1j)–(5.1s), which are defined after their formulation through sets $\Xi^{\mathrm{Primal}}_{o\omega}$ and $\Xi^{\mathrm{Dual}}_{o\omega}$.

The objective function (5.1a) is minus the expected profit (investment cost minus expected operational revenue) of the strategic producer. In detail, the objective function (5.1a) consists of the following terms:

- $\sum\limits_{n} K^{\mathrm{C}}_{n}\, x^{\mathrm{C}}_{n}$ is the annualized investment cost of candidate conventional units of the strategic producer.
- $\sum\limits_{n} K^{\mathrm{S}}_{n}\, x^{\mathrm{S}}_{n}$ is the annualized investment cost of candidate stochastic units of the strategic producer.
- $\sum\limits_{n} p^{\mathrm{C}}_{no\omega}\, \lambda_{no\omega}$ is the revenue of the strategic producer obtained from selling the production of candidate conventional units.
- $\sum\limits_{n} p^{\mathrm{E}}_{no\omega}\, \lambda_{no\omega}$ is the revenue of the strategic producer obtained from selling the production of existing conventional units.
- $\sum\limits_{n} p^{\mathrm{S}}_{no\omega}\, \lambda_{no\omega}$ is the revenue of the strategic producer obtained from selling the production of candidate stochastic units.
- $\sum\limits_{n} p^{\mathrm{C}}_{no\omega}\, C^{\mathrm{C}}_{n}$ is the production cost of candidate conventional units of the strategic producer.
- $\sum\limits_{n} p^{\mathrm{E}}_{no\omega}\, C^{\mathrm{E}}_{n}$ is the production cost of existing conventional units of the strategic producer.

The production cost of stochastic units is assumed to be zero. Furthermore, other potential incomes including capacity payments and incentives (e.g., feed-in tariff or premium) are not considered. However, those incomes can be easily incorporated into the model [28]. It is also assumed that all existing units available in the initial year are conventional, i.e., there is no stochastic production unit within the initial production capacity portfolio of the producers.

For each available conventional technology (e.g., nuclear, coal, gas), constraints (5.1b)–(5.1d) allow the strategic producer to choose among the available investment options including no investment, e.g., the set of investment options are 0, 200, 500, and 1000 MW.

Unlike candidate conventional units that include discrete investment options, the production capacity investment options for investing in stochastic production units are assumed continuous for simplicity, and their bounds are enforced by (5.1e). In addition, upper-level constraint (5.1f) imposes a cap on the available annualized investment budget of the strategic producer that reflects its limited financial resources.

Finally, upper-level constraints (5.1g) and (5.1h) enforce the nonnegativity of offer prices associated with the candidate and existing conventional units of the strategic producer, respectively. For simplicity, the offering prices of stochastic units are assumed to be zero. However, it is not complex to model strategic offering by stochastic production units [5, 27, 36].

Specifically, the strategic producer makes the following decisions in the upper-level problem (5.1a)–(5.1h):

1. Strategic production capacity investment decisions including investment in conventional units ($x_n^{\rm C}, \forall n$) and investment in stochastic units ($x_n^{\rm S}, \forall n$).
2. Strategic offering decisions including offer prices of candidate conventional units ($\alpha_{no\omega}^{\rm C}, \forall n, \forall o, \forall \omega$) and offer prices of existing conventional units ($\alpha_{no\omega}^{\rm E}$, $\forall n, \forall o, \forall \omega$).

We consider that each conventional generating unit submits its installed capacity as its offer quantity. Likewise, the quantity offer of each stochastic unit is equal to its available stochastic production.

The market-clearing prices ($\lambda_{no\omega}$) and the production quantities ($p_{no\omega}^{\rm C}$, $p_{no\omega}^{\rm E}$, and $p_{no\omega}^{\rm S}$) belong to the feasible region defined by lower-level problems (5.1j)–(5.1s). Each lower-level problem, one per operating condition o and scenario ω, is formulated below. The dual variable of each lower-level constraint is indicated following a colon:

$$\Bigg\{ \min_{\Xi_{o\omega}^{\rm Primal}} \quad \sum_n \left[\alpha_{no\omega}^{\rm C} \, p_{no\omega}^{\rm C} + \alpha_{no\omega}^{\rm E} \, p_{no\omega}^{\rm E} + C_{no}^{\rm R} \, p_{no\omega}^{\rm R} - U_{no}^{\rm D} \, p_{no\omega}^{\rm D} \right] \tag{5.1j}$$

subject to

$$p_{no\omega}^{\rm D} + \sum_{m \in \Omega_n} B_{nm} \left(\theta_{no\omega} - \theta_{mo\omega} \right) - p_{no\omega}^{\rm C} - p_{no\omega}^{\rm S} - p_{no\omega}^{\rm E}$$

$$- \, p_{no\omega}^{\rm R} = 0 \quad : \lambda_{no\omega} \qquad \forall n \tag{5.1k}$$

$$0 \leq p_{no\omega}^{\rm C} \leq x_n^{\rm C} \quad : \mu_{no\omega}^{\rm C^{min}}, \mu_{no\omega}^{\rm C^{max}} \qquad \forall n \tag{5.1l}$$

$$0 \leq p_{no\omega}^{\rm S} \leq Q_{no}^{\rm S} \, x_n^{\rm S} \quad : \mu_{no\omega}^{\rm S^{min}}, \mu_{no\omega}^{\rm S^{max}} \qquad \forall n \tag{5.1m}$$

$$0 \leq p_{no\omega}^{\rm E} \leq P_n^{\rm E^{max}} \quad : \mu_{no\omega}^{\rm E^{min}}, \mu_{no\omega}^{\rm E^{max}} \qquad \forall n \tag{5.1n}$$

$$0 \leq p_{no\omega}^{\rm R} \leq P_{n\omega}^{\rm R^{max}} \quad : \mu_{no\omega}^{\rm R^{min}}, \mu_{no\omega}^{\rm R^{max}} \qquad \forall n \tag{5.1o}$$

$$0 \leq p_{no\omega}^{\rm D} \leq Q_{no}^{\rm D} \, P_n^{\rm D^{max}} \quad : \mu_{no\omega}^{\rm D^{min}}, \mu_{no\omega}^{\rm D^{max}} \qquad \forall n \tag{5.1p}$$

$$B_{nm} \left(\theta_{no\omega} - \theta_{mo\omega} \right) \leq F_{nm}^{\rm max} \quad : \mu_{nmo\omega}^{\rm F} \qquad \forall n, \forall m \in \Omega_n \tag{5.1q}$$

$$- \pi \leq \theta_{no\omega} \leq \pi \quad : \mu_{no\omega}^{\theta^{\rm min}}, \mu_{no\omega}^{\theta^{\rm max}} \qquad \forall n \tag{5.1r}$$

$$\theta_{no\omega} = 0 \quad : \mu_{o\omega}^{\theta^{\rm ref}} \qquad n = \text{ref}. \tag{5.1s}$$

$$\Bigg\} \; \forall o, \forall \omega.$$

The primal optimization variables of each lower-level problem (5.1j)–(5.1s) are in the set $\Xi_{o\omega}^{\rm Primal} = \{p_{no\omega}^{\rm C}, p_{no\omega}^{\rm S}, p_{no\omega}^{\rm E}, p_{no\omega}^{\rm R}, p_{no\omega}^{\rm D}, \theta_{no\omega}\}$. Likewise, variable set $\Xi_{o\omega}^{\rm Dual}$ contains the dual optimization variables of each lower-level problem (5.1j)–(5.1s), i.e., $\Xi^{\text{Dual}}_{o\omega} = \{\lambda_{no\omega}, \mu^{\text{C}^{\min}}_{no\omega}, \mu^{\text{C}^{\max}}_{no\omega}, \mu^{\text{S}^{\min}}_{no\omega}, \mu^{\text{S}^{\max}}_{no\omega}, \mu^{\text{E}^{\min}}_{no\omega}, \mu^{\text{E}^{\max}}_{no\omega}, \mu^{\text{R}^{\min}}_{no\omega}, \mu^{\text{R}^{\max}}_{no\omega}, \mu^{\text{D}^{\min}}_{no\omega}, \mu^{\text{D}^{\max}}_{no\omega}, \mu^{\text{F}}_{nmo\omega}, \mu^{\theta^{\min}}_{no\omega}, \mu^{\theta^{\max}}_{no\omega}, \mu^{\theta^{\text{ref}}}_{o\omega}\}$.

Each lower-level problem (5.1j)–(5.1s) represents the clearing of the market for given investment and offering decisions made at the upper-level problem. Accordingly, $x^{\text{C}}_n, \forall n, x^{\text{S}}_n, \forall n, \alpha^{\text{C}}_{no\omega}, \forall n, \forall o, \forall \omega$, and $\alpha^{\text{E}}_{no\omega}, \forall n, \forall o, \forall \omega$ are variables in the upper-level problem, while they are parameters (fixed values) in the lower-level problems (5.1j)–(5.1s). This makes the lower-level problems (5.1j)–(5.1s) linear and thus convex.

The minimization of minus the social welfare is expressed by (5.1j). Since we assume that the stochastic production units always offer at zero price, there is no term in (5.1j) related to the offer prices of those units.

Constraints (5.1k) enforce power balance at every node, and their dual variables provide market-clearing prices.

Constraints (5.1l), (5.1m), and (5.1n) enforce production capacity limits for the candidate conventional, candidate stochastic, and existing units of the strategic producer, respectively, while equations (5.1o) enforce similar constraints for the units of rival producers. Note that parameter Q^{S}_{no} included in the upper bound of Eq. (5.1m) is the power capacity factor of the stochastic unit located at node n in operating condition o (see Sect. 5.1.4). Furthermore, constraints (5.1m) implicitly allow the stochastic productions to be spilled.

Constraints (5.1p) bound the power consumed by each consumer. Parameter Q^{D}_{no} is the demand factor of the consumer located at node n in operating condition o (see Sect. 5.1.4).

Constraints (5.1q) enforce the transmission capacity limits of each line. Note that the left-hand side of constraints (5.1q) provides the power flow from node n to node m. Constraints (5.1r) enforce voltage angle bounds for each node and constraints (5.1s) identify the reference node, whose voltage angle is equal to zero.

Regarding uncertainty characterization, parameter $P^{\text{R}^{\max}}_{n\omega}$, i.e., the upper bound of rival producers' production levels in (5.1o), is indexed by ω. In this way, the production capacity investment uncertainty of the rival producers is characterized. Analogously, other sources of uncertainty can be represented through additional scenarios.

Next, we present four illustrative examples for the static investment bilevel model: from a very basic model based on a deterministic single-node example to more complex examples considering transmission constraints, stochastic production investment options, and uncertainties. Note that their solution procedures are explained later, in Sects. 5.4 and 5.5.

Illustrative Example 5.1 *Deterministic single-node static investment bilevel model*

The considered power system includes one single node (n_1) and two generating units. The first unit belongs to the strategic producer, while the second one is a rival unit. The capacity and production cost of the strategic unit are 150 MW and \$10/MWh, respectively, while those of the rival unit are 100 MW and \$15/MWh, respectively. It is assumed that the rival unit offers its capacity to the market at its production cost. In addition, the rival producer does not invest in new production units. Thus, no rival investment uncertainty needs to be taken into account.

As production capacity investment options, a single technology (conventional) with two capacity options are considered: $h_1 = 0$ MW and $h_2 = 100$ MW. The annualized investment cost of this technology is \$55,000/MW, and its production cost is \$12/MWh.

A single consumer is considered, whose maximum load is equal to 300 MW. In addition, a single operating condition (o_1) is considered that embodies the consumer's demand factor, which is 1 p.u. The weighting factor corresponding to condition o_1 is 8760. The consumer bids in this condition at \$35/MWh.

Finally, the available annualized investment budget of the strategic consumer is assumed to be \$10 million.

The bilevel model below is formulated to derive the most beneficial investment and offering decisions of the strategic producer:

$$\begin{aligned}
&\min_{\Xi^{\text{UL,Ex1}} \cup \Xi^{\text{P,Ex1}}_{o_1} \cup \Xi^{\text{D,Ex1}}_{o_1}} \\
&\qquad 55000\, x^{\text{C}}_{n_1} - 8760 \left[\lambda_{n_1 o_1} \left(p^{\text{C}}_{n_1 o_1} + p^{\text{E}}_{n_1 o_1} \right) - 12\, p^{\text{C}}_{n_1 o_1} - 10\, p^{\text{E}}_{n_1 o_1} \right] \\
&\text{subject to} \\
&x^{\text{C}}_{n_1} = 100\, u^{\text{C}}_{n_1 h_2} \\
&u^{\text{C}}_{n_1 h_1} + u^{\text{C}}_{n_1 h_2} = 1 \\
&u^{\text{C}}_{n_1 h_1},\ u^{\text{C}}_{n_1 h_2} \in \{0, 1\} \\
&55000\, x^{\text{C}}_{n_1} \le 10^7 \\
&\alpha^{\text{C}}_{n_1 o_1} \ge 0 \\
&\alpha^{\text{E}}_{n_1 o_1} \ge 0 \\
&\text{Lower-level problem associated with operating condition } o_1,
\end{aligned}$$

where the lower-level problem pertaining to the market clearing in operating condition o_1 is:

$$\begin{aligned}
&\min_{\Xi^{\text{P,Ex1}}_{o_1}} \quad \alpha^{\text{C}}_{n_1 o_1}\, p^{\text{C}}_{n_1 o_1} + \alpha^{\text{E}}_{n_1 o_1}\, p^{\text{E}}_{n_1 o_1} + 15\, p^{\text{R}}_{n_1 o_1} - 35\, p^{\text{D}}_{n_1 o_1} \\
&\text{subject to} \\
&p^{\text{D}}_{n_1 o_1} - p^{\text{C}}_{n_1 o_1} - p^{\text{E}}_{n_1 o_1} - p^{\text{R}}_{n_1 o_1} = 0 \qquad : \lambda_{n_1 o_1} \\
&0 \le p^{\text{C}}_{n_1 o_1} \le x^{\text{C}}_{n_1} \qquad : \mu^{\text{C}^{\min}}_{n_1 o_1}, \mu^{\text{C}^{\max}}_{n_1 o_1} \\
&0 \le p^{\text{E}}_{n_1 o_1} \le 150 \qquad : \mu^{\text{E}^{\min}}_{n_1 o_1}, \mu^{\text{E}^{\max}}_{n_1 o_1} \\
&0 \le p^{\text{R}}_{n_1 o_1} \le 100 \qquad : \mu^{\text{R}^{\min}}_{n_1 o_1}, \mu^{\text{R}^{\max}}_{n_1 o_1} \\
&0 \le p^{\text{D}}_{n_1 o_1} \le 300 \qquad : \mu^{\text{D}^{\min}}_{n_1 o_1}, \mu^{\text{D}^{\max}}_{n_1 o_1}.
\end{aligned}$$

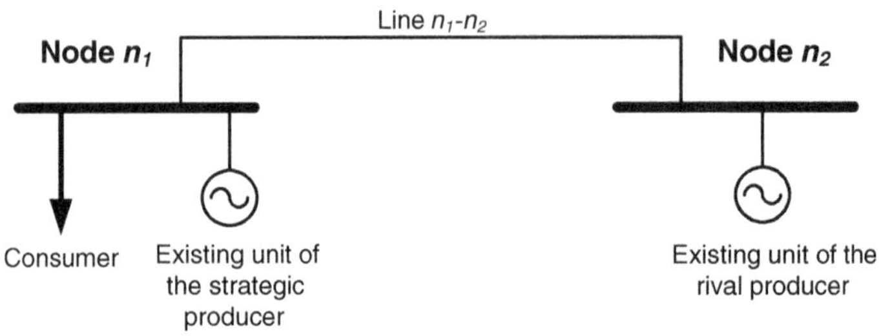

Fig. 5.2 Illustrative Example 5.2: Two-node network

The primal optimization variables of the lower-level problem are included in set $\Xi^{\mathrm{P,Ex1}}_{o_1} = \{p^{\mathrm{C}}_{n_1o_1}, p^{\mathrm{E}}_{n_1o_1}, p^{\mathrm{R}}_{n_1o_1}, p^{\mathrm{D}}_{n_1o_1}\}$, while the dual optimization variables of this problem are in set $\Xi^{\mathrm{D,Ex1}}_{o_1} = \{\lambda_{n_1o_1}, \mu^{\mathrm{C^{min}}}_{n_1o_1}, \mu^{\mathrm{C^{max}}}_{n_1o_1}, \mu^{\mathrm{E^{min}}}_{n_1o_1}, \mu^{\mathrm{E^{max}}}_{n_1o_1}, \mu^{\mathrm{R^{min}}}_{n_1o_1}, \mu^{\mathrm{R^{max}}}_{n_1o_1}, \mu^{\mathrm{D^{min}}}_{n_1o_1}, \mu^{\mathrm{D^{max}}}_{n_1o_1}\}$. Finally, the primal variables of the upper-level problem are those in set $\Xi^{\mathrm{U,Ex1}} = \{x^{\mathrm{C}}_{n_1}, u^{\mathrm{C}}_{n_1h_1}, u^{\mathrm{C}}_{n_1h_2}, \alpha^{\mathrm{C}}_{n_1o_1}, \alpha^{\mathrm{E}}_{n_1o_1}\}$ plus $\Xi^{\mathrm{P,Ex1}}_{o_1}$ and $\Xi^{\mathrm{D,Ex1}}_{o_1}$. □

Illustrative Example 5.2 *Deterministic network-constrained static investment bilevel model*

A power system with two nodes (n_1 and n_2), as illustrated in Fig. 5.2, is considered in this example. These two nodes are connected by transmission line $n_1 - n_2$ with a capacity of 200 MW and a susceptance of 1000 S. Node n_1 is the reference node and is considered the only candidate location for investing in new conventional units. Stochastic units are not considered as investment options in this example. Other input data are identical to those in Illustrative Example 5.1.

The bilevel model below allows the derivation of the optimal investment and offering decisions of the strategic producer:

$$\min_{\Xi^{\mathrm{UL,Ex2}} \cup \Xi^{\mathrm{P,Ex2}}_{o_1} \cup \Xi^{\mathrm{D,Ex2}}_{o_1}}$$

$$55000\, x^{\mathrm{C}}_{n_1} - 8760 \left[\lambda_{n_1o_1} \left(p^{\mathrm{E}}_{n_1o_1} + p^{\mathrm{C}}_{n_1o_1} \right) - 12\, p^{\mathrm{C}}_{n_1o_1} - 10\, p^{\mathrm{E}}_{n_1o_1} \right]$$

subject to

$$x^{\mathrm{C}}_{n_1} = 100\, u^{\mathrm{C}}_{n_1h_2}$$

$$u^{\mathrm{C}}_{n_1h_1} + u^{\mathrm{C}}_{n_1h_2} = 1$$

$$u^{\mathrm{C}}_{n_1h_1},\ u^{\mathrm{C}}_{n_1h_2} \in \{0, 1\}$$

$$55000\, x^{\mathrm{C}}_{n_1} \leq 10^7$$

$$\alpha^{\mathrm{C}}_{n_1o_1} \geq 0$$

$$\alpha^{\mathrm{E}}_{n_1o_1} \geq 0$$

Lower-level problem associated with operating condition o_1, where the lower-level problem pertaining to the market clearing in operating condition o_1 is:

$$
\begin{aligned}
&\min_{\Xi^{\mathrm{P,Ex2}}_{o_1}} \quad \alpha^{\mathrm{C}}_{n_1 o_1} \, p^{\mathrm{C}}_{n_1 o_1} + \alpha^{\mathrm{E}}_{n_1 o_1} \, p^{\mathrm{E}}_{n_1 o_1} + 15 \, p^{\mathrm{R}}_{n_2 o_1} - 35 \, p^{\mathrm{D}}_{n_1 o_1} \\
&\text{subject to} \\
&p^{\mathrm{D}}_{n_1 o_1} + 1000 \left(\theta_{n_1 o_1} - \theta_{n_2 o_1}\right) - p^{\mathrm{E}}_{n_1 o_1} - p^{\mathrm{C}}_{n_1 o_1} = 0 \qquad : \lambda_{n_1 o_1} \\
&1000 \left(\theta_{n_2 o_1} - \theta_{n_1 o_1}\right) - p^{\mathrm{R}}_{n_2 o_1} = 0 \qquad : \lambda_{n_2 o_1} \\
&0 \leq p^{\mathrm{C}}_{n_1 o_1} \leq x^{\mathrm{C}}_{n_1} \qquad : \mu^{\mathrm{C^{min}}}_{n_1 o_1}, \mu^{\mathrm{C^{max}}}_{n_1 o_1} \\
&0 \leq p^{\mathrm{E}}_{n_1 o_1} \leq 150 \qquad : \mu^{\mathrm{E^{min}}}_{n_1 o_1}, \mu^{\mathrm{E^{max}}}_{n_1 o_1} \\
&0 \leq p^{\mathrm{R}}_{n_2 o_1} \leq 100 \qquad : \mu^{\mathrm{R^{min}}}_{n_2 o_1}, \mu^{\mathrm{R^{max}}}_{n_2 o_1} \\
&0 \leq p^{\mathrm{D}}_{n_1 o_1} \leq 300 \qquad : \mu^{\mathrm{D^{min}}}_{n_1 o_1}, \mu^{\mathrm{D^{max}}}_{n_1 o_1} \\
&1000 \left(\theta_{n_1 o_1} - \theta_{n_2 o_1}\right) \leq 200 \qquad : \mu^{\mathrm{F}}_{n_1 n_2 o_1} \\
&1000 \left(\theta_{n_2 o_1} - \theta_{n_1 o_1}\right) \leq 200 \qquad : \mu^{\mathrm{F}}_{n_2 n_1 o_1} \\
&-\pi \leq \theta_{n_1 o_1} \leq \pi \qquad : \mu^{\theta^{\min}}_{n_1 o_1}, \mu^{\theta^{\max}}_{n_1 o_1} \\
&-\pi \leq \theta_{n_2 o_1} \leq \pi \qquad : \mu^{\theta^{\min}}_{n_2 o_1}, \mu^{\theta^{\max}}_{n_2 o_1} \\
&\theta_{n_1 o_1} = 0 \qquad : \mu^{\theta^{\mathrm{ref}}}_{o_1}.
\end{aligned}
$$

The primal optimization variables of the lower-level problem are included in set $\Xi^{\mathrm{P,Ex2}}_{o_1} = \{p^{\mathrm{C}}_{n_1 o_1}, p^{\mathrm{E}}_{n_1 o_1}, p^{\mathrm{R}}_{n_2 o_1}, p^{\mathrm{D}}_{n_1 o_1}, \theta_{n_1 o_1}, \theta_{n_2 o_1}\}$, while the dual optimization variables of this problem are those included in set $\Xi^{\mathrm{D,Ex2}}_{o_1} = \{\lambda_{n_1 o_1}, \lambda_{n_2 o_1}\, \mu^{\mathrm{C^{min}}}_{n_1 o_1}, \mu^{\mathrm{C^{max}}}_{n_1 o_1}, \mu^{\mathrm{E^{min}}}_{n_1 o_1}, \mu^{\mathrm{E^{max}}}_{n_1 o_1}, \mu^{\mathrm{R^{min}}}_{n_2 o_1}, \mu^{\mathrm{R^{max}}}_{n_2 o_1}, \mu^{\mathrm{D^{min}}}_{n_1 o_1}, \mu^{\mathrm{D^{max}}}_{n_1 o_1}, \mu^{\mathrm{F}}_{n_1 n_2 o_1}, \mu^{\mathrm{F}}_{n_2 n_1 o_1}, \mu^{\theta^{\min}}_{n_1 o_1}, \mu^{\theta^{\max}}_{n_1 o_1}, \mu^{\theta^{\min}}_{n_2 o_1}, \mu^{\theta^{\max}}_{n_2 o_1}, \mu^{\theta^{\mathrm{ref}}}_{o_1}\}$. Finally, the primal variables of the upper-level problem are those included in set $\Xi^{\mathrm{U,Ex2}} = \{x^{\mathrm{C}}_{n_1}, u^{\mathrm{C}}_{n_1 h_1}, u^{\mathrm{C}}_{n_1 h_2}, \alpha^{\mathrm{C}}_{n_1 o_1}, \alpha^{\mathrm{E}}_{n_1 o_1}\}$ plus $\Xi^{\mathrm{P,Ex2}}_{o_1}$ and $\Xi^{\mathrm{D,Ex2}}_{o_1}$. □

Illustrative Example 5.3 *Network-constrained static investment bilevel model considering stochastic units as investment options*

This example is similar to Illustrative Example 5.2, but in addition to the candidate conventional unit at node n_1, a stochastic production unit (wind power) at node n_2 is considered as an investment option. The maximum capacity of the stochastic production unit to be built is 200 MW and its annualized investment cost is \$66,000/MW. Two operating conditions (o_1 and o_2) are considered with the following characteristics:

- o_1: demand factor equals 1.00 p.u. and wind power capacity factor equals 0.35 p.u.
- o_2: demand factor equals 0.80 p.u. and wind power capacity factor equals 0.70 p.u.

The weighting factor associated with condition o_1 is 3530, while that of condition o_2 is 5230. Note that the summation of those two factors is 8760, i.e., the number of hours in a year. The consumer bids in conditions o_1 and o_2 at \$35/MWh and \$32/MWh, respectively. Other input data are identical to those in Illustrative Example 5.2.

Accordingly, we formulate below a bilevel problem including an upper-level problem and two lower-level problems, one corresponding to the market clearing in operating condition o_1, and another pertaining to the market clearing in operating condition o_2:

$$\min_{\Xi^{\mathrm{UL,Ex3}} \cup \Xi_{o_1}^{\mathrm{P,Ex3}} \cup \Xi_{o_1}^{\mathrm{D,Ex3}} \cup \Xi_{o_2}^{\mathrm{P,Ex3}} \cup \Xi_{o_2}^{\mathrm{D,Ex3}}} \quad 55000\, x_{n_1}^{\mathrm{C}} + 66000\, x_{n_2}^{\mathrm{S}}$$
$$- 3530 \left[\lambda_{n_1 o_1} \left(p_{n_1 o_1}^{\mathrm{E}} + p_{n_1 o_1}^{\mathrm{C}} \right) + \lambda_{n_2 o_1}\, p_{n_2 o_1}^{\mathrm{S}} - 12\, p_{n_1 o_1}^{\mathrm{C}} - 10\, p_{n_1 o_1}^{\mathrm{E}} \right]$$
$$- 5230 \left[\lambda_{n_1 o_2} \left(p_{n_1 o_2}^{\mathrm{E}} + p_{n_1 o_2}^{\mathrm{C}} \right) + \lambda_{n_2 o_2}\, p_{n_2 o_2}^{\mathrm{S}} - 12\, p_{n_1 o_2}^{\mathrm{C}} - 10\, p_{n_1 o_2}^{\mathrm{E}} \right]$$

subject to

$$x_{n_1}^{\mathrm{C}} = 100\, u_{n_1 h_2}^{\mathrm{C}}$$
$$u_{n_1 h_1}^{\mathrm{C}} + u_{n_1 h_2}^{\mathrm{C}} = 1$$
$$u_{n_1 h_1}^{\mathrm{C}},\ u_{n_1 h_2}^{\mathrm{C}} \in \{0, 1\}$$
$$0 \le x_{n_2}^{\mathrm{S}} \le 200$$
$$\left(55000\, x_{n_1}^{\mathrm{C}} + 66000\, x_{n_2}^{\mathrm{S}}\right) \le 10^7$$
$$\alpha_{n_1 o_1}^{\mathrm{C}} \ge 0$$
$$\alpha_{n_1 o_1}^{\mathrm{E}} \ge 0$$
$$\alpha_{n_1 o_2}^{\mathrm{C}} \ge 0$$
$$\alpha_{n_1 o_2}^{\mathrm{E}} \ge 0$$

Lower-level problems associated with operating conditions o_1 and o_2,

where the lower-level problem referring to the market clearing in operating condition o_1 is:

$$\min_{\Xi_{o_1}^{\mathrm{P,Ex3}}} \quad \alpha_{n_1 o_1}^{\mathrm{C}}\, p_{n_1 o_1}^{\mathrm{C}} + \alpha_{n_1 o_1}^{\mathrm{E}}\, p_{n_1 o_1}^{\mathrm{E}} + 15\, p_{n_2 o_1}^{\mathrm{R}} - 35\, p_{n_1 o_1}^{\mathrm{D}}$$

subject to

$$p_{n_1 o_1}^{\mathrm{D}} + 1000 \left(\theta_{n_1 o_1} - \theta_{n_2 o_1}\right) - p_{n_1 o_1}^{\mathrm{E}} - p_{n_1 o_1}^{\mathrm{C}} = 0 \quad : \lambda_{n_1 o_1}$$
$$1000 \left(\theta_{n_2 o_1} - \theta_{n_1 o_1}\right) - p_{n_2 o_1}^{\mathrm{S}} - p_{n_2 o_1}^{\mathrm{R}} = 0 \quad : \lambda_{n_2 o_1}$$
$$0 \le p_{n_1 o_1}^{\mathrm{C}} \le x_{n_1}^{\mathrm{C}} \quad : \mu_{n_1 o_1}^{\mathrm{C^{min}}}, \mu_{n_1 o_1}^{\mathrm{C^{max}}}$$
$$0 \le p_{n_2 o_1}^{\mathrm{S}} \le 0.35\, x_{n_2}^{\mathrm{S}} \quad : \mu_{n_2 o_1}^{\mathrm{S^{min}}}, \mu_{n_2 o_1}^{\mathrm{S^{max}}}$$
$$0 \le p_{n_1 o_1}^{\mathrm{E}} \le 150 \quad : \mu_{n_1 o_1}^{\mathrm{E^{min}}}, \mu_{n_1 o_1}^{\mathrm{E^{max}}}$$
$$0 \le p_{n_2 o_1}^{\mathrm{R}} \le 100 \quad : \mu_{n_2 o_1}^{\mathrm{R^{min}}}, \mu_{n_2 o_1}^{\mathrm{R^{max}}}$$
$$0 \le p_{n_1 o_1}^{\mathrm{D}} \le 300 \quad : \mu_{n_1 o_1}^{\mathrm{D^{min}}}, \mu_{n_1 o_1}^{\mathrm{D^{max}}}$$
$$1000 \left(\theta_{n_1 o_1} - \theta_{n_2 o_1}\right) \le 200 \quad : \mu_{n_1 n_2 o_1}^{\mathrm{F}}$$

$$
\begin{aligned}
&1000\left(\theta_{n_2 o_1} - \theta_{n_1 o_1}\right) \leq 200 \quad : \mu^{\mathrm{F}}_{n_2 n_1 o_1}\\
&-\pi \leq \theta_{n_1 o_1} \leq \pi \quad : \mu^{\theta^{\min}}_{n_1 o_1}, \mu^{\theta^{\max}}_{n_1 o_1}\\
&-\pi \leq \theta_{n_2 o_1} \leq \pi \quad : \mu^{\theta^{\min}}_{n_2 o_1}, \mu^{\theta^{\max}}_{n_2 o_1}\\
&\theta_{n_1 o_1} = 0 \quad : \mu^{\theta^{\mathrm{ref}}}_{o_1}.
\end{aligned}
$$

In addition, the lower-level problem pertaining to the market clearing in operating condition o_2 is:

$$
\begin{aligned}
&\min_{\Xi^{\mathrm{P,Ex3}}_{o_2}} \quad \alpha^{\mathrm{C}}_{n_1 o_2} p^{\mathrm{C}}_{n_1 o_2} + \alpha^{\mathrm{E}}_{n_1 o_2} p^{\mathrm{E}}_{n_1 o_2} + 15\, p^{\mathrm{R}}_{n_2 o_2} - 32\, p^{\mathrm{D}}_{n_1 o_2}\\
&\text{subject to}\\
&p^{\mathrm{D}}_{n_1 o_2} + 1000\left(\theta_{n_1 o_2} - \theta_{n_2 o_2}\right) - p^{\mathrm{E}}_{n_1 o_2} - p^{\mathrm{C}}_{n_1 o_2} = 0 \quad : \lambda_{n_1 o_2}\\
&1000\left(\theta_{n_2 o_2} - \theta_{n_1 o_2}\right) - p^{\mathrm{S}}_{n_2 o_2} - p^{\mathrm{R}}_{n_2 o_2} = 0 \quad : \lambda_{n_2 o_2}\\
&0 \leq p^{\mathrm{C}}_{n_1 o_2} \leq x^{\mathrm{C}}_{n_1} \quad : \mu^{\mathrm{C}^{\min}}_{n_1 o_2}, \mu^{\mathrm{C}^{\max}}_{n_1 o_2}\\
&0 \leq p^{\mathrm{S}}_{n_2 o_2} \leq 0.70\, x^{\mathrm{S}}_{n_2} \quad : \mu^{\mathrm{S}^{\min}}_{n_2 o_2}, \mu^{\mathrm{S}^{\max}}_{n_2 o_2}\\
&0 \leq p^{\mathrm{E}}_{n_1 o_2} \leq 150 \quad : \mu^{\mathrm{E}^{\min}}_{n_1 o_2}, \mu^{\mathrm{E}^{\max}}_{n_1 o_2}\\
&0 \leq p^{\mathrm{R}}_{n_2 o_2} \leq 100 \quad : \mu^{\mathrm{R}^{\min}}_{n_2 o_2}, \mu^{\mathrm{R}^{\max}}_{n_2 o_2}\\
&0 \leq p^{\mathrm{D}}_{n_1 o_2} \leq (0.80 \times 300) \quad : \mu^{\mathrm{D}^{\min}}_{n_1 o_2}, \mu^{\mathrm{D}^{\max}}_{n_1 o_2}\\
&1000\left(\theta_{n_1 o_2} - \theta_{n_2 o_2}\right) \leq 200 \quad : \mu^{\mathrm{F}}_{n_1 n_2 o_2}\\
&1000\left(\theta_{n_2 o_2} - \theta_{n_1 o_2}\right) \leq 200 \quad : \mu^{\mathrm{F}}_{n_2 n_1 o_2}\\
&-\pi \leq \theta_{n_1 o_2} \leq \pi \quad : \mu^{\theta^{\min}}_{n_1 o_2}, \mu^{\theta^{\max}}_{n_1 o_2}\\
&-\pi \leq \theta_{n_2 o_2} \leq \pi \quad : \mu^{\theta^{\min}}_{n_2 o_2}, \mu^{\theta^{\max}}_{n_2 o_2}\\
&\theta_{n_1 o_2} = 0 \quad : \mu^{\theta^{\mathrm{ref}}}_{o_2}.
\end{aligned}
$$

The primal optimization variables of the lower-level problem for operating condition o_1 are included in set $\Xi^{\mathrm{P,Ex3}}_{o_1} = \{p^{\mathrm{C}}_{n_1 o_1}, p^{\mathrm{S}}_{n_2 o_1}, p^{\mathrm{E}}_{n_1 o_1}, p^{\mathrm{R}}_{n_2 o_1}, p^{\mathrm{D}}_{n_1 o_1}, \theta_{n_1 o_1}, \theta_{n_2 o_1}\}$, while the dual optimization variables of this problem are in set $\Xi^{\mathrm{D,Ex3}}_{o_1} = \{\lambda_{n_1 o_1}, \lambda_{n_2 o_1}$ $\mu^{\mathrm{C}^{\min}}_{n_1 o_1}, \mu^{\mathrm{C}^{\max}}_{n_1 o_1}, \mu^{\mathrm{S}^{\min}}_{n_2 o_1}, \mu^{\mathrm{S}^{\max}}_{n_2 o_1}, \mu^{\mathrm{E}^{\min}}_{n_1 o_1}, \mu^{\mathrm{E}^{\max}}_{n_1 o_1}, \mu^{\mathrm{R}^{\min}}_{n_2 o_1}, \mu^{\mathrm{R}^{\max}}_{n_2 o_1}, \mu^{\mathrm{D}^{\min}}_{n_1 o_1}, \mu^{\mathrm{D}^{\max}}_{n_1 o_1}, \mu^{\mathrm{F}}_{n_1 n_2 o_1}, \mu^{\mathrm{F}}_{n_2 n_1 o_1},$ $\mu^{\theta^{\min}}_{n_1 o_1}, \mu^{\theta^{\max}}_{n_1 o_1}, \mu^{\theta^{\min}}_{n_2 o_1}, \mu^{\theta^{\max}}_{n_2 o_1}, \mu^{\theta^{\mathrm{ref}}}_{o_1}\}$. Likewise, the primal optimization variables of the lower-level problem for operating condition o_2 are included in set $\Xi^{\mathrm{P,Ex3}}_{o_2} = \{p^{\mathrm{C}}_{n_1 o_2},$ $p^{\mathrm{S}}_{n_2 o_2}, p^{\mathrm{E}}_{n_1 o_2}, p^{\mathrm{R}}_{n_2 o_2}, p^{\mathrm{D}}_{n_1 o_2}, \theta_{n_1 o_2}, \theta_{n_2 o_2}\}$, while the dual optimization variables of this problem are in set $\Xi^{\mathrm{D,Ex3}}_{o_2} = \{\lambda_{n_1 o_2}, \lambda_{n_2 o_2}$ $\mu^{\mathrm{C}^{\min}}_{n_1 o_2}, \mu^{\mathrm{C}^{\max}}_{n_1 o_2}, \mu^{\mathrm{S}^{\min}}_{n_2 o_2}, \mu^{\mathrm{S}^{\max}}_{n_2 o_2}, \mu^{\mathrm{E}^{\min}}_{n_1 o_2}, \mu^{\mathrm{E}^{\max}}_{n_1 o_2},$ $\mu^{\mathrm{R}^{\min}}_{n_2 o_2}, \mu^{\mathrm{R}^{\max}}_{n_2 o_2}, \mu^{\mathrm{D}^{\min}}_{n_1 o_2}, \mu^{\mathrm{D}^{\max}}_{n_1 o_2}, \mu^{\mathrm{F}}_{n_1 n_2 o_2}, \mu^{\mathrm{F}}_{n_2 n_1 o_2}, \mu^{\theta^{\min}}_{n_1 o_2}, \mu^{\theta^{\max}}_{n_1 o_2}, \mu^{\theta^{\min}}_{n_2 o_2}, \mu^{\theta^{\max}}_{n_2 o_2}, \mu^{\theta^{\mathrm{ref}}}_{o_2}\}$. Finally, the primal variables of the upper-level problem include those in set $\Xi^{\mathrm{U,Ex3}} =$ $\{x^{\mathrm{C}}_{n_1}, x^{\mathrm{S}}_{n_2}, u^{\mathrm{C}}_{n_1 h_1}, u^{\mathrm{C}}_{n_1 h_2}, \alpha^{\mathrm{C}}_{n_1 o_1}, \alpha^{\mathrm{C}}_{n_1 o_2}, \alpha^{\mathrm{E}}_{n_1 o_1}, \alpha^{\mathrm{E}}_{n_1 o_2}\}$ plus $\Xi^{\mathrm{P,Ex3}}_{o_1}$, $\Xi^{\mathrm{D,Ex3}}_{o_1}$, $\Xi^{\mathrm{P,Ex3}}_{o_2}$, and $\Xi^{\mathrm{D,Ex3}}_{o_2}$. □

Illustrative Example 5.4 *Stochastic network-constrained static investment bilevel model considering stochastic production and rival producer's investment uncertainties*

This example is similar to Illustrative Example 5.3, but we consider two scenarios (ω_1 and ω_2) to characterize the rival producer's investment uncertainty. These scenarios are listed below:

- ω_1: the rival producer does not invest in new units. The probability of this scenario is 0.6.
- ω_2: the rival producer builds a 40-MW new conventional unit at node n_1 and offers its production at \$13/MWh. The probability of this scenario is 0.4.

Other input data are identical to those in Illustrative Example 5.3. Among Illustrative Examples 5.1–5.4, this example is the most general since different operating conditions and scenarios are considered.

We formulate the bilevel model below including an upper-level problem and four lower-level problems, one per operating condition and scenario:

$$
\begin{aligned}
\min_{\Xi^{\text{UL,Ex4}} \cup \Xi^{\text{P,Ex4}}_{o_1\omega_1} \cup \Xi^{\text{D,Ex4}}_{o_1\omega_1} \cup \Xi^{\text{P,Ex4}}_{o_2\omega_1} \cup \Xi^{\text{D,Ex4}}_{o_2\omega_1} \cup \Xi^{\text{P,Ex4}}_{o_1\omega_2} \cup \Xi^{\text{D,Ex4}}_{o_1\omega_2} \cup \Xi^{\text{P,Ex4}}_{o_2\omega_2} \cup \Xi^{\text{D,Ex4}}_{o_2\omega_2}} \quad & 55000\, x^{\text{C}}_{n_1} + 66000\, x^{\text{S}}_{n_2} \\
& - 0.6 \left\{ 3530 \left[\lambda_{n_1 o_1 \omega_1} \left(p^{\text{C}}_{n_1 o_1 \omega_1} + p^{\text{E}}_{n_1 o_1 \omega_1} \right) + \lambda_{n_2 o_1 \omega_1}\, p^{\text{S}}_{n_2 o_1 \omega_1} \right. \right. \\
& \left. - 12\, p^{\text{C}}_{n_1 o_1 \omega_1} - 10\, p^{\text{E}}_{n_1 o_1 \omega_1} \right] \\
& + 5230 \left[\lambda_{n_1 o_2 \omega_1} \left(p^{\text{C}}_{n_1 o_2 \omega_1} + p^{\text{E}}_{n_1 o_2 \omega_1} \right) + \lambda_{n_2 o_2 \omega_1}\, p^{\text{S}}_{n_2 o_2 \omega_1} \right. \\
& \left. \left. - 12\, p^{\text{C}}_{n_1 o_2 \omega_1} - 10\, p^{\text{E}}_{n_1 o_2 \omega_1} \right] \right\} \\
& - 0.4 \left\{ 3530 \left[\lambda_{n_1 o_1 \omega_2} \left(p^{\text{C}}_{n_1 o_1 \omega_2} + p^{\text{E}}_{n_1 o_1 \omega_2} \right) + \lambda_{n_2 o_1 \omega_2}\, p^{\text{S}}_{n_2 o_1 \omega_2} \right. \right. \\
& \left. - 12\, p^{\text{C}}_{n_1 o_1 \omega_2} - 10\, p^{\text{E}}_{n_1 o_1 \omega_2} \right] \\
& + 5230 \left[\lambda_{n_1 o_2 \omega_2} \left(p^{\text{C}}_{n_1 o_2 \omega_2} + p^{\text{E}}_{n_1 o_2 \omega_2} \right) + \lambda_{n_2 o_2 \omega_2}\, p^{\text{S}}_{n_2 o_2 \omega_2} \right. \\
& \left. \left. - 12\, p^{\text{C}}_{n_1 o_2 \omega_2} - 10\, p^{\text{E}}_{n_1 o_2 \omega_2} \right] \right\}
\end{aligned}
$$

subject to

$$
\begin{aligned}
& x^{\text{C}}_{n_1} = 100\, u^{\text{C}}_{n_1 h_2} \\
& u^{\text{C}}_{n_1 h_1} + u^{\text{C}}_{n_1 h_2} = 1 \\
& u^{\text{C}}_{n_1 h_1},\ u^{\text{C}}_{n_1 h_2} \in \{0, 1\}
\end{aligned}
$$

$$0 \leq x^{\mathrm{S}}_{n_2} \leq 200$$
$$\left(55000\, x^{\mathrm{C}}_{n_1} + 66000\, x^{\mathrm{S}}_{n_2}\right) \leq 10^7$$
$$\alpha^{\mathrm{C}}_{n_1 o_1 \omega_1} \geq 0$$
$$\alpha^{\mathrm{E}}_{n_1 o_1 \omega_1} \geq 0$$
$$\alpha^{\mathrm{C}}_{n_1 o_2 \omega_1} \geq 0$$
$$\alpha^{\mathrm{E}}_{n_1 o_2 \omega_1} \geq 0$$
$$\alpha^{\mathrm{C}}_{n_1 o_1 \omega_2} \geq 0$$
$$\alpha^{\mathrm{E}}_{n_1 o_1 \omega_2} \geq 0$$
$$\alpha^{\mathrm{C}}_{n_1 o_2 \omega_2} \geq 0$$
$$\alpha^{\mathrm{E}}_{n_1 o_2 \omega_2} \geq 0$$

Four lower-level problems associated with operating conditions and scenarios,

where the first lower-level problem corresponds to the market clearing in operating condition o_1 under scenario ω_1 and is given below:

$$\min_{\Xi^{\mathrm{P.Ex4}}_{o_1 \omega_1}} \left[\alpha^{\mathrm{C}}_{n_1 o_1 \omega_1}\, p^{\mathrm{C}}_{n_1 o_1 \omega_1} + \alpha^{\mathrm{E}}_{n_1 o_1 \omega_1}\, p^{\mathrm{E}}_{n_1 o_1 \omega_1} + 13\, p^{\mathrm{R}}_{n_1 o_1 \omega_1} + 15\, p^{\mathrm{R}}_{n_2 o_1 \omega_1} - 35\, p^{\mathrm{D}}_{n_1 o_1 \omega_1}\right]$$

subject to

$$p^{\mathrm{D}}_{n_1 o_1 \omega_1} + 1000\left(\theta_{n_1 o_1 \omega_1} - \theta_{n_2 o_1 \omega_1}\right) - p^{\mathrm{E}}_{n_1 o_1 \omega_1} - p^{\mathrm{R}}_{n_1 o_1 \omega_1} - p^{\mathrm{C}}_{n_1 o_1 \omega_1} = 0 \quad : \lambda_{n_1 o_1 \omega_1}$$
$$1000\left(\theta_{n_2 o_1 \omega_1} - \theta_{n_1 o_1 \omega_1}\right) - p^{\mathrm{S}}_{n_2 o_1 \omega_1} - p^{\mathrm{R}}_{n_2 o_1 \omega_1} = 0 \quad : \lambda_{n_2 o_1 \omega_1}$$
$$0 \leq p^{\mathrm{C}}_{n_1 o_1 \omega_1} \leq x^{\mathrm{C}}_{n_1} \quad : \mu^{\mathrm{C^{min}}}_{n_1 o_1 \omega_1}, \mu^{\mathrm{C^{max}}}_{n_1 o_1 \omega_1}$$
$$0 \leq p^{\mathrm{S}}_{n_2 o_1 \omega_1} \leq 0.35\, x^{\mathrm{S}}_{n_2} \quad : \mu^{\mathrm{S^{min}}}_{n_2 o_1 \omega_1}, \mu^{\mathrm{S^{max}}}_{n_2 o_1 \omega_1}$$
$$0 \leq p^{\mathrm{E}}_{n_1 o_1 \omega_1} \leq 150 \quad : \mu^{\mathrm{E^{min}}}_{n_1 o_1 \omega_1}, \mu^{\mathrm{E^{max}}}_{n_1 o_1 \omega_1}$$
$$0 \leq p^{\mathrm{R}}_{n_1 o_1 \omega_1} \leq 0 \quad : \mu^{\mathrm{R^{min}}}_{n_1 o_1 \omega_1}, \mu^{\mathrm{R^{max}}}_{n_1 o_1 \omega_1}$$
$$0 \leq p^{\mathrm{R}}_{n_2 o_1 \omega_1} \leq 100 \quad : \mu^{\mathrm{R^{min}}}_{n_2 o_1 \omega_1}, \mu^{\mathrm{R^{max}}}_{n_2 o_1 \omega_1}$$
$$0 \leq p^{\mathrm{D}}_{n_1 o_1 \omega_1} \leq 300 \quad : \mu^{\mathrm{D^{min}}}_{n_1 o_1 \omega_1}, \mu^{\mathrm{D^{max}}}_{n_1 o_1 \omega_1}$$
$$1000\left(\theta_{n_1 o_1 \omega_1} - \theta_{n_2 o_1 \omega_1}\right) \leq 200 \quad : \mu^{\mathrm{F}}_{n_1 n_2 o_1 \omega_1}$$
$$1000\left(\theta_{n_2 o_1 \omega_1} - \theta_{n_1 o_1 \omega_1}\right) \leq 200 \quad : \mu^{\mathrm{F}}_{n_2 n_1 o_1 \omega_1}$$
$$-\pi \leq \theta_{n_1 o_1 \omega_1} \leq \pi \quad : \mu^{\theta^{\min}}_{n_1 o_1 \omega_1}, \mu^{\theta^{\max}}_{n_1 o_1 \omega_1}$$
$$-\pi \leq \theta_{n_2 o_1 \omega_1} \leq \pi \quad : \mu^{\theta^{\min}}_{n_2 o_1 \omega_1}, \mu^{\theta^{\max}}_{n_2 o_1 \omega_1}$$
$$\theta_{n_1 o_1 \omega_1} = 0 \quad : \mu^{\theta^{\mathrm{ref}}}_{o_1 \omega_1}.$$

In addition, the lower-level problem corresponding to the market clearing in operating condition o_2 under scenario ω_1 is:

$$\begin{aligned}
&\min_{\Xi^{\text{P,Ex4}}_{o_2\omega_1}} \quad \left[\alpha^{\text{C}}_{n_1o_2\omega_1}\ p^{\text{C}}_{n_1o_2\omega_1} + \alpha^{\text{E}}_{n_1o_2\omega_1}\ p^{\text{E}}_{n_1o_2\omega_1} + 13\ p^{\text{R}}_{n_1o_2\omega_1}\right.\\
&\qquad\qquad \left. +15\ p^{\text{R}}_{n_2o_2\omega_1} - 32\ p^{\text{D}}_{n_1o_2\omega_1}\right]\\
&\text{subject to}\\
&p^{\text{D}}_{n_1o_2\omega_1} + 1000\left(\theta_{n_1o_2\omega_1} - \theta_{n_2o_2\omega_1}\right)\\
&\qquad - p^{\text{E}}_{n_1o_2\omega_1} - p^{\text{R}}_{n_1o_2\omega_1} - p^{\text{C}}_{n_1o_2\omega_1} = 0 \quad : \lambda_{n_1o_2\omega_1}\\
&1000\left(\theta_{n_2o_2\omega_1} - \theta_{n_1o_2\omega_1}\right) - p^{\text{S}}_{n_2o_2\omega_1} - p^{\text{R}}_{n_2o_2\omega_1} = 0 \quad : \lambda_{n_2o_2\omega_1}\\
&0 \leq p^{\text{C}}_{n_1o_2\omega_1} \leq x^{\text{C}}_{n_1} \quad : \mu^{\text{C}^{\min}}_{n_1o_2\omega_1}, \mu^{\text{C}^{\max}}_{n_1o_2\omega_1}\\
&0 \leq p^{\text{S}}_{n_2o_2\omega_1} \leq 0.70\ x^{\text{S}}_{n_2} \quad : \mu^{\text{S}^{\min}}_{n_2o_2\omega_1}, \mu^{\text{S}^{\max}}_{n_2o_2\omega_1}\\
&0 \leq p^{\text{E}}_{n_1o_2\omega_1} \leq 150 \quad : \mu^{\text{E}^{\min}}_{n_1o_2\omega_1}, \mu^{\text{E}^{\max}}_{n_1o_2\omega_1}\\
&0 \leq p^{\text{R}}_{n_1o_2\omega_1} \leq 0 \quad : \mu^{\text{R}^{\min}}_{n_1o_2\omega_1}, \mu^{\text{R}^{\max}}_{n_1o_2\omega_1}\\
&0 \leq p^{\text{R}}_{n_2o_2\omega_1} \leq 100 \quad : \mu^{\text{R}^{\min}}_{n_2o_2\omega_1}, \mu^{\text{R}^{\max}}_{n_2o_2\omega_1}\\
&0 \leq p^{\text{D}}_{n_1o_2\omega_1} \leq (0.80 \times 300) \quad : \mu^{\text{D}^{\min}}_{n_1o_2\omega_1}, \mu^{\text{D}^{\max}}_{n_1o_2\omega_1}\\
&1000\left(\theta_{n_1o_2\omega_1} - \theta_{n_2o_2\omega_1}\right) \leq 200 \quad : \mu^{\text{F}}_{n_1n_2o_2\omega_1}\\
&1000\left(\theta_{n_2o_2\omega_1} - \theta_{n_1o_2\omega_1}\right) \leq 200 \quad : \mu^{\text{F}}_{n_2n_1o_2\omega_1}\\
&-\pi \leq \theta_{n_1o_2\omega_1} \leq \pi \quad : \mu^{\theta^{\min}}_{n_1o_2\omega_1}, \mu^{\theta^{\max}}_{n_1o_2\omega_1}\\
&-\pi \leq \theta_{n_2o_2\omega_1} \leq \pi \quad : \mu^{\theta^{\min}}_{n_2o_2\omega_1}, \mu^{\theta^{\max}}_{n_2o_2\omega_1}\\
&\theta_{n_1o_2\omega_1} = 0 \quad : \mu^{\theta^{\text{ref}}}_{o_2\omega_1}.
\end{aligned}$$

Additionally, the lower-level problem corresponding to the market clearing in operating condition o_1 under scenario ω_2 is:

$$\begin{aligned}
&\min_{\Xi^{\text{P,Ex4}}_{o_1\omega_2}} \quad \left[\alpha^{\text{C}}_{n_1o_1\omega_2}\ p^{\text{C}}_{n_1o_1\omega_2} + \alpha^{\text{E}}_{n_1o_1\omega_2}\ p^{\text{E}}_{n_1o_1\omega_2} + 13\ p^{\text{R}}_{n_1o_1\omega_2}\right.\\
&\qquad\qquad \left. +15\ p^{\text{R}}_{n_2o_1\omega_2} - 35\ p^{\text{D}}_{n_1o_1\omega_2}\right]\\
&\text{subject to}\\
&p^{\text{D}}_{n_1o_1\omega_2} + 1000\left(\theta_{n_1o_1\omega_2} - \theta_{n_2o_1\omega_2}\right)\\
&\qquad - p^{\text{E}}_{n_1o_1\omega_2} - p^{\text{R}}_{n_1o_1\omega_2} - p^{\text{C}}_{n_1o_1\omega_2} = 0 \quad : \lambda_{n_1o_1\omega_2}\\
&1000\left(\theta_{n_2o_1\omega_2} - \theta_{n_1o_1\omega_2}\right) - p^{\text{S}}_{n_2o_1\omega_2} - p^{\text{R}}_{n_2o_1\omega_2} = 0 \quad : \lambda_{n_2o_1\omega_2}\\
&0 \leq p^{\text{C}}_{n_1o_1\omega_2} \leq x^{\text{C}}_{n_1} \quad : \mu^{\text{C}^{\min}}_{n_1o_1\omega_2}, \mu^{\text{C}^{\max}}_{n_1o_1\omega_2}\\
&0 \leq p^{\text{S}}_{n_2o_1\omega_2} \leq 0.35\ x^{\text{S}}_{n_2} \quad : \mu^{\text{S}^{\min}}_{n_2o_1\omega_2}, \mu^{\text{S}^{\max}}_{n_2o_1\omega_2}\\
&0 \leq p^{\text{E}}_{n_1o_1\omega_2} \leq 150 \quad : \mu^{\text{E}^{\min}}_{n_1o_1\omega_2}, \mu^{\text{E}^{\max}}_{n_1o_1\omega_2}\\
&0 \leq p^{\text{R}}_{n_1o_1\omega_2} \leq 40 \quad : \mu^{\text{R}^{\min}}_{n_1o_1\omega_2}, \mu^{\text{R}^{\max}}_{n_1o_1\omega_2}
\end{aligned}$$

$$0 \leq p^{\text{R}}_{n_2 o_1 \omega_2} \leq 100 \quad : \mu^{\text{R}^{\min}}_{n_2 o_1 \omega_2}, \mu^{\text{R}^{\max}}_{n_2 o_1 \omega_2}$$
$$0 \leq p^{\text{D}}_{n_1 o_1 \omega_2} \leq 300 \quad : \mu^{\text{D}^{\min}}_{n_1 o_1 \omega_2}, \mu^{\text{D}^{\max}}_{n_1 o_1 \omega_2}$$
$$1000 \left(\theta_{n_1 o_1 \omega_2} - \theta_{n_2 o_1 \omega_2}\right) \leq 200 \quad : \mu^{\text{F}}_{n_1 n_2 o_1 \omega_2}$$
$$1000 \left(\theta_{n_2 o_1 \omega_2} - \theta_{n_1 o_1 \omega_2}\right) \leq 200 \quad : \mu^{\text{F}}_{n_2 n_1 o_1 \omega_2}$$
$$-\pi \leq \theta_{n_1 o_1 \omega_2} \leq \pi \quad : \mu^{\theta^{\min}}_{n_1 o_1 \omega_2}, \mu^{\theta^{\max}}_{n_1 o_1 \omega_2}$$
$$-\pi \leq \theta_{n_2 o_1 \omega_2} \leq \pi \quad : \mu^{\theta^{\min}}_{n_2 o_1 \omega_2}, \mu^{\theta^{\max}}_{n_2 o_1 \omega_2}$$
$$\theta_{n_1 o_1 \omega_2} = 0 \quad : \mu^{\theta^{\text{ref}}}_{o_1 \omega_2}.$$

Finally, the lower-level problem corresponding to the market clearing in operating condition o_2 under scenario ω_2 is:

$$\min_{\Xi^{\text{P,Ex4}}_{o_2 \omega_2}} \left[\alpha^{\text{C}}_{n_1 o_2 \omega_2} p^{\text{C}}_{n_1 o_2 \omega_2} + \alpha^{\text{E}}_{n_1 o_2 \omega_2} p^{\text{E}}_{n_1 o_2 \omega_2} + 13\, p^{\text{R}}_{n_1 o_2 \omega_2} + 15\, p^{\text{R}}_{n_2 o_2 \omega_2} - 32\, p^{\text{D}}_{n_1 o_2 \omega_2}\right]$$

subject to

$$p^{\text{D}}_{n_1 o_2 \omega_2} + 1000 \left(\theta_{n_1 o_2 \omega_2} - \theta_{n_2 o_2 \omega_2}\right) - p^{\text{E}}_{n_1 o_2 \omega_2} - p^{\text{R}}_{n_1 o_2 \omega_2} - p^{\text{C}}_{n_1 o_2 \omega_2} = 0 \quad : \lambda_{n_1 o_2 \omega_2}$$
$$1000 \left(\theta_{n_2 o_2 \omega_2} - \theta_{n_1 o_2 \omega_2}\right) - p^{\text{S}}_{n_2 o_2 \omega_2} - p^{\text{R}}_{n_2 o_2 \omega_2} = 0 \quad : \lambda_{n_2 o_2 \omega_2}$$
$$0 \leq p^{\text{C}}_{n_1 o_2 \omega_2} \leq x^{\text{C}}_{n_1} \quad : \mu^{\text{C}^{\min}}_{n_1 o_2 \omega_2}, \mu^{\text{C}^{\max}}_{n_1 o_2 \omega_2}$$
$$0 \leq p^{\text{S}}_{n_2 o_2 \omega_2} \leq 0.70\, x^{\text{S}}_{n_2} \quad : \mu^{\text{S}^{\min}}_{n_2 o_2 \omega_2}, \mu^{\text{S}^{\max}}_{n_2 o_2 \omega_2}$$
$$0 \leq p^{\text{E}}_{n_1 o_2 \omega_2} \leq 150 \quad : \mu^{\text{E}^{\min}}_{n_1 o_2 \omega_2}, \mu^{\text{E}^{\max}}_{n_1 o_2 \omega_2}$$
$$0 \leq p^{\text{R}}_{n_1 o_2 \omega_2} \leq 40 \quad : \mu^{\text{R}^{\min}}_{n_1 o_2 \omega_2}, \mu^{\text{R}^{\max}}_{n_1 o_2 \omega_2}$$
$$0 \leq p^{\text{R}}_{n_2 o_2 \omega_2} \leq 100 \quad : \mu^{\text{R}^{\min}}_{n_2 o_2 \omega_2}, \mu^{\text{R}^{\max}}_{n_2 o_2 \omega_2}$$
$$0 \leq p^{\text{D}}_{n_1 o_2 \omega_2} \leq (0.80 \times 300) \quad : \mu^{\text{D}^{\min}}_{n_1 o_2 \omega_2}, \mu^{\text{D}^{\max}}_{n_1 o_2 \omega_2}$$
$$1000 \left(\theta_{n_1 o_2 \omega_2} - \theta_{n_2 o_2 \omega_2}\right) \leq 200 \quad : \mu^{\text{F}}_{n_1 n_2 o_2 \omega_2}$$
$$1000 \left(\theta_{n_2 o_2 \omega_2} - \theta_{n_1 o_2 \omega_2}\right) \leq 200 \quad : \mu^{\text{F}}_{n_2 n_1 o_2 \omega_2}$$
$$-\pi \leq \theta_{n_1 o_2 \omega_2} \leq \pi \quad : \mu^{\theta^{\min}}_{n_1 o_2 \omega_2}, \mu^{\theta^{\max}}_{n_1 o_2 \omega_2}$$
$$-\pi \leq \theta_{n_2 o_2 \omega_2} \leq \pi \quad : \mu^{\theta^{\min}}_{n_2 o_2 \omega_2}, \mu^{\theta^{\max}}_{n_2 o_2 \omega_2}$$
$$\theta_{n_1 o_2 \omega_2} = 0 \quad : \mu^{\theta^{\text{ref}}}_{o_2 \omega_2}.$$

Note that in Illustrative Example 5.4, the production capacity investment decisions (i.e., $x^{\text{C}}_{n_1}$ and $x^{\text{S}}_{n_2}$) do not change over scenarios. In addition, similar to Illustrative Example 5.3, those decisions are unique for all operating conditions. □

5.3 Dynamic Production Capacity Investment Model

The previous section describes a static model to determine the optimal capacity investment by a strategic producer. As explained in Sect. 5.1.3, a static model constitutes an appropriate tradeoff between modeling accuracy and computational tractability. However, a dynamic (multistage) model allows making investment decisions at different points in time. Thus, this model generally results in better decisions since the investments can be adapted to changes in future system conditions.

Unlike a static model, a dynamic model considers a planning horizon comprising a specific number of time periods indexed by t and running from 1 to T. In turn, each time period spans a specified number of years from Y_1 to Y_N. The strategic producer makes its capacity investment decisions at the initial year (i.e., year Y_1) of each time period t, as illustrated in Fig. 5.3.

The two main advantages of a dynamic investment model are as follows:

- It allows adapting the investment decisions to the uncertainty realizations over the planning horizon. For example, the dynamic model provides different investment decisions at the beginning of time period $t = 2$, one per uncertainty realization at the end of time period $t = 1$.
- It allows considering how uncertainty unfolds over the planning period. For example, it is possible to consider uncertain values for investment costs of production technologies throughout the planning horizon.

Accordingly, the decision sequence for dynamic models is explained below, and the corresponding scenario tree is depicted in Fig. 5.4.

1. At the beginning of the planning horizon, i.e., at the beginning of the first time period ($t = 1$), the strategic producer makes its capacity investment decisions, which are *here-and-now* decisions since they do not depend on any future scenario realization. Note that these investment decisions affect the whole planning horizon

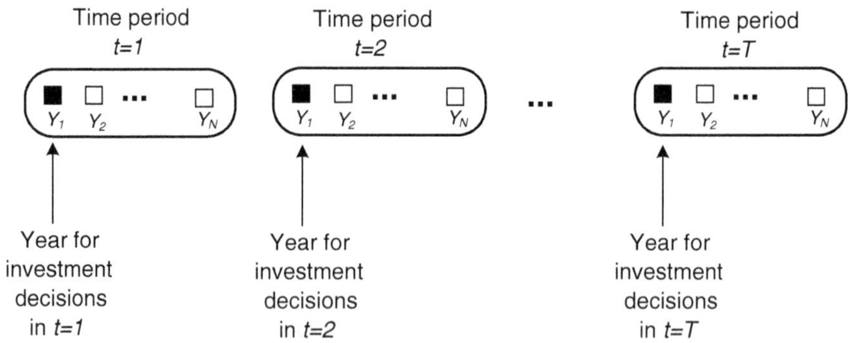

Fig. 5.3 Time model for dynamic production capacity investment

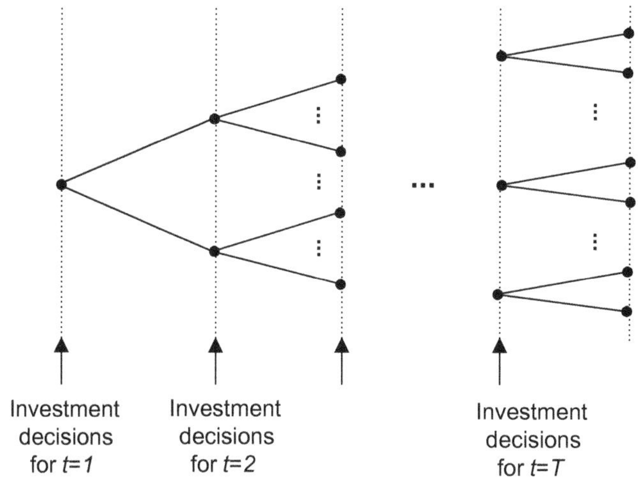

Fig. 5.4 Scenario tree for dynamic production capacity investment

because the installed units at this point in time are available throughout the entire planning horizon.

2. For each scenario realization in period $t = 1$ and for each operating condition, the pool is cleared, and the outcomes are market-clearing prices along with production and consumption levels.
3. Once time period $t = 1$ concludes, the strategic producer knows which scenario in this period is actually realized. Then, it makes its capacity investment decisions for the second time period ($t = 2$), which are *wait-and-see* decisions with respect to the first time period since they depend on the scenario realization throughout $t = 1$. However, they are *here-and-now* decisions with respect to the second and subsequent time periods.
4. For each potential scenario realization in time period $t = 2$ and for each operating condition, the market is cleared.

Steps 3 and 4 above are repeated for each remaining time period of the planning horizon.

The differences in formulation of a dynamic capacity investment model with respect to a static one, i.e., bilevel model (5.1), are as follows:

1. All variables are time-dependent, i.e., they are indexed by t.
2. In addition to each operating condition o and scenario ω, the market needs to be cleared for each time period t.
3. To avoid anticipating information, a set of *nonanticipativity* constraints are required in the upper-level problem. These constraints impose that investments depend on the scenario realizations in previous time periods but they are unique for all scenario realizations in the future.

4. Instead of the *annualized* investment costs, i.e., $K_n^{\mathrm{C}}, \forall n$, and $K_n^{\mathrm{S}}, \forall n$, the *yearly* investment costs of candidate units need to be considered in the dynamic model, i.e., $I_{tn\omega}^{\mathrm{C}}, \forall t, \forall n, \forall \omega$, and $I_{tn\omega}^{\mathrm{S}}, \forall t, \forall n, \forall \omega$ since these investment costs are time-variant (indexed by t). In addition, the values of such costs across time periods are uncertain. Thus, index ω is also incorporated into parameters $I_{tn\omega}^{\mathrm{C}}$ and $I_{tn\omega}^{\mathrm{S}}$.
5. Each investment cost is multiplied in each time period by an amortization rate ($a_t, \forall t$) to make investment costs and profits comparable across time periods. The approach for calculating the amortization rates of a production unit is explained in Appendix A.
6. In addition to investment costs, the maximum demand is considered time-variant, and index t is incorporated into parameter $P_{tn}^{\mathrm{D}^{\max}}$.
7. For the sake of simplicity, other parameters are assumed time-invariant. For example, the operating conditions including the demand factor of each consumer ($Q_{no}^{\mathrm{D}}, \forall o, \forall n$) and the power capacity factor of each stochastic unit ($Q_{no}^{\mathrm{S}}, \forall o, \forall n$) are considered not to change across time periods. In addition, the conventional technologies to be built ($X_{nh}^{\mathrm{C}}, \forall n, \forall h$) are considered to be identical across time periods.
8. For the sake of simplicity, we consider the investment costs of the candidate units as the only source of uncertainty. Other potential sources of uncertainty, e.g., production capacity investment of rival producers and demand growth, can also be incorporated into the model through additional scenarios.
9. The production level of a conventional candidate unit located at node n in time period t and condition o under scenario ω is less than or equal to its cumulative capacity added in that node during time periods $\tau \leq t$, i.e., $\sum_{\tau \leq t} x_{\tau n\omega}^{\mathrm{C}}$. Similarly, the upper bound for the stochastic production at node n in time period t and condition o under scenario ω is $Q_{no}^{\mathrm{S}} \sum_{\tau \leq t} x_{\tau n\omega}^{\mathrm{S}}$.

Considering the assumptions above, the dynamic investment bilevel model is given by (5.2) below:

$$\begin{aligned}\min_{\Xi^{\mathrm{UL}} \cup \Xi_{to\omega}^{\mathrm{Primal}} \cup \Xi_{to\omega}^{\mathrm{Dual}}} \quad & \sum_t \sum_\omega \varphi_\omega \Bigg[a_t \sum_n \left(I_{tn\omega}^{\mathrm{C}} x_{tn\omega}^{\mathrm{C}} + I_{tn\omega}^{\mathrm{S}} x_{tn\omega}^{\mathrm{S}} \right) \\ & - \sum_o \rho_o \sum_n \lambda_{tno\omega} \left(p_{tno\omega}^{\mathrm{C}} + p_{tno\omega}^{\mathrm{E}} + p_{tno\omega}^{\mathrm{S}} \right) \\ & - p_{tno\omega}^{\mathrm{C}} C_n^{\mathrm{C}} - p_{tno\omega}^{\mathrm{E}} C_n^{\mathrm{E}} \Bigg] \end{aligned} \tag{5.2a}$$

subject to

$$x_{tn\omega}^{\mathrm{C}} = \sum_h u_{tn\omega h}^{\mathrm{C}} X_{nh}^{\mathrm{C}} \quad \forall t, \forall n, \forall \omega \tag{5.2b}$$

$$\sum_h u_{tn\omega h}^{\mathrm{C}} = 1 \quad \forall t, \forall n, \forall \omega \tag{5.2c}$$

$$u^{\text{C}}_{tn\omega h} \in \{0, 1\} \quad \forall t, \forall n, \forall \omega, \forall h \tag{5.2d}$$

$$0 \leq \sum_{\tau \leq t} x^{\text{S}}_{\tau n\omega} \leq X^{\text{S}^{\max}}_{n} \quad \forall t, \forall n, \forall \omega \tag{5.2e}$$

$$\sum_{n} \left(I^{\text{C}}_{tn\omega} \, x^{\text{C}}_{tn\omega} + I^{\text{S}}_{tn\omega} \, x^{\text{S}}_{tn\omega} \right) \leq I^{\max}_{t} \quad \forall t, \forall \omega \tag{5.2f}$$

$$\alpha^{\text{C}}_{tno\omega} \geq 0 \quad \forall t, \forall n, \forall o, \forall \omega \tag{5.2g}$$

$$\alpha^{\text{E}}_{tno\omega} \geq 0 \quad \forall t, \forall n, \forall o, \forall \omega \tag{5.2h}$$

$$x^{\text{C}}_{tn\omega} = x^{\text{C}}_{tn\widetilde{\omega}} \quad \forall t, \forall n, \forall \omega, \forall \widetilde{\omega} \; : \; \Theta_{\tau\omega} = \Theta_{\tau\widetilde{\omega}} \quad \forall \tau \leq t \tag{5.2i}$$

$$x^{\text{S}}_{tn\omega} = x^{\text{S}}_{tn\widetilde{\omega}} \quad \forall t, \forall n, \forall \omega, \forall \widetilde{\omega} \; : \; \Theta_{\tau\omega} = \Theta_{\tau\widetilde{\omega}} \quad \forall \tau \leq t \tag{5.2j}$$

$$\text{Lower-level problems (5.2l)–(5.2u)} \quad \forall t, \forall o, \forall \omega. \tag{5.2k}$$

The structure of the bilevel model (5.2) is similar to that of (5.1), except that (5.2i)–(5.2j) are the nonanticipativity constraints. Note that $\Theta_{t\omega}$ is a set of those parameters indexed by scenario ω in time period t, i.e., parameters $I^{\text{C}}_{tn\omega}$ and $I^{\text{S}}_{tn\omega}$. Nonanticipativity constraints impose that investment decisions in time period t and node n are identical over those scenarios (e.g., scenarios ω and $\widetilde{\omega}$) in which the values of parameters $I^{\text{C}}_{tn\omega}$ and $I^{\text{S}}_{tn\omega}$ did not change during the considered time period and the previous ones. In other words, the nonanticipativity constraints enforce the investment decisions in time t under scenarios ω and $\widetilde{\omega}$ to be identical if $I^{\text{C}}_{\tau n\omega} = I^{\text{C}}_{\tau n\widetilde{\omega}}$ and $I^{\text{S}}_{\tau n\omega} = I^{\text{S}}_{\tau n\widetilde{\omega}}$, $\forall \tau \leq t, \forall n$.

Each lower-level problem (5.2l)–(5.2u) included in (5.2k), one per time period t, operating condition o, and scenario ω, is given below. The dual variable of each lower-level constraint is indicated following a colon:

$$\left\{ \min_{\Xi^{\text{Primal}}_{to\omega}} \quad \sum_{n} \left[\alpha^{\text{C}}_{tno\omega} \, p^{\text{C}}_{tno\omega} + \alpha^{\text{E}}_{tno\omega} \, p^{\text{E}}_{tno\omega} + C^{\text{R}}_{no} \, p^{\text{R}}_{tno\omega} - U^{\text{D}}_{no} \, p^{\text{D}}_{tno\omega} \right] \right. \tag{5.2l}$$

subject to

$$p^{\text{D}}_{tno\omega} + \sum_{m \in \Omega_n} B_{nm} \left(\theta_{tno\omega} - \theta_{tmo\omega} \right) - p^{\text{C}}_{tno\omega} - p^{\text{S}}_{tno\omega} - p^{\text{E}}_{tno\omega}$$
$$- p^{\text{R}}_{tno\omega} = 0 \quad : \lambda_{tno\omega} \quad \forall n \tag{5.2m}$$

$$0 \leq p^{\text{C}}_{tno\omega} \leq \sum_{\tau \leq t} x^{\text{C}}_{\tau n\omega} \quad : \mu^{\text{C}^{\min}}_{tno\omega}, \mu^{\text{C}^{\max}}_{tno\omega} \quad \forall n \tag{5.2n}$$

$$0 \leq p^{\text{S}}_{tno\omega} \leq Q^{\text{S}}_{no} \sum_{\tau \leq t} x^{\text{S}}_{\tau n\omega} \quad : \mu^{\text{S}^{\min}}_{tno\omega}, \mu^{\text{S}^{\max}}_{tno\omega} \quad \forall n \tag{5.2o}$$

$$0 \leq p^{\text{E}}_{tno\omega} \leq P^{\text{E}^{\max}}_{n} \quad : \mu^{\text{E}^{\min}}_{tno\omega}, \mu^{\text{E}^{\max}}_{tno\omega} \quad \forall n \tag{5.2p}$$

$$0 \leq p^{\text{R}}_{tno\omega} \leq P^{\text{R}^{\max}}_{n} \quad : \mu^{\text{R}^{\min}}_{tno\omega}, \mu^{\text{R}^{\max}}_{tno\omega} \quad \forall n \tag{5.2q}$$

$$0 \leq p^{\text{D}}_{tno\omega} \leq Q^{\text{D}}_{no} \, P^{\text{D}^{\max}}_{tn} \quad : \mu^{\text{D}^{\min}}_{tno\omega}, \mu^{\text{D}^{\max}}_{tno\omega} \quad \forall n \tag{5.2r}$$

$$B_{nm} (\theta_{tno\omega} - \theta_{tmo\omega}) \leq F^{\max}_{nm} \quad : \mu^{\text{F}}_{tnmo\omega} \quad \forall n, \forall m \in \Omega_n \tag{5.2s}$$

$$\left.\begin{array}{ll} -\pi \leq \theta_{tno\omega} \leq \pi \quad : \mu^{\theta^{\min}}_{tno\omega}, \mu^{\theta^{\max}}_{tno\omega} \quad \forall n & (5.2t) \\ \theta_{tno\omega} = 0 \quad : \mu^{\theta^{\text{ref}}}_{to\omega} \quad n = \text{ref}. & (5.2u) \end{array}\right\} \quad \forall t, \forall o, \forall \omega.$$

Next, we present an illustrative example for this dynamic investment bilevel model. Note that its solution procedure is explained later, in Sects. 5.4 and 5.5.

Illustrative Example 5.5 *Stochastic single-node dynamic investment bilevel model*

This example is similar to Illustrative Example 5.1, but it considers a dynamic production capacity investment model instead of the static one. A ten-year planning horizon including two five-year time periods (t_1 and t_2) is considered. Therefore, new production units can be built at the beginning of the first and the sixth years. Similar to Illustrative Example 5.1, a single technology (conventional) is considered as production capacity investment candidate, but three sizing options are available: $h_1 = 0$ MW, $h_2 = 50$ MW, and $h_3 = 100$ MW. The actual investment cost in the first time period is \$700,000/MW. However, there are two possible investment cost realizations for the second time period, which are characterized by scenarios ω_1 and ω_2. Based on scenario ω_1 with probability 0.4, the investment cost in the second time period is equal to that in the first time period. However, it is 20 % higher in scenario ω_2 with probability 0.6. Recall that the strategic producer should make an investment decision at the beginning of the first time period that does not depend on any scenario realization in the future. On the other hand, it makes two alternative investment decisions at the beginning of the second time period that depend on the investment cost realization at the end of the first time period. Moreover, the amortization rates (a_t) are equal to 30 % and 15 % in the first and second time periods, respectively. In addition, the maximum loads of the consumer in the first and second time periods are 300 MW and 330 MW, respectively. The available investment budget of the strategic producer in each time period is \$50 million. Other input data are identical to those in Illustrative Example 5.1.

The strategic producer solves the following bilevel model at the beginning of time period t_1 to make the most beneficial investment and offering decisions. This bilevel model includes an upper-level problem, and four lower-level problems, one per time period, operating condition, and scenario. Note that the number of lower-level problems is equal to the number of time periods, operating conditions, and scenarios, i.e., 2 × 1 × 2:

$$\min_{\Xi^{\text{UL,Ex5}} \cup \Xi^{\text{P,Ex5}}_{t_1 o_1 \omega_1} \cup \Xi^{\text{D,Ex5}}_{t_1 o_1 \omega_1} \cup \Xi^{\text{P,Ex5}}_{t_1 o_1 \omega_2} \cup \Xi^{\text{D,Ex5}}_{t_1 o_1 \omega_2} \cup \Xi^{\text{P,Ex5}}_{t_2 o_1 \omega_1} \cup \Xi^{\text{D,Ex5}}_{t_2 o_1 \omega_1} \cup \Xi^{\text{P,Ex5}}_{t_2 o_1 \omega_2} \cup \Xi^{\text{D,Ex5}}_{t_2 o_1 \omega_2}}$$

$$0.4 \Bigg\{ (0.30 \times 700000)\, x^{\text{C}}_{t_1 n_1 \omega_1}$$

$$- 8760 \Bigg[\lambda_{t_1 n_1 o_1 \omega_1} \left(p^{\text{C}}_{t_1 n_1 o_1 \omega_1} + p^{\text{E}}_{t_1 n_1 o_1 \omega_1} \right)$$

$$- 12\ p^{\mathrm{C}}_{t_1 n_1 o_1 \omega_1} - 10\ p^{\mathrm{E}}_{t_1 n_1 o_1 \omega_1} \Big] \Big\}$$

$$+ 0.6 \Big\{ (0.30 \times 700000)\ x^{\mathrm{C}}_{t_1 n_1 \omega_2}$$

$$- 8760 \Big[\lambda_{t_1 n_1 o_1 \omega_2} \left(p^{\mathrm{C}}_{t_1 n_1 o_1 \omega_2} + p^{\mathrm{E}}_{t_1 n_1 o_1 \omega_2} \right)$$

$$- 12\ p^{\mathrm{C}}_{t_1 n_1 o_1 \omega_2} - 10\ p^{\mathrm{E}}_{t_1 n_1 o_1 \omega_2} \Big] \Big\}$$

$$+ 0.4 \Big\{ (0.15 \times 700000)\ x^{\mathrm{C}}_{t_2 n_1 \omega_1}$$

$$- 8760 \Big[\lambda_{t_2 n_1 o_1 \omega_1} \left(p^{\mathrm{C}}_{t_2 n_1 o_1 \omega_1} + p^{\mathrm{E}}_{t_2 n_1 o_1 \omega_1} \right)$$

$$- 12\ p^{\mathrm{C}}_{t_2 n_1 o_1 \omega_1} - 10\ p^{\mathrm{E}}_{t_2 n_1 o_1 \omega_1} \Big] \Big\}$$

$$+ 0.6 \Big\{ (0.15 \times 840000)\ x^{\mathrm{C}}_{t_2 n_1 \omega_2}$$

$$- 8760 \Big[\lambda_{t_2 n_1 o_1 \omega_2} \left(p^{\mathrm{C}}_{t_2 n_1 o_1 \omega_2} + p^{\mathrm{E}}_{t_2 n_1 o_1 \omega_2} \right)$$

$$- 12\ p^{\mathrm{C}}_{t_2 n_1 o_1 \omega_2} - 10\ p^{\mathrm{E}}_{t_2 n_1 o_1 \omega_2} \Big] \Big\}$$

subject to

$$x^{\mathrm{C}}_{t_1 n_1 \omega_1} = 50\ u^{\mathrm{C}}_{t_1 n_1 \omega_1 h_2} + 100\ u^{\mathrm{C}}_{t_1 n_1 \omega_1 h_3}$$
$$u^{\mathrm{C}}_{t_1 n_1 \omega_1 h_1} + u^{\mathrm{C}}_{t_1 n_1 \omega_1 h_2} + u^{\mathrm{C}}_{t_1 n_1 \omega_1 h_3} = 1$$
$$u^{\mathrm{C}}_{t_1 n_1 \omega_1 h_1},\ u^{\mathrm{C}}_{t_1 n_1 \omega_1, h_2},\ u^{\mathrm{C}}_{t_1 n_1 \omega_1 h_3} \in \{0, 1\}$$
$$x^{\mathrm{C}}_{t_1 n_1 \omega_2} = 50\ u^{\mathrm{C}}_{t_1 n_1 \omega_2 h_2} + 100\ u^{\mathrm{C}}_{t_1 n_1 \omega_2 h_3}$$
$$u^{\mathrm{C}}_{t_1 n_1 \omega_2 h_1} + u^{\mathrm{C}}_{t_1 n_1 \omega_2 h_2} + u^{\mathrm{C}}_{t_1 n_1 \omega_2 h_3} = 1$$
$$u^{\mathrm{C}}_{t_1 n_1 \omega_2 h_1},\ u^{\mathrm{C}}_{t_1 n_1 \omega_2 h_2},\ u^{\mathrm{C}}_{t_1 n_1 \omega_2, h_3} \in \{0, 1\}$$
$$x^{\mathrm{C}}_{t_2 n_1 \omega_1} = 50\ u^{\mathrm{C}}_{t_2 n_1 \omega_1 h_2} + 100\ u^{\mathrm{C}}_{t_2 n_1 \omega_1 h_3}$$
$$u^{\mathrm{C}}_{t_2 n_1 \omega_1 h_1} + u^{\mathrm{C}}_{t_2 n_1 \omega_1 h_2} + u^{\mathrm{C}}_{t_2 n_1 \omega_1 h_3} = 1$$
$$u^{\mathrm{C}}_{t_2 n_1 \omega_1 h_1},\ u^{\mathrm{C}}_{t_2 n_1 \omega_1 h_2},\ u^{\mathrm{C}}_{t_2 n_1 \omega_1 h_3} \in \{0, 1\}$$
$$x^{\mathrm{C}}_{t_2 n_1 \omega_2} = 50\ u^{\mathrm{C}}_{t_2 n_1 \omega_2 h_2} + 100\ u^{\mathrm{C}}_{t_2 n_1 \omega_2 h_3}$$
$$u^{\mathrm{C}}_{t_2 n_1 \omega_2 h_1} + u^{\mathrm{C}}_{t_2 n_1 \omega_2 h_2} + u^{\mathrm{C}}_{t_2 n_1 \omega_2 h_3} = 1$$
$$u^{\mathrm{C}}_{t_2 n_1 \omega_2 h_1},\ u^{\mathrm{C}}_{t_2 n_1 \omega_2 h_2},\ u^{\mathrm{C}}_{t_2 n_1 \omega_2 h_3} \in \{0, 1\}$$
$$700000\ x^{\mathrm{C}}_{t_1 n_1 \omega_1} \le 50 \times 10^6$$
$$700000\ x^{\mathrm{C}}_{t_1 n_1 \omega_2} \le 50 \times 10^6$$
$$700000\ x^{\mathrm{C}}_{t_2 n_1 \omega_1} \le 50 \times 10^6$$

$$
\begin{aligned}
&840000\, x^{\mathrm{C}}_{t_2 n_1 \omega_2} \leq 50 \times 10^6\\
&\alpha^{\mathrm{C}}_{t_1 n_1 o_1 \omega_1} \geq 0\\
&\alpha^{\mathrm{C}}_{t_1 n_1 o_1 \omega_2} \geq 0\\
&\alpha^{\mathrm{C}}_{t_2 n_1 o_1 \omega_1} \geq 0\\
&\alpha^{\mathrm{C}}_{t_2 n_1 o_1 \omega_2} \geq 0\\
&\alpha^{\mathrm{E}}_{t_1 n_1 o_1 \omega_1} \geq 0\\
&\alpha^{\mathrm{E}}_{t_1 n_1 o_1 \omega_2} \geq 0\\
&\alpha^{\mathrm{E}}_{t_2 n_1 o_1 \omega_1} \geq 0\\
&\alpha^{\mathrm{E}}_{t_2 n_1 o_1 \omega_2} \geq 0\\
&\text{Four lower} - \text{level problems}\\
&x^{\mathrm{C}}_{t_1 n_1 \omega_1} = x^{\mathrm{C}}_{t_1 n_1 \omega_2}.
\end{aligned}
$$

Note that the amortization rates, i.e., 0.30 for the first time period and 0.15 for the second one, are included in the objective function. Note also that the last equation above is a nonanticipativity constraint and implies that the investment decision in the first time period does not depend on any scenario realization in the second time period.

The first lower-level problem referring to the market clearing at time period t_1, operating condition o_1, and scenario ω_1 is:

$$
\begin{aligned}
\min_{\Xi^{\mathrm{P,Ex5}}_{t_1 o_1 \omega_1}} \quad & \Big[\alpha^{\mathrm{C}}_{t_1 n_1 o_1 \omega_1}\, p^{\mathrm{C}}_{t_1 n_1 o_1 \omega_1} + \alpha^{\mathrm{E}}_{t_1 n_1 o_1 \omega_1}\, p^{\mathrm{E}}_{t_1 n_1 o_1 \omega_1} + 15\, p^{\mathrm{R}}_{t_1 n_1 o_1 \omega_1}\\
& - 35\, p^{\mathrm{D}}_{t_1 n_1 o_1 \omega_1}\Big]\\
\text{subject to} \quad &\\
& p^{\mathrm{D}}_{t_1 n_1 o_1 \omega_1} - p^{\mathrm{C}}_{t_1 n_1 o_1 \omega_1} - p^{\mathrm{E}}_{t_1 n_1 o_1 \omega_1} - p^{\mathrm{R}}_{t_1 n_1 o_1 \omega_1} = 0 \qquad : \lambda_{t_1 n_1 o_1 \omega_1}\\
& 0 \leq p^{\mathrm{C}}_{t_1 n_1 o_1 \omega_1} \leq x^{\mathrm{C}}_{t_1 n_1 \omega_1} \qquad : \mu^{\mathrm{C^{min}}}_{t_1 n_1 o_1 \omega_1}, \mu^{\mathrm{C^{max}}}_{t_1 n_1 o_1 \omega_1}\\
& 0 \leq p^{\mathrm{E}}_{t_1 n_1 o_1 \omega_1} \leq 150 \qquad : \mu^{\mathrm{E^{min}}}_{t_1 n_1 o_1 \omega_1}, \mu^{\mathrm{E^{max}}}_{t_1 n_1 o_1 \omega_1}\\
& 0 \leq p^{\mathrm{R}}_{t_1 n_1 o_1 \omega_1} \leq 100 \qquad : \mu^{\mathrm{R^{min}}}_{t_1 n_1 o_1 \omega_1}, \mu^{\mathrm{R^{max}}}_{t_1 n_1 o_1 \omega_1}\\
& 0 \leq p^{\mathrm{D}}_{t_1 n_1 o_1 \omega_1} \leq 300 \qquad : \mu^{\mathrm{D^{min}}}_{t_1 n_1 o_1 \omega_1}, \mu^{\mathrm{D^{max}}}_{t_1 n_1 o_1 \omega_1}.
\end{aligned}
$$

In addition, the second lower-level problem pertaining to the market clearing at time period t_1, operating condition o_1, and scenario ω_2 is:

$$\min_{\Xi^{\text{P.Ex5}}_{t_1 o_1 \omega_2}} \left[\alpha^{\text{C}}_{t_1 n_1 o_1 \omega_2}\, p^{\text{C}}_{t_1 n_1 o_1 \omega_2} + \alpha^{\text{E}}_{t_1 n_1 o_1 \omega_2}\, p^{\text{E}}_{t_1 n_1 o_1 \omega_2} + 15\, p^{\text{R}}_{t_1 n_1 o_1 \omega_2} - 35\, p^{\text{D}}_{t_1 n_1 o_1 \omega_2} \right]$$

subject to

$$p^{\text{D}}_{t_1 n_1 o_1 \omega_2} - p^{\text{C}}_{t_1 n_1 o_1 \omega_2} - p^{\text{E}}_{t_1 n_1 o_1 \omega_2} - p^{\text{R}}_{t_1 n_1 o_1 \omega_2} = 0 \quad : \lambda_{t_1 n_1 o_1 \omega_2}$$

$$0 \le p^{\text{C}}_{t_1 n_1 o_1 \omega_2} \le x^{\text{C}}_{t_1 n_1 \omega_2} \quad : \mu^{\text{C}^{\min}}_{t_1 n_1 o_1 \omega_2}, \mu^{\text{C}^{\max}}_{t_1 n_1 o_1 \omega_2}$$

$$0 \le p^{\text{E}}_{t_1 n_1 o_1 \omega_2} \le 150 \quad : \mu^{\text{E}^{\min}}_{t_1 n_1 o_1 \omega_2}, \mu^{\text{E}^{\max}}_{t_1 n_1 o_1 \omega_2}$$

$$0 \le p^{\text{R}}_{t_1 n_1 o_1 \omega_2} \le 100 \quad : \mu^{\text{R}^{\min}}_{t_1 n_1 o_1 \omega_2}, \mu^{\text{R}^{\max}}_{t_1 n_1 o_1 \omega_2}$$

$$0 \le p^{\text{D}}_{t_1 n_1 o_1 \omega_2} \le 300 \quad : \mu^{\text{D}^{\min}}_{t_1 n_1 o_1 \omega_2}, \mu^{\text{D}^{\max}}_{t_1 n_1 o_1 \omega_2}.$$

Likewise, the third lower-level problem referring to the market clearing at time period t_2, operating condition o_1, and scenario ω_1 is:

$$\min_{\Xi^{\text{P.Ex5}}_{t_2 o_1 \omega_1}} \left[\alpha^{\text{C}}_{t_2 n_1 o_1 \omega_1}\, p^{\text{C}}_{t_2 n_1 o_1 \omega_1} + \alpha^{\text{E}}_{t_2 n_1 o_1 \omega_1}\, p^{\text{E}}_{t_2 n_1 o_1 \omega_1} + 15\, p^{\text{R}}_{t_2 n_1 o_1 \omega_1} - 35\, p^{\text{D}}_{t_2 n_1 o_1 \omega_1} \right]$$

subject to

$$p^{\text{D}}_{t_2 n_1 o_1 \omega_1} - p^{\text{C}}_{t_2 n_1 o_1 \omega_1} - p^{\text{E}}_{t_2 n_1 o_1 \omega_1} - p^{\text{R}}_{t_2 n_1 o_1 \omega_1} = 0 \quad : \lambda_{t_2 n_1 o_1 \omega_1}$$

$$0 \le p^{\text{C}}_{t_2 n_1 o_1 \omega_1} \le \left[x^{\text{C}}_{t_1 n_1 \omega_1} + x^{\text{C}}_{t_2 n_1 \omega_1} \right] \quad : \mu^{\text{C}^{\min}}_{t_2 n_1 o_1 \omega_1}, \mu^{\text{C}^{\max}}_{t_2 n_1 o_1 \omega_1}$$

$$0 \le p^{\text{E}}_{t_2 n_1 o_1 \omega_1} \le 150 \quad : \mu^{\text{E}^{\min}}_{t_2 n_1 o_1 \omega_1}, \mu^{\text{E}^{\max}}_{t_2 n_1 o_1 \omega_1}$$

$$0 \le p^{\text{R}}_{t_2 n_1 o_1 \omega_1} \le 100 \quad : \mu^{\text{R}^{\min}}_{t_2 n_1 o_1 \omega_1}, \mu^{\text{R}^{\max}}_{t_2 n_1 o_1 \omega_1}$$

$$0 \le p^{\text{D}}_{t_2 n_1 o_1 \omega_1} \le 330 \quad : \mu^{\text{D}^{\min}}_{t_2 n_1 o_1 \omega_1}, \mu^{\text{D}^{\max}}_{t_2 n_1 o_1 \omega_1}.$$

Finally, the fourth lower-level problem related to the market clearing at time period t_2, operating condition o_1, and scenario ω_2 is:

$$\min_{\Xi^{\text{P.Ex5}}_{t_2 o_1 \omega_2}} \left[\alpha^{\text{C}}_{t_2 n_1 o_1 \omega_2}\, p^{\text{C}}_{t_2 n_1 o_1 \omega_2} + \alpha^{\text{E}}_{t_2 n_1 o_1 \omega_2}\, p^{\text{E}}_{t_2 n_1 o_1 \omega_2} + 15\, p^{\text{R}}_{t_2 n_1 o_1 \omega_2} - 35\, p^{\text{D}}_{t_2 n_1 o_1 \omega_2} \right]$$

subject to

$$p^{\text{D}}_{t_2 n_1 o_1 \omega_2} - p^{\text{C}}_{t_2 n_1 o_1 \omega_2} - p^{\text{E}}_{t_2 n_1 o_1 \omega_2} - p^{\text{R}}_{t_2 n_1 o_1 \omega_2} = 0 \quad : \lambda_{t_2 n_1 o_1 \omega_2}$$

$$0 \leq p^{\mathrm{C}}_{t_2 n_1 o_1 \omega_2} \leq \left[x^{\mathrm{C}}_{t_1 n_1 \omega_2} + x^{\mathrm{C}}_{t_2 n_1 \omega_2}\right] \quad : \mu^{\mathrm{C^{min}}}_{t_2 n_1 o_1 \omega_2}, \mu^{\mathrm{C^{max}}}_{t_2 n_1 o_1 \omega_2}$$
$$0 \leq p^{\mathrm{E}}_{t_2 n_1 o_1 \omega_2} \leq 150 \quad : \mu^{\mathrm{E^{min}}}_{t_2 n_1 o_1 \omega_2}, \mu^{\mathrm{E^{max}}}_{t_2 n_1 o_1 \omega_2}$$
$$0 \leq p^{\mathrm{R}}_{t_2 n_1 o_1 \omega_2} \leq 100 \quad : \mu^{\mathrm{R^{min}}}_{t_2 n_1 o_1 \omega_2}, \mu^{\mathrm{R^{max}}}_{t_2 n_1 o_1 \omega_2}$$
$$0 \leq p^{\mathrm{D}}_{t_2 n_1 o_1 \omega_2} \leq 330 \quad : \mu^{\mathrm{D^{min}}}_{t_2 n_1 o_1 \omega_2}, \mu^{\mathrm{D^{max}}}_{t_2 n_1 o_1 \omega_2}.$$

□

5.4 Direct Solution Approach

This section provides a direct solution approach, which is the first alternative solution approach mentioned in Sect. 5.1.7. This approach is used to solve bilevel models (5.1) and (5.2) presented in Sects. 5.2 and 5.3, respectively. To this end, the following two steps are carried out:

1. In both bilevel models (5.1) and (5.2), all lower-level problems are linear, continuous, and thus convex. Therefore, their Karush–Kuhn–Tucker (KKT) conditions are necessary and sufficient for optimality. This allows replacing these problems by their corresponding KKT conditions. In this way, each bilevel model is transformed into a single-level optimization problem, called an MPEC.
2. Both MPECs resulting from bilevel models (5.1) and (5.2) are mixed-integer and nonlinear. We recast each of these MPECs as an MILP problem through exact linearization techniques.

In Sect. 5.4.1, we derive the MPEC associated with bilevel model (5.1) corresponding to the static investment problem as well as the MPEC associated with Illustrative Example 5.1. A similar approach can also be used to derive the MPEC for bilevel model (5.2) corresponding to the dynamic production capacity investment problem and Illustrative Examples 5.2–5.5. In Sect. 5.4.2, the MILP problem equivalent to the MPEC associated with Illustrative Example 5.1 is derived. A similar approach can also be used to derive the MILP formulations of the static and dynamic investment models in Illustrative Examples 5.2–5.5.

5.4.1 MPEC

The MPEC associated with bilevel model (5.1) corresponding to the static investment problem is derived in this section. First, the KKT conditions associated with each lower-level problem (5.1j)–(5.1s), one per operating condition o and scenario ω, are obtained. To this end, the corresponding Lagrangian function $\mathscr{L}_{o\omega}$ is written below:

$$
\begin{aligned}
&\mathcal{L}_{o\omega} = \\
&\sum_{n}\Bigg[\alpha^{\mathrm{C}}_{no\omega}\, p^{\mathrm{C}}_{no\omega} + \alpha^{\mathrm{E}}_{no\omega}\, p^{\mathrm{E}}_{no\omega} + C^{\mathrm{R}}_{no}\, p^{\mathrm{R}}_{no\omega} - U^{\mathrm{D}}_{no}\, p^{\mathrm{D}}_{no\omega} \\
&+ \lambda_{no\omega}\left(p^{\mathrm{D}}_{no\omega} + \sum_{m\in\Omega_n} B_{nm}\left(\theta_{no\omega} - \theta_{mo\omega}\right) - p^{\mathrm{C}}_{no\omega} - p^{\mathrm{S}}_{no\omega} - p^{\mathrm{E}}_{no\omega} - p^{\mathrm{R}}_{no\omega}\right) \\
&+ \mu^{\mathrm{C}^{\max}}_{no\omega}\left(p^{\mathrm{C}}_{no\omega} - x^{\mathrm{C}}_{n}\right) - \mu^{\mathrm{C}^{\min}}_{no\omega}\, p^{\mathrm{C}}_{no\omega} + \mu^{\mathrm{S}^{\max}}_{no\omega}\left(p^{\mathrm{S}}_{no\omega} - Q^{\mathrm{S}}_{no}\, x^{\mathrm{S}}_{n}\right) - \mu^{\mathrm{S}^{\min}}_{no\omega}\, p^{\mathrm{S}}_{no\omega} \\
&+ \mu^{\mathrm{E}^{\max}}_{no\omega}\left(p^{\mathrm{E}}_{no\omega} - P^{\mathrm{E}^{\max}}_{n}\right) - \mu^{\mathrm{E}^{\min}}_{no\omega}\, p^{\mathrm{E}}_{no\omega} + \mu^{\mathrm{R}^{\max}}_{no\omega}\left(p^{\mathrm{R}}_{no\omega} - P^{\mathrm{R}^{\max}}_{n\omega}\right) - \mu^{\mathrm{R}^{\min}}_{no\omega}\, p^{\mathrm{R}}_{no\omega} \\
&+ \mu^{\mathrm{D}^{\max}}_{no\omega}\left(p^{\mathrm{D}}_{no\omega} - Q^{\mathrm{D}}_{no}\, P^{\mathrm{D}^{\max}}_{n}\right) - \mu^{\mathrm{D}^{\min}}_{no\omega}\, p^{\mathrm{D}}_{no\omega} \\
&+ \sum_{m\in\Omega_n} \mu^{\mathrm{F}}_{nmo\omega}\left(B_{nm}(\theta_{no\omega} - \theta_{mo\omega}) - F^{\max}_{nm}\right) \\
&+ \mu^{\theta^{\max}}_{no\omega}\left(\theta_{no\omega} - \pi\right) - \mu^{\theta^{\min}}_{no\omega}\left(\theta_{no\omega} + \pi\right)\Bigg] + \mu^{\theta^{\mathrm{ref}}}_{o\omega}\, \theta_{(n=\mathrm{ref.})o\omega} \quad \forall o, \forall \omega. \qquad (5.3)
\end{aligned}
$$

Considering the Lagrangian function (5.3), the KKT conditions associated with the lower-level problems (5.1j)–(5.1s) are given by (5.4). Note that equality constraints (5.4a)–(5.4f) are derived by differentiating the Lagrangian function $\mathcal{L}_{o\omega}$ with respect to the primal variables included in the set $\Xi^{\mathrm{Primal}}_{o\omega}$:

$$
\frac{\partial \mathcal{L}_{o\omega}}{\partial p^{\mathrm{C}}_{no\omega}} = \alpha^{\mathrm{C}}_{no\omega} - \lambda_{no\omega} + \mu^{\mathrm{C}^{\max}}_{no\omega} - \mu^{\mathrm{C}^{\min}}_{no\omega} = 0 \quad \forall n, \forall o, \forall \omega \qquad (5.4a)
$$

$$
\frac{\partial \mathcal{L}_{o\omega}}{\partial p^{\mathrm{S}}_{no\omega}} = - \lambda_{no\omega} + \mu^{\mathrm{S}^{\max}}_{no\omega} - \mu^{\mathrm{S}^{\min}}_{no\omega} = 0 \quad \forall n, \forall o, \forall \omega \qquad (5.4b)
$$

$$
\frac{\partial \mathcal{L}_{o\omega}}{\partial p^{\mathrm{E}}_{no\omega}} = \alpha^{\mathrm{E}}_{no\omega} - \lambda_{no\omega} + \mu^{\mathrm{E}^{\max}}_{no\omega} - \mu^{\mathrm{E}^{\min}}_{no\omega} = 0 \quad \forall n, \forall o, \forall \omega \qquad (5.4c)
$$

$$
\frac{\partial \mathcal{L}_{o\omega}}{\partial p^{\mathrm{R}}_{no\omega}} = C^{\mathrm{R}}_{no} - \lambda_{no\omega} + \mu^{\mathrm{R}^{\max}}_{no\omega} - \mu^{\mathrm{R}^{\min}}_{no\omega} = 0 \quad \forall n, \forall o, \forall \omega \qquad (5.4d)
$$

$$
\frac{\partial \mathcal{L}_{o\omega}}{\partial p^{\mathrm{D}}_{no\omega}} = - U^{\mathrm{D}}_{no} + \lambda_{no\omega} + \mu^{\mathrm{D}^{\max}}_{no\omega} - \mu^{\mathrm{D}^{\min}}_{no\omega} = 0 \quad \forall n, \forall o, \forall \omega \qquad (5.4e)
$$

$$
\frac{\partial \mathcal{L}_{o\omega}}{\partial \theta_{no\omega}} = \sum_{m\in\Omega_n} B_{nm}\left(\lambda_{no\omega} - \lambda_{mo\omega}\right) + \sum_{m\in\Omega_n} B_{nm}\left(\mu^{\mathrm{F}}_{nmo\omega} - \mu^{\mathrm{F}}_{mno\omega}\right)
$$

$$+ \mu^{\theta^{\max}}_{no\omega} - \mu^{\theta^{\min}}_{no\omega} + \left(\mu^{\theta^{\text{ref.}}}_{o\omega}\right)_{n=\text{ref.}} = 0 \quad \forall n, \forall o, \forall \omega. \tag{5.4f}$$

In addition, equality constraints (5.4g) and (5.4h) below are the primal equality constraints (5.1k) and (5.1s) in the lower-level problems:

$$p^{\text{D}}_{no\omega} + \sum_{m\in\Omega_n} B_{nm}\left(\theta_{no\omega} - \theta_{mo\omega}\right) - p^{\text{C}}_{no\omega} - p^{\text{S}}_{no\omega} - p^{\text{E}}_{no\omega}$$
$$- p^{\text{R}}_{no\omega} = 0 \quad \forall n, \forall o, \forall \omega \tag{5.4g}$$
$$\theta_{on\omega} = 0 \quad n = \text{ref}. \tag{5.4h}$$

Furthermore, complementarity conditions (5.4i)–(5.4u) below are part of the KKT conditions associated with the lower-level problems (5.1j)–(5.1s), and are related to the inequality constraints (5.1l)–(5.1r). Note that each complementarity condition of the form $0 \leq p \perp \mu \geq 0$ is equivalent to $p \geq 0$, $\mu \geq 0$, and $p\mu = 0$:

$$0 \leq p^{\text{C}}_{no\omega} \perp \mu^{\text{C}^{\min}}_{no\omega} \geq 0 \quad \forall n, \forall o, \forall \omega \tag{5.4i}$$
$$0 \leq \left(x^{\text{C}}_{n} - p^{\text{C}}_{no\omega}\right) \perp \mu^{\text{C}^{\max}}_{no\omega} \geq 0 \quad \forall n, \forall o, \forall \omega \tag{5.4j}$$
$$0 \leq p^{\text{S}}_{no\omega} \perp \mu^{\text{S}^{\min}}_{no\omega} \geq 0 \quad \forall n, \forall o, \forall \omega \tag{5.4k}$$
$$0 \leq \left(Q^{\text{S}}_{no}\, x^{\text{S}}_{n} - p^{\text{S}}_{no\omega}\right) \perp \mu^{\text{S}^{\max}}_{no\omega} \geq 0 \quad \forall n, \forall o, \forall \omega \tag{5.4l}$$
$$0 \leq p^{\text{E}}_{no\omega} \perp \mu^{\text{E}^{\min}}_{no\omega} \geq 0 \quad \forall n, \forall o, \forall \omega \tag{5.4m}$$
$$0 \leq \left(P^{\text{E}^{\max}}_{n} - p^{\text{E}}_{no\omega}\right) \perp \mu^{\text{E}^{\max}}_{no\omega} \geq 0 \quad \forall n, \forall o, \forall \omega \tag{5.4n}$$
$$0 \leq p^{\text{R}}_{no\omega} \perp \mu^{\text{R}^{\min}}_{no\omega} \geq 0 \quad \forall n, \forall o, \forall \omega \tag{5.4o}$$
$$0 \leq \left(P^{\text{R}^{\max}}_{n\omega} - p^{\text{R}}_{no\omega}\right) \perp \mu^{\text{R}^{\max}}_{no\omega} \geq 0 \quad \forall n, \forall o, \forall \omega \tag{5.4p}$$
$$0 \leq p^{\text{D}}_{no\omega} \perp \mu^{\text{D}^{\min}}_{no\omega} \geq 0 \quad \forall n, \forall o, \forall \omega \tag{5.4q}$$
$$0 \leq \left(Q^{\text{D}}_{no}\, P^{\text{D}^{\max}}_{n} - p^{\text{D}}_{no\omega}\right) \perp \mu^{\text{D}^{\max}}_{no\omega} \geq 0 \quad \forall n, \forall o, \forall \omega \tag{5.4r}$$
$$0 \leq \left[F^{\max}_{nm} - B_{nm}(\theta_{no\omega} - \theta_{mo\omega})\right] \perp \mu^{\text{F}}_{nmo\omega} \geq 0 \quad \forall n, \forall o, \forall m \in \Omega_n, \forall \omega \tag{5.4s}$$
$$0 \leq (\pi + \theta_{no\omega}) \perp \mu^{\theta^{\min}}_{no\omega} \geq 0 \quad \forall n, \forall o, \forall \omega \tag{5.4t}$$
$$0 \leq (\pi - \theta_{no\omega}) \perp \mu^{\theta^{\max}}_{no\omega} \geq 0 \quad \forall n, \forall o, \forall \omega. \tag{5.4u}$$

Finally, conditions (5.4v) and (5.4w) below state that the dual variables associated with the equality constraints (5.1k) and (5.1s) are free:

$$\lambda_{no\omega} \in \text{free} \quad \forall n, \forall o, \forall \omega \tag{5.4v}$$
$$\mu^{\theta^{\text{ref}}}_{o\omega} \in \text{free} \quad \forall o, \forall \omega. \tag{5.4w}$$

Using KKT conditions (5.4) instead of lower-level problems (5.1j)–(5.1s) within the bilevel model (5.1) yields the following MPEC:

$$\min_{\Xi^{\mathrm{UL}} \cup \Xi_{o\omega}^{\mathrm{Primal}} \cup \Xi_{o\omega}^{\mathrm{Dual}}} \quad (5.1\mathrm{a}) \tag{5.5a}$$

subject to

$$(5.1\mathrm{b})-(5.1\mathrm{h}) \tag{5.5b}$$

$$(5.4). \tag{5.5c}$$

Note that the MPEC associated with dynamic bilevel model (5.2) can be analogously derived. To this end, each lower-level problem (5.2l)–(5.2u) is replaced by its equivalent KKT optimality conditions. The resulting single-level optimization problem is an MPEC corresponding to the bilevel model (5.2).

Illustrative Example 5.6 *MPEC corresponding to the bilevel model in Illustrative Example 5.1*

The MPEC corresponding to the bilevel model in Illustrative Example 5.1 is given by (5.6) below:

$$\min_{\Xi^{\mathrm{UL,Ex1}} \cup \Xi_{o_1}^{\mathrm{P,Ex1}} \cup \Xi_{o_1}^{\mathrm{D,Ex1}}} \quad 55000\, x_{n_1}^{\mathrm{C}} - 8760 \left[\lambda_{n_1 o_1} \left(p_{n_1 o_1}^{\mathrm{C}} + p_{n_1 o_1}^{\mathrm{E}} \right) - 12\, p_{n_1 o_1}^{\mathrm{C}} - 10\, p_{n_1 o_1}^{\mathrm{E}} \right] \tag{5.6a}$$

subject to

The upper-level constraints:

$$x_{n_1}^{\mathrm{C}} = 100\, u_{n_1 h_2}^{\mathrm{C}} \tag{5.6b}$$

$$u_{n_1 h_1}^{\mathrm{C}} + u_{n_1 h_2}^{\mathrm{C}} = 1 \tag{5.6c}$$

$$u_{n_1 h_1}^{\mathrm{C}},\ u_{n_1 h_2}^{\mathrm{C}} \in \{0, 1\} \tag{5.6d}$$

$$55000\, x_{n_1}^{\mathrm{C}} \leq 10^7 \tag{5.6e}$$

$$\alpha_{n_1 o_1}^{\mathrm{C}} \geq 0 \tag{5.6f}$$

$$\alpha_{n_1 o_1}^{\mathrm{E}} \geq 0 \tag{5.6g}$$

The KKT conditions associated with the lower-level problems:

$$\frac{\partial \mathcal{L}_{o_1}}{\partial p_{n_1 o_1}^{\mathrm{C}}} = \alpha_{n_1 o_1}^{\mathrm{C}} - \lambda_{n_1 o_1} + \mu_{n_1 o_1}^{\mathrm{C^{max}}} - \mu_{n_1 o_1}^{\mathrm{C^{min}}} = 0 \tag{5.6h}$$

$$\frac{\partial \mathcal{L}_{o_1}}{\partial p_{n_1 o_1}^{\mathrm{E}}} = \alpha_{n_1 o_1}^{\mathrm{E}} - \lambda_{n_1 o_1} + \mu_{n_1 o_1}^{\mathrm{E^{max}}} - \mu_{n_1 o_1}^{\mathrm{E^{min}}} = 0 \tag{5.6i}$$

$$\frac{\partial \mathcal{L}_{o_1}}{\partial p_{n_1 o_1}^{\mathrm{R}}} = 15 - \lambda_{n_1 o_1} + \mu_{n_1 o_1}^{\mathrm{R^{max}}} - \mu_{n_1 o_1}^{\mathrm{R^{min}}} = 0 \tag{5.6j}$$

$$\frac{\partial \mathcal{L}_{o_1}}{\partial p_{n_1 o_1}^{\mathrm{D}}} = -35 + \lambda_{n_1 o_1} + \mu_{n_1 o_1}^{\mathrm{D^{max}}} - \mu_{n_1 o_1}^{\mathrm{D^{min}}} = 0 \tag{5.6k}$$

$$p^{\mathrm{D}}_{n_1o_1} - p^{\mathrm{C}}_{n_1o_1} - p^{\mathrm{E}}_{n_1o_1} - p^{\mathrm{R}}_{n_1o_1} = 0 \quad (5.6l)$$

$$0 \leq p^{\mathrm{C}}_{n_1o_1} \perp \mu^{\mathrm{C^{min}}}_{n_1o_1} \geq 0 \quad (5.6m)$$

$$0 \leq \left(x^{\mathrm{C}}_{n_1} - p^{\mathrm{C}}_{n_1o_1}\right) \perp \mu^{\mathrm{C^{max}}}_{n_1o_1} \geq 0 \quad (5.6n)$$

$$0 \leq p^{\mathrm{E}}_{n_1o_1} \perp \mu^{\mathrm{E^{min}}}_{n_1o_1} \geq 0 \quad (5.6o)$$

$$0 \leq \left(150 - p^{\mathrm{E}}_{n_1o_1}\right) \perp \mu^{\mathrm{E^{max}}}_{n_1o_1} \geq 0 \quad (5.6p)$$

$$0 \leq p^{\mathrm{R}}_{n_1o_1} \perp \mu^{\mathrm{R^{min}}}_{n_1o_1} \geq 0 \quad (5.6q)$$

$$0 \leq \left(100 - p^{\mathrm{R}}_{n_1o_1}\right) \perp \mu^{\mathrm{R^{max}}}_{n_1o_1} \geq 0 \quad (5.6r)$$

$$0 \leq p^{\mathrm{D}}_{n_1o_1} \perp \mu^{\mathrm{D^{min}}}_{n_1o_1} \geq 0 \quad (5.6s)$$

$$0 \leq \left(300 - p^{\mathrm{D}}_{n_1o_1}\right) \perp \mu^{\mathrm{D^{max}}}_{n_1o_1} \geq 0 \quad (5.6t)$$

$$\lambda_{n_1o_1} \in \text{free}. \quad (5.6u)$$

□

5.4.2 *MPEC Linearization*

This section describes the MILP problem equivalent of MPEC (5.6), obtained in Illustrative Example 5.6 and derived from the bilevel model in Illustrative Example 5.1. Note that the same linearization approach can be applied to the MPEC formulation of static and dynamic investment models (5.1) and (5.2). MPEC (5.6) includes the following nonlinearities:

1. The nonlinear term $\lambda_{n_1o_1}\left(p^{\mathrm{C}}_{n_1o_1} + p^{\mathrm{E}}_{n_1o_1}\right)$ in the objective function (5.6a). The reason for nonlinearity is the product of variables, i.e., production quantities $p^{\mathrm{C}}_{n_1o_1}$, $p^{\mathrm{E}}_{n_1o_1}$ and market-clearing price $\lambda_{n_1o_1}$. An exact linear expression for this nonlinear term can be obtained as explained in Sect. 5.4.2.1.
2. The complementarity conditions (5.6m)–(5.6t). Such conditions can be linearized without approximation through the approach explained in Sect. 5.4.2.2, which relies on auxiliary binary variables.

5.4.2.1 Linearizing the Nonlinear Term in the Objective Function

An exact linearization approach [34] is described in this section to obtain a linear expression for the nonlinear term $\lambda_{n_1o_1}\left(p^{\mathrm{C}}_{n_1o_1} + p^{\mathrm{E}}_{n_1o_1}\right)$ in objective function (5.6a). First, we derive the dual problem associated with the lower-level problem of the bilevel model in Illustrative Example 5.1. Then we formulate the strong duality equality, which states that the values of primal and dual objective functions are equal at the optimal solution.

The dual problem associated with the lower-level problem in Illustrative Example 5.1 is given by (5.7) below:

$$\max_{\Xi^{\mathrm{D,Ex1}}_{o_1}} \quad -x^{\mathrm{C}}_{n_1}\,\mu^{\mathrm{C^{max}}}_{n_1o_1} - 150\,\mu^{\mathrm{E^{max}}}_{n_1o_1} - 100\,\mu^{\mathrm{R^{max}}}_{n_1o_1} - 300\,\mu^{\mathrm{D^{max}}}_{n_1o_1} \tag{5.7a}$$

subject to

$$(5.6\mathrm{h}) - (5.6\mathrm{k}) \tag{5.7b}$$

$$\mu^{\mathrm{C^{min}}}_{n_1o_1},\ \mu^{\mathrm{C^{max}}}_{n_1o_1} \geq 0 \tag{5.7c}$$

$$\mu^{\mathrm{E^{min}}}_{n_1o_1},\ \mu^{\mathrm{E^{max}}}_{n_1o_1} \geq 0 \tag{5.7d}$$

$$\mu^{\mathrm{R^{min}}}_{n_1o_1},\ \mu^{\mathrm{R^{max}}}_{n_1o_1} \geq 0 \tag{5.7e}$$

$$\mu^{\mathrm{D^{min}}}_{n_1o_1},\ \mu^{\mathrm{D^{max}}}_{n_1o_1} \geq 0. \tag{5.7f}$$

The strong duality equality is:

$$\begin{aligned} &\alpha^{\mathrm{C}}_{n_1o_1}\,p^{\mathrm{C}}_{n_1o_1} + \alpha^{\mathrm{E}}_{n_1o_1}\,p^{\mathrm{E}}_{n_1o_1} + 15\,p^{\mathrm{R}}_{n_1o_1} - 35\,p^{\mathrm{D}}_{n_1o_1} = \\ &\qquad - x^{\mathrm{C}}_{n_1}\,\mu^{\mathrm{C^{max}}}_{n_1o_1} - 150\,\mu^{\mathrm{E^{max}}}_{n_1o_1} - 100\,\mu^{\mathrm{R^{max}}}_{n_1o_1} - 300\,\mu^{\mathrm{D^{max}}}_{n_1o_1}. \end{aligned} \tag{5.8}$$

In the second step of the linearization approach, we use some complementarity conditions obtained in Illustrative Example 5.6. The complementarity conditions (5.6n) and (5.6p) imply:

$$x^{\mathrm{C}}_{n_1}\,\mu^{\mathrm{C^{max}}}_{n_1o_1} = p^{\mathrm{C}}_{n_1o_1}\,\mu^{\mathrm{C^{max}}}_{n_1o_1} \tag{5.9a}$$

$$150\,\mu^{\mathrm{E^{max}}}_{n_1o_1} = p^{\mathrm{E}}_{n_1o_1}\,\mu^{\mathrm{E^{max}}}_{n_1o_1}. \tag{5.9b}$$

Substituting conditions (5.9a) and (5.9b) into (5.8) renders:

$$\begin{aligned} &p^{\mathrm{C}}_{n_1o_1}\left(\alpha^{\mathrm{C}}_{n_1o_1} + \mu^{\mathrm{C^{max}}}_{n_1o_1}\right) + p^{\mathrm{E}}_{n_1o_1}\left(\alpha^{\mathrm{E}}_{n_1o_1} + \mu^{\mathrm{E^{max}}}_{n_1o_1}\right) = \\ &\qquad - 15\,p^{\mathrm{R}}_{n_1o_1} + 35\,p^{\mathrm{D}}_{n_1o_1} - 100\,\mu^{\mathrm{R^{max}}}_{n_1o_1} - 300\,\mu^{\mathrm{D^{max}}}_{n_1o_1}. \end{aligned} \tag{5.9c}$$

One the other hand, KKT equality conditions (5.6h) and (5.6i) imply:

$$\lambda_{n_1o_1} = \alpha^{\mathrm{C}}_{n_1o_1} + \mu^{\mathrm{C^{max}}}_{n_1o_1} - \mu^{\mathrm{C^{min}}}_{n_1o_1} \tag{5.9d}$$

$$\lambda_{n_1o_1} = \alpha^{\mathrm{E}}_{n_1o_1} + \mu^{\mathrm{E^{max}}}_{n_1o_1} - \mu^{\mathrm{E^{min}}}_{n_1o_1}. \tag{5.9e}$$

Multiplying equalities (5.9d) and (5.9e) by variables $p^{\mathrm{C}}_{n_1o_1}$ and $p^{\mathrm{E}}_{n_1o_1}$, respectively, results in:

$$p^{\mathrm{C}}_{n_1o_1}\,\lambda_{n_1o_1} = p^{\mathrm{C}}_{n_1o_1}\,\alpha^{\mathrm{C}}_{n_1o_1} + p^{\mathrm{C}}_{n_1o_1}\,\mu^{\mathrm{C^{max}}}_{n_1o_1} - p^{\mathrm{C}}_{n_1o_1}\,\mu^{\mathrm{C^{min}}}_{n_1o_1} \tag{5.9f}$$

$$p^{\mathrm{E}}_{n_1o_1}\,\lambda_{n_1o_1} = p^{\mathrm{E}}_{n_1o_1}\,\alpha^{\mathrm{E}}_{n_1o_1} + p^{\mathrm{E}}_{n_1o_1}\,\mu^{\mathrm{E^{max}}}_{n_1o_1} - p^{\mathrm{E}}_{n_1o_1}\,\mu^{\mathrm{E^{min}}}_{n_1o_1}. \tag{5.9g}$$

From complementarity conditions (5.6m) and (5.6o), it can be concluded that the last terms of equalities (5.9f) and (5.9g) are zero, i.e., $p^{\mathrm{C}}_{n_1o_1}\,\mu^{\mathrm{C^{min}}}_{n_1o_1} = 0$ and $p^{\mathrm{E}}_{n_1o_1}\,\mu^{\mathrm{E^{min}}}_{n_1o_1} = 0$. Thus, summation of equalities (5.9f) and (5.9g) yields:

$$\lambda_{n_1 o_1} \left(p^{\mathrm{C}}_{n_1 o_1} + p^{\mathrm{E}}_{n_1 o_1}\right) = p^{\mathrm{C}}_{n_1 o_1} \left(\alpha^{\mathrm{C}}_{n_1 o_1} + \mu^{\mathrm{C^{max}}}_{n_1 o_1}\right) + p^{\mathrm{E}}_{n_1 o_1} \left(\alpha^{\mathrm{E}}_{n_1 o_1} + \mu^{\mathrm{E^{max}}}_{n_1 o_1}\right). \tag{5.9h}$$

Finally, comparing (5.9c) and (5.9h) yields:

$$\lambda_{n_1 o_1} \left(p^{\mathrm{C}}_{n_1 o_1} + p^{\mathrm{E}}_{n_1 o_1}\right) = -15\ p^{\mathrm{R}}_{n_1 o_1} + 35\ p^{\mathrm{D}}_{n_1 o_1} - 100\ \mu^{\mathrm{R^{max}}}_{n_1 o_1} - 300\ \mu^{\mathrm{D^{max}}}_{n_1 o_1}. \tag{5.9i}$$

Note that all terms on the right-hand side of (5.9i) are linear. □

5.4.2.2 Linearizing Complementarity Conditions

Each complementarity condition (5.6m)–(5.6t) of the form $0 \le p \perp \mu \ge 0$ can be linearized without approximation using a set of auxiliary binary variables [15]. Note that p is a primal variable, whereas μ is a dual one. The complementarity condition $0 \le p \perp \mu \ge 0$ is equivalent to the following constraints:

$$p \ge 0 \tag{5.10a}$$
$$\mu \ge 0 \tag{5.10b}$$
$$p \le \psi\ M^P \tag{5.10c}$$
$$\mu \le (1 - \psi)\ M^\mu \tag{5.10d}$$
$$\psi \in \{0, 1\}, \tag{5.10e}$$

where M^P and M^μ are large enough positive constants, and ψ is an auxiliary binary variable. The values of constants M^P and M^μ need to be appropriately selected. Note that values that are too large may result in complementarity not being satisfied, while values that are too small may result in numerical ill-conditioning.

The following strategy is proposed to select appropriate values for M^P and M^μ:

1. Arbitrarily select large values for M^P and M^μ, e.g., 10^4 for M^P and 10^5 for M^μ. In general, a comparatively lower value is needed for M^P than for M^μ because M^P limits the value of primal variable p in (5.10c), whose upper bound is usually known, while M^μ limits the value of dual variable μ in (5.10d) with unknown upper bound.
2. Solve the MILP problem and then obtain the optimal values for p and μ.
3. Check whether $p\mu = 0$. If not, the values selected for M^P and M^μ need to be reduced.

Steps 1–3 are repeated until the complementarity condition holds. In an MILP problem with several complementarity conditions, we may need to tune up separately the values of the large constants of each complementarity condition.

For example, the mixed-integer linear equivalent of complementarity condition (5.6m) is:

$$p^{\mathrm{C}}_{n_1 o_1} \geq 0 \quad (5.11a)$$
$$\mu^{\mathrm{C^{min}}}_{n_1 o_1} \geq 0 \quad (5.11b)$$
$$p^{\mathrm{C}}_{n_1 o_1} \leq \psi^{\mathrm{C^{min}}}_{n_1 o_1} M^{P1} \quad (5.11c)$$
$$\mu^{\mathrm{C^{min}}}_{n_1 o_1} \leq \left(1 - \psi^{\mathrm{C^{min}}}_{n_1 o_1}\right) M^{\mu 1} \quad (5.11d)$$
$$\psi^{\mathrm{C^{min}}}_{n_1 o_1} \in \{0, 1\}, \quad (5.11e)$$

where M^{P1} and $M^{\mu 1}$ are large enough positive constants.

In addition, the mixed-integer linear equivalent of complementarity condition (5.6n) is:

$$\left(x^{\mathrm{C}}_{n_1} - p^{\mathrm{C}}_{n_1 o_1}\right) \geq 0 \quad (5.12a)$$
$$\mu^{\mathrm{C^{max}}}_{n_1 o_1} \geq 0 \quad (5.12b)$$
$$\left(x^{\mathrm{C}}_{n_1} - p^{\mathrm{C}}_{n_1 o_1}\right) \leq \psi^{\mathrm{C^{max}}}_{n_1 o_1} M^{P2} \quad (5.12c)$$
$$\mu^{\mathrm{C^{max}}}_{n_1 o_1} \leq \left(1 - \psi^{\mathrm{C^{max}}}_{n_1 o_1}\right) M^{\mu 2} \quad (5.12d)$$
$$\psi^{\mathrm{C^{max}}}_{n_1 o_1} \in \{0, 1\}, \quad (5.12e)$$

where M^{P2} and $M^{\mu 2}$ are large enough positive constants.

5.4.3 *Numerical Results*

This section provides the numerical results corresponding to Illustrative Examples 5.1–5.5 presented in Sects. 5.2 and 5.3.

5.4.3.1 Results for Illustrative Example 5.1

In this static investment example, the strategic producer decides to build 100 MW of conventional capacity, i.e., $x^{\mathrm{C}}_{n_1}$ =100 MW. Thus, the annualized investment cost is \$5.5 million (i.e., 100 $\times$ 55000), which is lower than its annualized investment budget, i.e., \$10 million.

To maximize its profit, the strategic producer offers its two production units (existing and newly built) at a price of \$35/MWh, which is identical to the consumer's bid price. The market-clearing outcomes are provided in Table 5.1. Accordingly, the market-clearing price ($\lambda_{n_1 o_1}$) is \$35/MWh, which is equal to the offer price of the last production unit dispatched. In addition, the annual profit of the strategic producer is minus the upper-level objective function value, which is equal to \$37.42 million.

To highlight the impact of strategic behavior of the producer on investment results, we consider a *nonstrategic* case in which the producer submits its actual production costs as offer prices, i.e., $\alpha^{\mathrm{C}}_{n_1 o_1} = C^{\mathrm{C}}_{n_1}$ =\$12/MWh and $\alpha^{\mathrm{E}}_{n_1 o_1} = C^{\mathrm{E}}_{n_1}$ =\$10/MWh. This

Table 5.1 Illustrative Example 5.1: Market-clearing outcomes

Market participant	Offer/bid price [$/MWh]	Production/consumption level [MW]
Rival unit	$C^{R}_{n_1 o_1} = 15$	$p^{R}_{n_1 o_1} = 100$
Existing unit of the strategic producer	$\alpha^{E}_{n_1 o_1} = 35$	$p^{E}_{n_1 o_1} = 150$
Newly built unit of the strategic producer	$\alpha^{C}_{n_1 o_1} = 35$	$p^{C}_{n_1 o_1} = 50$
Consumer	$U^{D}_{n_1 o_1} = 35$	$p^{D}_{n_1 o_1} = 300$

case corresponds to a perfectly competitive market. In this case, no new production unit is built by the producer, i.e., $x^{C}_{n_1}$ =0 MW, and its annual profit decreases to $32.85 million, which is comparatively lower than that in the strategic case.

5.4.3.2 Results for Illustrative Example 5.2

This example is similar to Illustrative Example 5.1 but with consideration of the transmission line n_1-n_2. Since the transmission network is modeled, a market-clearing price at each node of the network, i.e., n_1 and n_2, is obtained, the so-called *locational marginal price (LMP)* of that node. The LMP of a given node represents the social welfare increment in the market as a result of a marginal demand increment at that node.

Since the capacity of line $n_1 - n_2$ is assumed to be large enough (200 MW), the transmission constraints do not change the market-clearing outcomes. Thus, the results obtained for this example are identical to those in Illustrative Example 5.1. In particular, the LMPs at both nodes are identical, and equal to $35/MWh. Note that the power flow in line $n_1 - n_2$ is 100 MW, which is equal to the rival unit's production, located at node n_2.

To gain insight into the impact of transmission constraints on investment decisions and profit, we consider a reduced transmission capacity of line $n_1 - n_2$; specifically, we limit the capacity of this line to 80 MW. We refer to this problem as the *congested case*. Similar to the uncongested case, the strategic producer invests in a 100 MW conventional unit at node n_1 and offers its production units at price $35/MWh. An important observation is that although the rival unit offers at a comparatively lower price, i.e., $15/MWh, it is not dispatched at its maximum (100 MW) due to transmission congestion. The market-clearing outcomes in this case are provided in Table 5.2. Accordingly, the production level of the rival unit decreases to 80 MW, while that of the newly built conventional unit of the strategic producer increases to 70 MW. Thus, the strategic producer's annual profit increases to $41.45 million.

Table 5.2 Illustrative Example 5.2: Market-clearing outcomes (congested case)

Market participant	Offer/bid price [$/MWh]	Production/consumption level [MW]
Rival unit	$C^{R}_{n_2 o_1} = 15$	$p^{R}_{n_2 o_1} = 80$
Existing unit of the strategic producer	$\alpha^{E}_{n_1 o_1} = 35$	$p^{E}_{n_1 o_1} = 150$
Newly built unit of the strategic producer	$\alpha^{C}_{n_1 o_1} = 35$	$p^{C}_{n_1 o_1} = 70$
Consumer	$U^{D}_{n_1 o_1} = 35$	$p^{D}_{n_1 o_1} = 300$

Another important impact of transmission congestion is that the LMPs at nodes n_1 and n_2 become different. In this case, the LMP at node n_1 is $35/MWh, while that of node n_2 is $15/MWh.

5.4.3.3 Results for Illustrative Example 5.3

In this example, two technologies as investment options are considered: conventional and stochastic (wind power) units. The strategic producer decides to build only a wind power unit at node n_2 with a capacity equal to $x^{S}_{n_2}$ =142.86 MW. Thus, the annualized investment cost of the strategic producer is $9.43 million (i.e., 142.86 × 66000), which is lower than its annualized investment budget, i.e., $10 million. The wind power capacity factors in operating conditions o_1 and o_2 are 0.35 and 0.70, respectively. Therefore, the available wind power production in operating conditions o_1 and o_2 are 50 MW (i.e., 142.86 × 0.35) and 100 MW (i.e., 142.86 × 0.70), respectively.

In analogy to the results obtained in Illustrative Examples 5.1 and 5.2, the strategic producer offers its existing conventional unit in each operating condition at a price identical to the consumer's bid price in that condition, i.e., $35/MWh in condition o_1 and $32/MWh in condition o_2. In this way, it forces the market to be cleared at a higher price. The market-clearing outcomes in both operating conditions o_1 and o_2 are provided in Table 5.3. The consumer is fully supplied in both conditions, and the line flows in operating conditions o_1 and o_2 are 200 MW and 150 MW, respectively. In this example, the transmission limit does not affect the market-clearing outcomes. Therefore, the LMPs corresponding to nodes n_1 and n_2 in each condition are identical, i.e., $35/MWh in operating condition o_1 and $32/MWh in operating condition o_2. Finally, the annual profit of the strategic producer in this example is $31.32 million.

Table 5.3 Illustrative Example 5.3: Market-clearing outcomes

Market participant	Production/consumption level in condition o_1 [MW]	Production/consumption level in condition o_2 [MW]
Rival unit	$p^{\mathrm{R}}_{n_2 o_1} = 100$	$p^{\mathrm{R}}_{n_2 o_2} = 100$
Existing unit of the strategic producer	$p^{\mathrm{E}}_{n_1 o_1} = 150$	$p^{\mathrm{E}}_{n_1 o_2} = 40$
Newly built wind unit of the strategic producer	$p^{\mathrm{S}}_{n_2 o_1} = 50$	$p^{\mathrm{S}}_{n_2 o_2} = 100$
Consumer	$p^{\mathrm{D}}_{n_1 o_1} = 300$	$p^{\mathrm{D}}_{n_1 o_2} = 240$

5.4.3.4 Results for Illustrative Example 5.4

The investment and offering decisions of the strategic producer are identical to those in Illustrative Example 5.3. However, its *expected* annual profit decreases with respect to its annual profit in Illustrative Example 5.3 due to the rival's investment uncertainty. The expected annual profit of the strategic producer is \$28.07 million.

Regarding market-clearing outcomes, the production/consumption levels under scenario ω_1 (no rival investment) are identical to those in Table 5.3. However in scenario ω_2, the rival producer builds a 40-MW conventional unit. Thus, the production of the strategic producer's existing unit decreases to 110 MW in operating condition o_1 and to zero in condition o_2. The LMPs in operating conditions o_1 and o_2 are \$35/MWh and \$32/MWh, respectively. Note that although the strategic producer's existing unit is not dispatched in condition o_2 under scenario ω_2, the LMP at both nodes is \$32/MWh because all other units are fully dispatched. Thus, the cost for increasing 1 MW demand is equal to the offer price of the strategic producer's existing unit, i.e., \$32/MWh.

The GAMS code for solving the MILP problem corresponding to Illustrative Example 5.4 is provided in Sect. 5.8.

5.4.3.5 Results for Illustrative Example 5.5

In this dynamic investment example, the decisions are made at the beginning of time periods t_1 and t_2. The strategic producer builds a 50 MW conventional unit at the beginning of time period t_1, i.e., $x^{\mathrm{C}}_{t_1 n_1 \omega_1} = x^{\mathrm{C}}_{t_1 n_1 \omega_2} = 50$ MW. This decision is identical over scenarios ω_1 and ω_2 as enforced by the nonanticipativity constraint. However, the investment decision at the beginning of time period t_2 depends on the scenario realized at the end of time period t_1. If the investment cost in time period t_2 remains equal to that in period t_1, i.e., if scenario ω_1 is realized, then the strategic producer builds an additional 50-MW conventional unit in time period t_2, i.e., $x^{\mathrm{C}}_{t_2 n_1 \omega_1}$ =50 MW. Therefore, the conventional unit capacity increases to 100 MW. One the other hand, if scenario ω_2 is realized, then the strategic producer does not invest in time period t_2, i.e., $x^{\mathrm{C}}_{t_2 n_1 \omega_2}$ =0. Therefore, the capacity remains at 50 MW. The expected profit of the strategic producer to be obtained during the considered ten-year time horizon is \$75.66 million. Similarly to the static investment examples, the strategic producer offers in each time period, operating condition, and scenario at a price identical to the consumer's bid price, which is assumed to be \$35/MWh for both time periods under any scenario. Thus, the market-clearing price in this example in both time periods under any scenario is \$35/MWh.

5.5 Benders Solution Approach

As explained in Sect. 5.1.7, the computational burden for solving bilevel models (5.1) and (5.2) through a direct solution approach (Sect. 5.4) increases significantly with the number of operating conditions and scenarios, which is in fact the main drawback of this approach. To overcome this computational issue and eventual intractability, another solution approach based on a Benders decomposition technique [7, 12] is described in this section. This solution approach provides a computationally tractable formulation that allows us to solve a problem with many operating conditions and scenarios.

5.5.1 *Complicating Variables*

Solving the static investment model (5.1) presented in Sect. 5.2 and the dynamic investment model (5.2) presented in Sect. 5.3 requires considering all involved operating conditions and scenarios simultaneously. Therefore, a large number of operating conditions and scenarios may result in high computational burden and eventual intractability.

If production capacity investment decision variables within the static investment model (5.1), i.e., x_n^{C} and x_n^{S}, are considered to be *complicating variables*, then model (5.1) can be solved using a Benders decomposition approach. The reason for selecting x_n^{C} and x_n^{S} as complicating variables is that fixing these variables to given values yields one decomposed problem per operating condition o and scenario ω.

Analogously, the dynamic investment model (5.2) can be solved using Benders decomposition, provided that the production capacity investment variables $x_{tn\omega}^{\mathrm{C}}$ and $x_{tn\omega}^{\mathrm{S}}$ are fixed to given values. In this case, the original bilevel model (5.2) decomposes into a number of smaller problems, one per time period t, operating condition o, and scenario ω.

Illustrative Example 5.7 *Decomposing Illustrative Example 5.4 through fixing complicating variables to given values*

To clarify the idea behind Benders decomposition, we analyze Illustrative Example 5.4 in Sect. 5.2 and its corresponding bilevel model. This example refers to a static production capacity investment model including two operating conditions, o_1 and o_2, and two scenarios, ω_1 and ω_2.

Note that if we consider the two investment variables $x^{\mathrm{C}}_{n_1}$ and $x^{\mathrm{S}}_{n_2}$ as complicating variables and fix them to given values $x^{\mathrm{C,fixed}}_{n_1}$ and $x^{\mathrm{S,fixed}}_{n_2}$, respectively, then the bilevel model of Illustrative Example 5.4 decomposes into four smaller bilevel models, one per operating condition and scenario. The first decomposed bilevel model corresponding to operating condition o_1 and scenario ω_1 is given below:

$$\min_{\Xi^{\mathrm{P,Ex4}}_{o_1\omega_1}, x^{\mathrm{C}}_{n_1}, x^{\mathrm{S}}_{n_2}} \quad -(3530 \times 0.6)\Big[\lambda_{n_1o_1\omega_1}\left(p^{\mathrm{C}}_{n_1o_1\omega_1} + p^{\mathrm{E}}_{n_1o_1\omega_1}\right) + \lambda_{n_2o_1\omega_1}\, p^{\mathrm{S}}_{n_2o_1\omega_1}$$
$$- 12\, p^{\mathrm{C}}_{n_1o_1\omega_1} - 10\, p^{\mathrm{E}}_{n_1o_1\omega_1}\Big]$$

subject to

$$x^{\mathrm{C}}_{n_1} = x^{\mathrm{C,fixed}}_{n_1}$$
$$x^{\mathrm{S}}_{n_2} = x^{\mathrm{S,fixed}}_{n_2}$$
$$\alpha^{\mathrm{C}}_{n_1o_1\omega_1} \geq 0$$
$$\alpha^{\mathrm{E}}_{n_1o_1\omega_1} \geq 0$$

Lower-level problem for operating condition o_1 and scenario ω_1.

The second decomposed bilevel model corresponding to operating condition o_2 and scenario ω_1 is given below:

$$\min_{\Xi^{\mathrm{P,Ex4}}_{o_2\omega_1}, x^{\mathrm{C}}_{n_1}, x^{\mathrm{S}}_{n_2}} \quad -(5230 \times 0.6)\Big[\lambda_{n_1o_2\omega_1}\left(p^{\mathrm{C}}_{n_1o_2\omega_1} + p^{\mathrm{E}}_{n_1o_2\omega_1}\right) + \lambda_{n_2o_2\omega_1}\, p^{\mathrm{S}}_{n_2o_2\omega_1}$$
$$- 12\, p^{\mathrm{C}}_{n_1o_2\omega_1} - 10\, p^{\mathrm{E}}_{n_1o_2\omega_1}\Big]$$

subject to

$$x^{\mathrm{C}}_{n_1} = x^{\mathrm{C,fixed}}_{n_1}$$
$$x^{\mathrm{S}}_{n_2} = x^{\mathrm{S,fixed}}_{n_2}$$
$$\alpha^{\mathrm{C}}_{n_1o_2\omega_1} \geq 0$$
$$\alpha^{\mathrm{E}}_{n_1o_2\omega_1} \geq 0$$

Lower-level problem for operating condition o_2 and scenario ω_1.

The third decomposed bilevel model corresponding to operating condition o_1 and scenario ω_2 is given below:

$$\min_{\Xi^{\mathrm{P,Ex4}}_{o_1\omega_2}, x^{\mathrm{C}}_{n_1}, x^{\mathrm{S}}_{n_2}} \quad -(3530 \times 0.4)\Big[\lambda_{n_1o_1\omega_2}\left(p^{\mathrm{C}}_{n_1o_1\omega_2} + p^{\mathrm{E}}_{n_1o_1\omega_2}\right) + \lambda_{n_2o_1\omega_2}\, p^{\mathrm{S}}_{n_2o_1\omega_2}$$
$$- 12\, p^{\mathrm{C}}_{n_1o_1\omega_2} - 10\, p^{\mathrm{E}}_{n_1o_1\omega_2}\Big]$$

subject to

$$x^{\mathrm{C}}_{n_1} = x^{\mathrm{C,fixed}}_{n_1}$$
$$x^{\mathrm{S}}_{n_2} = x^{\mathrm{S,fixed}}_{n_2}$$
$$\alpha^{\mathrm{C}}_{n_1o_1\omega_2} \geq 0$$
$$\alpha^{\mathrm{E}}_{n_1o_1\omega_2} \geq 0$$

Lower-level problem for operating condition o_1 and scenario ω_2.

Finally, the fourth decomposed bilevel model corresponding to operating condition o_2 and scenario ω_2 is given below:

$$\min_{\Xi^{\mathrm{P,Ex4}}_{o_2\omega_2}, x^{\mathrm{C}}_{n_1}, x^{\mathrm{S}}_{n_2}} \quad -(5230 \times 0.4)\Big[\lambda_{n_1o_2\omega_2}\left(p^{\mathrm{C}}_{n_1o_2\omega_2} + p^{\mathrm{E}}_{n_1o_2\omega_2}\right) + \lambda_{n_2o_2\omega_2}\, p^{\mathrm{S}}_{n_2o_2\omega_2}$$
$$- 12\, p^{\mathrm{C}}_{n_1o_2\omega_2} - 10\, p^{\mathrm{E}}_{n_1o_2\omega_2}\Big]$$

subject to

$$x^{\mathrm{C}}_{n_1} = x^{\mathrm{C,fixed}}_{n_1}$$
$$x^{\mathrm{S}}_{n_2} = x^{\mathrm{S,fixed}}_{n_2}$$
$$\alpha^{\mathrm{C}}_{n_1o_2\omega_2} \geq 0$$
$$\alpha^{\mathrm{E}}_{n_1o_2\omega_2} \geq 0$$

Lower-level problem for operating condition o_2 and scenario ω_2.

Note that each decomposed bilevel model is smaller than the original undecomposed bilevel model, and thus it is easier to solve. Each of these decomposed bilevel models is called a *subproblem*. □

5.5.2 Convexity Analysis

For bilevel models (5.1) and (5.2), an effective implementation of the Benders decomposition technique is possible if the expected profit of the strategic producer as a function of the complicating variables (i.e., the production capacity investment variables) has a convex envelope. Although bilevel models are generally nonconvex and do not meet such a requirement, if the number of operating conditions and scenarios is large enough, then the objective function of such bilevel models as a function of each complicating variable becomes *sufficiently convex* [8, 26]. In other words, the objective function of a stochastic programming problem becomes sufficiently convex as the number of operating conditions and scenarios increases. The reason for this is that the objective function represents the expectation over a number of operating conditions and scenarios. Thus, as the number of operating conditions and scenarios increases, the diversity of objective functions increases, while the weight of each single operating condition and scenario decreases. This results in a smoothing effect leading to the convexification of the expected value objective function. This convexification generally allows a successful implementation of Benders decomposition. Nevertheless, convergence cannot be generally guaranteed for the bilevel models considered.

5.5.3 Functioning of Benders Decomposition

Conceptually, the Benders algorithm works as explained through the following three steps:

1. Given fixed investment decisions, i.e., x_n^{C} and x_n^{S} in the static investment model (5.1), and $x_{tn\omega}^{\mathrm{C}}$ and $x_{tn\omega}^{\mathrm{S}}$ in the dynamic investment model (5.2), the resulting decomposed bilevel models (subproblems) are solved, and their solutions provide (i) offering and operating decisions for the strategic producer and (ii) sensitivities of the strategic producer's expected profit with respect to the investment decisions. In general, these sensitivities are derived through the dual variables associated with the constraints fixing the investment decisions.
2. The sensitivities obtained in step 1 above allow the formulation of a so-called *Benders master problem*, whose solution provides updated investment decisions.

Steps 1 and 2 above are repeated until no improvement in expected profit is achieved.

In the case of the static production capacity investment model (5.1), Fig. 5.5 further illustrates the Benders algorithm. Box A corresponds to the original undecomposed bilevel model (5.1). Fixing the complicating investment variables x_n^{C} and x_n^{S} (Box B) renders one decomposed bilevel model (subproblem) per operating condition and scenario (Box C), which can be recast as an MPEC (Box D). The sensitivities obtained from the subproblems allow formulating the master problem (Box E) to update the production capacity investment variables. The algorithm continues until no improvement in expected profit is achieved. The detailed steps of the Benders algorithm to solve the production capacity investment problem are presented in the next section.

For the dynamic production capacity investment model (5.2), the Benders algorithm is similar to that in Fig. 5.5; however, the original problem decomposes by time period as well.

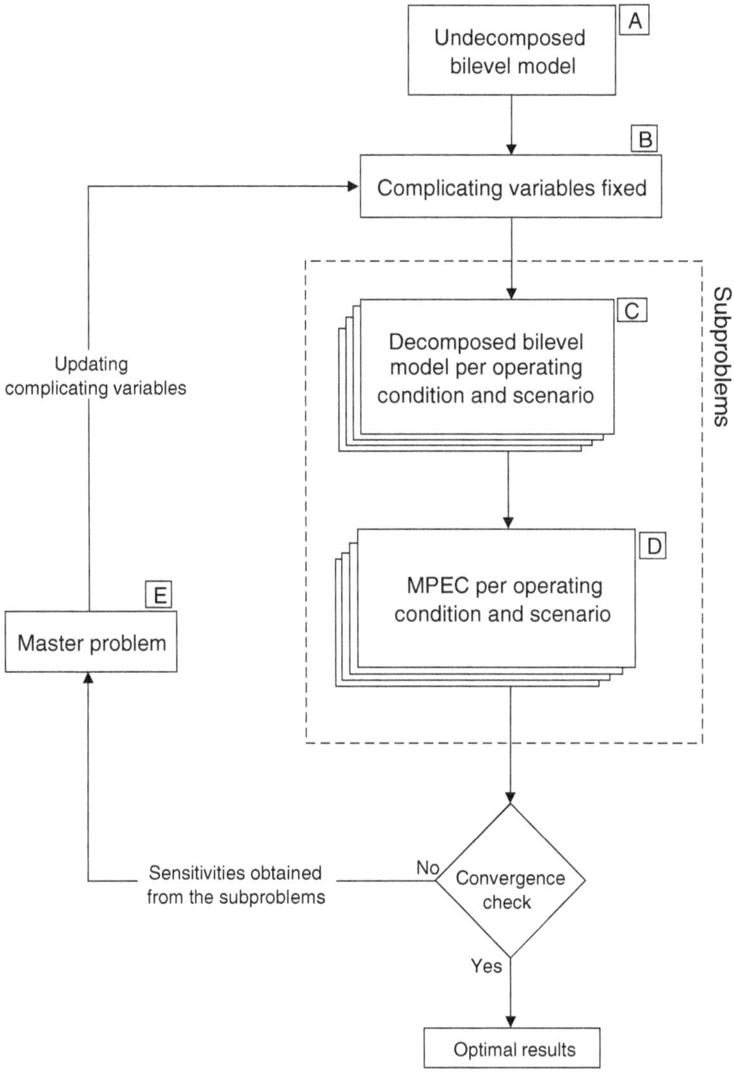

Fig. 5.5 Flowchart of the Benders solution approach to solving the static production capacity investment model

5.5.4 *The Benders Algorithm*

A detailed description of the Benders algorithm to solve the production capacity investment problem is presented below:

Step 0 Input data for the Benders algorithm including a tolerance ε, initial guesses for the complicating variables (e.g., no investment), operating conditions, and scenario data. In addition, this step sets Benders iteration $v = 1$ and considers an initial lower bound for the strategic producer's profit.

Step 1 Select the first operating condition and scenario (and time period in the case of a dynamic investment model).

Step 2 Solve the subproblem for operating condition o and scenario ω (and time period t in the case of a dynamic investment model).

Step 3 Repeat Steps 1 and 2 for all involved operating conditions and scenarios (and time periods in the case of a dynamic investment model). This step provides the values for sensitivities as well as an upper bound for the strategic producer's profit.

Step 4 Check convergence by comparing the values of the upper bound of the profit obtained from the subproblems and the initial lower bound of the profit as given in Step 0. Note that this lower bound in subsequent iterations (i.e., $v > 1$) is obtained from the master problem (Step 5). If the difference between these two bounds is smaller than the tolerance ε, then the solution at iteration v is optimal with a level of accuracy ε. Otherwise, the iteration counter is updated and the next iteration is considered.

Step 5 Solve the master problem based on the sensitivities obtained in the subproblems. This problem updates the values of complicating variables and of the lower bound profit.

The algorithm continues in Step 1.

Illustrative Example 5.8 *Benders solution of Illustrative Example 5.4*

In this example, we consider Illustrative Example 5.4 as presented in Sect. 5.2 and solve its corresponding bilevel model using Benders decomposition. Recall that Illustrative Example 5.4 has already been solved using a direct solution approach in Sect. 5.4.3. Note also that four subproblems have already been derived in Illustrative Example 5.7 from the original bilevel model corresponding to Illustrative Example 5.4 through fixing the investment decisions $x^{\mathrm{C}}_{n_1}$ and $x^{\mathrm{S}}_{n_2}$ to given values $x^{\mathrm{C,fixed}}_{n_1}$ and $x^{\mathrm{S,fixed}}_{n_2}$, respectively. To solve this example through the Benders solution approach, the following steps are examined:

Initialization: As an initial step (explained in Sect. 5.5.4), we set the tolerance $\varepsilon = 0.1$, the initial guesses of the complicating variables $x^{\mathrm{C}}_{n_1} = 0$ and $x^{\mathrm{S}}_{n_2} = 0$, and the Benders iteration $v = v_1$. Finally, we consider $-\$10^{10}$ as the initial lower bound for minus the strategic producer's expected annual profit.

Benders iteration $v = v_1$: According to the initial inputs, we need to solve separately four decomposed bilevel models derived in Illustrative Example 5.7 as subproblems. To this end, the MPEC and then the MILP equivalent of each subproblem are formulated and solved. However, the binary variables in the mixed-integer subproblems that are involved due to complementarity linearization may hinder obtaining appropriate values for sensitivities. To overcome such a problem, we first solve the MILP equivalent of each subproblem as an *auxiliary problem*. Then, a continuous and linear form of each subproblem is formulated in which each complementarity condition is replaced by its corresponding strong duality equality, and some variables

Table 5.4 Illustrative Example 5.8: Values obtained for sensitivities [$/MW] and the strategic producer's minus profit [$] in each subproblem (iteration $\nu = \nu_1$)

Subproblem	Variable	The value obtained
Subproblem 1 related to the operating condition o_1 and scenario ω_1	Minus the strategic producer's profit	-7.94×10^6
	Sensitivity with respect to $x_{n_1}^{C(\nu_1)}$	0
	Sensitivity with respect to $x_{n_2}^{S(\nu_1)}$	−25945.5
Subproblem 2 related to the operating condition o_1 and scenario ω_2	Minus the strategic producer's profit	-5.29×10^6
	Sensitivity with respect to $x_{n_1}^{C(\nu_1)}$	0
	Sensitivity with respect to $x_{n_2}^{S(\nu_1)}$	−17297.0
Subproblem 3 related to the operating condition o_2 and scenario ω_1	Minus the strategic producer's profit	-9.67×10^6
	Sensitivity with respect to $x_{n_1}^{C(\nu_1)}$	0
	Sensitivity with respect to $x_{n_2}^{S(\nu_1)}$	−21966.0
Subproblem 4 related to the operating condition o_2 and scenario ω_2	Minus the strategic producer's profit	-4.60×10^6
	Sensitivity with respect to $x_{n_1}^{C(\nu_1)}$	0
	Sensitivity with respect to $x_{n_2}^{S(\nu_1)}$	−14644.0

within the bilinear terms are substituted by their optimal values obtained from the auxiliary problem. The outcomes of this step are operational decisions as well as the values for sensitivities, which are dual variables associated with the constraints fixing investment decisions. Further details on this approach can be found in [26]. The values obtained for sensitivities and minus the strategic producer's annual profit in each subproblem are provided in Table 5.4. According to the values given in this table, minus the strategic producer's expected profit is $-\$27.50 \times 10^6$, which is the summation of minus its profit over the four subproblems. In addition, the expected values obtained for sensitivities in this iteration with respect to complicating variables $x_{n_1}^{C(\nu_1)}$ and $x_{n_2}^{S(\nu_1)}$ are 0 and −79852.5, respectively.

Convergence check: The value obtained in the first iteration for the upper bound of minus the strategic producer's expected profit is $-\$27.50 \times 10^6$ plus the investment cost, which is zero. On the other hand, the initial lower bound of minus the strategic producer's expected profit is $-\$10^{10}$. Thus, the absolute difference between these two bounds is greater than the value ε considered for tolerance. Therefore, the next iteration (i.e., $\nu = \nu_2$) needs to be considered.

Iteration $\nu = \nu_2$: As the first step in the second iteration, the Benders master problem is formulated and solved. The aim of the master problem is to update the values for complicating variables through sensitivities obtained in the first iteration. The formulation of the Benders master problem in iteration $\nu = \nu_2$ is given below. Note that the superscript (ν_2) for the variables illustrates the Benders iteration:

$$\min_{\Xi^{\mathrm{MP}(\nu_2)}} \quad \left(55000\, x_{n_1}^{C(\nu_2)} + 66000\, x_{n_2}^{S(\nu_2)}\right) + \beta^{(\nu_2)}$$

subject to

$$x_{n_1}^{\mathrm{C}(\nu_2)} = 100\, u_{n_1h_2}^{\mathrm{C}(\nu_2)}$$
$$u_{n_1h_1}^{\mathrm{C}(\nu_2)} + u_{n_1h_2}^{\mathrm{C}(\nu_2)} = 1$$
$$u_{n_1h_1}^{\mathrm{C}(\nu_2)},\ u_{n_1h_2}^{\mathrm{C}(\nu_2)} \in \{0, 1\}$$
$$0 \le x_{n_2}^{\mathrm{S}(\nu_2)} \le 200$$
$$\left(55000\, x_{n_1}^{\mathrm{C}(\nu_2)} + 66000\, x_{n_2}^{\mathrm{S}(\nu_2)}\right) \le 10^7$$
$$\beta^{(\nu_2)} \ge -10^{10}$$
$$\beta^{(\nu_2)} \ge \left(-27.50 \times 10^6\right) + 0\left(x_{n_1}^{\mathrm{C}(\nu_2)} - 0\right) - 79852.5\left(x_{n_2}^{\mathrm{S}(\nu_2)} - 0\right).$$

The optimization variables of the master problem above in iteration $\nu = \nu_2$ are those in the set $\Xi^{\mathrm{MP}(\nu_2)} = \{x_{n_1}^{\mathrm{C}(\nu_2)}, x_{n_2}^{\mathrm{S}(\nu_2)}, \beta^{(\nu_2)}, u_{n_1h_1}^{\mathrm{C}(\nu_2)}, u_{n_1h_2}^{\mathrm{C}(\nu_2)}\}$. This problem is mixed-integer and linear.

Note that the variable $\beta^{(\nu_2)}$ in the objective function represents minus the operational profit of the strategic producer. The first five constraints enforce the available capacity options and the investment budget. The sixth constraint imposes a lower bound on $\beta^{(\nu_2)}$ to accelerate convergence. Finally, the last constraint is a Benders cut, in which the values of minus the strategic producers' operational profit and sensitivities obtained in iteration $\nu = \nu_1$ are used. This constraint provides feedback to the master problem on the impact of the complicating variables' values on the outcomes of the subproblems in the previous iteration. Note that a new Benders cut is incorporated into the master problem in every Benders iteration.

The solution of the master problem updates the values of the complicating variables as $x_{n_1}^{\mathrm{C}(\nu_2)}$ =100 MW and $x_{n_2}^{\mathrm{S}(\nu_2)}$ =68.182 MW. In addition, the value obtained for the objective function provides the updated lower bound for minus the strategic producer's profit, which is $-\$31.07 \times 10^6$.

Then, the four subproblems are solved in iteration $\nu = \nu_2$ based on the updated values obtained for the complicating variables. Next, the convergence needs to be checked. If required, iteration $\nu = \nu_3$ is considered.

The next iterations: Table 5.5 provides the Benders procedure for solving this example, which converges in iteration 4. Note that the results obtained through the Benders solution approach given in the last row of Table 5.5 are identical to those obtained through the direct solution approach.

□

5.6 Summary

This chapter provides a mathematical framework based on a stochastic bilevel model to be used by a strategic producer to make production capacity investment decisions. Both conventional and stochastic production units are considered as investment candidates. Diverse sources of uncertainty, e.g., uncertainties on rival investment and investment costs, are appropriately characterized through a set of scenarios. In addition, the variability of demand and the production level of stochastic units are

Table 5.5 Illustrative Example 5.8: Benders iterations

	Iteration 1	Iteration 2	Iteration 3	Iteration 4
The value obtained for $x^{\mathrm{C}}_{n_1}$ [MW]	0	100	0	0
The value obtained for $x^{\mathrm{S}}_{n_2}$ [MW]	0	68.182	94.838	142.857
The lower bound for minus the profit [\$]	-10^{10}	-31.07×10^6	-28.82×10^6	-28.07×10^6
The upper bound for minus the profit [\$]	-27.50×10^6	-23.73×10^6	-28.00×10^6	-28.07×10^6
Convergence error [\$]	99.72×10^8	7.33×10^6	0.82×10^6	3.72×10^{-9}

represented through a set of operating conditions. Then, two investment models are studied: static and dynamic (multistage). In the static investment model, the strategic producer considers a single future year and decides its optimal production capacity portfolio for that target year. On the other hand, in the dynamic model, the investment decisions are made at several points in time throughout the planning horizon. In general, the dynamic model results in more accurate investment decisions than the static model but at the cost of a high computational burden and potential intractability.

In addition, two solution approaches are described. The first is a direct solution, which may result in intractability if a large number of operating conditions and scenarios is used. Another solution approach based on Benders decomposition, valid for very large problems, is provided as well.

5.7 End-of-Chapter Exercises

5.1 Reformulate the static investment model (5.1) and the dynamic investment model (5.2) to include several units per node, piecewise production costs of units, and piecewise bidding curves of consumers.

5.2 Reformulate the static investment model (5.1) and the dynamic investment model (5.2) to include the strategic offering of candidate stochastic units (instead of zero offer price).

5.3 Consider the lower-level problem of the static investment model (5.1). Write the dual problem corresponding to this lower-level problem and its KKT conditions. Verify that the solutions of the primal problem, the dual problem, and the optimality conditions are identical.

5.4 Derive the MPEC of the generic bilevel model (5.2) corresponding to the dynamic investment model. Then recast this generic MPEC as an MILP problem.

5.5 Consider investment decisions as complicating variables in Illustrative Example 5.5 corresponding to the dynamic investment model. Write the decomposed subproblems. Then, solve it by the Benders decomposition algorithm described in Sect. 5.5 and compare the results to those obtained from the direct solution approach.

5.6 Determine the most beneficial investment decisions using the static investment model (5.1) for a strategic producer competing in the power system depicted in Fig. 5.6. Note that to solve this exercise, the reformulated form of model (5.1) derived in Exercise 5.1 is needed, in which several units per node can be considered. This power system includes three nodes (n_1, n_2, and n_3) and three transmission lines ($n_1 - n_2, n_1 - n_3$, and $n_2 - n_3$). The transmission capacity and susceptance of each line are 900 MW and 1000 S, respectively.

According to Fig. 5.6, the strategic producer owns an existing unit at node n_1 with capacity of 400 MW and production cost of \$15.00/MWh. In addition, two rival production units (R_1 and R_2) are considered at node n_3. One of the rival units (R_1) exists in the target year, while another one (R_2) may or may not be built by the rival producer. In fact, unit R_2 characterizes the rival investment uncertainty. The capacity of unit R_1 is 450 MW, while that of unit R_2 is uncertain and characterized by three scenarios ω_1, ω_2, and ω_3. The capacities of rival unit R_2 corresponding to scenarios ω_1, ω_2, and ω_3 are 0, 300 MW, and 450 MW, respectively, and the corresponding

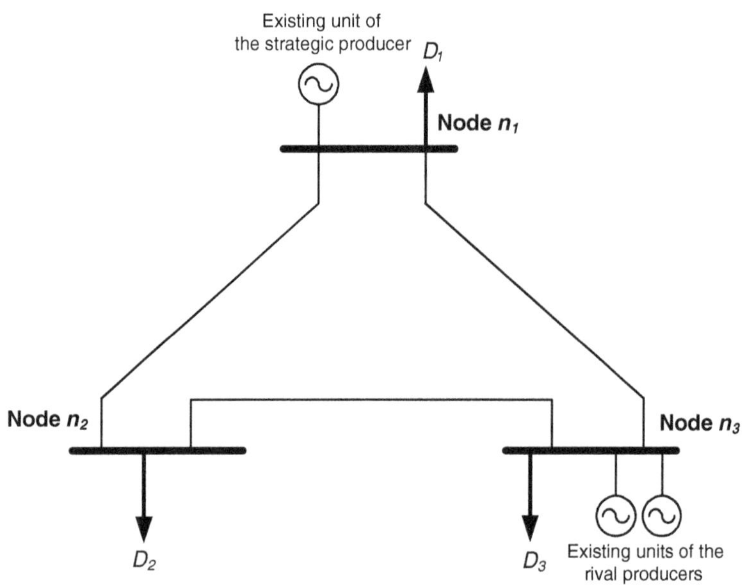

Fig. 5.6 Exercise 5.6: three-node test system

probabilities are 0.3, 0.4, and 0.3, respectively. Note that the production cost of both rival units R_1 and R_2 is $14.00/MWh.

Regarding the production capacity investment options, three different technologies are considered as follows:

1. Conventional base technology (e.g., nuclear) with high investment cost but low production cost.
2. Conventional peak technology (e.g., CCGTs) with low investment cost but high production cost.
3. Stochastic production technology (e.g., wind turbines) with high investment cost but zero production cost.

Two potential locations are considered to build stochastic units, i.e., nodes n_1 and n_2. On the other hand, nodes n_1 and n_3 are candidate locations to build new conventional (base and peak) units. The characteristics of production capacity investment candidates including their annualized investment costs and available capacity options are presented in Table 5.6. The investment budget is assumed to be unlimited. The maximum consumption levels of consumers D_1, D_2, and D_3 are 250 MW, 850 MW, and 1000 MW, respectively. In addition, five operating conditions (o_1 to o_5) are considered, whose weighting factors are 1465, 1277, 2377, 2155, and 1486, respectively. We assume that the demand factors of all consumers are identical. In addition, different power capacity factors are considered for the two candidate stochastic production units since their geographic locations are different. Table 5.7 presents the values considered for demand and stochastic production factors. Finally, Table 5.8 presents the bid prices of consumers under different operating conditions.

Table 5.6 Exercise 5.6: data for production capacity investment options

Candidate unit (technology)	Annualized investment cost [$/MW]	Available capacity for investment options [MW]	Production cost [$/MWh]
Conventional base	55000	0, 275, 500, 625, 750	12.00
Conventional peak	15000	0, 125, 250, 375, 500	20.00
Stochastic	78000	Continuous options between 0 and 500	0.00

Table 5.7 Exercise 5.6: data for operating conditions considered

Operating conditions	Demand factor for each consumer [p.u.]	Production factor for candidate stochastic unit located at node n_1 [p.u.]	Production factor for candidate stochastic unit located at node n_2 [p.u.]
o_1	0.86	0.13	0.43
o_2	0.78	0.13	0.82
o_3	0.70	0.62	0.66
o_4	0.63	0.10	0.13
o_5	0.54	0.59	0.36

Table 5.8 Exercise 5.6: data for bid price of consumers in different operating conditions

Operating conditions	Bid price of D_1 [\$/MWh]	Bid price of D_2 [\$/MWh]	Bid price of D_3 [\$/MWh]
o_1	30.00	32.00	36.00
o_2	27.00	30.00	35.00
o_3	26.00	28.00	33.00
o_4	23.00	25.00	30.00
o_5	22.00	24.00	28.00

5.7 Solve Exercise 5.6 for the case in which the capacity of each transmission line is 300 MW. Interpret why transmission congestion alters the investment decisions of the strategic producer.

5.8 Solve Exercise 5.6 using the dynamic investment model (5.2). Consider a planning horizon comprising two five-year time periods (t_1 and t_2). Amortization rates are 20 % and 10 % in the first and second periods, respectively. In addition, the investment technologies are identical to those in Table 5.6; however, their actual investment costs are \$600,000/MW, \$180,000/MW, and \$800,000/MW, respectively. The maximum consumption levels of each consumer in the first and the second time periods are 80 % and 100 % of those in Exercise 5.6, respectively. We assume that the bid prices of consumers (Table 5.8) do not change across time periods.

5.8 GAMS Code

This section provides the GAMS code for solving the MILP problem corresponding to Illustrative Example 5.4, which is more complex than Illustrative Examples 5.1–5.3. Therefore, this code can be easily adapted to those examples. Likewise, it is not difficult to change the code for use in a dynamic investment example, e.g., Illustrative Example 5.5. Note that this code is written in a general form, and thus it is straightforward to adapt it to any static production capacity investment case.

```
SETS
o            operating conditions /o1*o2/
h            conventional technologies /h1*h2/
w            scenarios /w1*w2/
n            nodes /n1*n2/
s(n)         reference node /n1/
Omega(n,n)   transmission lines /n1.n2,n2.n1/
ALIAS (n,m);

PARAMETERS
C_C(n) production cost of candidate conventional
    units /
n1   12/
```

```
C_E(n) production cost of existing conventional units /
n1   10/

K_C(n) annualized investment cost of candidate conventional units /
n1   55000/

K_S(n) annualized investment cost of candidate stochastic units /
n2   66000/

P_Emax(n) capacity of existing conventional units /
n1   150/

P_Dmax(n) maximum load of consumers /
n1   300/

X_Smax(n) maximum capacity investment of candidate stochastic units /
n2   200/

phi(w)   probability of scenarios /
w1   0.6
w2   0.4/

rho(o)   weighting factor of operating conditions /
o1   3530
o2   5230/;

TABLE  B(n,n) susceptance of transmission lines
       n1     n2
n1     0      1e3
n2     1e3    0;

TABLE  C_R(n,o) offer price of rival units
       o1     o2
n1     13     13
n2     15     15;

TABLE Fmax(n,n) transmission capacity of lines
       n1     n2
n1     0      200
n2     200    0;

TABLE P_Rmax(n,w) capacity of rival units
       w1     w2
n1     0      40
n2     100    100;

TABLE Q_S(n,o) power capacity factor of candidate stochastic units
```

```
      o1      o2
n2    0.35    0.70;

TABLE Q_D(n,o) demand factor of consumers
      o1      o2
n1    1.0    0.8;

TABLE U_D(n,o) bid price of consumers
      o1      o2
n1    35      32;

TABLE X_Option(n,h) conventional investment alternatives
      h1      h2
n1    0       100;

SCALAR Kmax available investment budget /1e7/
SCALAR BigM1   a large value /1e4/
SCALAR BigM2   a large value /5e4/
SCALAR PI    pi   /3.1416/;

FREE VARIABLES
minus_profit              minus expected profit of the producer
linear_term(n,o,w)        linear equivalent of the revenue term
theta(n,o,w)              nodal voltage angles
lambda(n,o,w)             locational marginal prices (LMPs);

POSITIVE VARIABLES
p_C(n,o,w)                production of candidate conventional units
p_D(n,o,w)                Consumption of consumers
p_E(n,o,w)                production of existing conventional units
p_R(n,o,w)                production of rival units
p_S(n,o,w)                production of candidate stochastic units
x_C(n)                    capacity investment in conventional units
x_S(n)                    capacity investment in stochastic units
alpha_C(n,o,w)            offer price by candidate conventional units
alpha_E(n,o,w)            offer price by existing conventional units
mu_Cmin(n,o,w)            dual
mu_Cmax(n,o,w)            dual
mu_Dmin(n,o,w)            dual
mu_Dmax(n,o,w)            dual
mu_Emin(n,o,w)            dual
mu_Emax(n,o,w)            dual
```

```
mu_F(n,m,o,w)          dual
mu_Rmin(n,o,w)         dual
mu_Rmax(n,o,w)         dual
mu_Smin(n,o,w)         dual
mu_Smax(n,o,w)         dual
mu_theta_min(n,o,w)    dual
mu_theta_max(n,o,w)    dual
mu_theta_ref(o,w)      dual

BINARY VARIABLES
u_C(n,h)               conventional investment
   decisions
u_mu_Smin(n,o,w)       auxiliary
u_mu_Smax(n,o,w)       auxiliary
u_mu_Cmin(n,o,w)       auxiliary
u_mu_Cmax(n,o,w)       auxiliary
u_mu_Rmin(n,o,w)       auxiliary
u_mu_Rmax(n,o,w)       auxiliary
u_mu_Emin(n,o,w)       auxiliary
u_mu_Emax(n,o,w)       auxiliary
u_mu_Dmin(n,o,w)       auxiliary
u_mu_Dmax(n,o,w)       auxiliary
u_mu_F(n,m,o,w)        auxiliary
u_mu_theta_min(n,o,w) auxiliary
u_mu_theta_max(n,o,w) auxiliary;

x_S.up(n)=X_Smax(n);
p_E.up(n,o,w)=P_Emax(n);
p_R.up(n,o,w)=P_Rmax(n,w);
p_D.up(n,o,w)=P_Dmax(n)*Q_D(n,o);
theta.lo(n,o,w)=-PI;
theta.up(n,o,w)=PI;
theta.fx(s,o,w)=0;

EQUATIONS
OF , EQ1, EQ2, EQ3, EQ4, EQ5, EQ6, EQ7, EQ8, EQ9,
   EQ10, EQ11, EQ12, EQ13, EQ14, EQ15, EQ16, EQ17,
    EQ18, EQ19, EQ20,
EQ21, EQ22, EQ23, EQ24, EQ25, EQ26, EQ27, EQ28,
   EQ29, EQ30, EQ31, EQ32, EQ33, EQ34, EQ35, EQ36,
    EQ37, EQ38, EQ39, EQ40;

OF..minus_profit=E=SUM(n,K_C(n)*x_C(n)+K_S(n)*x_S(n
   )-SUM(w,phi(w)*(SUM(o,rho(o)*[linear_term(n,o,w
   )-p_C(n,o,w)*C_C(n)-p_E(n,o,w)*C_E(n)]))));

EQ1(n,o,w)..linear_term(n,o,w)=E=-[C_R(n,o)*p_R(n,o
   ,w)]+[U_D(n,o)*p_D(n,o,w)]-[mu_Rmax(n,o,w)*
   P_Rmax(n,w)]-[mu_Dmax(n,o,w)*P_Dmax(n)*Q_D(n,o)
   ]-SUM(m$Omega(n,m),Fmax(n,m)*mu_F(n,m,o,w))-PI
   *[mu_theta_max(n,o,w)+mu_theta_min(n,o,w)];

EQ2(n)..x_C(n)=E=SUM(h,u_C(n,h)*X_Option(n,h));
```

```
EQ3(n)..SUM(h,u_C(n,h))=E=1;

EQ4..SUM(n,x_C(n)*K_C(n)+x_S(n)*K_S(n))=L=Kmax;

EQ5(n,o,w)..p_D(n,o,w)+SUM(m$Omega(n,m),B(n,m)*[
    theta(n,o,w)-theta(m,o,w)])-p_C(n,o,w)-p_S(n,o,
    w)-p_E(n,o,w)-p_R(n,o,w)=E=0;

EQ6(n,o,w)..p_C(n,o,w)=L=x_C(n);

EQ7(n,o,w)..p_S(n,o,w)=L=Q_S(n,o)*x_S(n);

EQ8(n,m,o,w)$Omega(n,m)..B(n,m)*[theta(n,o,w)-theta
    (m,o,w)]=L=Fmax(n,m);

EQ9(n,o,w)..alpha_C(n,o,w)-lambda(n,o,w)+mu_Cmax(n,
    o,w)-mu_Cmin(n,o,w)=E=0;

EQ10(n,o,w)..alpha_E(n,o,w)-lambda(n,o,w)+mu_Emax(n
    ,o,w)-mu_Emin(n,o,w)=E=0;

EQ11(n,o,w)..C_R(n,o)-lambda(n,o,w)+mu_Rmax(n,o,w)-
    mu_Rmin(n,o,w)=E=0;

EQ12(n,o,w)..-lambda(n,o,w)+mu_Smax(n,o,w)-mu_Smin(
    n,o,w)=E=0;

EQ13(n,o,w)..-U_D(n,o)+lambda(n,o,w)+mu_Dmax(n,o,w)
    -mu_Dmin(n,o,w)=E=0;

EQ14(n,o,w)..SUM(m$Omega(n,m),B(n,m)*[lambda(n,o,w)
    -lambda(m,o,w)])+SUM(m$Omega(n,m),B(n,m)*[mu_F(
    n,m,o,w)-mu_F(m,n,o,w)])-mu_theta_min(n,o,w)+
    mu_theta_max(n,o,w)+[mu_theta_ref(o,w)$s(n)]=E
    =0;

EQ15(n,o,w)..p_C(n,o,w)=L=u_mu_Cmin(n,o,w)*BigM1;
EQ16(n,o,w)..mu_Cmin(n,o,w)=L=[1-u_mu_Cmin(n,o,w)]*
    BigM2;

EQ17(n,o,w)..x_C(n)-p_C(n,o,w)=L=u_mu_Cmax(n,o,w)*
    BigM1;
EQ18(n,o,w)..mu_Cmax(n,o,w)=L=[1-u_mu_Cmax(n,o,w)]*
    BigM2;

EQ19(n,o,w)..p_S(n,o,w)=L=u_mu_Smin(n,o,w)*BigM1;
EQ20(n,o,w)..mu_Smin(n,o,w)=L=[1-u_mu_Smin(n,o,w)]*
    BigM2;

EQ21(n,o,w)..(Q_S(n,o)*x_S(n))-p_S(n,o,w)=L=
    u_mu_Smax(n,o,w)*BigM1;
```

```
EQ22(n,o,w)..mu_Smax(n,o,w)=L=[1-u_mu_Smax(n,o,w)]*
    BigM2;
EQ23(n,o,w)..p_E(n,o,w)=L=u_mu_Emin(n,o,w)*BigM1;
EQ24(n,o,w)..mu_Emin(n,o,w)=L=[1-u_mu_Emin(n,o,w)]*
    BigM2;

EQ25(n,o,w)..P_Emax(n)-p_E(n,o,w)=L=u_mu_Emax(n,o,w
    )*BigM1;
EQ26(n,o,w)..mu_Emax(n,o,w)=L=[1-u_mu_Emax(n,o,w)]*
    BigM2;

EQ27(n,o,w)..p_R(n,o,w)=L=u_mu_Rmin(n,o,w)*BigM1;
EQ28(n,o,w)..mu_Rmin(n,o,w)=L=[1-u_mu_Rmin(n,o,w)]*
    BigM2;

EQ29(n,o,w)..P_Rmax(n,w)-p_R(n,o,w)=L=u_mu_Rmax(n,o
    ,w)*BigM1;
EQ30(n,o,w)..mu_Rmax(n,o,w)=L=[1-u_mu_Rmax(n,o,w)]*
    BigM2;

EQ31(n,o,w)..p_D(n,o,w)=L=u_mu_Dmin(n,o,w)*BigM1;
EQ32(n,o,w)..mu_Dmin(n,o,w)=L=[1-u_mu_Dmin(n,o,w)]*
    BigM2;

EQ33(n,o,w)..(Q_D(n,o)*P_Dmax(n))-p_D(n,o,w)=L=
    u_mu_Dmax(n,o,w)*BigM1;
EQ34(n,o,w)..mu_Dmax(n,o,w)=L=[1-u_mu_Dmax(n,o,w)]*
    BigM2;

EQ35(n,m,o,w)$Omega(n,m)..Fmax(n,m)-[B(n,m)*(theta(
    n,o,w)-theta(m,o,w))]=L=u_mu_F(n,m,o,w)*BigM1;
EQ36(n,m,o,w)$Omega(n,m)..mu_F(n,m,o,w)=L=[1-u_mu_F
    (n,m,o,w)]*BigM2;

EQ37(n,o,w)..PI+theta(n,o,w)=L=[1-u_mu_theta_min(n,
    o,w)]*BigM1;
EQ38(n,o,w)..mu_theta_min(n,o,w)=L=u_mu_theta_min(n
    ,o,w)*BigM2;

EQ39(n,o,w)..PI-theta(n,o,w)=L=[1-u_mu_theta_max(n,
    o,w)]*BigM1;
EQ40(n,o,w)..mu_theta_max(n,o,w)=L=u_mu_theta_max(n
    ,o,w)*BigM2;

MODEL EXAMPLE4  /ALL/ ;
SOLVE EXAMPLE4 USING MIP MINIMIZING minus_profit;
```

References

1. Baringo, L., Conejo, A.J.: Wind power investment within a market environment. Appl. Energy **88**(9), 3239–3247 (2011)
2. Baringo, L., Conejo, A.J.: Wind power investment: A Benders' decomposition approach. IEEE Trans. Power Syst. **27**(1), 433–441 (2012)
3. Baringo, L., Conejo, A.J.: Risk-constrained multi-stage wind power investment. IEEE Trans. Power Syst. **28**(1), 401–411 (2013)
4. Baringo, L., Conejo, A.J.: Correlated wind-power production and electric load scenarios for investment decisions. Appl. Energy **101**, 475–482 (2013)
5. Baringo, L., Conejo, A.J.: Strategic offering for a wind power producer. IEEE Trans. Power Syst. **28**(4), 4645–4654 (2013)
6. Baringo, L., Conejo, A.J.: Strategic wind power investment. IEEE Trans. Power Syst. **29**(3), 1250–1260 (2014)
7. Benders, J.F.: Partitioning procedures for solving mixed variable programming problems. Numerische Mathematik **4**(1), 238–252 (1962)
8. Bertsekas, D.P., Sandell, N.R.: Estimates of the duality gap for large-scale separable nonconvex optimization problems. In: IEEE Conference on Decision and Control, pp. 782–785 (1982)
9. Bertsimas, D., Sim, M.: Robust discrete optimization and network flows. Math. Program. Ser. B **98**(1–3): 49–71 (2002)
10. Birge, J.R., Louveaux, F.: Introduction to Stochastic Programming. Springer, New York (1997)
11. Botterud, A., Ilic, M.D., Wangensteen, I.: Optimal investment in power generation under centralized and decentralized decision making. IEEE Trans. Power Syst. **20**(1), 254–263 (2005)
12. Conejo, A.J., Castillo, E., Minguez, R., Garcia-Bertrand, R.: Decomposition Techniques in Mathematical Programming. Engineering and Science Applications. Springer, Heidelberg, Germany (2006)
13. Conejo, A.J., Carrión, M., Morales, J.M.: Decision Making Under Uncertainty in Electricity Markets. Springer, New York (2010)
14. Ehrenmann, A., Smeers, Y.: Generation capacity expansion in a risky environment: A stochastic equilibrium analysis. Oper. Res. **59**(6), 1332–1346 (2011)
15. Fortuny-Amat, J., McCarl, B.: A representation and economic interpretation of a two-level programming problem. J. Oper. Res. Soc. **32**(9), 783–792 (1981)
16. Gabriel, S., Conejo, A.J., Hobbs, B., Fuller, D., Ruiz, C.: Complementarity Modeling in Energy Markets. Springer, New York (2012)
17. Garcia, A., Shen, Z.: Equilibrium capacity expansion under stochastic demand growth. Oper. Res. **58**(1), 30–42 (2010)
18. Garcia, A., Stacchetti, E.: Investment dynamics in electricity markets. Econ. Theory **46**(2), 149–187 (2011)
19. Genc, T.S., Thille, H.: Investment in electricity markets with asymmetric technologies. Energy Econ. **33**(3), 379–387 (2011)
20. Gómez-Expósito, A., Conejo, A.J., Cañizares, C.: Electric Energy Systems: Analysis and Operation. CRC, Boca Raton (2008)
21. Hansen, E., Walster, G.W.: Global Optimization Using Interval Analysis. Marcel Dekker Inc, New York (2004)
22. Kazempour, S.J., Conejo, A.J., Ruiz, C.: Strategic generation investment using a complementarity approach. IEEE Trans. Power Syst. **26**(2), 940–948 (2011)
23. Kazempour, S.J., Conejo, A.J., Ruiz, C.: Strategic generation investment considering futures and spot markets. IEEE Trans. Power Syst. **27**(3), 1467–1476 (2012)
24. Kazempour, S.J., Conejo, A.J., Ruiz, C.: Generation investment equilibria with strategic producers—Part I: formulation. IEEE Trans. Power Syst. **28**(3), 2613–2622 (2013)
25. Kazempour, S.J., Conejo, A.J., Ruiz, C.: Generation investment equilibria with strategic producers—Part II: case studies. IEEE Trans. Power Syst. **28**(3), 2623–2631 (2013)
26. Kazempour, S.J., Conejo, A.J.: Strategic generation investment under uncertainty via Benders' decomposition. IEEE Trans. Power Syst. **27**(1), 424–432 (2012)

27. Kazempour, S.J., Zareipour, H.: Equilibria in an oligopolistic market with wind power production. IEEE Trans. Power Syst. **29**(2), 686–697 (2014)
28. Martín, S., Smeers, Y., Aguado, J.A.: A stochastic two settlement equilibrium model for electricity markets with wind generation. IEEE Trans. Power Syst. **30**(1), 233–245 (2015)
29. Morales, J.M., Conejo, A.J., Perez-Ruiz, J.: Economic valuation of reserves in power systems with high penetration of wind power. IEEE Trans. Power Syst. **24**(2), 900–910 (2009)
30. Morales, J.M., Conejo, A.J., Madsen, H., Pinson, P., Zugno, M.: Integrating Renewables in Electricity Markets. Springer, New York (2013)
31. Motto, A.L., Galiana, F.D., Conejo, A.J., Arroyo, J.M.: Network-constrained multiperiod auction for a pool-based electricity market. IEEE Trans. Power Syst. **17**(3), 646–653 (2002)
32. Murphy, F.H., Smeers, Y.: Generation capacity expansion in imperfectly competitive restructured electricity markets. Oper. Res. **53**(4), 646–661 (2005)
33. Murphy, F.H., Smeers, Y.: On the impact of forward markets on investment in oligopolistic markets wtih reference to electricity. Oper. Res. **58**(3), 515–528 (2010)
34. Ruiz, C., Conejo, A.J.: Pool strategy of a producer with endogenous formation of locational marginal prices. IEEE Trans. Power Syst. **24**(4), 1855–1866 (2009)
35. Wogrin, S., Centeno, E., Barquín, J.: Generation capacity expansion in liberalized electricity markets: A stochastic MPEC approach. IEEE Trans. Power Syst. **26**(4), 2526–2532 (2011)
36. Zugno, M., Morales, J.M., Pinson, P., Madsen, H.: Pool strategy of a price-maker wind power producer. IEEE Trans. Power Syst. **28**(3), 3440–3450 (2013)

Chapter 6
Investment Equilibria

Investment equilibrium analysis constitutes a useful framework for regulators to gain insights into the behavior of strategic producers and the evolution of generation investment in an electricity market. Such a perspective enables regulators to design better market rules, which in turn may contribute to increasing the competitiveness of the market and to stimulating investment in generation capacity. This chapter provides a methodology based on optimization and complementarity modeling for identifying generation investment equilibria in a network-constrained electricity market.

6.1 Introduction

The objective of a producer competing in an electricity market is to maximize its profit. To this purpose, such a producer makes its own decisions through its investment strategies (long-term decisions) and operational strategies (short-term decisions). However, the strategic decisions of each producer are related to those of other producers (rivals) due to market interactions. In fact, decisions made by each producer may influence the strategies of other producers. Within this framework, a number of investment equilibria generally exist, whereby each producer cannot increase its profit by changing its strategies unilaterally [7, 8, 10, 11, 23, 25]. The objective of this chapter is to identify such investment equilibria mathematically.

Investment equilibrium analysis is particularly useful for a regulator to gain insights into the investment behavior of producers and the evolution of the total production capacity. As a result, the regulator may be able to design better market rules, which in turn may enhance the competitiveness of the market and stimulate investment in production capacity.

In contrast to Chap. 5, in which a single strategic producer is considered, *all* producers considered in this chapter are strategic, thereby creating an oligopoly. This means that all producers can alter the market outcomes, i.e., market-clearing prices and production quantities, through their strategies. One important observation is that the feasibility region for the investment decision-making problem of each

A.J. Conejo et al., *Investment in Electricity Generation and Transmission*,
DOI 10.1007/978-3-319-29501-5_6

producer is interrelated with those of other producers. Thus, the production capacity investment equilibria problem is a *generalized Nash equilibrium (GNE)* problem [2, 4, 12].

Several treatments of operational and investment equilibria in the extant literature address oligopolistic energy markets, e.g., [1, 3, 9–11, 13–15, 19, 21, 23, 24].

The remainder of this chapter is organized as follows. Section 6.2 describes the available approaches for solving an equilibrium problem. Section 6.3 presents modeling features and assumptions. Section 6.4 provides a bilevel model for a single producer to make its investment decisions, which renders to a mathematical program with equilibrium constraint (MPEC). Section 6.5 presents the investment decision-making problem of multiple producers, which results in an equilibrium problem with equilibrium constraints (EPEC). Section 6.6 summarizes the chapter and discusses the main conclusions of the models and results reported in the chapter. Section 6.7 proposes some exercises to enable a deeper understanding of the models and concepts described in the chapter. Finally, Sect. 6.8 includes the GAMS code for an illustrative example.

6.2 Solution Approach

Similar to Chap. 5, in which a bilevel model is considered to represent the investment and offering decisions of a strategic producer, we consider in this chapter such a model for each strategic producer. Within the bilevel model of each strategic producer, the upper-level problem determines its optimal investment and strategic offer prices with the aim of maximizing its profit. In addition, a number of lower-level problems represent the clearing of the market under different operating conditions. As in Chap. 5, each lower-level problem is replaced by its optimality conditions, which yields an MPEC. This transformation is schematically illustrated in Fig. 6.1.

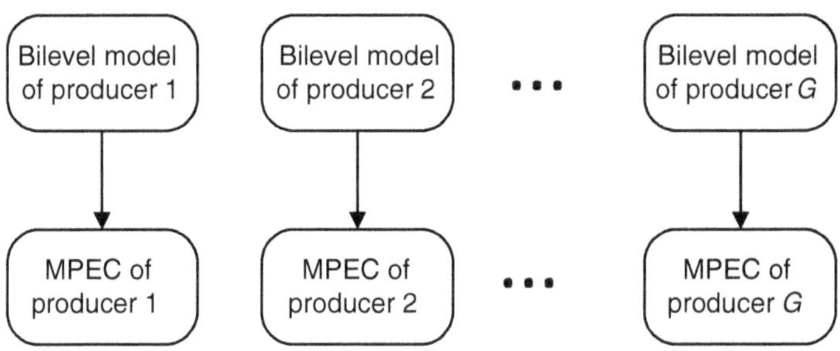

Fig. 6.1 Transforming the bilevel models into MPECs in an oligopolistic market with several strategic producers

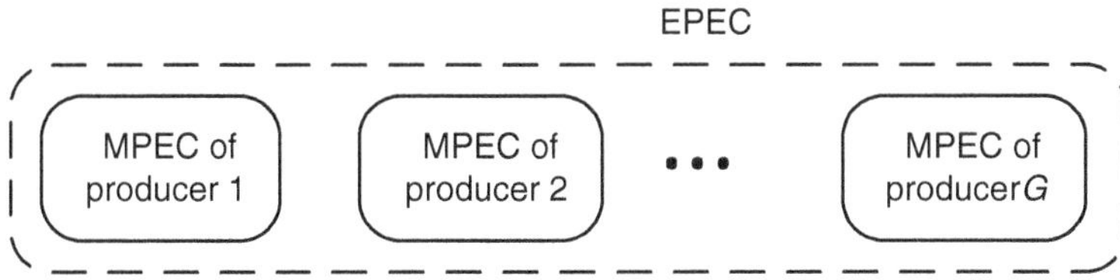

Fig. 6.2 MPECs and EPEC

Note that since several strategic producers are considered in this chapter, several MPECs are obtained, one per strategic producer. The joint consideration of all these MPECs constitutes an EPEC, as depicted in Fig. 6.2. The general mathematical structure of an EPEC is explained in Appendix C. Note that the EPEC solution identifies the market equilibria.

In general, two solution alternatives are available to solve an EPEC and thus to identify the market equilibria:

1. Diagonalization (iterative) approach [18, 22].
2. Simultaneous (noniterative) approach [9–11, 19].

The first solution alternative, i.e., the diagonalization approach, is an iterative technique, in which a single MPEC is solved in each iteration, while the strategic decisions of other producers are fixed. For example, consider a duopoly with two strategic producers 1 and 2, whose bilevel models are transformed into two MPECs 1 and 2, respectively. In the first iteration, MPEC 1 is solved, while the strategic decisions of producer 2 are fixed to some initial guesses. Then, in the second iteration, MPEC 2 is solved, while the strategic decisions of producer 1 are fixed to those obtained in the first iteration. This iterative process is continued until no decision is changed in the two subsequent iterations. The solution obtained is a Nash equilibrium since no producer desires to deviate from its decisions. Note that this approach is generally inefficient since it is iterative and provides, if convergence is achieved, at most a single equilibrium point. In addition, it may require a large number of iterations in the case of markets with many producers. Besides, it is not straightforward to find appropriate initial guesses, and suboptimal guesses may greatly affect the functioning of this approach.

The second solution alternative, i.e., the simultaneous approach, is a noniterative technique, in which all producers' MPECs are solved together. Therefore, it generally yields a complex mathematical problem. In this approach, each MPEC is replaced by its Karush–Kuhn–Tucker (KKT) conditions, which provide its strong stationary conditions. A collection of all those conditions for all producers results in the strong stationary conditions of the EPEC, whose solutions identify the market equilibria. This latter approach is the focus of this chapter.

6.3 Modeling Features and Assumptions

The technical features and assumptions of the investment equilibria model presented in this chapter are stated below:

1. An electricity pool is considered in which the market operator clears the pool once a day, one day ahead, and on an hourly basis.
2. A dc transmission network representation is considered.
3. Pursuing simplicity, a static investment model is used, i.e., a single target year is considered. The target year represents the final stage of the planning horizon, and the model uses annualized costs for this target year. Further details on investment models are available in Chap. 5. Note that a dynamic (multistage) model can also be considered within the investment equilibria problem [23] but at the cost of increased computational complexity.
4. A set of operating conditions is considered to represent the potential levels of the consumers' demands and the production of stochastic units during the target year. Accordingly, we define a set of demand and power capacity factors. Further details on operating conditions are available in Chap. 5.
5. For the sake of simplicity, uncertainties are not considered in this chapter. However, note that the investment equilibria problem is generally subject to several uncertainties, e.g., demand growth, investment costs for different technologies, and regulatory changes, which may be modeled through a set of plausible scenarios [3, 8, 21].

The notation used in this chapter is defined below:

Indices

- g Index for producers.
- n, m Indices for nodes.
- o Index for operating conditions.

Sets

- Ω_n Set of nodes connected to node n.

Parameters

- B_{nm} Susceptance of the transmission line connecting nodes n and m [S].
- C^{C}_{gn} Production cost of the candidate conventional unit of producer g located at node n [\$/MWh].
- C^{E}_{gn} Production cost of the existing conventional unit of producer g located at node n [\$/MWh].
- $F^{\max}_{nm}$ Transmission capacity of the line connecting nodes n and m [MW].
- K^{C}_{gn} Annualized investment cost of the candidate conventional unit of producer g located at node n [\$/MW].
- K^{S}_{gn} Annualized investment cost of the candidate stochastic unit of producer g located at node n [\$/MW].

$K_g^{\max}$ Available annualized investment budget of producer g [$].
$P_{gn}^{E^{\max}}$ Capacity of the existing conventional unit of producer g located at node n [MW].
$P_n^{D^{\max}}$ Maximum load of the consumer located at node n [MW].
Q_{no}^{S} Power capacity factor of the candidate stochastic unit located at node n in operating condition o [p.u.].
Q_{no}^{D} Demand factor of the consumer located at node n in operating condition o [p.u.].
U_{no}^{D} Bid price of the consumer located at node n in operating condition o [$/MWh].
$X_n^{C^{\max}}$ Maximum production capacity of the candidate conventional unit located at node n [MW].
$X_n^{S^{\max}}$ Maximum production capacity of the candidate stochastic unit located at node n [MW].
ρ_o Number of hours (weight) corresponding to operating condition o [h].

Variables

p_{gno}^{C} Power produced by the candidate conventional unit of producer g located at node n in operating condition o [MW].
p_{no}^{D} Power consumed by the consumer located at node n in operating condition o [MW].
p_{gno}^{E} Power produced by the existing conventional unit of producer g located at node n in operating condition o [MW].
p_{gno}^{S} Power produced by the candidate stochastic unit of producer g located at node n in operating condition o [MW].
x_{gn}^{C} Capacity of the candidate conventional unit of producer g located at node n [MW].
x_{gn}^{S} Capacity of the candidate stochastic unit of producer g located at node n [MW].
α_{gno}^{C} Offer price by the candidate conventional unit of producer g located at node n in operating condition o [$/MWh].
α_{gno}^{E} Offer price by the existing conventional unit of producer g located at node n in operating condition o [$/MWh].
λ_{no} Market-clearing price at node n in operating condition o [$/MWh].
θ_{no} Voltage angle at node n in operating condition o [rad].

6.4 Single-Producer Problem

The bilevel model for each single strategic producer is similar to one presented in Chap. 5. We consider two types of generating units: candidate (conventional and stochastic) and existing (conventional) units. These units belong to different strategic producers (i.e., producers $g = 1, \ldots, G$) and offer at strategic prices, except the candidate stochastic units, which always offer at zero. It is also assumed that all existing units (available at the initial year) are conventional, i.e., there is no stochastic production unit within the initial production portfolio of the producers.

The formulation of the bilevel model for a particular strategic producer, e.g., producer G, is given by (6.1). Note that (6.1a)–(6.1f) refer to the upper-level problem of producer G, whereas (6.1g) pertains to the lower-level problems, one per operating condition o. Note that a similar bilevel problem can be considered for any other producers, i.e., producers $g = 1, \ldots, G-1$. The bilevel problem for producer G is formulated below:

$$\left\{ \min_{\Xi_g^{\text{UL}} \cup \Xi_o^{\text{Primal}} \cup \Xi_o^{\text{Dual}}} \quad \sum_n \left\{ K_{gn}^{\text{C}} \, x_{gn}^{\text{C}} + K_{gn}^{\text{S}} \, x_{gn}^{\text{S}} \right. \right.$$

$$\left. - \sum_o \rho_o \left[\lambda_{no} \left(p_{gno}^{\text{C}} + p_{gno}^{\text{S}} + p_{gno}^{\text{E}} \right) - p_{gno}^{\text{C}} \, C_{gn}^{\text{C}} - p_{gno}^{\text{E}} \, C_{gn}^{\text{E}} \right] \right\} \tag{6.1a}$$

subject to

$$0 \leq x_{gn}^{\text{C}} \leq X_n^{\text{C}^{\max}} \quad \forall n \tag{6.1b}$$

$$0 \leq x_{gn}^{\text{S}} \leq X_n^{\text{S}^{\max}} \quad \forall n \tag{6.1c}$$

$$\sum_n \left(K_{gn}^{\text{C}} \, x_{gn}^{\text{C}} + K_{gn}^{\text{S}} \, x_{gn}^{\text{S}} \right) \leq K_g^{\max} \tag{6.1d}$$

$$\alpha_{gno}^{\text{C}} \geq 0 \quad \forall o, \forall n \tag{6.1e}$$

$$\alpha_{gno}^{\text{E}} \geq 0 \quad \forall o, \forall n \tag{6.1f}$$

$$\text{Lower-level problems (6.1h)–(6.1p)} \quad \forall o \tag{6.1g}$$

$$\left. \right\} \quad g = G.$$

The primal variables of the upper-level problem (6.1a)–(6.1f) are those in set $\Xi_g^{\text{UL}} = \{\alpha_{gno}^{\text{C}}, \alpha_{gno}^{\text{E}}, x_{gn}^{\text{C}}, x_{gn}^{\text{S}}\}$ plus all primal and dual variables of the lower-level problems (6.1g), which are defined after their formulation through sets Ξ_o^{Primal} and Ξ_o^{Dual}.

The objective function (6.1a) refers to minus the expected annual profit of the considered producer, i.e., annualized investment cost minus expected annual operational profit. Note that the market-clearing price λ_{no} is the dual variable of the power balance constraint at node n and operating condition o obtained endogenously from the corresponding lower-level problem. The objective function (6.1a) comprises the following terms:

- $\sum_n K_{gn}^{\text{C}} \, x_{gn}^{\text{C}}$ is the annualized investment cost of candidate conventional units of producer g.
- $\sum_n K_{gn}^{\text{S}} \, x_{gn}^{\text{S}}$ is the annualized investment cost of candidate stochastic units of producer g.

- $\sum_n \sum_o \rho_o \, p^{\rm C}_{gno} \, \lambda_{no}$ is the annualized revenue of producer g obtained from selling the production of candidate conventional units (production quantity multiplied by market-clearing price).
- $\sum_n \sum_o \rho_o \, p^{\rm E}_{gno} \, \lambda_{no}$ is the annualized revenue of producer g obtained from selling the production of existing conventional units (production quantity multiplied by market-clearing price).
- $\sum_n \sum_o \rho_o \, p^{\rm S}_{gno} \, \lambda_{no}$ is the annualized revenue of producer g obtained from selling the production of candidate stochastic units (production quantity multiplied by market-clearing price).
- $\sum_n \sum_o \rho_o \, p^{\rm C}_{gno} \, C^{\rm C}_{gn}$ is the annualized production cost of candidate conventional units of producer g (production quantity multiplied by marginal cost).
- $\sum_n \sum_o \rho_o \, p^{\rm E}_{gno} \, C^{\rm E}_{gn}$ is the annualized production cost of existing conventional units of producer g (production quantity multiplied by marginal cost).

The production cost of stochastic units is assumed to be zero. Note that the market-clearing prices (λ_{no}) and the production quantities ($p^{\rm C}_{gno}$, $p^{\rm E}_{gno}$, and $p^{\rm S}_{gno}$) belong to the feasible region defined by lower-level problems (6.1g).

For the sake of simplicity, the capacity options for investing in both conventional and stochastic units are assumed continuous. The capacity bounds for such options are enforced by (6.1b) and (6.1c). In addition, a cap on the available annualized investment budget of producer g is enforced by (6.1d). Finally, the upper-level constraints (6.1e)–(6.1f) enforce the nonnegativity of the offer prices associated with the candidate and existing conventional units, respectively, of producer G.

Each lower-level problem, one per operating condition o, is formulated below. The dual variable of each lower-level constraint is indicated following a colon:

$$\left\{ \min_{\Xi_o^{\rm Primal}} \quad \sum_n \left[\sum_g \left(\alpha^{\rm C}_{gno} \, p^{\rm C}_{gno} + \alpha^{\rm E}_{gno} \, p^{\rm E}_{gno} \right) - U^{\rm D}_{no} \, p^{\rm D}_{no} \right] \right. \tag{6.1h}$$

subject to

$$p^{\rm D}_{no} + \sum_{m \in \Omega_n} B_{nm} \left(\theta_{no} - \theta_{mo} \right) - \sum_g p^{\rm C}_{gno} - \sum_g p^{\rm S}_{gno} - \sum_g p^{\rm E}_{gno} = 0 \quad : \lambda_{no} \qquad \forall n \tag{6.1i}$$

$$0 \le p^{\rm C}_{gno} \le x^{\rm C}_{gn} \quad : \mu^{\rm C^{min}}_{gno}, \mu^{\rm C^{max}}_{gno} \qquad \forall g, \forall n \tag{6.1j}$$

$$0 \le p^{\rm S}_{gno} \le Q^{\rm S}_{no} \, x^{\rm S}_{gn} \quad : \mu^{\rm S^{min}}_{gno}, \mu^{\rm S^{max}}_{gno} \qquad \forall g, \forall n \tag{6.1k}$$

$$0 \le p^{\rm E}_{gno} \le P^{\rm E^{max}}_{gn} \quad : \mu^{\rm E^{min}}_{gno}, \mu^{\rm E^{max}}_{gno} \qquad \forall g, \forall n \tag{6.1l}$$

$$0 \le p^{\rm D}_{no} \le Q^{\rm D}_{no} \, P^{\rm D^{max}}_{n} \quad : \mu^{\rm D^{min}}_{no}, \mu^{\rm D^{max}}_{no} \qquad \forall n \tag{6.1m}$$

$$B_{nm}\left(\theta_{no}-\theta_{mo}\right) \leq F_{nm}^{\max} \quad : \mu_{nmo}^{\mathrm{F}} \qquad \forall n, \forall m \in \Omega_n \tag{6.1n}$$

$$-\pi \leq \theta_{no} \leq \pi \quad : \mu_{no}^{\theta^{\min}}, \mu_{no}^{\theta^{\max}} \qquad \forall n \tag{6.1o}$$

$$\theta_{no}=0 \quad : \mu_{o}^{\theta^{\mathrm{ref}}} \qquad n=\mathrm{ref}. \tag{6.1p}$$

$$\Bigg\} \forall o.$$

The primal optimization variables of each lower-level problem (6.1h)–(6.1p) are included in set $\Xi_o^{\text{Primal}} = \{p^{\mathrm{C}}_{gno}, p^{\mathrm{S}}_{gno}, p^{\mathrm{E}}_{gno}, p^{\mathrm{D}}_{no}, \theta_{no}\}$. Additionally, the dual optimization variables of each lower-level problem (6.1h)–(6.1p) are those included in set $\Xi_o^{\text{Dual}} = \{\lambda_{no}, \mu^{\mathrm{C}^{\min}}_{gno}, \mu^{\mathrm{C}^{\max}}_{gno}, \mu^{\mathrm{S}^{\min}}_{gno}, \mu^{\mathrm{S}^{\max}}_{gno}, \mu^{\mathrm{E}^{\min}}_{gno}, \mu^{\mathrm{E}^{\max}}_{gno}, \mu^{\mathrm{D}^{\min}}_{no}, \mu^{\mathrm{D}^{\max}}_{no}, \mu^{\mathrm{F}}_{nmo}, \mu^{\theta^{\min}}_{no}, \mu^{\theta^{\max}}_{no}, \mu^{\theta^{\mathrm{ref}}}_{o}\}$.

Lower-level problems (6.1h)–(6.1p) represent the clearing of the market for each operating condition and for given investment and offering decisions made in the upper-level problems by different producers. Accordingly, x^{C}_{gn}, x^{S}_{gn}, $\alpha^{\mathrm{C}}_{gno}$, and $\alpha^{\mathrm{E}}_{gno}$ are variables in the upper-level problem (6.1a)–(6.1f), but they are fixed values (parameters) in the lower-level problems (6.1h)–(6.1p). This makes the lower-level problems (6.1h)–(6.1p) linear and convex since there is no term containing the product of variables within the lower-level problems. The objective function (6.1h) minimizes minus the social welfare considering the offer prices of all strategic producers $g = 1, \ldots, G$, and bid prices of all demands. The power balance at every node is enforced by (6.1i), and its dual variable provides the market-clearing price at that node under operating condition o. Equations (6.1j), (6.1k), and (6.1l) impose production capacity limits for candidate conventional, candidate stochastic, and existing units, respectively. In addition, Eqs. (6.1m) bounds the power consumption of each demand. Equations (6.1n) enforce the transmission capacity limits of each line. Finally, Eqs. (6.1o) enforce voltage angle bounds for each node, and constraints (6.1p) fix the voltage angle to zero at the reference node.

Note that the market-clearing problems, i.e., the lower-level problems (6.1h)–(6.1p), are common to all producers $g = 1, \ldots, G$. Thus, the investment equilibria model presented in this chapter is in fact a *GNE problem with shared constraints*.

Illustrative Example 6.1 *A two-node electricity market with two strategic producers (duopoly)*

A power system with two nodes (n_1 and n_2) is considered as illustrated in Fig. 6.3. The capacity of transmission line $n_1 - n_2$ is 400 MW, and its susceptance is 1000 S. Node n_1 is the reference node. Two strategic producers (g_1 and g_2) compete together, creating a duopoly. Producer g_1 owns an existing unit located at node n_1 with capacity of 150 MW and production cost of \$10/MWh. On the other hand, producer g_2 owns an existing unit located at node n_2 with capacity of 100 MW and production cost of \$15/MWh.

Both producers g_1 and g_2 desire to build new production units. The available annualized investment budget for each producer is \$20 million. In addition, the investment options for each producer are identical and stated below:

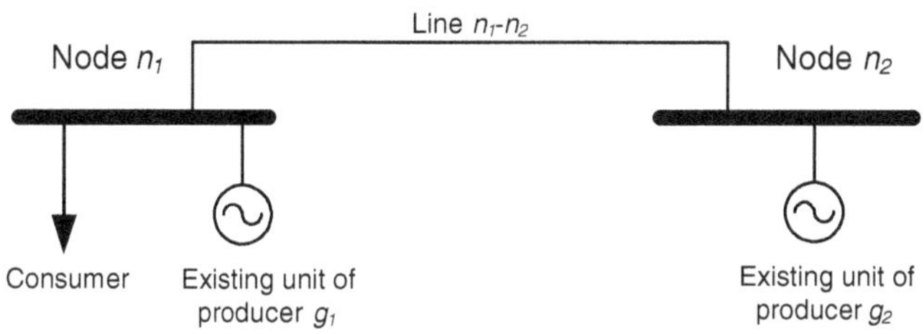

Fig. 6.3 Illustrative Example 6.1: two-node network

- A conventional unit to be built at node n_1. The maximum capacity of this candidate unit is 200 MW, and its annualized investment cost is \$55,000/MW. The production cost of this candidate conventional unit is \$12/MWh.
- A stochastic (wind-power) unit to be built at node n_2. The maximum capacity of this candidate stochastic unit is 200 MW, and its annualized investment cost is \$66,000/MW.

As depicted in Fig. 6.3, a single consumer is considered at node n_1, whose maximum load is equal to 400 MW.

In addition, two operating conditions (o_1 and o_2) are considered, whose characteristics are stated below:

- o_1: Demand factor equals 1.00 p.u. and wind power capacity factor equals 0.35 p.u.
- o_2: Demand factor equals 0.80 p.u. and wind power capacity factor equals 0.70 p.u.

The weight of condition o_1 is 3530 h and that of condition o_2 is 5230 h. The consumer bids in conditions o_1 and o_2 at \$35/MWh and \$32/MWh, respectively.

According to the data above, two bilevel problems, one per producer, are formulated. The bilevel problem for producer g_1 is given by (6.2) including upper-level problem (6.2a)–(6.2e) and lower-level problems (6.2f):

$$\begin{aligned}
\min_{\Xi_{g_1}^{\mathrm{UL,Ex}} \cup \Xi_{o_1}^{\mathrm{P,Ex}} \cup \Xi_{o_1}^{\mathrm{D,Ex}} \cup \Xi_{o_2}^{\mathrm{P,Ex}} \cup \Xi_{o_2}^{\mathrm{D,Ex}}} \quad & 55000\, x_{g_1 n_1}^{\mathrm{C}} + 66000\, x_{g_1 n_2}^{\mathrm{S}} \\
& - 3530 \Big[\lambda_{n_1 o_1} \left(p_{g_1 n_1 o_1}^{\mathrm{C}} + p_{g_1 n_1 o_1}^{\mathrm{E}} \right) + \lambda_{n_2 o_1}\, p_{g_1 n_2 o_1}^{\mathrm{S}} \\
& - 12\, p_{g_1 n_1 o_1}^{\mathrm{C}} - 10\, p_{g_1 n_1 o_1}^{\mathrm{E}} \Big] \\
& - 5230 \Big[\lambda_{n_1 o_2} \left(p_{g_1 n_1 o_2}^{\mathrm{C}} + p_{g_1 n_1 o_2}^{\mathrm{E}} \right) + \lambda_{n_2 o_2}\, p_{g_1 n_2 o_2}^{\mathrm{S}} \\
& - 12\, p_{g_1 n_1 o_2}^{\mathrm{C}} - 10\, p_{g_1 n_1 o_2}^{\mathrm{E}} \Big] \qquad (6.2a)
\end{aligned}$$

subject to

$$0 \le x^{\mathrm{C}}_{g_1 n_1} \le 200 \tag{6.2b}$$

$$0 \le x^{\mathrm{S}}_{g_1 n_2} \le 200 \tag{6.2c}$$

$$55000\, x^{\mathrm{C}}_{g_1 n_1} + 66000\, x^{\mathrm{S}}_{g_1 n_2} \le 2 \times 10^7 \tag{6.2d}$$

$$\alpha^{\mathrm{C}}_{g_1 n_1 o_1},\ \alpha^{\mathrm{E}}_{g_1 n_1 o_1},\ \alpha^{\mathrm{C}}_{g_1 n_1 o_2},\ \alpha^{\mathrm{E}}_{g_1 n_1 o_2} \ge 0 \tag{6.2e}$$

$$\text{Lower-level problems (6.4)–(6.5).} \tag{6.2f}$$

Similarly, the bilevel problem for producer g_2 is given by (6.3) including upper-level problem (6.3a)–(6.3e) and lower-level problems (6.3f):

$$\begin{aligned}\min_{\Xi^{\mathrm{UL,Ex}}_{g_2} \cup \Xi^{\mathrm{P,Ex}}_{o_1} \cup \Xi^{\mathrm{D,Ex}}_{o_1} \cup \Xi^{\mathrm{P,Ex}}_{o_2} \cup \Xi^{\mathrm{D,Ex}}_{o_2}} \quad & 55000\, x^{\mathrm{C}}_{g_2 n_1} + 66000\, x^{\mathrm{S}}_{g_2 n_2} \\ & - 3530 \Big[\lambda_{n_1 o_1}\, p^{\mathrm{C}}_{g_2 n_1 o_1} + \lambda_{n_2 o_1} \left(p^{\mathrm{S}}_{g_2 n_2 o_1} + p^{\mathrm{E}}_{g_2 n_2 o_1} \right) \\ & - 12\, p^{\mathrm{C}}_{g_2 n_1 o_1} - 15\, p^{\mathrm{E}}_{g_2 n_2 o_1} \Big] \\ & - 5230 \Big[\lambda_{n_1 o_2}\, p^{\mathrm{C}}_{g_2 n_1 o_2} + \lambda_{n_2 o_2} \left(p^{\mathrm{S}}_{g_2 n_2 o_2} + p^{\mathrm{E}}_{g_2 n_2 o_2} \right) \\ & - 12\, p^{\mathrm{C}}_{g_2 n_1 o_2} - 15\, p^{\mathrm{E}}_{g_2 n_2 o_2} \Big] \end{aligned} \tag{6.3a}$$

subject to

$$0 \le x^{\mathrm{C}}_{g_2 n_1} \le 200 \tag{6.3b}$$

$$0 \le x^{\mathrm{S}}_{g_2 n_2} \le 200 \tag{6.3c}$$

$$55000\, x^{\mathrm{C}}_{g_2 n_1} + 66000\, x^{\mathrm{S}}_{g_2 n_2} \le 2 \times 10^7 \tag{6.3d}$$

$$\alpha^{\mathrm{C}}_{g_2 n_1 o_1},\ \alpha^{\mathrm{E}}_{g_2 n_2 o_1},\ \alpha^{\mathrm{C}}_{g_2 n_1 o_2},\ \alpha^{\mathrm{E}}_{g_2 n_2 o_2} \ge 0 \tag{6.3e}$$

$$\text{Lower-level problems (6.4)–(6.5).} \tag{6.3f}$$

Within bilevel problems (6.2) and (6.3) associated with producers g_1 and g_2, lower-level problems (one per operating condition) are common. The lower-level problem referring to the operating condition o_1 is given by (6.4) below:

$$\begin{aligned}\min_{\Xi^{\mathrm{P,Ex}}_{o_1}} \quad & \alpha^{\mathrm{C}}_{g_1 n_1 o_1}\, p^{\mathrm{C}}_{g_1 n_1 o_1} + \alpha^{\mathrm{E}}_{g_1 n_1 o_1}\, p^{\mathrm{E}}_{g_1 n_1 o_1} \\ & + \alpha^{\mathrm{C}}_{g_2 n_1 o_1}\, p^{\mathrm{C}}_{g_2 n_1 o_1} + \alpha^{\mathrm{E}}_{g_2 n_2 o_1}\, p^{\mathrm{E}}_{g_2 n_2 o_1} - 35\, p^{\mathrm{D}}_{n_1 o_1} \end{aligned} \tag{6.4a}$$

subject to

$$\begin{aligned} & p^{\mathrm{D}}_{n_1 o_1} + 1000 \left(\theta_{n_1 o_1} - \theta_{n_2 o_1} \right) - p^{\mathrm{E}}_{g_1 n_1 o_1} - p^{\mathrm{C}}_{g_1 n_1 o_1} \\ & \quad - p^{\mathrm{C}}_{g_2 n_1 o_1} = 0 \quad : \lambda_{n_1 o_1} \end{aligned} \tag{6.4b}$$

$$1000 \left(\theta_{n_2 o_1} - \theta_{n_1 o_1} \right) - p^{\mathrm{E}}_{g_2 n_2 o_1} - p^{\mathrm{S}}_{g_1 n_2 o_1} - p^{\mathrm{S}}_{g_2 n_2 o_1} = 0 \quad : \lambda_{n_2 o_1} \tag{6.4c}$$

$$0 \le p^{\mathrm{C}}_{g_1 n_1 o_1} \le x^{\mathrm{C}}_{g_1 n_1} \quad : \mu^{\mathrm{C^{min}}}_{g_1 n_1 o_1}, \mu^{\mathrm{C^{max}}}_{g_1 n_1 o_1} \tag{6.4d}$$

$$0 \leq p^{\text{C}}_{g_2 n_1 o_1} \leq x^{\text{C}}_{g_2 n_1} \quad : \mu^{\text{C}^{\min}}_{g_2 n_1 o_1}, \mu^{\text{C}^{\max}}_{g_2 n_1 o_1} \tag{6.4e}$$

$$0 \leq p^{\text{S}}_{g_1 n_2 o_1} \leq 0.35\, x^{\text{S}}_{g_1 n_2} \quad : \mu^{\text{S}^{\min}}_{g_1 n_2 o_1}, \mu^{\text{S}^{\max}}_{g_1 n_2 o_1} \tag{6.4f}$$

$$0 \leq p^{\text{S}}_{g_2 n_2 o_1} \leq 0.35\, x^{\text{S}}_{g_2 n_2} \quad : \mu^{\text{S}^{\min}}_{g_2 n_2 o_1}, \mu^{\text{S}^{\max}}_{g_2 n_2 o_1} \tag{6.4g}$$

$$0 \leq p^{\text{E}}_{g_1 n_1 o_1} \leq 150 \quad : \mu^{\text{E}^{\min}}_{g_1 n_1 o_1}, \mu^{\text{E}^{\max}}_{g_1 n_1 o_1} \tag{6.4h}$$

$$0 \leq p^{\text{E}}_{g_2 n_2 o_1} \leq 100 \quad : \mu^{\text{E}^{\min}}_{g_2 n_2 o_1}, \mu^{\text{E}^{\max}}_{g_2 n_2 o_1} \tag{6.4i}$$

$$0 \leq p^{\text{D}}_{n_1 o_1} \leq 1 \times 400 \quad : \mu^{\text{D}^{\min}}_{n_1 o_1}, \mu^{\text{D}^{\max}}_{n_1 o_1} \tag{6.4j}$$

$$1000\left(\theta_{n_1 o_1} - \theta_{n_2 o_1}\right) \leq 400 \quad : \mu^{\text{F}}_{n_1 n_2 o_1} \tag{6.4k}$$

$$1000\left(\theta_{n_2 o_1} - \theta_{n_1 o_1}\right) \leq 400 \quad : \mu^{\text{F}}_{n_2 n_1 o_1} \tag{6.4l}$$

$$-\pi \leq \theta_{n_1 o_1} \leq \pi \quad : \mu^{\theta^{\min}}_{n_1 o_1}, \mu^{\theta^{\max}}_{n_1 o_1} \tag{6.4m}$$

$$-\pi \leq \theta_{n_2 o_1} \leq \pi \quad : \mu^{\theta^{\min}}_{n_2 o_1}, \mu^{\theta^{\max}}_{n_2 o_1} \tag{6.4n}$$

$$\theta_{n_1 o_1} = 0 \quad : \mu^{\theta^{\text{ref}}}_{o_1}. \tag{6.4o}$$

In addition, the lower-level problem referring to the operating condition o_2 (common to both producers) is given by (6.5) below:

$$\min_{\Xi^{\text{P,Ex}}_{o_2}} \quad \alpha^{\text{C}}_{g_1 n_1 o_2}\, p^{\text{C}}_{g_1 n_1 o_2} + \alpha^{\text{E}}_{g_1 n_1 o_2}\, p^{\text{E}}_{g_1 n_1 o_2} + \alpha^{\text{C}}_{g_2 n_1 o_2}\, p^{\text{C}}_{g_2 n_1 o_2} + \alpha^{\text{E}}_{g_2 n_2 o_2}\, p^{\text{E}}_{g_2 n_2 o_2} - 32\, p^{\text{D}}_{n_1 o_2} \tag{6.5a}$$

subject to

$$p^{\text{D}}_{n_1 o_2} + 1000\left(\theta_{n_1 o_2} - \theta_{n_2 o_2}\right) - p^{\text{E}}_{g_1 n_1 o_2} - p^{\text{C}}_{g_1 n_1 o_2} - p^{\text{C}}_{g_2 n_1 o_2} = 0 \quad : \lambda_{n_1 o_2} \tag{6.5b}$$

$$1000\left(\theta_{n_2 o_2} - \theta_{n_1 o_2}\right) - p^{\text{E}}_{g_2 n_2 o_2} - p^{\text{S}}_{g_1 n_2 o_2} - p^{\text{S}}_{g_2 n_2 o_2} = 0 \quad : \lambda_{n_2 o_2} \tag{6.5c}$$

$$0 \leq p^{\text{C}}_{g_1 n_1 o_2} \leq x^{\text{C}}_{g_1 n_1} \quad : \mu^{\text{C}^{\min}}_{g_1 n_1 o_2}, \mu^{\text{C}^{\max}}_{g_1 n_1 o_2} \tag{6.5d}$$

$$0 \leq p^{\text{C}}_{g_2 n_1 o_2} \leq x^{\text{C}}_{g_2 n_1} \quad : \mu^{\text{C}^{\min}}_{g_2 n_1 o_2}, \mu^{\text{C}^{\max}}_{g_2 n_1 o_2} \tag{6.5e}$$

$$0 \leq p^{\text{S}}_{g_1 n_2 o_2} \leq 0.70\, x^{\text{S}}_{g_1 n_2} \quad : \mu^{\text{S}^{\min}}_{g_1 n_2 o_2}, \mu^{\text{S}^{\max}}_{g_1 n_2 o_2} \tag{6.5f}$$

$$0 \leq p^{\text{S}}_{g_2 n_2 o_2} \leq 0.70\, x^{\text{S}}_{g_2 n_2} \quad : \mu^{\text{S}^{\min}}_{g_2 n_2 o_2}, \mu^{\text{S}^{\max}}_{g_2 n_2 o_2} \tag{6.5g}$$

$$0 \leq p^{\text{E}}_{g_1 n_1 o_2} \leq 150 \quad : \mu^{\text{E}^{\min}}_{g_1 n_1 o_2}, \mu^{\text{E}^{\max}}_{g_1 n_1 o_2} \tag{6.5h}$$

$$0 \leq p^{\text{E}}_{g_2 n_2 o_2} \leq 100 \quad : \mu^{\text{E}^{\min}}_{g_2 n_2 o_2}, \mu^{\text{E}^{\max}}_{g_2 n_2 o_2} \tag{6.5i}$$

$$0 \leq p^{\text{D}}_{n_1 o_2} \leq 0.8 \times 400 \quad : \mu^{\text{D}^{\min}}_{n_1 o_2}, \mu^{\text{D}^{\max}}_{n_1 o_2} \tag{6.5j}$$

$$1000\left(\theta_{n_1 o_2} - \theta_{n_2 o_2}\right) \leq 400 \quad : \mu^{\text{F}}_{n_1 n_2 o_2} \tag{6.5k}$$

$$1000\left(\theta_{n_2 o_2} - \theta_{n_1 o_2}\right) \leq 400 \quad : \mu^{\text{F}}_{n_2 n_1 o_2} \tag{6.5l}$$

$$-\pi \leq \theta_{n_1 o_2} \leq \pi \quad : \mu^{\theta^{\min}}_{n_1 o_2}, \mu^{\theta^{\max}}_{n_1 o_2} \tag{6.5m}$$

$$-\pi \leq \theta_{n_2o_2} \leq \pi \quad : \mu^{\theta^{\min}}_{n_2o_2}, \mu^{\theta^{\max}}_{n_2o_2} \tag{6.5n}$$

$$\theta_{n_1o_2} = 0 \quad : \mu^{\theta^{\text{ref.}}}_{o_2}. \tag{6.5o}$$

The primal optimization variables of lower-level problem (6.4) associated with operating condition o_1 are included in set $\Xi^{\text{P,Ex}}_{o_1} = \{p^{\text{C}}_{g_1n_1o_1}, p^{\text{E}}_{g_1n_1o_1}, p^{\text{S}}_{g_1n_2o_1}, p^{\text{C}}_{g_2n_1o_1}, p^{\text{E}}_{g_2n_2o_1}, p^{\text{S}}_{g_2n_2o_1}, p^{\text{D}}_{n_1o_1}, \theta_{n_1o_1}, \theta_{n_2o_1}\}$. Additionally, its dual variables are those included in set $\Xi^{\text{D,Ex}}_{o_1} = \{\lambda_{n_1o_1}, \lambda_{n_2o_1}, \mu^{\text{C}^{\min}}_{g_1n_1o_1}, \mu^{\text{C}^{\max}}_{g_1n_1o_1}, \mu^{\text{C}^{\min}}_{g_2n_1o_1}, \mu^{\text{C}^{\max}}_{g_2n_1o_1}, \mu^{\text{S}^{\min}}_{g_1n_2o_1}, \mu^{\text{S}^{\max}}_{g_1n_2o_1}, \mu^{\text{S}^{\min}}_{g_2n_2o_1}, \mu^{\text{S}^{\max}}_{g_2n_2o_1}, \mu^{\text{E}^{\min}}_{g_1n_1o_1}, \mu^{\text{E}^{\max}}_{g_1n_1o_1}, \mu^{\text{E}^{\min}}_{g_2n_2o_1}, \mu^{\text{E}^{\max}}_{g_2n_2o_1}, \mu^{\text{D}^{\min}}_{n_1o_1}, \mu^{\text{D}^{\max}}_{n_1o_1}, \mu^{\text{F}}_{n_1n_2o_1}, \mu^{\text{F}}_{n_2n_1o_1}, \mu^{\theta^{\min}}_{n_1o_1}, \mu^{\theta^{\max}}_{n_1o_1}, \mu^{\theta^{\min}}_{n_2o_1}, \mu^{\theta^{\max}}_{n_2o_1}, \mu^{\theta^{\text{ref}}}_{o_1}\}$. In addition, the primal optimization variables of lower-level problem (6.5) associated with operating condition o_2 are included in set $\Xi^{\text{P,Ex}}_{o_2} = \{p^{\text{C}}_{g_1n_1o_2}, p^{\text{E}}_{g_1n_1o_2}, p^{\text{S}}_{g_1n_2o_2}, p^{\text{C}}_{g_2n_1o_2}, p^{\text{E}}_{g_2n_2o_2}, p^{\text{S}}_{g_2n_2o_2}, p^{\text{D}}_{n_1o_2}, \theta_{n_1o_2}, \theta_{n_2o_2}\}$. Likewise, its dual variables are those included in set $\Xi^{\text{D,Ex}}_{o_2} = \{\lambda_{n_1o_2}, \lambda_{n_2o_2}, \mu^{\text{C}^{\min}}_{g_1n_1o_2}, \mu^{\text{C}^{\max}}_{g_1n_1o_2}, \mu^{\text{C}^{\min}}_{g_2n_1o_2}, \mu^{\text{C}^{\max}}_{g_2n_1o_2}, \mu^{\text{S}^{\min}}_{g_1n_2o_2}, \mu^{\text{S}^{\max}}_{g_1n_2o_2}, \mu^{\text{S}^{\min}}_{g_2n_2o_2}, \mu^{\text{S}^{\max}}_{g_2n_2o_2}, \mu^{\text{E}^{\min}}_{g_1n_1o_2}, \mu^{\text{E}^{\max}}_{g_1n_1o_2}, \mu^{\text{E}^{\min}}_{g_2n_2o_2}, \mu^{\text{E}^{\max}}_{g_2n_2o_2}, \mu^{\text{D}^{\min}}_{n_1o_2}, \mu^{\text{D}^{\max}}_{n_1o_2}, \mu^{\text{F}}_{n_1n_2o_2}, \mu^{\text{F}}_{n_2n_1o_2}, \mu^{\theta^{\min}}_{n_1o_2}, \mu^{\theta^{\max}}_{n_1o_2}, \mu^{\theta^{\min}}_{n_2o_2}, \mu^{\theta^{\max}}_{n_2o_2}, \mu^{\theta^{\text{ref}}}_{o_2}\}$. The primal optimization variables of upper-level problem (6.2a)–(6.2e) pertaining to producer g_1 are included in set $\Xi^{\text{UL,Ex}}_{g_1} = \{x^{\text{C}}_{g_1n_1}, x^{\text{S}}_{g_1n_2}, \alpha^{\text{C}}_{g_1n_1o_1}, \alpha^{\text{E}}_{g_1n_1o_1}, \alpha^{\text{C}}_{g_1n_1o_2}, \alpha^{\text{E}}_{g_1n_1o_2}\}$ plus $\Xi^{\text{P,Ex}}_{o_1}$, $\Xi^{\text{D,Ex}}_{o_1}$, $\Xi^{\text{P,Ex}}_{o_2}$, and $\Xi^{\text{D,Ex}}_{o_2}$. Finally, the primal optimization variables of upper-level problem (6.3a)–(6.3e) pertaining to producer g_2 are included in set $\Xi^{\text{UL,Ex}}_{g_2} = \{x^{\text{C}}_{g_2n_1}, x^{\text{S}}_{g_2n_2}, \alpha^{\text{C}}_{g_2n_1o_1}, \alpha^{\text{E}}_{g_2n_2o_1}, \alpha^{\text{C}}_{g_2n_1o_2}, \alpha^{\text{E}}_{g_2n_2o_2}\}$ plus $\Xi^{\text{P,Ex}}_{o_1}$, $\Xi^{\text{D,Ex}}_{o_1}$, $\Xi^{\text{P,Ex}}_{o_2}$, and $\Xi^{\text{D,Ex}}_{o_2}$. □

6.4.1 MPEC

As stated in the previous section, each strategic producer solves its own bilevel model to derive the most beneficial investment and offering decisions. To this end, each lower-level problem within the bilevel model of each producer needs to be replaced by its equivalent optimality conditions. In general, two alternative approaches are available to derive those conditions for a continuous and linear problem: (i) the KKT conditions and (ii) the primal–dual transformation [7].

In contrast to Chap. 5, in which the first approach, i.e., KKT conditions, is used, we use in this chapter the second approach, i.e., primal–dual transformation, which includes the strong duality equality instead of all complementarity conditions. However, the primal–dual transformation introduces some nonlinearities due to bilinear terms within the strong duality equality.

Pursuing further clarity, we derive the MPECs corresponding to the strategic producers g_1 and g_2 in Illustrative Example 6.1 presented in the previous section. Recall that lower-level problems (market-clearing problems for different operating conditions) are common within the bilevel models of both producers. Thus, we derive the optimality conditions corresponding to the lower-level problems (6.4) and (6.5) using the primal–dual transformation. Figure 6.4 schematically illustrates this transformation.

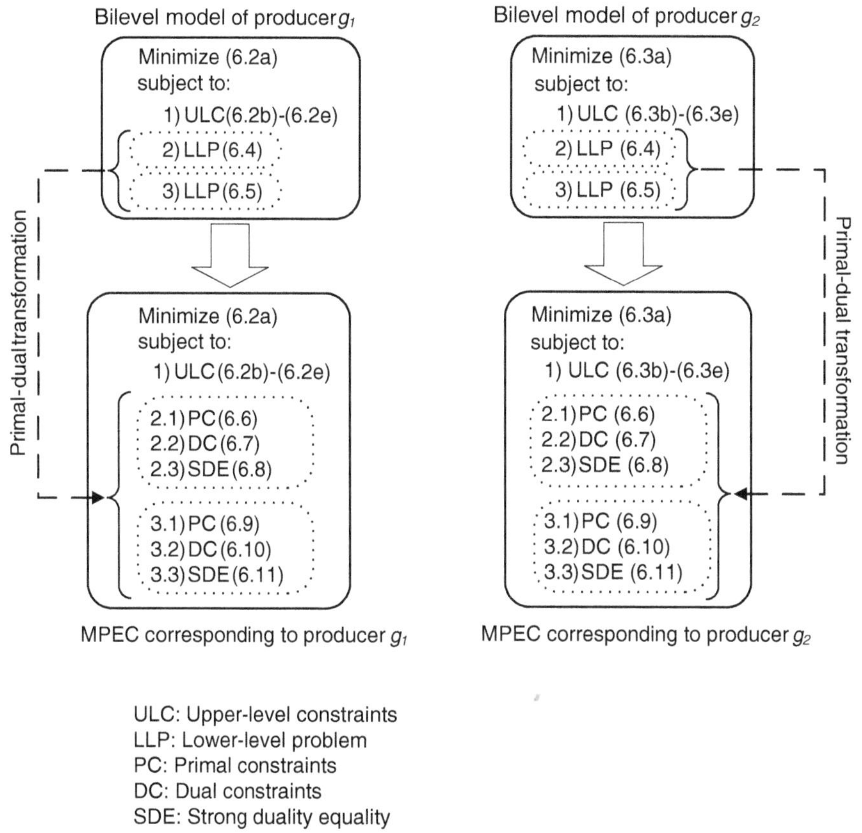

Fig. 6.4 Illustrative Example 6.1: transformation of the bilevel models of strategic producers g_1 and g_2 into their corresponding MPECs (primal–dual transformation)

First, we derive the optimality conditions corresponding to the lower-level problem (6.4), which include the primal constraints (6.6), the dual constraints (6.7), and the strong duality equality (6.8). The primal constraints of lower-level problem (6.4) are given by (6.6) below:

$$p^{\mathrm{D}}_{n_1o_1} + 1000\left(\theta_{n_1o_1} - \theta_{n_2o_1}\right) - p^{\mathrm{E}}_{g_1n_1o_1} - p^{\mathrm{C}}_{g_1n_1o_1} - p^{\mathrm{C}}_{g_2n_1o_1} = 0 \tag{6.6a}$$

$$1000\left(\theta_{n_2o_1} - \theta_{n_1o_1}\right) - p^{\mathrm{E}}_{g_2n_2o_1} - p^{\mathrm{S}}_{g_1n_2o_1} - p^{\mathrm{S}}_{g_2n_2o_1} = 0 \tag{6.6b}$$

$$0 \le p^{\mathrm{C}}_{g_1n_1o_1} \le x^{\mathrm{C}}_{g_1n_1} \tag{6.6c}$$

$$0 \le p^{\mathrm{C}}_{g_2n_1o_1} \le x^{\mathrm{C}}_{g_2n_1} \tag{6.6d}$$

$$0 \le p^{\mathrm{S}}_{g_1n_2o_1} \le 0.35\, x^{\mathrm{S}}_{g_1n_2} \tag{6.6e}$$

$$0 \le p^{\mathrm{S}}_{g_2n_2o_1} \le 0.35\, x^{\mathrm{S}}_{g_2n_2} \tag{6.6f}$$

$$0 \le p^{\mathrm{E}}_{g_1n_1o_1} \le 150 \tag{6.6g}$$

$$0 \leq p^{\mathrm{E}}_{g_2 n_2 o_1} \leq 100 \tag{6.6h}$$
$$0 \leq p^{\mathrm{D}}_{n_1 o_1} \leq 1 \times 400 \tag{6.6i}$$
$$1000\left(\theta_{n_1 o_1} - \theta_{n_2 o_1}\right) \leq 400 \tag{6.6j}$$
$$1000\left(\theta_{n_2 o_1} - \theta_{n_1 o_1}\right) \leq 400 \tag{6.6k}$$
$$-\pi \leq \theta_{n_1 o_1} \leq \pi \tag{6.6l}$$
$$-\pi \leq \theta_{n_2 o_1} \leq \pi \tag{6.6m}$$
$$\theta_{n_1 o_1} = 0. \tag{6.6n}$$

The dual constraints of lower-level problem (6.4) are given by (6.7) below:

$$\alpha^{\mathrm{C}}_{g_1 n_1 o_1} - \lambda_{n_1 o_1} + \mu^{\mathrm{C^{max}}}_{g_1 n_1 o_1} - \mu^{\mathrm{C^{min}}}_{g_1 n_1 o_1} = 0 \tag{6.7a}$$
$$\alpha^{\mathrm{E}}_{g_1 n_1 o_1} - \lambda_{n_1 o_1} + \mu^{\mathrm{E^{max}}}_{g_1 n_1 o_1} - \mu^{\mathrm{E^{min}}}_{g_1 n_1 o_1} = 0 \tag{6.7b}$$
$$\alpha^{\mathrm{C}}_{g_2 n_1 o_1} - \lambda_{n_1 o_1} + \mu^{\mathrm{C^{max}}}_{g_2 n_1 o_1} - \mu^{\mathrm{C^{min}}}_{g_2 n_1 o_1} = 0 \tag{6.7c}$$
$$\alpha^{\mathrm{E}}_{g_2 n_2 o_1} - \lambda_{n_2 o_1} + \mu^{\mathrm{E^{max}}}_{g_2 n_2 o_1} - \mu^{\mathrm{E^{min}}}_{g_2 n_2 o_1} = 0 \tag{6.7d}$$
$$-\lambda_{n_2 o_1} + \mu^{\mathrm{S^{max}}}_{g_1 n_2 o_1} - \mu^{\mathrm{S^{min}}}_{g_1 n_2 o_1} = 0 \tag{6.7e}$$
$$-\lambda_{n_2 o_1} + \mu^{\mathrm{S^{max}}}_{g_2 n_2 o_1} - \mu^{\mathrm{S^{min}}}_{g_2 n_2 o_1} = 0 \tag{6.7f}$$
$$-35 + \lambda_{n_1 o_1} + \mu^{\mathrm{D^{max}}}_{n_1 o_1} - \mu^{\mathrm{D^{min}}}_{n_1 o_1} = 0 \tag{6.7g}$$
$$1000\left(\lambda_{n_1 o_1} - \lambda_{n_2 o_1} + \mu^{\mathrm{F}}_{n_1 n_2 o_1} - \mu^{\mathrm{F}}_{n_2 n_1 o_1}\right) + \mu^{\theta^{\max}}_{n_1 o_1} - \mu^{\theta^{\min}}_{n_1 o_1} + \mu^{\theta^{\mathrm{ref}}}_{o_1} = 0 \tag{6.7h}$$
$$1000\left(\lambda_{n_2 o_1} - \lambda_{n_1 o_1} + \mu^{\mathrm{F}}_{n_2 n_1 o_1} - \mu^{\mathrm{F}}_{n_1 n_2 o_1}\right) + \mu^{\theta^{\max}}_{n_2 o_1} - \mu^{\theta^{\min}}_{n_2 o_1} = 0 \tag{6.7i}$$
$$\mu^{\mathrm{C^{min}}}_{g_1 n_1 o_1}, \ \mu^{\mathrm{C^{max}}}_{g_1 n_1 o_1} \geq 0 \tag{6.7j}$$
$$\mu^{\mathrm{C^{min}}}_{g_2 n_1 o_1}, \ \mu^{\mathrm{C^{max}}}_{g_2 n_1 o_1} \geq 0 \tag{6.7k}$$
$$\mu^{\mathrm{E^{min}}}_{g_1 n_1 o_1}, \ \mu^{\mathrm{E^{max}}}_{g_1 n_1 o_1} \geq 0 \tag{6.7l}$$
$$\mu^{\mathrm{E^{min}}}_{g_2 n_2 o_1}, \ \mu^{\mathrm{E^{max}}}_{g_2 n_2 o_1} \geq 0 \tag{6.7m}$$
$$\mu^{\mathrm{S^{min}}}_{g_1 n_2 o_1}, \ \mu^{\mathrm{S^{max}}}_{g_1 n_2 o_1} \geq 0 \tag{6.7n}$$
$$\mu^{\mathrm{S^{min}}}_{g_2 n_2 o_1}, \ \mu^{\mathrm{S^{max}}}_{g_2 n_2 o_1} \geq 0 \tag{6.7o}$$
$$\mu^{\mathrm{D^{min}}}_{n_1 o_1} \geq 0, \ \mu^{\mathrm{D^{max}}}_{n_1 o_1} \geq 0 \tag{6.7p}$$
$$\mu^{\mathrm{F}}_{n_1 n_2 o_1}, \ \mu^{\mathrm{F}}_{n_2 n_1 o_1} \geq 0 \tag{6.7q}$$
$$\mu^{\theta^{\min}}_{n_1 o_1}, \ \mu^{\theta^{\max}}_{n_1 o_1} \geq 0 \tag{6.7r}$$
$$\mu^{\theta^{\min}}_{n_2 o_1}, \ \mu^{\theta^{\max}}_{n_2 o_1} \geq 0. \tag{6.7s}$$

Finally, the strong duality equality corresponding to lower-level problem (6.4) is (6.8), which enforces the equality of its primal and dual objective function values at the optimal solution:

$$\alpha^{\mathrm{C}}_{g_1n_1o_1}\, p^{\mathrm{C}}_{g_1n_1o_1} + \alpha^{\mathrm{E}}_{g_1n_1o_1}\, p^{\mathrm{E}}_{g_1n_1o_1} + \alpha^{\mathrm{C}}_{g_2n_1o_1}\, p^{\mathrm{C}}_{g_2n_1o_1} + \alpha^{\mathrm{E}}_{g_2n_2o_1}\, p^{\mathrm{E}}_{g_2n_2o_1}$$
$$- 35\, p^{\mathrm{D}}_{n_1o_1} = -x^{\mathrm{C}}_{g_1n_1}\, \mu^{\mathrm{C^{max}}}_{g_1n_1o_1} - x^{\mathrm{C}}_{g_2n_1}\, \mu^{\mathrm{C^{max}}}_{g_2n_1o_1} - 0.35\, x^{\mathrm{S}}_{g_1n_2}\, \mu^{\mathrm{S^{max}}}_{g_1n_2o_1}$$
$$- 0.35\, x^{\mathrm{S}}_{g_2n_2}\, \mu^{\mathrm{S^{max}}}_{g_2n_2o_1} - 150\, \mu^{\mathrm{E^{max}}}_{g_1n_1o_1} - 100\, \mu^{\mathrm{E^{max}}}_{g_2n_2o_1} - 400\, \mu^{\mathrm{D^{max}}}_{n_1o_1}$$
$$- 400\left(\mu^{\mathrm{F}}_{n_1n_2o_1} + \mu^{\mathrm{F}}_{n_2n_1o_1}\right) - \pi\left(\mu^{\theta^{\min}}_{n_1o_1} + \mu^{\theta^{\max}}_{n_1o_1} + \mu^{\theta^{\min}}_{n_2o_1} + \mu^{\theta^{\max}}_{n_2o_1}\right). \quad (6.8)$$

Next, we derive the optimality conditions corresponding to lower-level problem (6.5), which consist of primal constraints (6.9), dual constraints (6.10), and strong duality equality (6.11). The primal constraints of lower-level problem (6.5) are given by (6.9) below:

$$p^{\mathrm{D}}_{n_1o_2} + 1000\left(\theta_{n_1o_2} - \theta_{n_2o_2}\right) - p^{\mathrm{E}}_{g_1n_1o_2} - p^{\mathrm{C}}_{g_1n_1o_2} - p^{\mathrm{C}}_{g_2n_1o_2} = 0 \quad (6.9a)$$
$$1000\left(\theta_{n_2o_2} - \theta_{n_1o_2}\right) - p^{\mathrm{E}}_{g_2n_2o_2} - p^{\mathrm{S}}_{g_1n_2o_2} - p^{\mathrm{S}}_{g_2n_2o_2} = 0 \quad (6.9b)$$
$$0 \le p^{\mathrm{C}}_{g_1n_1o_2} \le x^{\mathrm{C}}_{g_1n_1} \quad (6.9c)$$
$$0 \le p^{\mathrm{C}}_{g_2n_1o_2} \le x^{\mathrm{C}}_{g_2n_1} \quad (6.9d)$$
$$0 \le p^{\mathrm{S}}_{g_1n_2o_2} \le 0.70\, x^{\mathrm{S}}_{g_1n_2} \quad (6.9e)$$
$$0 \le p^{\mathrm{S}}_{g_2n_2o_2} \le 0.70\, x^{\mathrm{S}}_{g_2n_2} \quad (6.9f)$$
$$0 \le p^{\mathrm{E}}_{g_1n_1o_2} \le 150 \quad (6.9g)$$
$$0 \le p^{\mathrm{E}}_{g_2n_2o_2} \le 100 \quad (6.9h)$$
$$0 \le p^{\mathrm{D}}_{n_1o_2} \le 0.8 \times 400 \quad (6.9i)$$
$$1000\left(\theta_{n_1o_2} - \theta_{n_2o_2}\right) \le 400 \quad (6.9j)$$
$$1000\left(\theta_{n_2o_2} - \theta_{n_1o_2}\right) \le 400 \quad (6.9k)$$
$$-\pi \le \theta_{n_1o_2} \le \pi \quad (6.9l)$$
$$-\pi \le \theta_{n_2o_2} \le \pi \quad (6.9m)$$
$$\theta_{n_1o_2} = 0. \quad (6.9n)$$

The dual constraints of lower-level problem (6.5) are given by (6.10) below:

$$\alpha^{\mathrm{C}}_{g_1n_1o_2} - \lambda_{n_1o_2} + \mu^{\mathrm{C^{max}}}_{g_1n_1o_2} - \mu^{\mathrm{C^{min}}}_{g_1n_1o_2} = 0 \quad (6.10a)$$
$$\alpha^{\mathrm{E}}_{g_1n_1o_2} - \lambda_{n_1o_2} + \mu^{\mathrm{E^{max}}}_{g_1n_1o_2} - \mu^{\mathrm{E^{min}}}_{g_1n_1o_2} = 0 \quad (6.10b)$$
$$\alpha^{\mathrm{C}}_{g_2n_1o_2} - \lambda_{n_1o_2} + \mu^{\mathrm{C^{max}}}_{g_2n_1o_2} - \mu^{\mathrm{C^{min}}}_{g_2n_1o_2} = 0 \quad (6.10c)$$
$$\alpha^{\mathrm{E}}_{g_2n_2o_2} - \lambda_{n_2o_2} + \mu^{\mathrm{E^{max}}}_{g_2n_2o_2} - \mu^{\mathrm{E^{min}}}_{g_2n_2o_2} = 0 \quad (6.10d)$$
$$-\lambda_{n_2o_2} + \mu^{\mathrm{S^{max}}}_{g_1n_2o_2} - \mu^{\mathrm{S^{min}}}_{g_1n_2o_2} = 0 \quad (6.10e)$$
$$-\lambda_{n_2o_2} + \mu^{\mathrm{S^{max}}}_{g_2n_2o_2} - \mu^{\mathrm{S^{min}}}_{g_2n_2o_2} = 0 \quad (6.10f)$$
$$-32 + \lambda_{n_1o_2} + \mu^{\mathrm{D^{max}}}_{n_1o_2} - \mu^{\mathrm{D^{min}}}_{n_1o_2} = 0 \quad (6.10g)$$
$$1000\left(\lambda_{n_1o_2} - \lambda_{n_2o_2} + \mu^{\mathrm{F}}_{n_1n_2o_2} - \mu^{\mathrm{F}}_{n_2n_1o_2}\right) + \mu^{\theta^{\max}}_{n_1o_2} - \mu^{\theta^{\min}}_{n_1o_2} + \mu^{\theta^{\mathrm{ref}}}_{o_2} = 0 \quad (6.10h)$$

$$1000\left(\lambda_{n_2o_2}-\lambda_{n_1o_2}+\mu^{\mathrm{F}}_{n_2n_1o_2}-\mu^{\mathrm{F}}_{n_1n_2o_2}\right)+\mu^{\theta^{\max}}_{n_2o_2}-\mu^{\theta^{\min}}_{n_2o_2}=0 \quad (6.10i)$$

$$\mu^{\mathrm{C}^{\min}}_{g_1n_1o_2},\ \mu^{\mathrm{C}^{\max}}_{g_1n_1o_2}\geq 0 \quad (6.10j)$$

$$\mu^{\mathrm{C}^{\min}}_{g_2n_1o_2},\ \mu^{\mathrm{C}^{\max}}_{g_2n_1o_2}\geq 0 \quad (6.10k)$$

$$\mu^{\mathrm{E}^{\min}}_{g_1n_1o_2},\ \mu^{\mathrm{E}^{\max}}_{g_1n_1o_2}\geq 0 \quad (6.10l)$$

$$\mu^{\mathrm{E}^{\min}}_{g_2n_2o_2},\ \mu^{\mathrm{E}^{\max}}_{g_2n_2o_2}\geq 0 \quad (6.10m)$$

$$\mu^{\mathrm{S}^{\min}}_{g_1n_2o_2},\ \mu^{\mathrm{S}^{\max}}_{g_1n_2o_2}\geq 0 \quad (6.10n)$$

$$\mu^{\mathrm{S}^{\min}}_{g_2n_2o_2},\ \mu^{\mathrm{S}^{\max}}_{g_2n_2o_2}\geq 0 \quad (6.10o)$$

$$\mu^{\mathrm{D}^{\min}}_{n_1o_2},\ \mu^{\mathrm{D}^{\max}}_{n_1o_2}\geq 0 \quad (6.10p)$$

$$\mu^{\mathrm{F}}_{n_1n_2o_2},\ \mu^{\mathrm{F}}_{n_2n_1o_2}\geq 0 \quad (6.10q)$$

$$\mu^{\theta^{\min}}_{n_1o_2},\ \mu^{\theta^{\max}}_{n_1o_2}\geq 0 \quad (6.10r)$$

$$\mu^{\theta^{\min}}_{n_2o_2},\ \mu^{\theta^{\max}}_{n_2o_2}\geq 0. \quad (6.10s)$$

Finally, the strong duality equality corresponding to lower-level problem (6.5) is (6.11) below:

$$\begin{aligned}\alpha^{\mathrm{C}}_{g_1n_1o_2}\ p^{\mathrm{C}}_{g_1n_1o_2}&+\alpha^{\mathrm{E}}_{g_1n_1o_2}\ p^{\mathrm{E}}_{g_1n_1o_2}+\alpha^{\mathrm{C}}_{g_2n_1o_2}\ p^{\mathrm{C}}_{g_2n_1o_2}+\alpha^{\mathrm{E}}_{g_2n_2o_2}\ p^{\mathrm{E}}_{g_2n_2o_2}\\&-32\ p^{\mathrm{D}}_{n_1o_2}=-x^{\mathrm{C}}_{g_1n_1}\ \mu^{\mathrm{C}^{\max}}_{g_1n_1o_2}-x^{\mathrm{C}}_{g_2n_1}\ \mu^{\mathrm{C}^{\max}}_{g_2n_1o_2}-0.70\ x^{\mathrm{S}}_{g_1n_2}\ \mu^{\mathrm{S}^{\max}}_{g_1n_2o_2}\\&-0.70\ x^{\mathrm{S}}_{g_2n_2}\ \mu^{\mathrm{S}^{\max}}_{g_2n_2o_2}-150\ \mu^{\mathrm{E}^{\max}}_{g_1n_1o_2}-100\ \mu^{\mathrm{E}^{\max}}_{g_2n_2o_2}-320\ \mu^{\mathrm{D}^{\max}}_{n_1o_2}\\&-400\left(\mu^{\mathrm{F}}_{n_1n_2o_2}+\mu^{\mathrm{F}}_{n_2n_1o_2}\right)-\pi\left(\mu^{\theta^{\min}}_{n_1o_2}+\mu^{\theta^{\max}}_{n_1o_2}+\mu^{\theta^{\min}}_{n_2o_2}+\mu^{\theta^{\max}}_{n_2o_2}\right). \quad (6.11)\end{aligned}$$

Accordingly, the MPEC corresponding to the strategic producer g_1 includes the upper-level problem (6.2a)–(6.2e) and the optimality conditions (6.6)–(6.11). Hereinafter, this MPEC will be called MPEC 1. Similarly, the MPEC corresponding to the strategic producer g_2 comprises the upper-level problem (6.3a)–(6.3e) and the optimality conditions (6.6)–(6.11). Hereinafter, this MPEC will be called MPEC 2.

One important observation is that both MPECs 1 and 2 are continuous, but nonlinear. The reason for nonlinearities is the existence of bilinear terms within objective functions (6.2a) and (6.3a) and strong duality equalities (6.8) and (6.11).

Note that the dual variables associated with the constraints of MPECs 1 and 2 are needed in the next section. Pursuing notational clarity, we use the equation numbers to indicate the dual variables of the MPECs. The following notional examples are provided:

1. In MPEC 1 corresponding to producer g_1, the dual variable associated with strong duality equality (6.8) is $\eta^{(6.8)}_{g_1}$. However, the dual variable of that equality within MPEC 2 corresponding to producer g_2 is $\eta^{(6.8)}_{g_2}$.

2. In MPEC 1 corresponding to producer g_1, the dual variables associated with lower and upper bounds in inequality (6.9c) are $\underline{\eta}_{g_1}^{(6.9c)}$ and $\overline{\eta}_{g_1}^{(6.9c)}$, respectively.
3. In MPEC 2 corresponding to producer g_2, the dual variables associated with nonnegativity conditions (6.10j) are $\underline{\eta}_{g_2}^{(6.10j)}$ and $\overline{\eta}_{g_2}^{(6.10j)}$, respectively.

6.5 Multiple-Producer Problem: EPEC

The joint consideration of all MPECs, one per producer, constitutes an EPEC. Figure 6.5 illustrates the EPEC corresponding to Illustrative Example 6.1 presented in Sect. 6.4. Accordingly, this EPEC includes both MPECs 1 and 2 corresponding to the strategic producers g_1 and g_2, respectively. Note that the EPEC solution identifies the market equilibria.

6.5.1 EPEC Solution

To obtain the EPEC solution, we first need to derive the KKT conditions associated with each MPEC. However, it is important to recall that MPECs are continuous, nonlinear, and thus nonconvex. Therefore, the KKT conditions associated with each MPEC provide its strong stationarity conditions. A collection of all those conditions corresponding to all MPECs constitutes the strong stationarity conditions associated with the EPEC, whose solution identifies the equilibria. For Illustrative Example 6.1 of Sect. 6.4, this transformation is depicted in Fig. 6.6.

It is important to note that the solutions obtained from this procedure can be Nash equilibria, local equilibria, and saddle points. To detect the Nash equilibria among

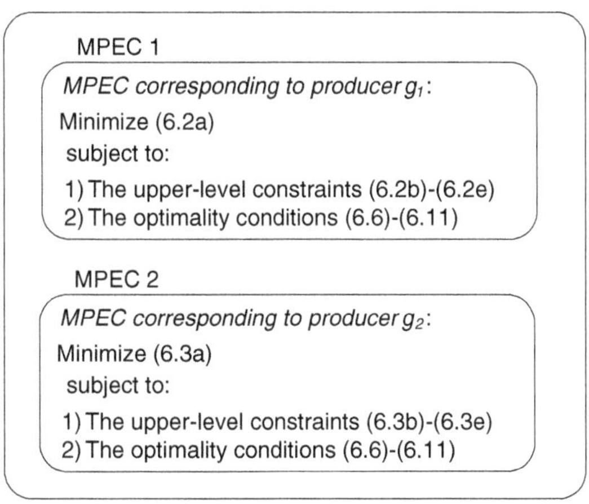

Fig. 6.5 Illustrative Example 6.1: EPEC

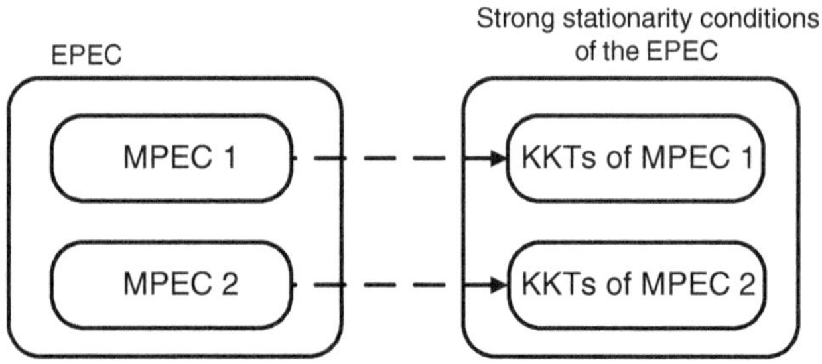

Fig. 6.6 Illustrative Example 6.1: strong stationarity conditions of the EPEC

the solutions obtained, an ex-post algorithm is provided in Sect. 6.5.3. The next two sections present the KKT conditions of both MPECs.

6.5.1.1 KKT Conditions of MPEC 1

The KKT conditions of MPEC 1 corresponding to producer g_1 include the constraints below:

1. Primal equality constraints of MPEC 1 including (6.6a)–(6.6b), (6.6n), (6.7a)–(6.7i), (6.8), (6.9a)–(6.9b), (6.9n), (6.10a)–(6.10i), and (6.11). We refer to these equality constraints as the set Γ_1.
2. Equality constraints obtained from differentiating the corresponding Lagrangian associated with MPEC 1 with respect to its variables. We refer to these equality constraints as the set Γ_2. Four examples of the members of this set are stated below. Note that $\mathscr{L}^{\text{MPEC1}}_{g_1}$ is the Lagrangian function of the MPEC 1 pertaining to the strategic producer g_1:

$$\frac{\partial \mathscr{L}^{\text{MPEC1}}_{g_1}}{\partial x^{\text{C}}_{g_1 n_1}} = 55000 + \overline{\eta}^{(6.2\text{b})}_{g_1} - \underline{\eta}^{(6.2\text{b})}_{g_1} + 55000\, \eta^{(6.2\text{d})}_{g_1} - \overline{\eta}^{(6.6\text{c})}_{g_1} + \mu^{\text{C}^{\max}}_{g_1 n_1 o_1}\, \eta^{(6.8)}_{g_1} - \overline{\eta}^{(6.9\text{c})}_{g_1} + \mu^{\text{C}^{\max}}_{g_1 n_1 o_2}\, \eta^{(6.11)}_{g_1} = 0 \tag{6.12a}$$

$$\frac{\partial \mathscr{L}^{\text{MPEC1}}_{g_1}}{\partial x^{\text{C}}_{g_2 n_1}} = -\overline{\eta}^{(6.6\text{d})}_{g_1} + \mu^{\text{C}^{\max}}_{g_2 n_1 o_1}\, \eta^{(6.8)}_{g_1} - \overline{\eta}^{(6.9\text{d})}_{g_1} + \mu^{\text{C}^{\max}}_{g_2 n_1 o_2}\, \eta^{(6.11)}_{g_1} = 0 \tag{6.12b}$$

$$\frac{\partial \mathscr{L}^{\text{MPEC1}}_{g_1}}{\partial \alpha^{\text{E}}_{g_1 n_1 o_2}} = -\, \eta^{(6.2\text{e})}_{g_1} + \eta^{(6.10\text{b})}_{g_1} + p^{\text{E}}_{g_1 n_1 o_2}\, \eta^{(6.11)}_{g_1} = 0 \tag{6.12c}$$

$$\frac{\partial \mathscr{L}^{\text{MPEC1}}_{g_1}}{\partial \mu^{\text{S}^{\max}}_{g_2 n_2 o_1}} = \eta^{(6.7\text{f})}_{g_1} - \overline{\eta}^{(6.7\text{o})}_{g_1} + 0.35\, x^{\text{S}}_{g_2 n_2}\, \eta^{(6.8)}_{g_1} = 0. \tag{6.12d}$$

3. Complementarity conditions related to the inequality constraints of MPEC 1. We refer to these inequality constraints as the Γ_3. Four examples of the members of this set are as follows:

$$0 \leq x^{\mathrm{C}}_{g_1 n_1} \perp \underline{\eta}^{(6.2\mathrm{b})}_{g_1} \geq 0 \tag{6.13a}$$

$$0 \leq \left[200 - x^{\mathrm{C}}_{g_1 n_1}\right] \perp \overline{\eta}^{(6.2\mathrm{b})}_{g_1} \geq 0 \tag{6.13b}$$

$$0 \leq p^{\mathrm{C}}_{g_2 n_1 o_2} \perp \underline{\eta}^{(6.9\mathrm{d})}_{g_1} \geq 0 \tag{6.13c}$$

$$0 \leq \left[x^{\mathrm{C}}_{g_2 n_1} - p^{\mathrm{C}}_{g_2 n_1 o_2}\right] \perp \overline{\eta}^{(6.9\mathrm{d})}_{g_1} \geq 0. \tag{6.13d}$$

6.5.1.2 KKT Conditions of MPEC 2

The KKT conditions of MPEC 2 corresponding to producer g_2 consist of the following three sets of constraints:

1. Primal equality constraints of MPEC 2 that are identical to those included in the constraint set Γ_1.
2. Equality constraints resulting from differentiating the corresponding Lagrangian of MPEC 2 with respect to its variables. These equality constraints are referred to as the set Γ_4. Four examples of the members of this set are stated below, in which $\mathscr{L}^{\mathrm{MPEC2}}_{g_2}$ is the Lagrangian function of the MPEC 2 pertaining to the strategic producer g_2:

$$\begin{aligned}\frac{\partial \mathscr{L}^{\mathrm{MPEC2}}_{g_2}}{\partial x^{\mathrm{S}}_{g_2 n_2}} = {} & 66000 + \overline{\eta}^{(6.3\mathrm{c})}_{g_2} - \underline{\eta}^{(6.3\mathrm{c})}_{g_2} + 66000\, \eta^{(6.3\mathrm{d})}_{g_2} - 0.35\, \overline{\eta}^{(6.6\mathrm{f})}_{g_2} \\ & + 0.35\, \mu^{\mathrm{S^{max}}}_{g_2 n_2 o_1}\, \eta^{(6.8)}_{g_2} - 0.70\, \overline{\eta}^{(6.9\mathrm{f})}_{g_2} \\ & + 0.70\, \mu^{\mathrm{S^{max}}}_{g_2 n_2 o_2}\, \eta^{(6.11)}_{g_2} = 0\end{aligned} \tag{6.14a}$$

$$\begin{aligned}\frac{\partial \mathscr{L}^{\mathrm{MPEC2}}_{g_2}}{\partial x^{\mathrm{S}}_{g_1 n_2}} = {} & -0.35\, \overline{\eta}^{(6.6\mathrm{e})}_{g_2} + 0.35\, \mu^{\mathrm{S^{max}}}_{g_1 n_2 o_1}\, \eta^{(6.8)}_{g_2} \\ & - 0.70\, \overline{\eta}^{(6.9\mathrm{e})}_{g_2} + 0.70\, \mu^{\mathrm{S^{max}}}_{g_1 n_2 o_2}\, \eta^{(6.11)}_{g_2} = 0\end{aligned} \tag{6.14b}$$

$$\begin{aligned}\frac{\partial \mathscr{L}^{\mathrm{MPEC2}}_{g_2}}{\partial p^{\mathrm{E}}_{g_2 n_2 o_2}} = {} & -5230\, \lambda_{n_2 o_2} - \eta^{(6.9\mathrm{b})}_{g_2} + \overline{\eta}^{(6.9\mathrm{h})}_{g_2} - \underline{\eta}^{(6.9\mathrm{h})}_{g_2} \\ & + \alpha^{\mathrm{E}}_{g_2 n_2 o_2}\, \eta^{(6.11)}_{g_2} = 0\end{aligned} \tag{6.14c}$$

$$\frac{\partial \mathscr{L}^{\mathrm{MPEC2}}_{g_2}}{\partial \mu^{\mathrm{D^{max}}}_{n_1 o_1}} = \eta^{(6.7\mathrm{g})}_{g_2} - \overline{\eta}^{(6.7\mathrm{p})}_{g_2} + 400\, \eta^{(6.8)}_{g_2} = 0. \tag{6.14d}$$

3. Complementarity conditions related to the inequality constraints of MPEC 2. We refer to these inequality constraints as the set Γ_5. Four examples of the members of this set are as follows:

$$0 \le x^{\mathrm{S}}_{g_2 n_2} \perp \underline{\eta}^{(6.3\mathrm{c})}_{g_2} \ge 0 \quad (6.15\mathrm{a})$$

$$0 \le \left[200 - x^{\mathrm{S}}_{g_2 n_2}\right] \perp \overline{\eta}^{(6.3\mathrm{c})}_{g_2} \ge 0 \quad (6.15\mathrm{b})$$

$$0 \le p^{\mathrm{E}}_{g_2 n_2 o_2} \perp \underline{\eta}^{(6.9\mathrm{h})}_{g_2} \ge 0 \quad (6.15\mathrm{c})$$

$$0 \le \left[100 - p^{\mathrm{E}}_{g_2 n_2 o_2}\right] \perp \overline{\eta}^{(6.9\mathrm{h})}_{g_2} \ge 0. \quad (6.15\mathrm{d})$$

6.5.1.3 Strong Stationarity Conditions of the EPEC: Linearization

The strong stationarity conditions of the EPEC (the right-hand box of Fig. 6.6) is a system of equalities and inequalities included in Γ_1–Γ_5. Although the solution to this system identifies the investment equilibria, this system includes the following three nonlinearities:

1. The complementarity conditions included in Γ_3 and Γ_5. Such conditions can be exactly linearized through the approach explained in Chap. 5 using auxiliary binary variables and large enough positive constants [6]. For example, the mixed-integer linear equivalent of complementarity condition (6.15a) is provided by (6.16) below:

$$x^{\mathrm{S}}_{g_2 n_2} \ge 0 \quad (6.16\mathrm{a})$$

$$\underline{\eta}^{(6.3\mathrm{c})}_{g_2} \ge 0 \quad (6.16\mathrm{b})$$

$$x^{\mathrm{S}}_{g_2 n_2} \le \psi \, M^{x} \quad (6.16\mathrm{c})$$

$$\underline{\eta}^{(6.3\mathrm{c})}_{g_2} \le (1 - \psi) \, M^{\eta} \quad (6.16\mathrm{d})$$

$$\psi \in \{0, 1\}, \quad (6.16\mathrm{e})$$

   where M^x and M^η are large enough positive constants. A method for appropriate value selection for those constants is provided in Chap. 5.
2. The products of variables involved in the strong duality equalities (6.8) and (6.11) included in Γ_1. Unlike the complementarity conditions included in Γ_3 and Γ_5, which can be exactly linearized through auxiliary binary variables, the strong duality equalities (6.8) and (6.11) included in Γ_1 cannot be linearized straightforwardly due to the nature of the nonlinearities, i.e., the product of continuous variables. However, we take advantage of the fact that the strong duality equality resulting from the primal–dual transformation is equivalent to the set of complementarity conditions obtained from the KKT conditions [7]. Hence, pursuing linearity, the strong duality equality (6.8) is replaced by its equivalent complementarity conditions (6.17) below:

$$0 \le p^{\mathrm{C}}_{g_1 n_1 o_1} \perp \mu^{\mathrm{C^{min}}}_{g_1 n_1 o_1} \ge 0 \quad (6.17\mathrm{a})$$

$$0 \le \left(x^{\mathrm{C}}_{g_1 n_1} - p^{\mathrm{C}}_{g_1 n_1 o_1}\right) \perp \mu^{\mathrm{C^{max}}}_{g_1 n_1 o_1} \ge 0 \quad (6.17\mathrm{b})$$

$$0 \le p^{\mathrm{C}}_{g_2 n_1 o_1} \perp \mu^{\mathrm{C^{min}}}_{g_2 n_1 o_1} \ge 0 \quad (6.17\mathrm{c})$$

$$0 \leq \left(x^{\mathrm{C}}_{g_2 n_1} - p^{\mathrm{C}}_{g_2 n_1 o_1}\right) \perp \mu^{\mathrm{C}^{\max}}_{g_2 n_1 o_1} \geq 0 \tag{6.17d}$$

$$0 \leq p^{\mathrm{S}}_{g_1 n_2 o_1} \perp \mu^{\mathrm{S}^{\min}}_{g_1 n_2 o_1} \geq 0 \tag{6.17e}$$

$$0 \leq \left(0.35\, x^{\mathrm{S}}_{g_1 n_2} - p^{\mathrm{S}}_{g_1 n_2 o_1}\right) \perp \mu^{\mathrm{S}^{\max}}_{g_1 n_2 o_1} \geq 0 \tag{6.17f}$$

$$0 \leq p^{\mathrm{S}}_{g_2 n_2 o_1} \perp \mu^{\mathrm{S}^{\min}}_{g_2 n_2 o_1} \geq 0 \tag{6.17g}$$

$$0 \leq \left(0.35\, x^{\mathrm{S}}_{g_2 n_2} - p^{\mathrm{S}}_{g_2 n_2 o_1}\right) \perp \mu^{\mathrm{S}^{\max}}_{g_2 n_2 o_1} \geq 0 \tag{6.17h}$$

$$0 \leq p^{\mathrm{E}}_{g_1 n_1 o_1} \perp \mu^{\mathrm{E}^{\min}}_{g_1 n_1 o_1} \geq 0 \tag{6.17i}$$

$$0 \leq \left(150 - p^{\mathrm{E}}_{g_1 n_1 o_1}\right) \perp \mu^{\mathrm{E}^{\max}}_{g_1 n_1 o_1} \geq 0 \tag{6.17j}$$

$$0 \leq p^{\mathrm{E}}_{g_2 n_2 o_1} \perp \mu^{\mathrm{E}^{\min}}_{g_2 n_2 o_1} \geq 0 \tag{6.17k}$$

$$0 \leq \left(100 - p^{\mathrm{E}}_{g_2 n_2 o_1}\right) \perp \mu^{\mathrm{E}^{\max}}_{g_2 n_2 o_1} \geq 0 \tag{6.17l}$$

$$0 \leq p^{\mathrm{D}}_{n_1 o_1} \perp \mu^{\mathrm{D}^{\min}}_{n_1 o_1} \geq 0 \tag{6.17m}$$

$$0 \leq \left(400 - p^{\mathrm{D}}_{n_1 o_1}\right) \perp \mu^{\mathrm{D}^{\max}}_{n_1 o_1} \geq 0 \tag{6.17n}$$

$$0 \leq \left[400 - 1000\left(\theta_{n_1 o_1} - \theta_{n_2 o_1}\right)\right] \perp \mu^{\mathrm{F}}_{n_1 n_2 o_1} \geq 0 \tag{6.17o}$$

$$0 \leq \left[400 - 1000\left(\theta_{n_2 o_1} - \theta_{n_1 o_1}\right)\right] \perp \mu^{\mathrm{F}}_{n_2 n_1 o_1} \geq 0 \tag{6.17p}$$

$$0 \leq \left(\pi + \theta_{n_1 o_1}\right) \perp \mu^{\theta^{\min}}_{n_1 o_1} \geq 0 \tag{6.17q}$$

$$0 \leq \left(\pi - \theta_{n_1 o_1}\right) \perp \mu^{\theta^{\max}}_{n_1 o_1} \geq 0 \tag{6.17r}$$

$$0 \leq \left(\pi + \theta_{n_2 o_1}\right) \perp \mu^{\theta^{\min}}_{n_2 o_1} \geq 0 \tag{6.17s}$$

$$0 \leq \left(\pi - \theta_{n_2 o_1}\right) \perp \mu^{\theta^{\max}}_{n_2 o_1} \geq 0. \tag{6.17t}$$

Similarly, the strong duality equality (6.11) can be replaced by its equivalent complementarity conditions. Recall that these complementarity conditions can be linearized using the auxiliary binary variables and large enough positive constants.

3. The ones arising from the product of variables in Γ_2 and Γ_4, e.g., the bilinear term $\mu^{\mathrm{C}^{\max}}_{g_1 n_1 o_1}\ \eta^{(6.8)}_{g_1}$ in condition (6.12a). Observe that the common variables of such nonlinear terms are either dual variables $\eta^{(6.8)}_{g_1}$ and $\eta^{(6.8)}_{g_2}$ or dual variables $\eta^{(6.11)}_{g_1}$ and $\eta^{(6.11)}_{g_2}$. From a mathematical point of view, the nonconvex nature of the MPECs 1 and 2 implies that the Mangasarian–Fromovitz constraint qualification (MFCQ) [5] does not hold at any feasible solution, i.e., the set of dual variables associated with the MPECs (all dual variables denoted by η) is unbounded. In other words, the values of these dual variables are not unique, and thus there are some degrees of freedom in the choice of values for those dual variables at any solution [5, 19, 20]. This redundancy allows the parameterization of dual variables $\eta^{(6.8)}_{g_1}$, $\eta^{(6.8)}_{g_2}$, $\eta^{(6.11)}_{g_1}$, and $\eta^{(6.11)}_{g_2}$. Hence, the bilinear terms in Γ_2 and Γ_4 become linear if the strong stationarity conditions of the EPEC are parameterized in dual variables $\eta^{(6.8)}_{g_1}$, $\eta^{(6.8)}_{g_2}$, $\eta^{(6.11)}_{g_1}$, and $\eta^{(6.11)}_{g_2}$.

6.5.2 Searching for Multiple Solutions

The mixed-integer linear form of the strong stationarity conditions of the EPEC, i.e., condition sets Γ_1–Γ_5, constitutes a system of mixed-integer linear equalities and inequalities that involves continuous and binary variables. This system generally has multiple solutions; however, recall that these solutions can be Nash equilibria, local equilibria, and saddle points. To detect the Nash equilibria among the solutions obtained, an ex-post algorithm is provided in Sect. 6.5.3.

To explore multiple solutions, it is straightforward to formulate an auxiliary optimization problem considering the mixed-integer linear condition sets Γ_1–Γ_5 as constraints. In addition, several auxiliary objective functions can be considered to identify different solutions [19]. For example, the following objectives can be maximized:

1. Total profit (TP).
2. Annual *true* social welfare (ATSW) considering the *actual* production costs of the generation units.
3. Annual social welfare considering the strategic offer prices of the generation units.
4. Minus the payment of the demands.
5. Profit of a given producer.
6. Minus the payment of a given demand.

In this chapter, the first two objectives are selected because (i) they can be formulated linearly and (ii) they refer to general market measures. Thus, the auxiliary optimization problem to find multiple solutions is formulated as follows:

$$\max \text{ TP or ATSW} \tag{6.18a}$$

$$\text{subject to the mixed-integer linear system } \Gamma_1 - \Gamma_5. \tag{6.18b}$$

The two linear objective functions selected, i.e., TP and ATSW, to be included in (6.18a) are described in the following two sections.

6.5.2.1 Objective Function (6.18a): TP Maximization

The summation of the MPEC's objective function for all producers provides minus the total profit of all producers, but this expression is nonlinear due to the products of continuous variables (i.e., production quantities and clearing prices). An identical linearization approach to one presented in Chap. 5 is used to linearize these bilinear terms. For Illustrative Example 6.1 presented in Sect. 6.4, the following exact linear expression can be obtained as an equivalent for the total profit of strategic producers g_1 and g_2:

$$\begin{aligned}
\text{TP} = 3530 \Big[& 35\, p^{\text{D}}_{n_1 o_1} - 400\, \mu^{\text{D}^{\max}}_{n_1 o_1} - 400\, (\mu^{\text{F}}_{n_1 n_2 o_1} + \mu^{\text{F}}_{n_2 n_1 o_1}) \\
& - \pi \left(\mu^{\theta^{\min}}_{n_1 o_1} + \mu^{\theta^{\max}}_{n_1 o_1} + \mu^{\theta^{\min}}_{n_2 o_1} + \mu^{\theta^{\max}}_{n_2 o_1} \right) \\
& - 12\, p^{\text{C}}_{g_1 n_1 o_1} - 10\, p^{\text{E}}_{g_1 n_1 o_1} - 12\, p^{\text{C}}_{g_2 n_1 o_1} - 15\, p^{\text{E}}_{g_2 n_2 o_1} \Big] \\
& + 5230 \Big[32\, p^{\text{D}}_{n_1 o_2} - 320\, \mu^{\text{D}^{\max}}_{n_1 o_2} - 400\, (\mu^{\text{F}}_{n_1 n_2 o_2} + \mu^{\text{F}}_{n_2 n_1 o_2}) \\
& - \pi \left(\mu^{\theta^{\min}}_{n_1 o_2} + \mu^{\theta^{\max}}_{n_1 o_2} + \mu^{\theta^{\min}}_{n_2 o_2} + \mu^{\theta^{\max}}_{n_2 o_2} \right) \\
& - 12\, p^{\text{C}}_{g_1 n_1 o_2} - 10\, p^{\text{E}}_{g_1 n_1 o_2} - 12\, p^{\text{C}}_{g_2 n_1 o_2} - 15\, p^{\text{E}}_{g_2 n_2 o_2} \Big] \\
& - 55000\, x^{\text{C}}_{g_1 n_1} - 66000\, x^{\text{S}}_{g_1 n_2} - 55000\, x^{\text{C}}_{g_2 n_1} - 66000\, x^{\text{S}}_{g_2 n_2}.
\end{aligned} \quad (6.19)$$

6.5.2.2 Objective Function (6.18a): ATSW Maximization

For Illustrative Example 6.1 presented in Sect. 6.4, the linear formulation of the ATSW to be included in (6.18a) is given by (6.20) below:

$$\begin{aligned}
\text{ATSW} = 3530 & \left[35\, p^{\text{D}}_{n_1 o_1} - 12\, p^{\text{C}}_{g_1 n_1 o_1} - 10\, p^{\text{E}}_{g_1 n_1 o_1} - 12\, p^{\text{C}}_{g_2 n_1 o_1} - 15\, p^{\text{E}}_{g_2 n_2 o_1} \right] \\
+ 5230 & \left[32\, p^{\text{D}}_{n_1 o_2} - 12\, p^{\text{C}}_{g_1 n_1 o_2} - 10\, p^{\text{E}}_{g_1 n_1 o_2} - 12\, p^{\text{C}}_{g_2 n_1 o_2} - 15\, p^{\text{E}}_{g_2 n_2 o_2} \right].
\end{aligned} \quad (6.20)$$

Note that to formulate the ATSW in (6.20), instead of the strategic offers of the generating units, their true production costs are considered.

6.5.3 Ex-Post Algorithm for Detecting Nash Equilibria

In this section, we provide an ex-post algorithm [10] based on a single-iteration diagonalization approach, which is the next step after solving problem (6.18). This algorithm allows us to check whether each solution of problem (6.18) obtained is, in fact, a Nash equilibrium. Note that if under the diagonalization framework, no producer desires to deviate from its actual strategy, then the set of strategies of all producers satisfies the definition of a Nash equilibrium [16, 17].

Let us consider the duopoly introduced in Illustrative Example 6.1 in Sect. 6.4 with two strategic producers g_1 and g_2. The strategic decisions of producer g_1 include its investment decisions, i.e., $x^{\text{C}}_{g_1 n_1}$ and $x^{\text{S}}_{g_1 n_2}$, and its offering decisions, i.e., $\alpha^{\text{C}}_{g_1 n_1 o_1}$, $\alpha^{\text{E}}_{g_1 n_1 o_1}$, $\alpha^{\text{C}}_{g_1 n_1 o_2}$, and $\alpha^{\text{E}}_{g_1 n_1 o_2}$. We refer to these strategic decisions of producer g_1

as decision set S_{g_1}. A similar set can be defined including the strategic decisions of producer g_2, denoted by S_{g_2}. In order to verify that each solution obtained for problem (6.18) constitutes a Nash equilibrium, the following four steps are carried out:

1. Consider the mixed-integer linear form of MPEC 1 pertaining to producer g_1.
2. Set the investment decisions of producer g_2, i.e., S_{g_2}, to those obtained by the equilibrium model through solving problem (6.18). Then, solve MPEC 1. Note that its solution provides the strategic decisions of producer g_1, which we denote by $\widehat{S}_{g_1}$.
3. Repeat the two steps above for producer g_2 through solving MPEC 2, while strategic decisions S_{g_1} are fixed to those values obtained from the equilibrium model. This step results in deriving the strategic decisions $\widehat{S}_{g_2}$ pertaining to producer g_2.
4. Compare the results obtained from the previous steps of the diagonalization algorithm, i.e., $\widehat{S}_{g_1}$ and $\widehat{S}_{g_2}$, with those achieved from the equilibrium model, i.e., S_{g_1} and S_{g_2}. If the investment results of each strategic producer obtained from the single-iteration diagonalization algorithm are identical to those attained by the equilibrium model, i.e., $\widehat{S}_{g_1}=S_{g_1}$ and $\widehat{S}_{g_2}=S_{g_2}$, then such a solution is a Nash equilibrium because each producer cannot increase its profit by changing its strategy unilaterally.

6.5.4 *Numerical Results*

This section provides the numerical results corresponding to Illustrative Example 6.1 presented in Sect. 6.4. Table 6.1 presents the investment equilibrium results. Note that these results are obtained by solving the auxiliary optimization problem (6.18) considering two different terms as objective function (6.18a), i.e., (i) maximizing TP and (ii) maximizing ATSW. The GAMS code for solving this MILP problem maximizing TP is provided in Sect. 6.8. Note that all results reported in Table 6.1 are verified to be Nash equilibria through the ex-post algorithm provided in Sect. 6.5.3.

As described in Sect. 6.5.1.3, the equilibrium model is parameterized in dual variables $\eta^{(6.8)}_{g_1}$, $\eta^{(6.8)}_{g_2}$, $\eta^{(6.11)}_{g_1}$, and $\eta^{(6.11)}_{g_2}$, which makes it a linear problem. The value considered for those parameterized variables, i.e., $\eta^{(6.8)}_{g_1}$, $\eta^{(6.8)}_{g_2}$, $\eta^{(6.11)}_{g_1}$, and $\eta^{(6.11)}_{g_2}$, are equal to the weights of the corresponding operating conditions, that is, 3530, 3530, 5230, and 5230, respectively. We have checked a variety of values for those parameterized dual variables, e.g., half of those values considered, but the numerical results obtained do not change.

According to the results presented in Table 6.1, several observations can be made, as stated below:

1. In the TP maximization case, the strategic producers g_1 and g_2 invest in both conventional and stochastic units. However, the total capacity of newly built units (280 MW) is comparatively lower than that in the ATSW maximization case (400 MW).

Table 6.1 Illustrative Example 6.1: production capacity investment equilibrium results

Objective function (6.18a)	max TP	max ATSW
Newly built conventional units	80 MW (node n_1)	No investment
Newly built stochastic units	200 MW (node n_2)	400 MW (node n_2)
Total newly built units [MW]	280	400
Total investment cost of the producers [$ million]	17.60	26.40
TP of the producers [$ million]	61.67	58.06
ATSW of the market [$ million]	79.27	84.46

2. As expected, the TP of producers g_1 and g_2 in the TP maximization case ($61.67 million) is comparatively higher than that in the ATSW maximization case ($58.06 million). However, the ATSW in the TP maximization case ($79.27 million) is comparatively lower than that in the ATSW maximization case ($84.46 million).
3. Since the stochastic units with zero offer prices lead to a higher ATSW, only those units are built in the ATSW maximization case.

Regarding the investment results for each producer in the TP maximization case, the new units can be built by each of the two producers. In other words, all new units (280 MW) may be built by producer g_1 or producer g_2. In addition, those units may be built by both producers, e.g., 80 MW conventional unit by producer g_1 and 200 MW stochastic unit by producer g_2. Therefore, several equilibrium points can be found in this case.

Regarding the investment results for each producer in the ATSW maximization case, the only possible equilibrium point is to invest in a 200-MW stochastic unit by each producer (i.e., 400 MW all together) since producers g_1 and g_2 cannot invest in such a unit with a capacity greater than 200 MW.

Regarding the LMPs obtained, as expected, at least one of the producers strategically offers at a price identical to the bid price of the demand. Therefore, the LMPs obtained in operation conditions o_1 and o_2 are $35 and $32/MWh, respectively. Note that in each condition, the LMPs at both nodes are the same since the transmission line is not congested.

6.6 Summary

This chapter provides a methodology to characterize generation investment equilibria in a pool-based network-constrained electricity market in which all producers behave strategically. To this end, the following steps are carried out:

Step (1) The investment problem of each strategic producer is represented using a bilevel model, whose upper-level problem determines the optimal production investment (capacity and location) and the offer prices to maximize its profit, and whose several lower-level problems represent the clearing of the market for different operating conditions.

Step (2) The single-producer bilevel models formulated in Step 1 are transformed into single-level MPECs by replacing the lower-level problems with their optimality conditions resulting from the primal–dual transformation. The resulting MPECs are continuous but nonlinear, due to the product of variables in the objective function and strong duality equalities.

Step (3) The joint consideration of all producer MPECs, one per producer, constitutes an EPEC, whose solution identifies the market equilibria.

Step (4) To identify EPEC solutions, the strong stationarity conditions associated with the EPEC, i.e., the strong stationarity conditions of all producer MPECs, are derived. To this end, each MPEC obtained in Step 2 is replaced by its KKT conditions. The set of resulting strong stationarity conditions of all MPECs, which are the strong stationarity conditions of the EPEC, is a collection of nonlinear systems of equalities and inequalities.

Step (5) The strong stationarity conditions associated with the EPEC obtained in Step 4 are linearized without approximation through three procedures: (i) linearizing the complementarity conditions, (ii) parameterizing the resulting conditions in the dual variables corresponding to the strong duality equalities, and (iii) replacing the strong duality equalities with their equivalent complementarity conditions. This linearization results in a mixed-integer and linear system of equalities and inequalities characterizing the EPEC.

Step (6) To explore multiple solutions, an auxiliary mixed-integer linear optimization problem is formulated, whose constraints are the mixed-integer linear conditions obtained in Step 5 and whose objective function is either a linear form of the total profit of all producers or a linear form of the annual true social welfare.

Step (7) The auxiliary mixed-integer linear optimization problem formulated in Step 6 is solved and a number of solutions are obtained.

Step (8) To detect Nash equilibria among the solutions achieved in Step 7, an ex-post algorithm based on a single-iteration diagonalization approach is provided. This algorithm checks whether each solution achieved in Step 7 is, in fact, a Nash equilibrium.

To validate numerically the methodology provided in this chapter, a two-node illustrative example with two strategic producers is examined and the equilibrium results obtained are reported and discussed.

6.7 End-of-Chapter Exercises

6.1 Reformulate the production capacity investment equilibrium model (6.1) to include several units per node and piecewise linear production costs.

6.2 Reformulate the production capacity investment equilibrium model (6.1) considering a multistage investment model instead of a static one.

6.3 Reformulate the production capacity investment equilibrium model (6.1) considering uncertainty in demand bid prices and investment costs of different technologies.

6.4 Solve Illustrative Example 6.1 presented in Sect. 6.4 considering the capacity of transmission line to be 200 MW (congested case) and then interpret the investment equilibrium results obtained.

6.5 Solve Illustrative Example 6.1 presented in Sect. 6.4 considering a single producer owning the entire production capacity portfolio (monopoly case) and then interpret the investment equilibrium results obtained.

6.6 Solve Illustrative Example 6.1 presented in Sect. 6.4 considering three strategic producers f_1, f_2, and f_3 (triopoly case), in which the capacity portfolio of each producer f_1 and f_3 is equal to half that of producer g_1 in the original example, while the capacity portfolio of producer f_2 is identical to that of producer g_2 is the original example. Then interpret the investment equilibrium results obtained.

6.7 Compare the investment equilibrium results obtained from the monopoly case (Exercise 6.5), the duopoly case (Illustrative Example 6.1 in Sect. 6.4), and the triopoly case (Exercise 6.6).

6.8 GAMS Code

This section provides the GAMS code for solving the MILP problem maximizing TP corresponding to Illustrative Example 6.1. Note that this code is written in a general form, and thus it is straightforward to adapt it to any investment equilibrium example.

```
SETS
o           operating conditions /o1*o2/
g           producers /g1*g2/
n           nodes /n1*n2/
s(n)        reference node /n1/
Omega(n,n)  transmission lines /n1.n2,n2.n1/
ALIAS (n,m)
ALIAS (g,y);

PARAMETERS
```

```
Kmax(g) available annualized investment budget of
   the producers /
g1   2e7
g2   2e7/

P_Dmax(n) maximum load of the consumers /
n1   400
n2     0/

X_Cmax(n) maximum capacity investment of the
   candidate conventional units /
n1   200/

X_Smax(n) maximum capacity investment of the
   candidate stochastic units /
n2   200/

rho(o)   weighting factor of operating conditions /
o1    3530
o2    5230/
;

TABLE  B(n,n) susceptance of the transmission lines
        n1      n2
n1    0       1e3
n2    1e3     0;

Table C_C(g,n) production cost of the candidate
   conventional units
        n1
g1    12
g2    12;

Table C_E(g,n) production cost of the existing
   conventional units
        n1     n2
g1    10     0
g2    0      15;

TABLE Fmax(n,n) capacity of the transmission lines
        n1      n2
n1    0       400
n2    400     0;

TABLE K_C(g,n) annualized investment cost of the
   candidate conventional units
        n1
g1    55000
g2    55000;

TABLE K_S(g,n) annualized investment cost of the
   candidate stochastic units
        n2
```

```
g1   66000
g2   66000;

TABLE P_Emax(g,n) capacity of the existing
    conventional units
        n1        n2
g1      150       0
g2      0         100;

TABLE Q_S(n,o) power capacity factor of the
    candidate stochastic units
      o1      o2
n2    0.35    0.70;

TABLE Q_D(n,o) demand factor of the consumers
      o1      o2
n1    1.0    0.8;

TABLE U_D(n,o) bid price of the consumers
      o1      o2
n1    35      32;

SCALAR BigM1  a large value /1e4/
SCALAR BigM2  a large value /1e6/
SCALAR BigM3  a large value /5e7/
SCALAR PI    pi  /3.1416/;

PARAMETERS
eta_parameterized(y,o);
eta_parameterized(y,o)=rho(o);

FREE VARIABLES
TP                      total profit of the producers
linear_term(o)          linear equivalent of the
    bilinear term
lambda(n,o)             locational marginal prices (
    LMPs)
theta(n,o)              nodal voltage angles
*dual variable associated with the lower-level
    problems
mu_theta_ref(o)
*dual variables associated with the MPEC of
    producer y
beta(y,n,o)
rho_C(y,g,n,o)
rho_S(y,g,n,o)
rho_E(y,g,n,o)
rho_D(y,n,o)
rho_theta(y,n,o)
eta_theta_ref(y,o);

POSITIVE VARIABLES
*primal variables
```

```
p_C(g,n,o)              power produced by the candidate conventional units
p_D(n,o)                power consumed by the consumers
p_E(g,n,o)              power produced by the existing conventional units
p_S(g,n,o)              power produced by the candidate stochastic units
x_C(g,n)                capacity of the candidate conventional units
x_S(g,n)                capacity of the candidate stochastic units
alpha_C(g,n,o)          offer price by the candidate conventional units
alpha_E(g,n,o)          offer price by the existing conventional units
*dual variables associated with the lower-level problems
mu_Cmin(g,n,o)
mu_Cmax(g,n,o)
mu_Dmin(n,o)
mu_Dmax(n,o)
mu_Emin(g,n,o)
mu_Emax(g,n,o)
mu_Smin(g,n,o)
mu_Smax(g,n,o)
mu_F(n,m,o)
mu_theta_min(n,o)
mu_theta_max(n,o)
*dual variables associated with the MPEC of producer y
eta_Cmax(y,g,n,o)
eta_Cmin(y,g,n,o)
eta_Emax(y,g,n,o)
eta_Emin(y,g,n,o)
eta_Smax(y,g,n,o)
eta_Smin(y,g,n,o)
eta_Dmax(y,n,o)
eta_Dmin(y,n,o)
eta_x_C_max(y,g,n)
eta_x_C_min(y,g,n)
eta_x_S_max(y,g,n)
eta_x_S_min(y,g,n)
eta_budget(y,g)
eta_alpha_C(y,g,n,o)
eta_alpha_E(y,g,n,o)
eta_F(y,n,m,o)
eta_theta_max(y,n,o)
eta_theta_min(y,n,o)
gamma_Cmin(y,g,n,o)
gamma_Cmax(y,g,n,o)
gamma_Dmin(y,n,o)
gamma_Dmax(y,n,o)
```

```
gamma_Emin(y,g,n,o)
gamma_Emax(y,g,n,o)
gamma_Smin(y,g,n,o)
gamma_Smax(y,g,n,o)
gamma_F(y,n,m,o)
gamma_theta_min(y,n,o)
gamma_theta_max(y,n,o);

BINARY VARIABLES
u_mu_Smin(g,n,o)
u_mu_Smax(g,n,o)
u_mu_Cmin(g,n,o)
u_mu_Cmax(g,n,o)
u_mu_Emin(g,n,o)
u_mu_Emax(g,n,o)
u_mu_Dmin(n,o)
u_mu_Dmax(n,o)
u_mu_F(n,m,o)
u_mu_theta_min(n,o)
u_mu_theta_max(n,o)
u_x_C_min(y,g,n)
u_x_C_max(y,g,n)
u_x_S_min(y,g,n)
u_x_S_max(y,g,n)
u_budget(y,g)
u_alpha_C(y,g,n,o)
u_alpha_E(y,g,n,o)
u_C_min(y,g,n,o)
u_C_max(y,g,n,o)
u_S_min(y,g,n,o)
u_S_max(y,g,n,o)
u_E_min(y,g,n,o)
u_E_max(y,g,n,o)
u_D_min(y,n,o)
u_D_max(y,n,o)
u_F(y,n,m,o)
u_theta_min(y,n,o)
u_theta_max(y,n,o)
u1(y,g,n,o)
u2(y,g,n,o)
u3(y,g,n,o)
u4(y,g,n,o)
u5(y,g,n,o)
u6(y,g,n,o)
u7(y,n,o)
u8(y,n,o)
u9(y,n,m,o)
u10(y,n,o)
u11(y,n,o);

x_S.UP(g,n)=X_Smax(n);
x_C.UP(g,n)=X_Cmax(n);
p_E.UP(g,n,o)=P_Emax(g,n);
```

```
p_D.UP(n,o)=P_Dmax(n)*Q_D(n,o);
theta.LO(n,o)=-PI;
theta.UP(n,o)=PI;
theta.FX('n1',o)=0;
EQUATIONS
OF , EQ1, EQ2, EQ3, EQ4, EQ5, EQ6, EQ7, EQ8, EQ9, EQ10, EQ11, EQ12, EQ13, EQ14, EQ15, EQ16, EQ17, EQ18, EQ19, EQ20,
EQ21, EQ22, EQ23, EQ24, EQ25, EQ26, EQ27, EQ28, EQ29, EQ30, EQ31, EQ32, EQ33, EQ34, EQ35, EQ36, EQ37, EQ38, EQ39, EQ40,
EQ41, EQ42, EQ43, EQ44, EQ45, EQ46, EQ47, EQ48, EQ49, EQ50, EQ51, EQ52, EQ53, EQ54, EQ55, EQ56, EQ57, EQ58, EQ59, EQ60,
EQ61, EQ62, EQ63, EQ64, EQ65, EQ66, EQ67, EQ68, EQ69, EQ70, EQ71, EQ72, EQ73, EQ74, EQ75, EQ76, EQ77, EQ78, EQ79, EQ80,
EQ81, EQ82, EQ83, EQ84, EQ85, EQ86, EQ87, EQ88, EQ89, EQ90, EQ91, EQ92, EQ93, EQ94, EQ95, EQ96, EQ97, EQ98, EQ99, EQ100,
EQ101, EQ102, EQ103, EQ104, EQ105, EQ106, EQ107, EQ108, EQ109, EQ110, EQ111, EQ112, EQ113, EQ114, EQ115, EQ116, EQ117,
EQ118, EQ119, EQ120;


OF..TP=E=SUM(o,rho(o)*linear_term(o))-SUM((n,g),x_C(g,n)*K_C(g,n))-SUM((n,g),x_S(g,n)*K_S(g,n))-SUM((g,n,o),rho(o)*p_C(g,n,o)*C_C(g,n))-SUM((g,n,o),rho(o)*p_E(g,n,o)*C_E(g,n));

EQ1(o)..linear_term(o)=E=SUM(n,U_D(n,o)*p_D(n,o))-SUM(n,mu_Dmax(n,o)*P_Dmax(n)*Q_D(n,o))-SUM((n,m)$Omega(n,m),Fmax(n,m)*mu_F(n,m,o))-SUM(n,PI*[mu_theta_max(n,o)+mu_theta_min(n,o)]);

EQ2(g)..SUM(n,x_C(g,n)*K_C(g,n))+SUM(n,x_S(g,n)*K_S(g,n))=L=Kmax(g);

EQ3(n,o)..p_D(n,o)+SUM(m$Omega(n,m),B(n,m)*[theta(n,o)-theta(m,o)])-SUM(g,p_C(g,n,o))-SUM(g,p_E(g,n,o))-SUM(g,p_S(g,n,o))=E=0;

EQ4(g,n,o)..p_C(g,n,o)=L=x_C(g,n);

EQ5(g,n,o)..p_S(g,n,o)=L=Q_S(n,o)*x_S(g,n);

EQ6(n,m,o)$Omega(n,m)..B(n,m)*[theta(n,o)-theta(m,o)]=L=Fmax(n,m);

EQ7(g,n,o)..alpha_C(g,n,o)-lambda(n,o)+mu_Cmax(g,n,o)-mu_Cmin(g,n,o)=E=0;
```

```
EQ8(g,n,o)..alpha_E(g,n,o)-lambda(n,o)+mu_Emax(g,n,
    o)-mu_Emin(g,n,o)=E=0;

EQ9(g,n,o)..-lambda(n,o)+mu_Smax(g,n,o)-mu_Smin(g,n
    ,o)=E=0;

EQ10(n,o)..-U_D(n,o)+lambda(n,o)+mu_Dmax(n,o)-
    mu_Dmin(n,o)=E=0;

EQ11(n,o).. SUM(m$Omega(n,m),B(n,m)*[lambda(n,o)-
    lambda(m,o)])+SUM(m$Omega(n,m),B(n,m)*[mu_F(n,m
    ,o)-mu_F(m,n,o)])+mu_theta_max(n,o)-
    mu_theta_min(n,o)+mu_theta_ref(o)$s(n)=E=0;

EQ12(g,n,o)..p_C(g,n,o)=L=u_mu_Cmin(g,n,o)*BigM1;
EQ13(g,n,o)..mu_Cmin(g,n,o)=L=[1-u_mu_Cmin(g,n,o)]*
    BigM2;

EQ14(g,n,o)..[x_C(g,n)-p_C(g,n,o)]=L=u_mu_Cmax(g,n,
    o)*BigM1;
EQ15(g,n,o)..mu_Cmax(g,n,o)=L=[1-u_mu_Cmax(g,n,o)]*
    BigM2;

EQ16(g,n,o)..p_S(g,n,o)=L=u_mu_Smin(g,n,o)*BigM1;
EQ17(g,n,o)..mu_Smin(g,n,o)=L=[1-u_mu_Smin(g,n,o)]*
    BigM2;

EQ18(g,n,o)..[(Q_S(n,o)*x_S(g,n))-p_S(g,n,o)]=L=
    u_mu_Smax(g,n,o)*BigM1;
EQ19(g,n,o)..mu_Smax(g,n,o)=L=[1-u_mu_Smax(g,n,o)]*
    BigM2;

EQ20(g,n,o)..p_E(g,n,o)=L=u_mu_Emin(g,n,o)*BigM1;
EQ21(g,n,o)..mu_Emin(g,n,o)=L=[1-u_mu_Emin(g,n,o)]*
    BigM2;

EQ22(g,n,o)..[P_Emax(g,n)-p_E(g,n,o)]=L=u_mu_Emax(g
    ,n,o)*BigM1;
EQ23(g,n,o)..mu_Emax(g,n,o)=L=[1-u_mu_Emax(g,n,o)]*
    BigM2;

EQ24(n,o)..p_D(n,o)=L=u_mu_Dmin(n,o)*BigM1;
EQ25(n,o)..mu_Dmin(n,o)=L=[1-u_mu_Dmin(n,o)]*BigM2;

EQ26(n,o)..[(Q_D(n,o)*P_Dmax(n))-p_D(n,o)]=L=
    u_mu_Dmax(n,o)*BigM1;
EQ27(n,o)..mu_Dmax(n,o)=L=[1-u_mu_Dmax(n,o)]*BigM2;

EQ28(n,m,o)$Omega(n,m)..Fmax(n,m)-[B(n,m)*(theta(n,
    o)-theta(m,o))]=L=u_mu_F(n,m,o)*BigM1;
EQ29(n,m,o)$Omega(n,m)..mu_F(n,m,o)=L=[1-u_mu_F(n,m
    ,o)]*BigM2;
```

```
EQ30(n,o)..PI+theta(n,o)=L=[1-u_mu_theta_min(n,o)]*BigM1;
EQ31(n,o)..mu_theta_min(n,o)=L=u_mu_theta_min(n,o)*BigM2;

EQ32(n,o)..PI-theta(n,o)=L=[1-u_mu_theta_max(n,o)]*BigM1;
EQ33(n,o)..mu_theta_max(n,o)=L=u_mu_theta_max(n,o)*BigM2;

EQ34(y,g,n,o)$[ORD(g) EQ ORD(y)]..-[rho(o)*(lambda(n,o)-C_C(g,n))]-beta(y,n,o)+eta_Cmax(y,g,n,o)-eta_Cmin(y,g,n,o)+[eta_parameterized(y,o)*alpha_C(g,n,o)]=E=0;
EQ35(y,g,n,o)$[ORD(g) NE ORD(y)]..-beta(y,n,o)+eta_Cmax(y,g,n,o)-eta_Cmin(y,g,n,o)+[eta_parameterized(y,o)*alpha_C(g,n,o)]=E=0;

EQ36(y,g,n,o)$[ORD(g) EQ ORD(y)]..-[rho(o)*(lambda(n,o)-C_E(g,n))]-beta(y,n,o)+eta_Emax(y,g,n,o)-eta_Emin(y,g,n,o)+[eta_parameterized(y,o)*alpha_E(g,n,o)]=E=0;
EQ37(y,g,n,o)$[ORD(g) NE ORD(y)]..-beta(y,n,o)+eta_Emax(y,g,n,o)-eta_Emin(y,g,n,o)+[eta_parameterized(y,o)*alpha_E(g,n,o)]=E=0;

EQ38(y,g,n,o)$[ORD(g) EQ ORD(y)]..-[rho(o)*lambda(n,o)]-beta(y,n,o)+eta_Smax(y,g,n,o)-eta_Smin(y,g,n,o)=E=0;
EQ39(y,g,n,o)$[ORD(g) NE ORD(y)]..-beta(y,n,o)+eta_Smax(y,g,n,o)-eta_Smin(y,g,n,o)=E=0;

EQ40(y,n,o)..beta(y,n,o)+eta_Dmax(y,n,o)-eta_Dmin(y,n,o)-[eta_parameterized(y,o)*U_D(n,o)]=E=0;

EQ41(y,g,n)$[ORD(g) EQ ORD(y)]..K_C(g,n)+eta_x_C_max(y,g,n)-eta_x_C_min(y,g,n)+[K_C(g,n)*eta_budget(y,g)]-SUM(o,eta_Cmax(y,g,n,o))+SUM(o,eta_parameterized(y,o)*mu_Cmax(g,n,o))=E=0;
EQ42(y,g,n)$[ORD(g) NE ORD(y)]..-SUM(o,eta_Cmax(y,g,n,o))+SUM(o,eta_parameterized(y,o)*mu_Cmax(g,n,o))=E=0;

EQ43(y,g,n)$[ORD(g) EQ ORD(y)]..K_S(g,n)+eta_x_S_max(y,g,n)-eta_x_S_min(y,g,n)+[K_S(g,n)*eta_budget(y,g)]-SUM(o,Q_S(n,o)*eta_Smax(y,g,n,o))+SUM(o,Q_S(n,o)*eta_parameterized(y,o)*mu_Smax(g,n,o))=E=0;
EQ44(y,g,n)$[ORD(g) NE ORD(y)]..-SUM(o,Q_S(n,o)*eta_Smax(y,g,n,o))+SUM(o,Q_S(n,o)*eta_parameterized(y,o)*mu_Smax(g,n,o))=E=0;
```

```
EQ45(y,g,n,o)$[ORD(g) EQ ORD(y)]..-eta_alpha_C(y,g,n,o)+rho_C(y,g,n,o)+[eta_parameterized(y,o)*p_C(g,n,o)]=E=0;
EQ46(y,g,n,o)$[ORD(g) NE ORD(y)]..rho_C(y,g,n,o)+[eta_parameterized(y,o)*p_C(g,n,o)]=E=0;

EQ47(y,g,n,o)$[ORD(g) EQ ORD(y)]..-eta_alpha_E(y,g,n,o)+rho_E(y,g,n,o)+[eta_parameterized(y,o)*p_E(g,n,o)]=E=0;
EQ48(y,g,n,o)$[ORD(g) NE ORD(y)]..rho_E(y,g,n,o)+[eta_parameterized(y,o)*p_E(g,n,o)]=E=0;

EQ49(y,n,o)..SUM(m$Omega(n,m),B(n,m)*[beta(y,n,o)-beta(y,m,o)])+SUM(m$Omega(n,m),B(n,m)*[eta_F(y,n,m,o)-eta_F(y,m,n,o)])+eta_theta_max(y,n,o)-eta_theta_min(y,n,o)+eta_theta_ref(y,o)$s(n)=E=0;

EQ50(y,n,o)..-[rho(o)*(p_C(y,n,o)+p_S(y,n,o)+p_E(y,n,o))]-SUM(g,rho_C(y,g,n,o))-SUM(g,rho_E(y,g,n,o))-SUM(g,rho_S(y,g,n,o))+rho_D(y,n,o)+SUM(m$Omega(n,m),B(n,m)*[rho_theta(y,n,o)-rho_theta(y,m,o)])=E=0;

EQ51(y,g,n,o)..-rho_C(y,g,n,o)-gamma_Cmin(y,g,n,o)=E=0;
EQ52(y,g,n,o)..+rho_C(y,g,n,o)-gamma_Cmax(y,g,n,o)+[eta_parameterized(y,o)*x_C(g,n)]=E=0;

EQ53(y,g,n,o)..-rho_S(y,g,n,o)-gamma_Smin(y,g,n,o)=E=0;
EQ54(y,g,n,o)..+rho_S(y,g,n,o)-gamma_Smax(y,g,n,o)+[eta_parameterized(y,o)*Q_S(n,o)*x_S(g,n)]=E=0;

EQ55(y,g,n,o)..-rho_E(y,g,n,o)-gamma_Emin(y,g,n,o)=E=0;
EQ56(y,g,n,o)..+rho_E(y,g,n,o)-gamma_Emax(y,g,n,o)+[eta_parameterized(y,o)*P_Emax(g,n)]=E=0;

EQ57(y,n,o)..-rho_D(y,n,o)-gamma_Dmin(y,n,o)=E=0;
EQ58(y,n,o)..+rho_D(y,n,o)-gamma_Dmax(y,n,o)+[eta_parameterized(y,o)*P_Dmax(n)*Q_D(n,o)]=E=0;

EQ59(y,n,m,o)$Omega(n,m)..[B(n,m)*(rho_theta(y,n,o)-rho_theta(y,m,o))]-gamma_F(y,n,m,o)+[eta_parameterized(y,o)*Fmax(n,m)]=E=0;

EQ60(y,n,o)..-rho_theta(y,n,o)-gamma_theta_min(y,n,o)+[eta_parameterized(y,o)*PI]=E=0;
EQ61(y,n,o)..+rho_theta(y,n,o)-gamma_theta_max(y,n,o)+[eta_parameterized(y,o)*PI]=E=0;
EQ62(y,n,o)..rho_theta(y,'n1',o)=E=0;
```

```
EQ63(y,g,n)$[ORD(g) EQ ORD(y)]..x_C(g,n)=L=[1-u_x_C_min(y,g,n)]*BigM1;
EQ64(y,g,n)$[ORD(g) EQ ORD(y)]..eta_x_C_min(y,g,n)=L=u_x_C_min(y,g,n)*BigM2;

EQ65(y,g,n)$[ORD(g) EQ ORD(y)]..[X_Cmax(n)-x_C(g,n)]=L=[1-u_x_C_max(y,g,n)]*BigM1;
EQ66(y,g,n)$[ORD(g) EQ ORD(y)]..eta_x_C_max(y,g,n)=L=u_x_C_max(y,g,n)*BigM2;

EQ67(y,g,n)$[ORD(g) EQ ORD(y)]..x_S(g,n)=L=[1-u_x_S_min(y,g,n)]*BigM1;
EQ68(y,g,n)$[ORD(g) EQ ORD(y)]..eta_x_S_min(y,g,n)=L=u_x_S_min(y,g,n)*BigM2;

EQ69(y,g,n)$[ORD(g) EQ ORD(y)]..[X_Smax(n)-x_S(g,n)]=L=[1-u_x_S_max(y,g,n)]*BigM1;
EQ70(y,g,n)$[ORD(g) EQ ORD(y)]..eta_x_S_max(y,g,n)=L=u_x_S_max(y,g,n)*BigM2;

EQ71(g,y)$[ORD(g) EQ ORD(y)]..Kmax(g)-SUM(n,x_C(g,n)*K_C(g,n))-SUM(n,x_S(g,n)*K_S(g,n))=L=[1-u_budget(g,y)]*BigM3;
EQ72(g,y)$[ORD(g) EQ ORD(y)]..eta_budget(y,g)=L=u_budget(g,y)*BigM3;

EQ73(y,g,n,o)$[ORD(g) EQ ORD(y)]..alpha_C(g,n,o)=L=[1-u_alpha_C(y,g,n,o)]*BigM1;
EQ74(y,g,n,o)$[ORD(g) EQ ORD(y)]..eta_alpha_C(y,g,n,o)=L=u_alpha_C(y,g,n,o)*BigM2;

EQ75(y,g,n,o)$[ORD(g) EQ ORD(y)]..alpha_E(g,n,o)=L=[1-u_alpha_E(y,g,n,o)]*BigM1;
EQ76(y,g,n,o)$[ORD(g) EQ ORD(y)]..eta_alpha_E(y,g,n,o)=L=u_alpha_E(y,g,n,o)*BigM2;

EQ77(y,g,n,o)..p_C(g,n,o)=L=(1-u_C_min(y,g,n,o))*BigM1;
EQ78(y,g,n,o)..eta_Cmin(y,g,n,o)=L=u_C_min(y,g,n,o)*BigM2;

EQ79(y,g,n,o)..[x_C(g,n)-p_C(g,n,o)]=L=(1-u_C_max(y,g,n,o))*BigM1;
EQ80(y,g,n,o)..eta_Cmax(y,g,n,o)=L=u_C_max(y,g,n,o)*BigM2;

EQ81(y,g,n,o)..p_S(g,n,o)=L=(1-u_S_min(y,g,n,o))*BigM1;
EQ82(y,g,n,o)..eta_Smin(y,g,n,o)=L=u_S_min(y,g,n,o)*BigM2;

EQ83(y,g,n,o)..[(Q_S(n,o)*x_S(g,n))-p_S(g,n,o)]=L=(1-u_S_max(y,g,n,o))*BigM1;
```

```
EQ84(y,g,n,o)..eta_Smax(y,g,n,o)=L=u_S_max(y,g,n,o)
    *BigM2;

EQ85(y,g,n,o)..p_E(g,n,o)=L=(1-u_E_min(y,g,n,o))*
    BigM1;
EQ86(y,g,n,o)..eta_Emin(y,g,n,o)=L=u_E_min(y,g,n,o)
    *BigM2;

EQ87(y,g,n,o)..[P_Emax(g,n)-p_E(g,n,o)]=L=(1-
    u_E_max(y,g,n,o))*BigM1;
EQ88(y,g,n,o)..eta_Emax(y,g,n,o)=L=u_E_max(y,g,n,o)
    *BigM2;

EQ89(y,n,o)..p_D(n,o)=L=(1-u_D_min(y,n,o))*BigM1;
EQ90(y,n,o)..eta_Dmin(y,n,o)=L=u_D_min(y,n,o)*BigM2
    ;

EQ91(y,n,o)..[(P_Dmax(n)*Q_D(n,o))-p_D(n,o)]=L=(1-
    u_D_max(y,n,o))*BigM1;
EQ92(y,n,o)..eta_Dmax(y,n,o)=L=u_D_max(y,n,o)*BigM2
    ;

EQ93(y,n,m,o)$Omega(n,m)..Fmax(n,m)-[B(n,m)*[theta(
    n,o)-theta(m,o)]]=L=[1-u_F(y,n,m,o)]*BigM1;
EQ94(y,n,m,o)$Omega(n,m)..eta_F(y,n,m,o)=L=u_F(y,n,
    m,o)*BigM1;

EQ95(y,n,o)..[PI+theta(n,o)]=L=[1-u_theta_min(y,n,o
    )]*BigM1;
EQ96(y,n,o)..eta_theta_min(y,n,o)=L=u_theta_min(y,n
    ,o)*BigM2;

EQ97(y,n,o)..[PI-theta(n,o)]=L=[1-u_theta_max(y,n,o
    )]*BigM1;
EQ98(y,n,o)..eta_theta_max(y,n,o)=L=u_theta_max(y,n
    ,o)*BigM2;

EQ99(y,g,n,o)..mu_Cmin(g,n,o)=L=[1-u1(y,g,n,o)]*
    BigM3;
EQ100(y,g,n,o)..gamma_Cmin(y,g,n,o)=L=u1(y,g,n,o)*
    BigM3;

EQ101(y,g,n,o)..mu_Cmax(g,n,o)=L=[1-u2(y,g,n,o)]*
    BigM3;
EQ102(y,g,n,o)..gamma_Cmax(y,g,n,o)=L=u2(y,g,n,o)*
    BigM3;

EQ103(y,g,n,o)..mu_Smin(g,n,o)=L=[1-u3(y,g,n,o)]*
    BigM3;
EQ104(y,g,n,o)..gamma_Smin(y,g,n,o)=L=u3(y,g,n,o)*
    BigM3;
```

```
EQ105(y,g,n,o)..mu_Smax(g,n,o)=L=[1-u4(y,g,n,o)]*
    BigM3;
EQ106(y,g,n,o)..gamma_Smax(y,g,n,o)=L=u4(y,g,n,o)*
    BigM3;

EQ107(y,g,n,o)..mu_Emin(g,n,o)=L=[1-u5(y,g,n,o)]*
    BigM3;
EQ108(y,g,n,o)..gamma_Emin(y,g,n,o)=L=u5(y,g,n,o)*
    BigM3;

EQ109(y,g,n,o)..mu_Emax(g,n,o)=L=[1-u6(y,g,n,o)]*
    BigM3;
EQ110(y,g,n,o)..gamma_Emax(y,g,n,o)=L=u6(y,g,n,o)*
    BigM3;

EQ111(y,n,o)..mu_Dmin(n,o)=L=[1-u7(y,n,o)]*BigM3;
EQ112(y,n,o)..gamma_Dmin(y,n,o)=L=u7(y,n,o)*BigM3;

EQ113(y,n,o)..mu_Dmax(n,o)=L=[1-u8(y,n,o)]*BigM3;
EQ114(y,n,o)..gamma_Dmax(y,n,o)=L=u8(y,n,o)*BigM3;

EQ115(y,n,m,o)$Omega(n,m)..mu_F(n,m,o)=L=[1-u9(y,n,
    m,o)]*BigM3;
EQ116(y,n,m,o)$Omega(n,m)..gamma_F(y,n,m,o)=L=u9(y,
    n,m,o)*BigM3;

EQ117(y,n,o)..mu_theta_min(n,o)=L=[1-u10(y,n,o)]*
    BigM3;
EQ118(y,n,o)..gamma_theta_min(y,n,o)=L=u10(y,n,o)*
    BigM3;

EQ119(y,n,o)..mu_theta_max(n,o)=L=[1-u11(y,n,o)]*
    BigM3;
EQ120(y,n,o)..gamma_theta_max(y,n,o)=L=u11(y,n,o)*
    BigM3;

MODEL EQUILIBRIA /ALL/;
OPTION OPTCR=0;
SOLVE EQUILIBRIA USING MIP MAXIMIZING TP;
```

References

1. Baldick, R., Hogan, W.W.: Stability of supply function equilibria implications for daily versus hourly bids in a poolco market. J. Regul. Econ. **30**(2), 119–139 (2006)
2. Debreu, G.: A social equilibrium existence theorem. Proc. Natl. Acad. Sci. U. S. A. **38**(10), 886–893 (1952)
3. De Wolf, D., Smeers, Y.: A stochastic version of a Stackelberg-Nash-Cournot equilibrium model. Manag. Sci. **43**(2), 190–197 (1997)
4. Facchinei, F., Kanzow, C.: Generalized Nash equilibrium problems. 4OR - A Q. J. Oper. Res. **5**(3), 173–173 (2007)

5. Fletcher, R., Leyffer, S., Ralph, D., Scholtes, S.: Local convergence of SQP methods for mathematical programs with equilibrium constraints. SIAM J. Optim. **17**(1), 259–286 (2006)
6. Fortuny-Amat, J., McCarl, B.: A representation and economic interpretation of a two-level programming problem. J. Oper. Res. Soc. **32**(9), 783–792 (1981)
7. Gabriel, S., Conejo, A.J., Hobbs, B.F., Fuller, D., Ruiz, C.: Complementarity Modeling in Energy Markets. Springer, New York (2012)
8. Garcia, A., Shen, Z.: Equilibrium capacity expansion under stochastic demand growth. Oper. Res. **58**(1), 30–42 (2010)
9. Hu, X., Ralph, D.: Using EPECs to model bilevel games in restructured electricity markets with locational prices. Oper. Res. **55**(5), 809–827 (2007)
10. Kazempour, S.J., Conejo, A.J., Ruiz, C.: Generation investment equilibria with strategic producers - Part I: formulation. IEEE Trans. Power Syst. **28**(3), 2613–2622 (2013)
11. Kazempour, S.J., Conejo, A.J., Ruiz, C.: Generation investment equilibria with strategic producers - Part II: case studies. IEEE Trans. Power Syst. **28**(3), 2623–2631 (2013)
12. Leyffer, S., Munson, T.: Solving multi-leader-common-follower games. Optim. Methods Softw. **25**(4), 601–623 (2010)
13. Martin, S., Smeers, Y., Aguado, J.A.: A stochastic two settlement equilibrium model for electricity markets with wind generation. IEEE Trans. Power Syst. **30**(1), 233–245 (2015)
14. Metzler, C., Hobbs, B.F., Pang, J.-S.: Nash-Cournot equilibria in power markets on a linearized DC network with arbitrage: formulations and properties. Netw. Spat. Econ. **3**(2), 123–150 (2003)
15. Murphy, F.H., Smeers, Y.: On the impact of forward markets on investment in oligopolistic markets wtih reference to electricity. Oper. Res. **58**(3), 515–528 (2010)
16. Nash, J.: Equilibrium points in n-person games. Proc. Natl. Acad. Sci. U. S. A. **36**(1), 48–49 (1950)
17. Nash, J.: Non-cooperative games. Ann. Math. **54**(2), 286–295 (1951)
18. Pandzic, H., Conejo, A.J., Kuzle, I.: An EPEC approach to the yearly maintenance scheduling of generating units. IEEE Trans. Power Syst. **28**(2), 922–930 (2013)
19. Ruiz, C., Conejo, A.J., Smeers, Y.: Equilibria in an oligopolistic electricity pool with stepwise offer curves. IEEE Trans. Power Syst. **27**(2), 752–761 (2012)
20. Scheel, H., Scholtes, S.: Mathematical programs with complementarity constraints: stationarity, optimality, and sensitivity. Math. Oper. Res. **25**(1), 1–22 (2000)
21. Shanbhag, U.V., Infanger, G., Glynn, P.W.: A complementarity framework for forward contracting under uncertainty. Oper. Res. **59**(4), 810–834 (2011)
22. Sonja, S., Micha, B.: Relaxation approach for equilibrium problems with equilibrium constraints. Comput. Oper. Res. **41**, 333–345 (2012)
23. Wogrin, S., Barquín, J., Centeno, E.: Capacity expansion equilibria in liberalized electricity markets: an EPEC approach. IEEE Trans. Power Syst. **28**(2), 1531–1539 (2013)
24. Yao, J., Adler, I., Oren, S.S.: Modeling and computing two-settlement oligopolistic equilibrium in a congested electricity network. Oper. Res. **56**(1), 34–47 (2008)
25. Zhang, X.-P.: Restructured Electric Power Systems: Analysis of Electricity Markets With Equilibrium Models. ser. Power Engineering. Wiley-IEEE Press, Hoboken, NJ (2010)

Chapter 7
Deciding on Alternative Investments: A Real Options Approach

Investment decisions in power plants and other assets are typically made under evolving uncertainties. Power companies often have managerial discretion over the timing of the investment as well as flexibility regarding the type of technology. By abstracting from some real-world details, the real options approach provides an elegant mathematical framework in which to assess the value of such flexibilities to provide both managerial and policy insights. In this chapter, we introduce the real options approach and contrast it with the now-or-never net present value perspective. Besides dealing with the issue of optimal timing, the real options approach also enables a power company to value operational flexibility, e.g., in the form of faster ramping, as compound options. Other flexibilities, such as modularized investment and endogenous capacity choice, are also amenable to analysis via this approach. Finally, the impact of risk aversion is explored, and the chapter concludes with extensions and exercises for further analysis.

7.1 Assumptions and the Need for Dynamic Programming

In previous chapters, we observed that it may be beneficial to delay investment in new technologies when there is uncertainty concerning prices or performance. For example, consider a small power company that may invest in a new power plant from which it will earn revenues by selling the generated electricity at the prevailing spot price of power and incur costs associated with fuel purchases.[1] At the time of investment, the company must pay a one-time capital cost to cover the expenses associated with purchasing the equipment and installing it. Subsequently, there may be operating and maintenance (O&M) expenses not related directly to fuel costs. The basic question in engineering economics is, "Is it profitable to proceed with the

[1]Of course, there may be other streams of revenue, e.g., from feed-in tariffs (FITs) or renewable energy certificates (RECs), and the possibility to sell the power in various types of markets, e.g., futures, day-ahead, and balancing. We neglect these possibilities for the sake of exposition.

A.J. Conejo et al., *Investment in Electricity Generation and Transmission*,
DOI 10.1007/978-3-319-29501-5_7

investment now?" For this purpose, it is straightforward to calculate the expected now-or-never net present value (NPV) of investment and to determine whether it warrants immediate investment.

However, if the power company has exclusive rights to invest at a particular location, e.g., because of licensing agreements, then it also has the discretion to consider investing at a later date. Indeed, given the trajectory of electricity and fuel prices, it may be beneficial to delay the adoption decision by a year. In doing so, the power company must trade off the following three aspects in determining the correct timing:

1. The marginal benefit from postponing the investment cost. Rather than paying the full investment cost now, delaying the project's start by a year would mean incurring the discounted investment cost.
2. The marginal benefit from starting the project with higher electricity prices or lower fuel prices. It may be profitable to invest immediately, but the trajectory of prices may be such that it is beneficial to delay adoption.
3. The marginal cost from forgone cash flows in the waiting period. In effect, the cash flows that the power company could have been earning are an opportunity cost that must be figured into its decision.

In general, if the power company has the discretion to defer investment perpetually, then what should be the optimal time to invest? For example, in Fig. 7.1, the NPV of a hypothetical power plant is given as a function of the current electricity price. In addition to being able to invest immediately, the power company may undertake the same project in five or ten years' time. Depending on the current electricity price and its growth rate, it may be optimal to invest immediately, wait five years, wait ten years, or never invest. Thus, the NPV of the overall investment opportunity is the upper envelope of the three NPV functions as well as zero. Furthermore, if underlying prices are uncertain, then how would the optimal investment decision be affected? How would characteristics of alternative technologies, e.g., operational flexibility or sizing, affect the investment decision? Would a modular investment strategy make sense in certain situations? Finally, how would the decisions change if the power company were risk averse? Tackling all of these features within an elegant mathematical framework would be desirable in order to elicit managerial insights.

While the timing question may be addressed adequately via the now-or-never NPV approach, the analysis becomes cumbersome and leads us to propose a more suitable framework for decision making: real options. Essentially, real options is a dynamic programming approach to making optimal decisions in which investment and operational opportunities are thought of as options on real (rather than financial) assets. In finance, an "option" refers to an instrument that provides the holder with the right, but not the obligation, to obtain an asset in exchange for a so-called strike price [20]. Given suitable simplifying assumptions, real options can provide powerful insights into the value of managerial flexibility in appraising alternative investment proposals. For example, the now-or-never NPV approach would not be able to distinguish between investing in a single 100 MW power plant and investing in two 50 MW modules constructed sequentially when there is uncertainty about

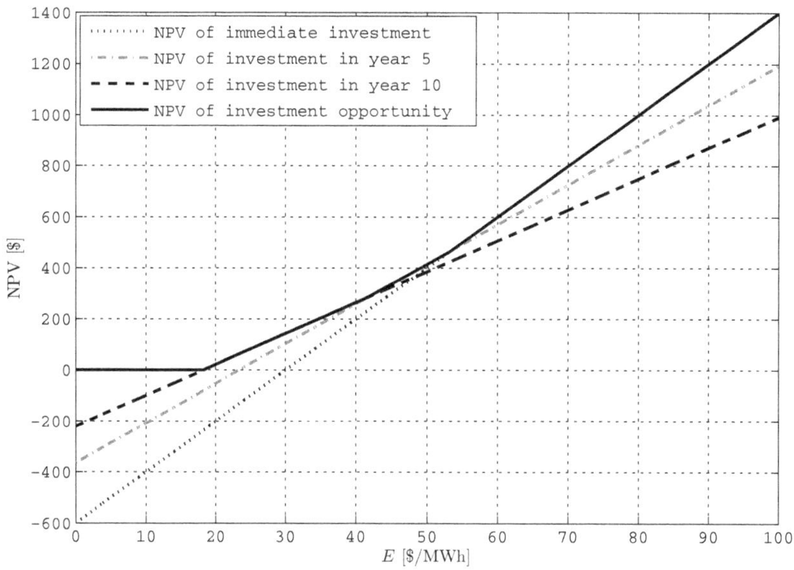

Fig. 7.1 NPV of power plant in different starting years

the electricity price. Yet intuitively, most managers would attribute more value to the modular approach, even though the now-or-never NPV approach would give the same expected value. Hence in this chapter, we abstract from some of the details of the previous chapters in order develop the intuition and methodology for applying real options analysis to investment and operational problems in the electricity industry.

Before proceeding to the exposition of the real options approach, we first state the assumptions and define the notation to be used for the rest of this chapter. Without loss of generality, we assume that the electricity price at time $t \geq 0$, E_t, follows a geometric Brownian motion (GBM). First, a Brownian motion (BM) may intuitively be thought of as a continuous-time analogue of a random walk with drift [35]. In other words, the absolute changes in the value of a random parameter following a BM are normally distributed. Second, in dealing with prices, it is convenient to limit the range of realizations to be nonnegative.[2] Thus, rather than considering absolute changes, it is expedient to deal with percentage changes, and a GBM is a stochastic process in which the percentage changes (rather than absolute changes) in the value are normally distributed. Consequently, if E_t follows a GBM, then:

[2]Because of nonconvexities in power plant operations such as startup costs and minimum uptimes, electricity prices may actually become negative during certain hours. For example, it may be cheaper for a power plant with high startup costs to remain online even during off-peak periods when demand is low. Thus, the power plant effectively pays to continue generation [22]. Nevertheless, we assume in this chapter for the sake of clarity that prices are nonnegative.

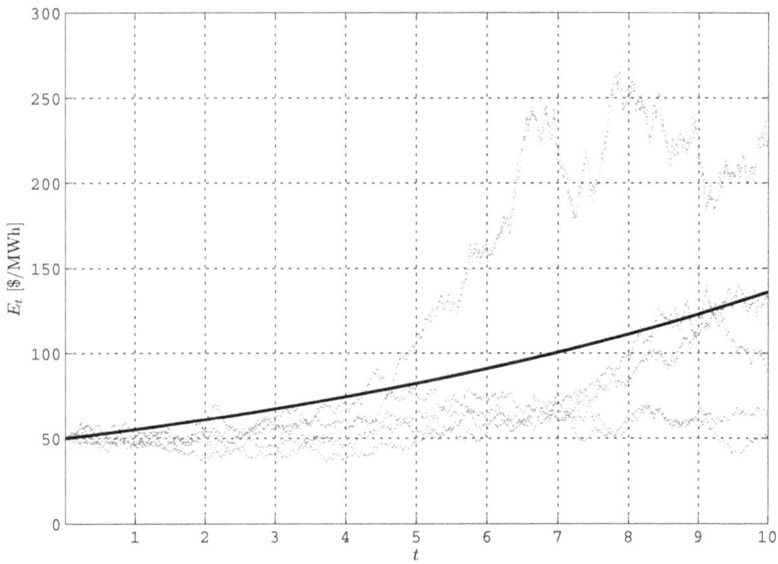

Fig. 7.2 Forecast value and sample paths of a GBM

$$dE_t = \alpha E_t dt + \sigma E_t dz_t, \tag{7.1}$$

where dz_t is the increment to a BM at time t, α is the annualized drift rate, $\sigma \geq 0$ is the annualized percentage volatility, and $E_0 \equiv E$.[3]

Illustrative Example 7.1 *GBM sample paths*

If we ignore the stochastic dz_t term, then we may derive the expected value of the GBM in year t conditional on E as $\mathbb{E}_E[E_t] = Ee^{\alpha t}$. In effect, the GBM, on average, exhibits exponential growth. Once the stochastic dz_t term is considered, sample paths for the GBM may be generated that evolve with uncertainty around the conditional expectation. In Fig. 7.2, the forecast value, $\mathbb{E}_E[E_t] = Ee^{\alpha t}$, is plotted (solid series) along with five sample paths (dotted series) from a GBM with parameters $\alpha = 0.10$, $\sigma = 0.20$, and $E = 50$ over a period of ten years. Note that the forecast value after ten years is $50e^{0.10 \times 10} = 135.91$. □

[3]There is considerable debate over whether energy prices follow BMs or mean-reverting processes. For example, Pindyck [32] analyzes 127 years of coal, natural gas, and oil prices to test for mean reversion. He finds that while such energy prices are indeed mean reverting, the rate of mean reversion is so low that using a GBM assumption for the purposes of investment analysis is "unlikely to lead to large errors." Combined with the fact that real options models in the BM family typically lead to closed-form solutions, we retain the GBM assumption for modeling long-term energy prices. However, this may not be valid for short-term operational analyses in which the electricity price is marked by stronger mean reversion and spikes [8].

Taking the dynamic programming approach (see Appendix E for a summary) to solving real options problems [12], we assume that all cash flows are real and the exogenous discount rate is $\rho > \alpha$. Without loss of generality, we assume that the power company holds a perpetual option to invest in a power plant that will last forever once constructed. The latter assumption about infinite lifetime may be easily relaxed. However, the former is necessary to ensure analytical solutions that will facilitate insights. Furthermore, it may be justified by the fact that typically, an investor in the electricity industry will have monopoly rights to build a facility at a particular location, e.g., through either an agreement with the municipal authority or rights to the land. Finally, the impact of a finite option to build on the optimal investment threshold price is weak when the time to expiration is relatively large.

At the optimal time, the power company pays a deterministic capital cost, I (in \$), to trigger the investment in the power plant. For now, we disregard the capacity size of the facility, i.e., we assume that it generates a notional 1 MWh of electricity per annum. Thus, I may be interpreted as a per-unit capacity cost. We assume that the power plant is constructed immediately once ordered by the power company[4] and starts generating electricity at heat rate H (in MWh_{th}/MWh), which is sold at price E_t (in \$/MWh). For now, we also assume that the fuel price at time t, F_t (in \$/MWh_{th}), is constant, i.e., $F_t = F$. In subsequent sections, we will explore the implications of relaxing some of these assumptions. However, in order to establish a benchmark and to gain intuition for how investment with a deferral option differs from a now-or-never NPV approach, we proceed to a stylized example in Sect. 7.2 with investment in a single power plant of given capacity. Next, in Sects. 7.3 and 7.4, we tackle flexibility in operations and modularity in investment, respectively. Sections 7.5 and 7.6 allow for the plant's capacity to be a decision variable in either a continuous or discrete setting. In order to examine the effect of risk aversion on investment timing, we expand the framework in Sect. 7.7 to incorporate concave utility functions. In Sect. 7.8, we summarize the chapter and provide an overview of the recent literature. End-of-chapter exercises are included in Sect. 7.9. Section 7.10 provides MATLAB codes for solving numerical examples.

The nomenclature for the rest of the chapter is as follows:

Indices

i, i'	Index for states.
j	Index for projects.
s, t	Index for time.

Parameters

A^j	Investment cost term for power plant j with endogenous sizing [\$/(MWh)^2].

---

[4]Depending on the type of facility, this may not be a reasonable assumption. For example, wind farms, solar plants, and run-of-river hydro plants can be constructed relatively quickly because they use standardized components. Nuclear power plants, however, are notorious for time and cost overruns because of the complexity of the task and the lack of suitable sites. Lead times for fossil-fueled installations are somewhere in the middle, e.g., taking up to two years for gas-fired plants and five years for coal plants. For how to handle this "time-to-build" problem, see [28].

B^j	Investment cost superscript for power plant j with endogenous sizing (unitless).
E_t	Electricity price at time t [\$/MWh].
F_t	Fuel price at time t [\$/MWh$_{th}$].
H^j	Heat rate of power plant j [MWh$_{th}$/MWh].
I^j	Investment cost of power plant j [\$].
K^j	Annual electricity output of power plant j [MWh].
$S^{i,i',j}$	Switching cost between states i and i' of power plant j [\$].

Constants and Rates

α	Percentage growth rate [1/year].
$\beta_1 (\beta_2)$	Positive (negative) root of the characteristic quadratic.
γ	Relative risk aversion parameter.
ρ	Discount rate [1/year].
σ	Percentage volatility [1/year].

Functions

$\mathscr{I}(K^j)$	Investment cost for power plant j with endogenous capacity sizing [\$].
$\mathscr{Q}(\beta)$	Characteristic quadratic function.
$\mathscr{U}(E)$	Utility function given electricity price E.
$\mathscr{V}^j(E)$	Expected now-or-never NPV of power plant j given electricity price E [\$].
$\mathscr{W}_i^j(E)$	Value of power plant j in state i given electricity price E [\$].

Variables

$a_{i,1} (a_{i,2})$	Coefficient for the positive (negative) branch of the option value function in state i.
$\kappa^j(E)$	Optimal size for power plant j given current electricity price E [MWh].
ξ^j	Optimal investment threshold price for power plant j [\$/MWh].
$\xi^{i,i',j}$	Optimal switching threshold from state i to state i' for power plant j [\$/MWh].
ξ^j_{NPV}	Now-or-never NPV investment threshold price for power plant j [\$/MWh].
τ^j	Optimal stopping time for investment in plant j.

7.2 Optimal Timing Versus Now-or-Never Net Present Value Approaches

In the now-or-never NPV approach, the expected discounted revenues of a project are compared with its investment cost. The instantaneous cash flows of a power plant with heat rate H at time t are $E_t - HF_t$. Assuming that the electricity price follows a GBM and that the fuel price is constant at F, the expected now-or-never NPV of such a power plant is:

$$\begin{aligned}\mathscr{V}(E) &= \mathbb{E}_E\left[\int_0^\infty (E_t - HF)\,e^{-\rho t}dt\right] - I\\ \Rightarrow \mathscr{V}(E) &= \int_0^\infty \left(Ee^{\alpha t} - HF\right)e^{-\rho t}dt - I\\ \Rightarrow \mathscr{V}(E) &= \frac{E}{\rho-\alpha} - \frac{HF}{\rho} - I.\end{aligned} \tag{7.2}$$

In Eq. (7.2), we use the fact that the conditional expectation of a continuous random variable involves taking an integral. Thus, the first line is effectively a double integration, and without loss of generality, the order of integration may be reversed. We use this fact in the second line by moving the conditional expectation operator inside the integral with respect to time. Since $\mathbb{E}_E[E_t] = Ee^{\alpha t}$, we obtain the expression in the second line, and evaluating this integral yields the result in the final line. Intuitively, it states that the expected now-or-never NPV of such a power plant is the difference between the present value of the operating cash flows (stemming from electricity sales and fuel purchases) and the up-front investment cost. Consequently, if the option to defer investment is ignored, then investment occurs immediately as long as $\mathscr{V}(E) \geq 0$. Otherwise, investment never occurs.

Rather than investing immediately, it may be desirable for the power company to postpone taking action. For example, the electricity price may be likely to increase in the next few years. The increase in expected revenues along with the delay in paying the investment cost could make deferral favorable. However, the forgone revenues from not having an active power plant in the intervening years are an opportunity cost of delaying that would have to be factored into the decision. Thus, the power company could consider waiting T years from now and then investing immediately in the power plant. The expected NPV of such a strategy is simply $e^{-\rho T}\mathbb{E}_E[\mathscr{V}(E_T)] = \frac{Ee^{-(\rho-\alpha)T}}{\rho-\alpha} - \frac{HFe^{-\rho T}}{\rho} - Ie^{-\rho T}$, i.e., it is the discounted expected NPV of a power plant that is constructed in T years when the electricity price is E_T. To elaborate, the expected NPV of a power plant that is constructed in T years is $\mathbb{E}_E\left[\int_T^\infty (E_t - HF)\,e^{-\rho t}dt\right] - Ie^{-\rho T}$. Note that the conditional expectation may be written as $\mathbb{E}_E\left[\mathbb{E}_{E_T}\left[\int_T^\infty (E_t - HF)\,e^{-\rho t}dt\right]\right]$ because of the law of iterated expectations, i.e., $\mathbb{E}[X] = \mathbb{E}[\mathbb{E}[X|Y]]$, where X and Y are random variables. Furthermore, since the GBM is a Markov process, i.e., the probabilistic structure of the future given the present is independent of the past, the inner conditional expectation may be rendered as $\int_0^\infty \left(E_Te^{\alpha t'} - HF\right)e^{-\rho(t'+T)}dt'$ after the change of variable $t' = t - T$. Thus, this integral becomes $e^{-\rho T}\mathscr{V}(E_T)$ after including the discounted investment cost.

Illustrative Example 7.2 *Investment timing at discrete points in time*

Using parameter values of $I = 100$, $\rho = 0.10$, $\alpha = 0.05$, $F = 20$, and $H = 2.5$, we obtain the value functions for $T = 0$, $T = 5$, and $T = 10$ in Fig. 7.1. We first note that since the power plant is to generate a notional 1 MWh of electricity per annum, this results in an investment cost of \$876/kW. Also, the heat rate implies a 40 % electrical conversion efficiency. These two parameters are in line with the characteristics of most gas-fired plants. Next, in terms of investment timing, if the power company can construct the plant at only these three points in time, then its optimal decision is dependent on the current electricity price. For example, if the current electricity price is \$25/MWh, then it is optimal to invest in ten years. This is because the electricity price will need a decade to increase to a level that makes the plant profitable. Plus, the investment cost will be discounted. On the other hand, a relatively high current electricity price, e.g., greater than \$60/MWh, warrants immediate investment. Intuitively, waiting is not worthwhile because the price is high enough to make the opportunity cost of not investing more than offset any benefit from deferral. An electricity price in the middle of the range makes it optimal to invest after five years, whereas a very low electricity price means that it is optimal never to invest. Therefore, the value of the entire investment opportunity to the power company is the upper envelope of the value functions as well as zero in Fig. 7.1. □

Suppose now that instead of having the option to invest in the power plant only at certain discrete points in time, i.e., $T = 0, 5, 10$, the power company may start the project at any point in time. In that case, the upper envelope in Fig. 7.1 reflecting the value of the investment opportunity should become a smooth curve and indicate the optimal investment threshold price. Using dynamic programming, we will derive this function and threshold rigorously. We assume that there are two states of the world: 0, in which the power company is waiting to invest, and 1, in which it has an active power plant. Working backward from state 1, we know that its value function is just the expected present value of an active power plant, i.e.:

$$\mathscr{W}_1(E) = \frac{E}{\rho - \alpha} - \frac{HF}{\rho}. \tag{7.3}$$

Now, in state 0, we begin with the Bellman equation in order to value $\mathscr{W}_0(E)$ and to determine the optimal investment threshold price, ξ:

$$\rho \mathscr{W}_0(E)\mathrm{d}t = \mathbb{E}_E \left[\mathrm{d}\mathscr{W}_0\right]. \tag{7.4}$$

This states that the instantaneous return on the option to invest is equal to its expected appreciation. Intuitively, an external assessor's required rate of return on the option to build the power plant, ρ, must equal the expected value from owning the right to build the power plant outright. Expanding the right-hand side of Eq. (7.4) via Itô's lemma and re-arranging, we obtain the following second-order ordinary differential equation (ODE):

$$\begin{aligned} \rho \mathscr{W}_0(E)\mathrm{d}t &= \mathbb{E}_E \left[\mathscr{W}_0'(E)\mathrm{d}E + \frac{1}{2}\mathscr{W}_0''(E)\,(\mathrm{d}E)^2\right] \\ \Rightarrow \frac{1}{2}\sigma^2 E^2 \mathscr{W}_0''(E) &+ \alpha E \mathscr{W}_0'(E) - \rho \mathscr{W}_0(E) = 0, \end{aligned} \tag{7.5}$$

where we use the fact that $\mathbb{E}_E\,[\mathrm{d}E] = \alpha E\mathrm{d}t$ and $(\mathrm{d}E)^2 = \sigma^2 E^2 \mathrm{d}t$ in going from the first line to the second.

The general solution to the ODE in Eq. (7.5) is of the power form, i.e.:

$$\mathscr{W}_0(E) = a_0 E^{\beta}, \tag{7.6}$$

which is subject to the following boundary conditions:

$$\lim_{E\to 0} \mathscr{W}_0(E) = 0 \tag{7.7a}$$
$$\mathscr{W}_0(\xi) = \mathscr{W}_1(\xi) - I \tag{7.7b}$$
$$\mathscr{W}_0'(\xi) = \mathscr{W}_1'(\xi). \tag{7.7c}$$

Intuitively, Eq. (7.7a) states that the option to invest in the power plant becomes worthless as the electricity price tends to zero. Since zero is an absorbing state for the GBM, it follows that there will be no value from either waiting to invest in such a plant or having an active one immediately. Next, Eq. (7.7b) is the value-matching condition, which requires the value of the investment opportunity to equal the expected NPV, $\mathscr{W}_1(E) - I$, at the optimal investment threshold price, ξ. Indeed, the value lost from killing the option must equal the value gained from an active power plant at this trigger price. Finally, Eq. (7.7c) is the smooth-pasting condition, which is actually a first-order condition for optimization that reflects the fact that the marginal benefit of waiting must equal the marginal cost of waiting at ξ.

Substituting the function from Eq. (7.6) and its derivatives into Eq. (7.5), we obtain:

$$\mathscr{Q}(\beta) \equiv \frac{1}{2}\sigma^2\beta(\beta - 1) + \alpha\beta - \rho = 0, \tag{7.8}$$

where β is taken as a generic parameter and $\mathscr{Q}(\beta)$ is the fundamental characteristic quadratic function, which implicitly defines β_1 and β_2 as its two roots. Although both of these roots can be solved for explicitly, it becomes apparent from the geometry of the problem that $\beta_1 > 1$ and $\beta_2 < 0$ as in Fig. 7.3. To observe this, note that the expression in Eq. (7.8) is an upward-facing parabola with $\mathscr{Q}(0) = -\rho < 0$ and $\mathscr{Q}(1) = \alpha - \rho < 0$. The latter inequality implies that $\beta_1 > 1$ because the parabola is still negative at $\beta = 1$. Likewise, the former inequality implies that $\beta_2 < 0$ because the parabola is still negative at $\beta = 0$.

Now, if $a_{0,1}E^{\beta_1}$ is a solution to the ODE in Eq. (7.5), then so is $a_{0,2}E^{\beta_2}$, where $a_{0,1}$ and $a_{0,2}$ are endogenous coefficients that depend on ξ. Thus, we have:

$$\mathscr{W}_0(E) = a_{0,1}E^{\beta_1} + a_{0,2}E^{\beta_2}. \tag{7.9}$$

Since the latter term in Eq. (7.9) goes to infinity as E goes to zero, it is inconsistent with the boundary condition in Eq. (7.7a). Therefore, it must be the case that $a_{0,2} = 0$, and, consequently, we obtain:

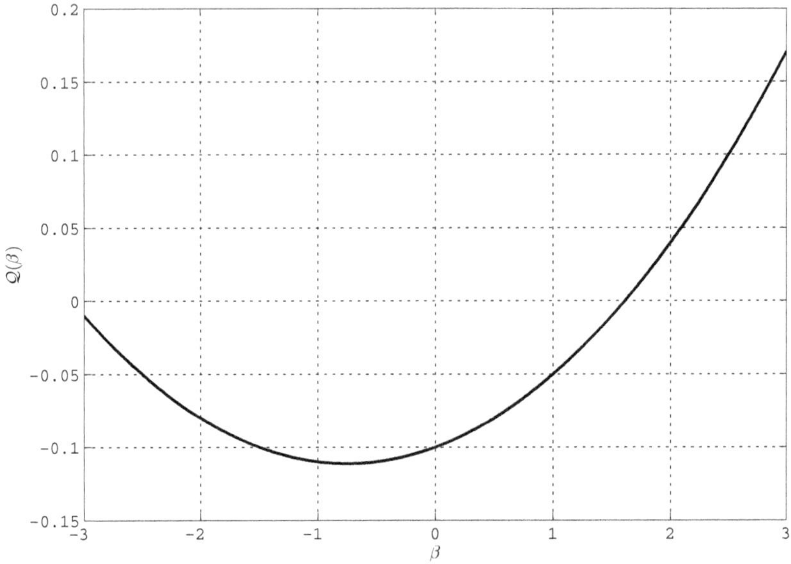

Fig. 7.3 Characteristic quadratic function, $\mathscr{Q}(\beta)$, for $\alpha = 0.05$, $\rho = 0.10$, and $\sigma = 0.20$

$$\mathscr{W}_0(E) = a_{0,1} E^{\beta_1}. \tag{7.10}$$

We now use the expression from Eq. (7.10) in Eqs. (7.7b) and (7.7c) to solve for ξ and $a_{0,1}$:

$$a_{0,1}\xi^{\beta_1} = \frac{\xi}{\rho - \alpha} - \frac{HF}{\rho} - I \tag{7.11}$$

$$\begin{aligned} \beta_1 a_{0,1}\xi^{\beta_1 - 1} &= \frac{1}{\rho - \alpha} \\ \Rightarrow a_{0,1} &= \frac{\xi^{1-\beta_1}}{\beta_1(\rho - \alpha)}. \end{aligned} \tag{7.12}$$

Substituting the solution for $a_{0,1}$ from Eq. (7.12) into Eq. (7.11), we obtain:

$$\begin{aligned} \tfrac{\xi}{\beta_1(\rho-\alpha)} &= \tfrac{\xi}{\rho-\alpha} - \tfrac{HF}{\rho} - I \\ \Rightarrow \xi &= \left(\tfrac{\beta_1}{\beta_1 - 1}\right)\left(\tfrac{\rho-\alpha}{\rho}\right)(HF + \rho I). \end{aligned} \tag{7.13}$$

For comparison with the now-or-never NPV, we set the expression for $\mathscr{W}_1(E) - I$ to zero in Eq. (7.3) and solve for E to obtain $\xi_{NPV} = \left(\frac{\rho-\alpha}{\rho}\right)(HF + \rho I)$. This means that investment should occur only if the current price is high enough to cover the amortized operating and investment costs. By contrast, via the real options approach, ξ in Eq. (7.13) results in a higher threshold price for investment because $\frac{\beta_1}{\beta_1-1} > 1$.[5] This discretion to wait for a higher threshold vis-à-vis the now-or-never NPV approach stems from a combination of positive drift and volatility in the electricity price. Note that even if $\sigma = 0$, then the optimal investment threshold in Eq. (7.13) is greater than the now-or-never one. In particular, $\lim_{\sigma \to 0} \beta_1 = \frac{\rho}{\alpha}$ from Eq. (7.8), which leads to $\lim_{\sigma \to 0} \xi = HF + \rho I > \xi_{NPV}$. Indeed, even without uncertainty in the electricity price, as long as the power company has the discretion to wait for a higher price at which to launch the power plant's operations, it will do so until the marginal benefit of waiting is just equal to the marginal cost of waiting, which results from the forgone cash flows in the waiting period. With greater uncertainty, the marginal benefit of waiting increases by more than the marginal cost because the latter depends only on the opportunity cost of lost cash flows in the immediate future. However, the marginal benefit of waiting is related to the possibility of a higher starting price for the power plant in the future, which is affected to a greater extent by uncertainty. Hence, with the deferral option, the value of the power plant project is higher, yet this also increases the opportunity cost of killing the option to wait, thereby leading to a higher investment threshold price.

Insights about investment under uncertainty may be facilitated via numerical examples. We first consider the investment decision in Illustrative Example 7.3. Next, we explore sensitivity analyses with respect to σ, α, and ρ in Illustrative Examples 7.4–7.6.

Illustrative Example 7.3 *Investment under uncertainty with continuous time*

Here, we use $\alpha = 0.05$, $\rho = 0.10$, $\sigma = 0.20$, $I = 100$, $H = 2.5$, and $F = 20$ as base parameters. First, we plot the expected NPV and value of the investment opportunity with respect to the electricity price, E, in Fig. 7.4. The expected NPV function, $\mathscr{W}_1(E) - I$, is the same as the immediate investment NPV in Fig. 7.1. Recall that when we allowed investment to occur only at discrete points in time, the value of the investment opportunity comprised the upper envelope of the NPV functions. By contrast, the real options approach enables such a comparison to be made at every infinitesimal point in time. Thus, as the time intervals between alternative investment opportunities go to zero, the kinked function in Fig. 7.1 becomes a smooth convex one, as indicated in Fig. 7.4. Note that the expected NPV of immediate investment, $\mathscr{W}_1(E) - I$, equals zero for $\xi_{NPV} = 30$. However, the value of the investment opportunity stemming from the real options approach, $\mathscr{W}_0(E)$, is strictly above the expected NPV, thereby revealing that there is positive value to waiting. In fact, it is optimal to wait until the electricity price hits the threshold $\xi = 79.30$. In the parlance of financial options, it is worthwhile retaining the option until it is deep "in the money." Furthermore, as the initial electricity price goes to zero, the value of the option to invest also goes to zero because investment never occurs in that case. □

[5]To see this, note that $\frac{\beta_1}{\beta_1-1} = \frac{1}{1-\frac{1}{\beta_1}}$. Since $\beta_1 > 1$, the denominator of the latter expression is strictly less than one.

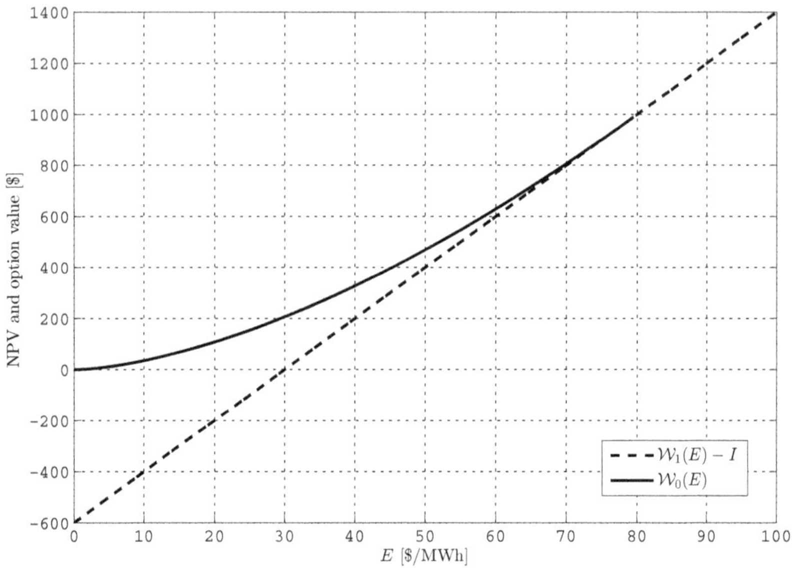

Fig. 7.4 Expected NPV, $\mathscr{W}_1(E) - I$, and value of the investment opportunity, $\mathscr{W}_0(E)$, for $\alpha = 0.05$, $\rho = 0.10$, $\sigma = 0.20$, $I = 100$, $H = 2.5$, and $F = 20$

Illustrative Example 7.4 *Sensitivity analysis of investment under uncertainty with respect to volatility*

Here, we vary σ while holding all other parameters constant in order to examine the sensitivity of ξ to uncertainty. In Fig. 7.5, we note that the now-or-never investment threshold, ξ_{NPV}, is not affected by uncertainty and remains constant at \$30/MWh. However, the real options threshold, ξ, increases from a value of \$60/MWh (for $\sigma = 0$) to nearly \$90/MWh (for $\sigma = 0.25$). Thus, greater uncertainty increases the investment threshold as the value of waiting becomes larger.[6] Consequently, the "wedge" between the now-or-never and real options thresholds, $\frac{\beta_1}{\beta_1 - 1}$, increases with uncertainty as well.[7]

An example of the value functions for $\sigma = 0$ is given in Fig. 7.6 to illustrate that there is a value in waiting since the electricity price is still going to increase. However, without the presence of uncertainty, its magnitude is reduced.

The effect of uncertainty on the value of the investment opportunity is summarized in Fig. 7.7, in which the relative value of $\mathscr{W}_0(E)$ to $\mathscr{W}_1(E) - I$ at $E = 50$ is plotted.

[6] Although it may be appealing to think of a higher investment threshold price as "delaying" the investment timing, in fact, as the volatility increases, so does the conditional probability that the threshold price will be reached from a given initial price. Intuitively, the higher volatility also increases the likelihood of extremely high (as well as extremely low) prices. Therefore, the overall impact on timing is ambiguous. For a rigorous analysis, see [27].

[7] The formal proof of this is left as an exercise at the end of this chapter.

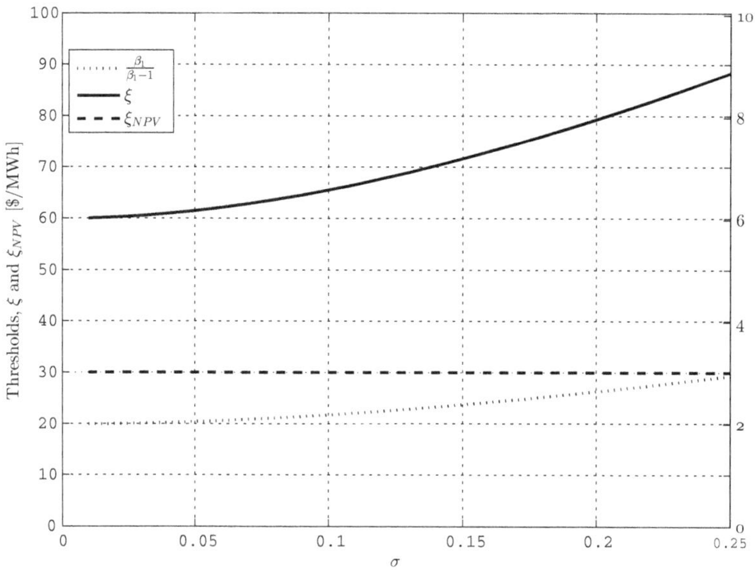

Fig. 7.5 Sensitivity of ξ, ξ_{NPV}, and $\frac{\beta_1}{\beta_1-1}$ with respect to σ for $\alpha = 0.05$, $\rho = 0.10$, $I = 100$, $H = 2.5$, and $F = 20$

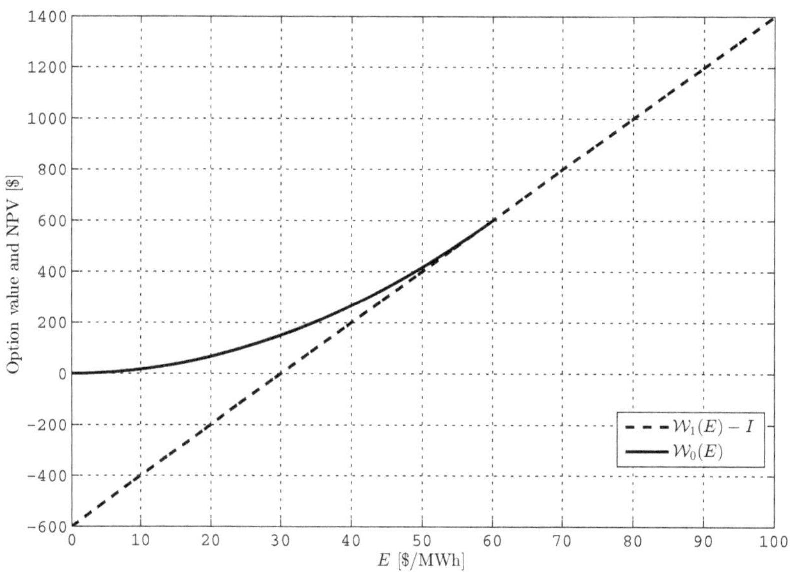

Fig. 7.6 Expected NPV, $\mathscr{W}_1(E) - I$, and value of the investment opportunity, $\mathscr{W}_0(E)$, for $\alpha = 0.05$, $\rho = 0.10$, $\sigma = 0$, $I = 100$, $H = 2.5$, and $F = 20$

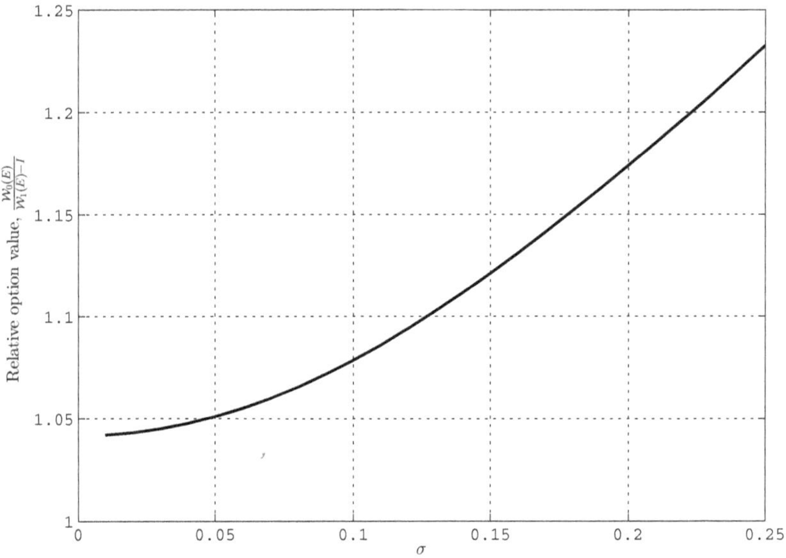

Fig. 7.7 Sensitivity of relative option value, $\frac{\mathscr{W}_0(E)}{\mathscr{W}_1(E)-I}$, with respect to σ for $\alpha = 0.05$, $\rho = 0.10$, $E = 50$, $I = 100$, $H = 2.5$, and $F = 20$

Hence, higher uncertainty increases the discretion to wait, which leads to a higher investment threshold. □

Illustrative Example 7.5 *Sensitivity analysis of investment timing with respect to the drift rate*

In Fig. 7.8, we vary α while holding all other parameters constant. Intuitively, we would postulate that an increase in the drift rate would increase the investment threshold since it would be desirable to wait for a higher future electricity price at which to launch the power plant's operations. Somewhat surprisingly, ξ actually decreases with α. In order to explain this seemingly counterintuitive outcome, we plot ξ_{NPV} and note that it is also decreasing but at a faster rate. Indeed, from Eq. (7.13), an increase in α leads to a decrease in the now-or-never threshold, ξ_{NPV}. But if we look at the expected NPV of the power plant upon investment at ξ, we obtain $\left(\frac{\beta_1}{\beta_1-1}\right)\left[\frac{HF}{\rho}+I\right]-\frac{HF}{\rho}-I$. In other words, it is the "wedge" between ξ and ξ_{NPV} that is affected by α. Plotting the wedge, i.e., $\frac{\beta_1}{\beta_1-1}$ in Fig. 7.8, we see that it is indeed increasing with respect to α, i.e., a higher drift rate increases the expected NPV of investment at the real options trigger. □

Illustrative Example 7.6 *Sensitivity analysis of investment timing with respect to the discount rate*

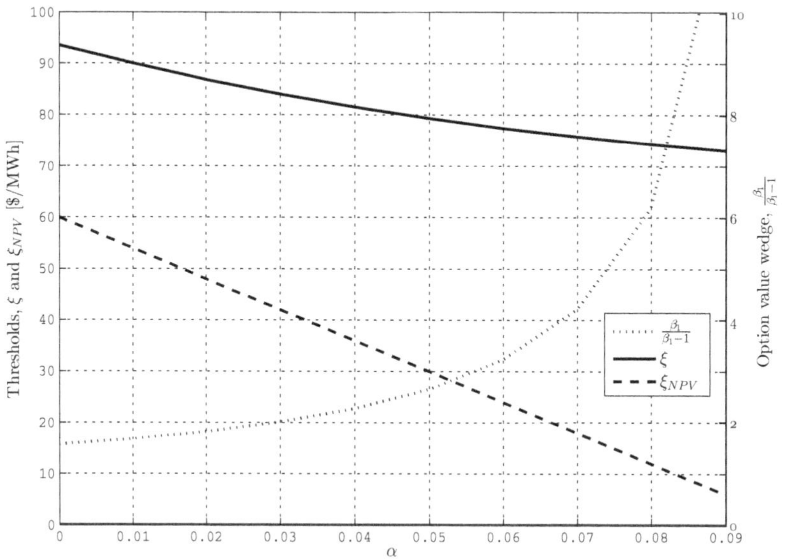

Fig. 7.8 Sensitivity of ξ, ξ_{NPV}, and $\frac{\beta_1}{\beta_1 - 1}$ with respect to α for $\sigma = 0.20$, $\rho = 0.10$, $I = 100$, $H = 2.5$, and $F = 20$

A similar breakdown of the behavior of ξ may be done with respect to ρ. Again, intuitively, we would imagine that an increase in the discount rate would make the future less important, thereby reducing the incentive to wait. Yet, Fig. 7.9 tells a story that seems to belief our understanding since ξ actually increases with ρ. However, as with the analysis with respect to α, it is important to note that the wedge, $\frac{\beta_1}{\beta_1 - 1}$, is the main driver of the result. Plotting it respect to ρ, we are able to reconcile the finding with our intuition: a higher discount rate facilitates investment even as it lowers the expected NPV of an operational power plant. □

7.3 Operational Flexibility

In Sect. 7.2, we focus on the optimal timing of the investment decision, assuming that the power plant operates forever. In other words, the power company's decision was completely irreversible. However, in many cases, there is at least partial reversibility in the form of subsequent managerial discretion to abandon, modify, or suspend temporarily the power plant. In this section, we focus on the latter aspect, i.e., treatment of the so-called compound option to turn the power plant on and off again after it has been constructed.

Given the volatile nature of energy prices, such flexibility may be highly valuable. Indeed, [9] showed that by accounting for the option value of such flexibility, plants

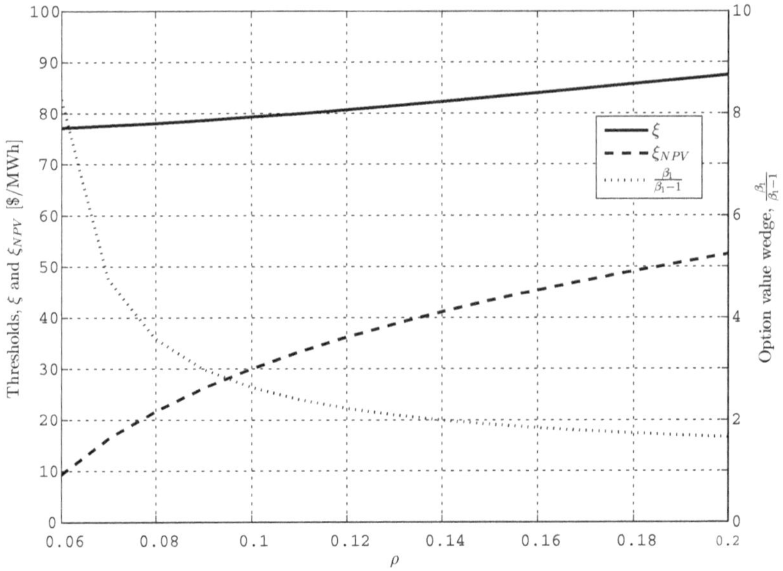

Fig. 7.9 Sensitivity of ξ, ξ_{NPV}, and $\frac{\beta_1}{\beta_1-1}$ with respect to ρ for $\sigma = 0.20$, $\alpha = 0.05$, $I = 100$, $H = 2.5$, and $F = 20$

being sold in California after deregulation in the late 1990s were valued at higher than their book values as indicated by the NPV approach. In particular, gas-fired power plants have the flexibility to ramp up and down relatively quickly, although this occurs at a cost and may be constrained by minimum uptimes. For example, a gas-fired power plant with a maximum power capacity of 431 MW requires 1200 GJ of fuel (or approximately 333 MWh_{th}) for a "hot start" (immediately one hour after the plant is shut down), which leads to a startup cost of almost $7000, assuming a fuel price of $20/$MWh_{th}$. These calculations are based on a combined-cycle gas turbine (CCGT) plant installed in 2010 in Aghada, Republic of Ireland [41]. Furthermore, according to the same source, CCGT plants have constraints on minimum uptimes and downtimes (typically four hours each).

Using the real options approach, we can analyze how the availability of such partial flexibility in the plant's operations influences not only the value of the investment opportunity but also the initial investment decision. Intuitively, any power plant operator would value a flexible power plant more. However, the value of this flexibility is difficult to quantify via a now-or-never NPV approach. For example, at what threshold electricity price would it be optimal to suspend or to resume operations? If the value of operational flexibility increases with electricity price volatility, then how does it affect the investment threshold vis-à-vis that from Sect. 7.2?

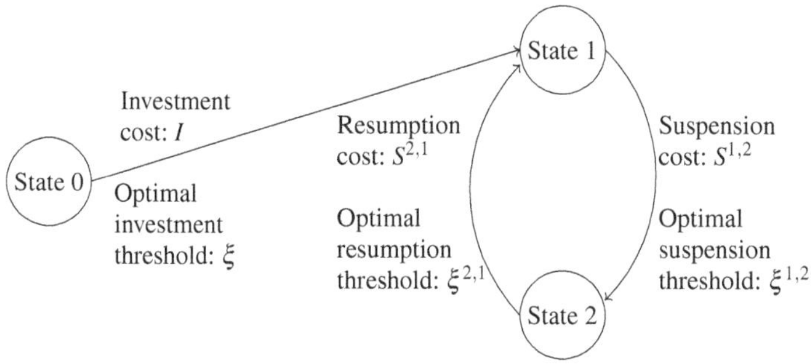

Fig. 7.10 State-transition diagram for a power plant with operational flexibility

In order to focus on the implications of operational flexibility, we assume that after investment, the plant may be in one of two states: on (state 1) or off (state 2).[8] Transitioning between these two states incurs fixed resumption and suspension costs, $S^{2,1}$ and $S^{1,2}$, respectively, as indicated in Fig. 7.10. Associated with these transitions are threshold prices, $\xi^{2,1}$ and $\xi^{1,2}$, which will be determined endogenously.[9] Thus, we abstract from technical constraints of actual power plants such as finite ramping rates and minimum uptimes and downtimes. Instead, we assume that these features may be captured via fixed switching costs.

Although the sequence of decisions depicted in Fig. 7.10 begins in state 0 with the option to invest and progresses to state 1 with the valuation of an active power plant with the suspension option, we solve the problem using backward induction. Specifically, we start by considering the operational decisions of the power plant given that the investment decision with associated threshold, ξ, has already been undertaken. We define $\mathscr{W}_i(E)$ as the value of the power plant in state i and use dynamic programming to find not only the value functions but also the optimal switching thresholds. First, we consider state 1 and note that the value of the plant should comprise both the expected PV of cash flows from indefinite operations and the option value to shut down. Intuitively, the latter component should increase in value as the electricity price decreases. We now formally determine the value function in state 1 by setting up the Bellman equation while keeping in mind that it should be adjusted from that in Eq. (7.4) to reflect cash flows from ongoing operations:

$$\rho \mathscr{W}_1(E)\mathrm{d}t = \mathbb{E}_E\left[\mathrm{d}\mathscr{W}_1\right] + (E - HF)\,\mathrm{d}t. \tag{7.14}$$

---

[8] Instead of two discrete on–off states, it may also be possible to have several operating states ranging from zero to full capacity. Alternatively, continuous adjustment of the plant's output may be handled by specifying a production function as in Chap. 6 of [12].

[9] In the limit as these fixed transition costs go to zero, the problem collapses to one of costless switching, e.g., as in [30]. The optimal switching thresholds then converge to the operating cost of the plant, i.e., HF. Intuitively, it is optimal to shut down (restart) the plant when the electricity price drops below (increases above) the operating cost.

The second term on the right-hand side of Eq. (7.14) is precisely the instantaneous cash flow from operations. Next, we expand $d\mathscr{W}_1$ via Itô's lemma as in Sect. 7.2 and rearrange it to obtain a second-order ODE similar to that in Eq. (7.5):

$$\frac{1}{2}\sigma^2 E^2 \mathscr{W}_1''(E) + \alpha E \mathscr{W}_1'(E) - \rho \mathscr{W}_1(E) + E - HF = 0. \tag{7.15}$$

The solution to the ODE in Eq. (7.15) is similar to that in Eq. (7.9) but with an extra term reflecting the expected PV of cash flows from a perpetually operating power plant:

$$\mathscr{W}_1(E) = a_{1,1}E^{\beta_1} + a_{1,2}E^{\beta_2} + \frac{E}{\rho - \alpha} - \frac{HF}{\rho}, \tag{7.16}$$

where β_1 and β_2 are still the positive and negative roots, respectively, of the characteristic quadratic function from Eq. (7.8). Here, $\frac{E}{\rho-\alpha} - \frac{HF}{\rho}$ represents the expected PV of a power plant that operates forever, and $a_{1,1}E^{\beta_1} + a_{1,2}E^{\beta_2}$ is the value of the option to suspend operations. Economically, we require $\lim_{E\to\infty} \mathscr{W}_1(E) = \frac{E}{\rho-\alpha} - \frac{HF}{\rho}$, i.e., the value of a power plant at very high electricity prices should be simply that of one that never shuts down. Indeed, it is only for relatively low electricity prices that the plant would ever shut down. Hence, we must have $a_{1,1} = 0$, thereby resulting in:

$$\mathscr{W}_1(E) = a_{1,2}E^{\beta_2} + \frac{E}{\rho - \alpha} - \frac{HF}{\rho}. \tag{7.17}$$

Second, we similarly tackle the value of a suspended power plant, i.e., one that is in state 2. Since there are no instantaneous cash flows, the Bellman equation becomes:

$$\rho \mathscr{W}_2(E)dt = \mathbb{E}_E\left[d\mathscr{W}_2\right]. \tag{7.18}$$

Again, by applying Itô's lemma to the right-hand side and rearranging, we obtain a second-order ODE:

$$\frac{1}{2}\sigma^2 E^2 \mathscr{W}_2''(E) + \alpha E \mathscr{W}_2'(E) - \rho \mathscr{W}_2(E) = 0. \tag{7.19}$$

The solution is $\mathscr{W}_2(E) = a_{2,1}E^{\beta_1} + a_{2,2}E^{\beta_2}$, which becomes the following after application of the boundary condition $\lim_{E\to 0} \mathscr{W}_2(E) = 0$:

$$\mathscr{W}_2(E) = a_{2,1}E^{\beta_1}. \tag{7.20}$$

Hence, the value of the power plant in state 2 is simply the value of the option to resume operations in the future, which is increasing with the electricity price.

From Eqs. (7.17) and (7.20), we have two endogenous variables, $a_{1,2}$ and $a_{2,1}$, as well as two thresholds, $\xi^{1,2}$ and $\xi^{2,1}$, to solve for. Thus, we need a total of four equations. We obtain these by writing a pair of value-matching and smooth-pasting conditions for each of the two operational transitions. First, in shutting down the power plant, i.e., going from state 1 to 2, we obtain:

$$\mathscr{W}_1(\xi^{1,2}) = \mathscr{W}_2(\xi^{1,2}) - S^{1,2}$$
$$\Rightarrow a_{1,2}\left(\xi^{1,2}\right)^{\beta_2} + \frac{\xi^{1,2}}{\rho-\alpha} - \frac{HF}{\rho} = a_{2,1}\left(\xi^{1,2}\right)^{\beta_1} - S^{1,2} \tag{7.21a}$$

$$\mathscr{W}_1'(\xi^{1,2}) = \mathscr{W}_2'(\xi^{1,2})$$
$$\Rightarrow \beta_2 a_{1,2}\left(\xi^{1,2}\right)^{\beta_2-1} + \frac{1}{\rho-\alpha} = \beta_1 a_{2,1}\left(\xi^{1,2}\right)^{\beta_1-1}. \tag{7.21b}$$

Second, we have a pair of such equations for the transition from state 2 to 1:

$$\mathscr{W}_2(\xi^{2,1}) = \mathscr{W}_1(\xi^{2,1}) - S^{2,1}$$
$$\Rightarrow a_{2,1}\left(\xi^{2,1}\right)^{\beta_1} = a_{1,2}\left(\xi^{2,1}\right)^{\beta_2} + \frac{\xi^{2,1}}{\rho-\alpha} - \frac{HF}{\rho} - S^{2,1} \tag{7.22a}$$

$$\mathscr{W}_2'(\xi^{2,1}) = \mathscr{W}_1'(\xi^{2,1})$$
$$\Rightarrow \beta_1 a_{2,1}\left(\xi^{2,1}\right)^{\beta_1-1} = \beta_2 a_{1,2}\left(\xi^{2,1}\right)^{\beta_2-1} + \frac{1}{\rho-\alpha}. \tag{7.22b}$$

Intuitively, Eqs. (7.21a)–(7.22b) state that the value gained must equal the value lost from switching operating modes and that the marginal benefit must equal the marginal cost from delaying any operational transitions. However, unlike Eqs. (7.11)–(7.12), the system of equations here with operational flexibility is highly nonlinear. Therefore, in general, it is not possible to obtain closed-form solutions for the four unknowns. Instead, we must resort to numerical methods to find solutions for specific parameter values. Most computational software packages like Mathematica, MATLAB, and Octave have functions, e.g., `fsolve` in MATLAB, that solve nonlinear systems if a guess for the solution is available.

We provide MATLAB code in Sect. 7.10 for solving the nonlinear system resulting from Illustrative Example 7.7. But, how should the guess for the solution be calculated? One way to proceed is to find analytical solutions to a simpler system, e.g., with a one-time abandonment option from state 1 or a one-time resumption option from state 2. Considering such a simplified model from state 1, we note that the resumption option from state 2 will not be available. Thus, the following system of two equations can be easily solved for guesses $\tilde{a}_{1,2}$ and $\tilde{\xi}^{1,2}$:

$$\tilde{a}_{1,2}\left(\tilde{\xi}^{1,2}\right)^{\beta_2} + \frac{\tilde{\xi}^{1,2}}{\rho-\alpha} - \frac{HF}{\rho} = -S^{1,2} \tag{7.23a}$$

$$\beta_2\tilde{a}_{1,2}\left(\tilde{\xi}^{1,2}\right)^{\beta_2-1} + \frac{1}{\rho-\alpha} = 0. \tag{7.23b}$$

Solving Eqs. (7.23a)–(7.23b), we obtain $\tilde{\xi}^{1,2} = \left(\frac{\beta_2}{\beta_2-1}\right)(\rho-\alpha)\left[\frac{HF}{\rho} - S^{1,2}\right]$ and $\tilde{a}_{1,2} = -\frac{\left(\tilde{\xi}^{1,2}\right)^{1-\beta_2}}{\beta_2(\rho-\alpha)}$ as the guesses for $\xi^{1,2}$ and $a_{1,2}$, respectively. By similarly simplifying Eqs. (7.22a) and (7.22b) to remove the $\left(\xi^{2,1}\right)^{\beta_2}$ terms, we obtain $\tilde{\xi}^{2,1} = \left(\frac{\beta_1}{\beta_1-1}\right)(\rho-\alpha)\left[\frac{HF}{\rho} + S^{2,1}\right]$ and $\tilde{a}_{2,1} = -\frac{\left(\tilde{\xi}^{2,1}\right)^{1-\beta_1}}{\beta_1(\rho-\alpha)}$ as the guesses for $\xi^{2,1}$ and $a_{2,1}$, respectively. Indeed, even without solving the nonlinear system numerically, we obtain the insight that with uncertainty and the option to make operational changes, the switching thresholds lead the decision-maker to be more cautious than the now-or-never NPV rule in which the plant would be shut down when the electricity price dropped below $\left(HF - \rho S^{1,2}\right) > \tilde{\xi}^{1,2}$ and restarted when the electricity price increased above $\left(HF + \rho S^{2,1}\right) < \tilde{\xi}^{2,1}$. These results follow because $\left(\frac{\beta_1}{\beta_1-1}\right)\left(\frac{\rho-\alpha}{\rho}\right) > 1$ and $\left(\frac{\beta_2}{\beta_2-1}\right)\left(\frac{\rho-\alpha}{\rho}\right) < 1$ for $\sigma > 0$. To see this, note that $\left(\frac{\beta_2}{\beta_2-1}\right) < 1$ and $\left(\frac{\rho-\alpha}{\rho}\right) < 1$. Thus, $\left(\frac{\beta_2}{\beta_2-1}\right)\left(\frac{\rho-\alpha}{\rho}\right) < 1$. On the other hand, $\left(\frac{\beta_1}{\beta_1-1}\right) > 1$ but $\left(\frac{\rho-\alpha}{\rho}\right) < 1$. Since $\left(\frac{\beta_1}{\beta_1-1}\right)$ is the lowest when $\sigma = 0$, we can show that even in this case, we have $\lim_{\sigma\to 0}\left(\frac{\beta_1}{\beta_1-1}\right) = \frac{\rho}{\rho-\alpha}$. Hence the product $\left(\frac{\beta_1}{\beta_1-1}\right)\left(\frac{\rho-\alpha}{\rho}\right)$ must be greater than 1 for all $\sigma > 0$.

As in Sect. 7.2, the optimal investment threshold, $\xi^{0,1}$, may be found via value-matching and smooth-pasting conditions between $\mathscr{W}_0(E)$ and $\mathscr{W}_1(E) - I$. We remark that the optimization must occur in going from state 0 to state 1 rather than state 2 because it would not make sense for the power company to invest I only to have an idle power plant. For this reason, the thresholds should have the ordering $\xi^{1,2} < \xi^{2,1} < \xi^{0,1}$. In order to obtain $\mathscr{W}_0(E)$, we follow the same procedure as in Eq. (7.5) to obtain:

$$\mathscr{W}_0(E) = a_{0,1}E^{\beta_1}. \tag{7.24}$$

Next, value-matching and smooth-pasting conditions yield the following system of equations:

$$\begin{aligned}&\mathscr{W}_0\left(\xi^{0,1}\right) = \mathscr{W}_1\left(\xi^{0,1}\right) - I\\ &\Rightarrow a_{0,1}\left(\xi^{0,1}\right)^{\beta_1} = \left(\xi^{0,1}\right)^{\beta_2} + \frac{\xi}{\rho-\alpha} - \frac{HF}{\rho} - I\end{aligned} \tag{7.25a}$$

$$\begin{aligned}&\mathscr{W}_0'\left(\xi^{0,1}\right) = \mathscr{W}_1'\left(\xi^{0,1}\right)\\ &\Rightarrow \beta_1 a_{0,1}\left(\xi^{0,1}\right)^{\beta_1-1} = \beta_2\left(\xi^{0,1}\right)^{\beta_2-1} + \frac{1}{\rho-\alpha}.\end{aligned} \tag{7.25b}$$

In contrast to Eqs. (7.11)–(7.12), here an analytical solution is impossible. However, it is possible to reduce Eqs. (7.25a)–(7.25b) to one nonlinear equation for $\xi^{0,1}$:

$$(\beta_1 - \beta_2)\, a_{1,2} \left(\xi^{0,1}\right)^{\beta_2} + (\beta_1 - 1)\, \frac{\xi^{0,1}}{\rho - \alpha} - \beta_1 \left(\frac{HF}{\rho} + I\right) = 0. \tag{7.26}$$

Using ξ from Eq. (7.13) as a guess, we can solve numerically for $\xi^{0,1}$ and consequently for $a_{0,1}$. Yet even without an analytical solution, it is possible to prove that $\xi^{0,1} < \xi$ by comparing the implicit definition of $\xi^{0,1}$ in Eq. (7.26) with the following for ξ:

$$(\beta_1 - 1)\, \frac{\xi}{\rho - \alpha} - \beta_1 \left(\frac{HF}{\rho} + I\right) = 0. \tag{7.27}$$

The two equations are identical except for the presence of the $(\beta_1 - \beta_2)\, a_{1,2} \left(\xi^{0,1}\right)^{\beta_2}$ term in Eq. (7.26), which is strictly positive. This adds to the linear term $(\beta_1 - 1)\, \frac{\xi^{0,1}}{\rho-\alpha}$, thereby ensuring that its intersection with the constant $\beta_1 \left(\frac{HF}{\rho} + I\right)$ is for a lower threshold price. In Illustrative Examples 7.7 and 7.8, we demonstrate the shapes of the value functions, perform sensitivity analyses on the thresholds with respect to the volatility, and provide MATLAB code in Sect. 7.10 for solving the nonlinear system of equations.

Illustrative Example 7.7 *Investment timing with operational flexibility*

Using $\alpha = 0.05$, $\rho = 0.10$, $\sigma = 0.20$, $I = 100$, $S^{1,2} = S^{2,1} = 10$, $H = 2.5$, and $F = 20$ as base parameters, we plot the value functions as in Fig. 7.11. Note that the functions $\mathscr{W}_1(E)$ and $\mathscr{W}_2(E)$ are defined over the ranges $(\xi^{1,2}, \infty)$ and $(0, \xi^{2,1})$, respectively. Thus, $\mathscr{W}_2(E)$ is a convex function that has the same gradient as $\mathscr{W}_1(E)$ at $\xi^{2,1}$ and differs from it by exactly $S^{2,1}$ at that point. Meanwhile, $\mathscr{W}_1(E)$ has a pronounced convex shape only for relatively low values of E, whereas it becomes asymptotically linear as $E \to \infty$. Indeed, for extremely high electricity prices, the value of the option to shut down the plant is nearly zero. Consequently, an active power plant's value converges to that of a plant that operates forever. However, for low electricity prices, it becomes optimal to suspend operations and switch to state 2. Therefore, at $\xi^{1,2}$, $\mathscr{W}_1(E)$ has the same gradient as $\mathscr{W}_2(E)$ and differs from it by $S^{1,2}$ at that point. Moreover, the suspension and resumption thresholds at \$37.99/MWh and \$61.38/MWh, respectively, are lower and higher, respectively, than the now-or-never NPV thresholds of \$49/MWh and \$51/MWh, respectively. Finally, the value function in state 0 satisfies the value-matching and smooth-pasting conditions with $\mathscr{W}_1(E) - I$ at $\xi^{0,1}$, which at \$76.23/MWh is lower than that of \$79.30/MWh for ξ. □

Illustrative Example 7.8 *Sensitivity analysis of operational flexibility with respect to volatility*

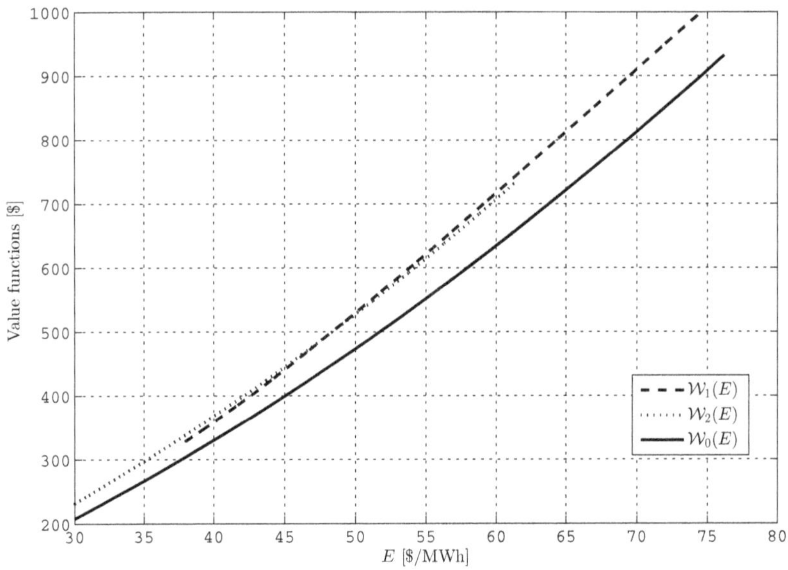

Fig. 7.11 Value functions with operational flexibility for $\alpha = 0.05$, $\rho = 0.10$, $\sigma = 0.20$, $I = 100$, $S^{1,2} = 10$, $S^{2,1} = 10$, $H = 2.5$, and $F = 20$

By varying the volatility parameter, e.g., between 0.15 and 0.35, we investigate the sensitivity of the thresholds and the relative value of flexibility. In Fig. 7.12, we show that higher volatility causes the thresholds to spread wider apart. Indeed, greater uncertainty induces more hesitancy as the value of suspension from an active state increases, but this value stems from the option value of keeping the discretion to suspend alive. Likewise, from state 2, higher volatility increases the value of the option to resume operations by moving to state 1. However, this also increases the opportunity cost of exercising the option to resume operations, and, as a result, the power company is more cautious in making the operational change. As for the relative value of flexibility, we plot in Fig. 7.13 the ratio of $a_{0,1}$ from Eq. (7.24) to that from Eq. (7.10) with respect to σ. It increases with volatility because, intuitively, more uncertainty gives more value to the flexibility option. Here, the power company would be willing to pay about 3 % more for a power plant with such flexibility. A similar analysis for a California-based distributed generation unit may be found in [39]. □

7.4 Modularity and Capacity Expansion

Rather than investing in a power plant all at once, it may be desirable to make incremental capacity additions. One motivation for modularizing adoption of the power plant is that the power company may prefer to observe how the electricity

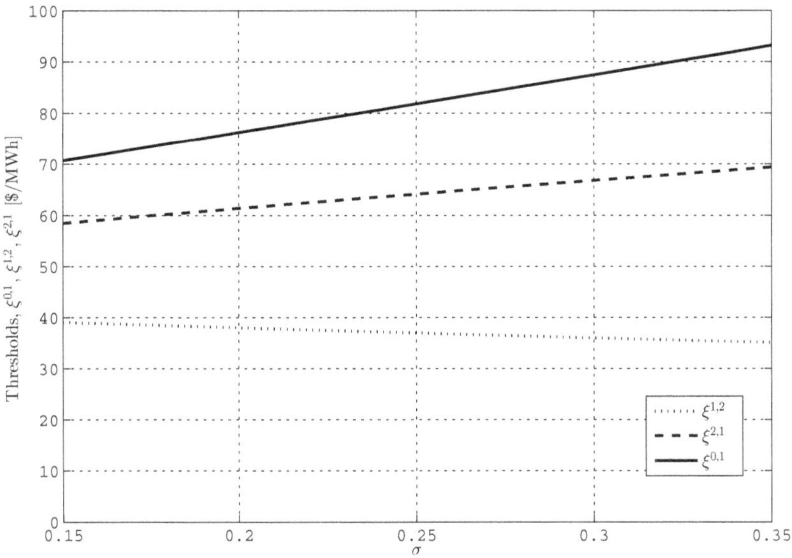

Fig. 7.12 Sensitivity of $\xi^{0,1}$, $\xi^{1,2}$, and $\xi^{2,1}$ with respect to σ with operational flexibility for $\alpha = 0.05$, $\rho = 0.10$, $I = 100$, $S^{1,2} = 10$, $S^{2,1} = 10$, $H = 2.5$, and $F = 20$

price is unfolding and to match capacity to the needs to the market. For example, technological advances have made it possible for small-scale modules, i.e., less than 300 MW, to be developed even for nuclear power plants [44]. By proceeding to add capacity in an incremental manner, the power company may benefit from starting cash flows sooner while adding larger modules later on [16]. Hence, although the total investment costs of modular units may be higher than that of a single large unit, these diseconomies of scale may be outweighed by the benefit from optimizing capacity additions.

In order to explore such modular capacity expansion, we assume that the power company may invest in a power plant of total capacity $K = K^1 + K^2$ either directly or sequentially. Without loss of generality, we assume that the capital cost from "lumpy" investment, I, will be the same as the total capital cost from the modular approach, i.e., $I^1 + I^2$, where I^j is the capital cost of module j. This may be extended to treat an arbitrary number of modules as well as operational flexibility. Thus, although we ignore total economies of scale, we nevertheless have $\frac{I^1}{K^1} < \frac{I^2}{K^2}$, i.e., relative diseconomies of scale in integrating the second module, which reflect difficulties associated with modifying fixed infrastructure. This is similar to the assumption made by [25, 40].

Figure 7.14 illustrates the sequence of decisions that are possible under the direct and modular investment strategies. In the former, the only possible transition is between states 0 and 2. Proceeding backward, state 2 is one in which both modules are active, i.e., the power company has a perpetually operating plant that outputs

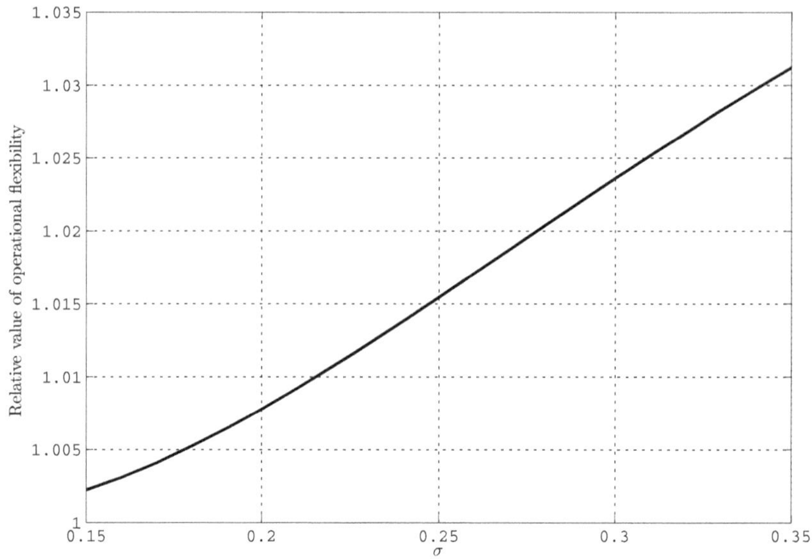

Fig. 7.13 Sensitivity of relative value of flexibility with respect to σ with operational flexibility for $\alpha = 0.05$, $\rho = 0.10$, $I = 100$, $S^{1,2} = 10$, $S^{2,1} = 10$, $H = 2.5$, and $F = 20$

K MWh of electricity per year and costs I to install. Thus, the value in state 2, assuming that the electricity price follows a GBM as in Eq. (7.1), a heat rate of H, a fuel price of F, and an exogenous discount rate of ρ, is:

$$\mathscr{W}_2(E) = \frac{KE}{\rho - \alpha} - \frac{KHF}{\rho}. \tag{7.28}$$

In state 0, the value function reflects simply the option to invest directly in such a power plant. Consequently, by following the same argument as in Eqs. (7.4)–(7.5), we have:

$$\mathscr{W}_0^d(E) = a_{0,1}^d E^{\beta_1}. \tag{7.29}$$

We let the ^{d} denote a "direct" investment strategy with the corresponding endogenous $a_{0,1}^d$. Via value-matching and smooth-pasting conditions between the functions in Eqs. (7.28) and (7.29), we obtain the optimal investment threshold price by following the direct investment strategy:

$$\xi^{0,2} = \left(\frac{\beta_1}{\beta_1 - 1}\right)(\rho - \alpha)\left[\frac{I}{K} + \frac{HF}{\rho}\right]. \tag{7.30}$$

Analogously, we have:

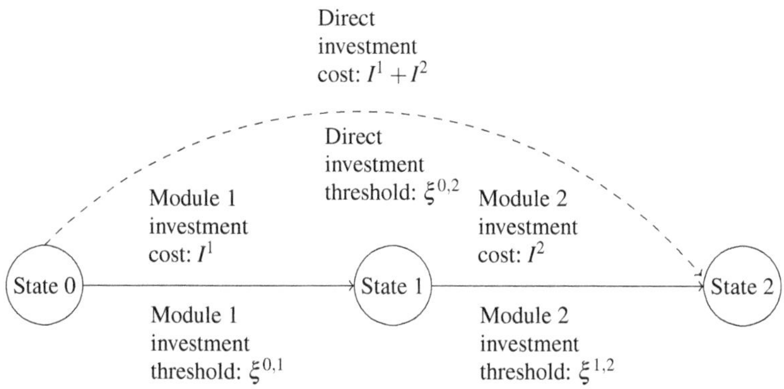

Fig. 7.14 State-transition diagram for a power plant with two modules

$$a_{0,1}^d = \frac{K\left(\xi^{0,2}\right)^{1-\beta_1}}{\beta_1(\rho-\alpha)}. \tag{7.31}$$

By contrast, with a modular investment strategy, the power company first invests in a module of annual output K^1, i.e., going from state 0 to state 1. In state 1, its value function includes not only the expected PV of cash flows from a module that operates forever but also the option to upgrade to the second module. Thus, the value function in state 1 is:

$$\mathscr{W}_1(E) = a_{1,1}E^{\beta_1} + \frac{K^1E}{\rho-\alpha} - \frac{K^1HF}{\rho}. \tag{7.32}$$

Finally, the value function in state 0 reflects the option value to invest in the first module with the subsequent option to acquire the second one:

$$\mathscr{W}_0(E) = a_{0,1}E^{\beta_1}. \tag{7.33}$$

Under the modular strategy, we need to solve for two investment thresholds, $\xi^{1,2}$ and $\xi^{0,1}$, as well as $a_{1,1}$ and $a_{0,1}$. We obtain these via four value-matching and smooth-pasting conditions:

$$\mathscr{W}_1\left(\xi^{1,2}\right) = \mathscr{W}_2\left(\xi^{1,2}\right) - I^2$$
$$\Rightarrow a_{1,1}\xi^{1,2\beta_1} + \frac{K^1\xi^{1,2}}{\rho-\alpha} - \frac{K^1HF}{\rho} = \frac{K\xi^{1,2}}{\rho-\alpha} - \frac{KHF}{\rho} - I^2 \tag{7.34a}$$

$$\mathscr{W}_1'\left(\xi^{1,2}\right) = \mathscr{W}_2'\left(\xi^{1,2}\right)$$
$$\Rightarrow \beta_1 a_{1,1}\xi^{1,2\beta_1-1} + \frac{K^1}{\rho-\alpha} = \frac{K}{\rho-\alpha} \tag{7.34b}$$

$$\mathscr{W}_0\left(\xi^{0,1}\right) = \mathscr{W}_1\left(\xi^{0,1}\right) - I^1$$

$$\Rightarrow a_{0,1}\xi^{0,1^{\beta_1}} = a_{1,1}\xi^{0,1^{\beta_1}} + \frac{K^1\xi^{0,1}}{\rho-\alpha} - \frac{K^1 HF}{\rho} - I^1 \tag{7.34c}$$

$$\mathscr{W}_0'\left(\xi^{0,1}\right) = \mathscr{W}_1'\left(\xi^{0,1}\right)$$

$$\Rightarrow \beta_1 a_{0,1}\xi^{0,1^{\beta_1-1}} = \beta_1 a_{1,1}\xi^{0,1^{\beta_1-1}} + \frac{K^1}{\rho-\alpha}. \tag{7.34d}$$

The analytical solutions are:

$$\xi^{1,2} = \left(\frac{\beta_1}{\beta_1-1}\right)(\rho-\alpha)\left[\frac{I^2}{K^2} + \frac{HF}{\rho}\right] \tag{7.35}$$

$$a_{1,1} = \frac{K^2\left(\xi^{1,2}\right)^{1-\beta_1}}{\beta_1(\rho-\alpha)} \tag{7.36}$$

$$\xi^{0,1} = \left(\frac{\beta_1}{\beta_1-1}\right)(\rho-\alpha)\left[\frac{I^1}{K^1} + \frac{HF}{\rho}\right] \tag{7.37}$$

$$a_{0,1} = a_{1,1} + \frac{K^1\left(\xi^{0,1}\right)^{1-\beta_1}}{\beta_1(\rho-\alpha)}. \tag{7.38}$$

In comparing the solutions, we note that $\xi^{0,1}$ is independent of $\xi^{1,2}$. Indeed, although the value in state 0 is affected by that of state 1 (since $a_{0,1}$ depends on $a_{1,1}$), the timing of the investment in the first module is myopic, i.e., it is as if the second module did not exist. This is due to the structure of the sequential decision-making problem. Recall that the investment is delayed up to the point that the marginal benefit of waiting equals the marginal cost of waiting. From Sect. 7.2, we know that the former quantity is related to starting the plant at a higher price and reducing the discounted investment cost. Meanwhile, the marginal cost of waiting is the opportunity cost of not earning cash flows from an active power plant. Now with a subsequent module, the marginal benefit of waiting additionally includes the discounted expected marginal benefit (from having to wait less until the second module is installed after the first one is adopted) and the discounted expected marginal cost (from having to wait longer from the initial point until the option to install the second module is available). These two extra marginal values cancel out, thereby rendering the effect of the second module on the timing inconsequential. In order to examine the properties of modularity, we next perform Illustrative Examples 7.9 and 7.10.

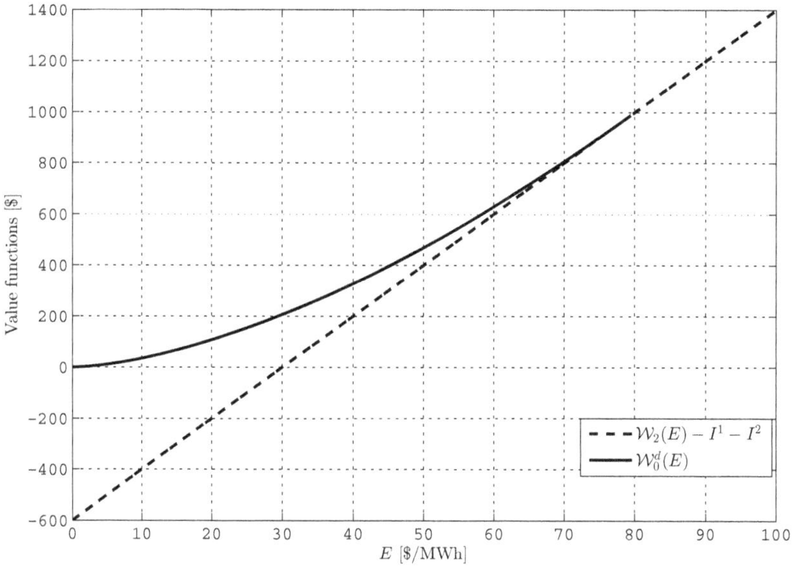

Fig. 7.15 Value functions with direct investment strategy for $\alpha = 0.05$, $\rho = 0.10$, $\sigma = 0.20$, $I = 100$, $K = 1$, $H = 2.5$, and $F = 20$

Illustrative Example 7.9 *Investment timing with modularity*

Let $\alpha = 0.05$, $\rho = 0.10$, $I^1 = 40$, $I^2 = 60$, $K^1 = 0.5$, $K^2 = 0.5$, $H = 2.5$, and $F = 20$. Thus, $I = 100$ and $K = 1.0$, and σ is allowed to vary between 0 and 0.35. First, Fig. 7.15 indicates the value functions with a direct investment strategy. Here, investment occurs when the electricity price reaches a threshold of \$79.30/MWh. Second, in Fig. 7.16, we have the modular investment strategy. As expected, the first module is adopted at a lower threshold, i.e., \$78.31/MWh, than in the direct investment strategy. A subsequent price increase to \$81.95/MWh is required to trigger adoption of the second module. □

Illustrative Example 7.10 *Sensitivity analysis of modularity with respect to volatility*

In performing sensitivity analysis, we examine how the thresholds change with uncertainty in Fig. 7.17. As anticipated, all thresholds increase with uncertainty, with those related to the modular investment strategy sandwiching the one for the direct investment strategy. The relative value of flexibility from following a modular approach is sketched out in Fig. 7.18. For the base case of $\alpha = 0.05$, this relative value is barely 0.1 %. However, with a lower annualized percentage growth rate, it can comprise nearly 5 % of the project's value. The reason is that a modular strategy enables the power company to take advantage of revenues from the more economic module even at relatively low prices before waiting for the right time to complete the project. Finally, this relative value of modularity decreases with uncertainty since

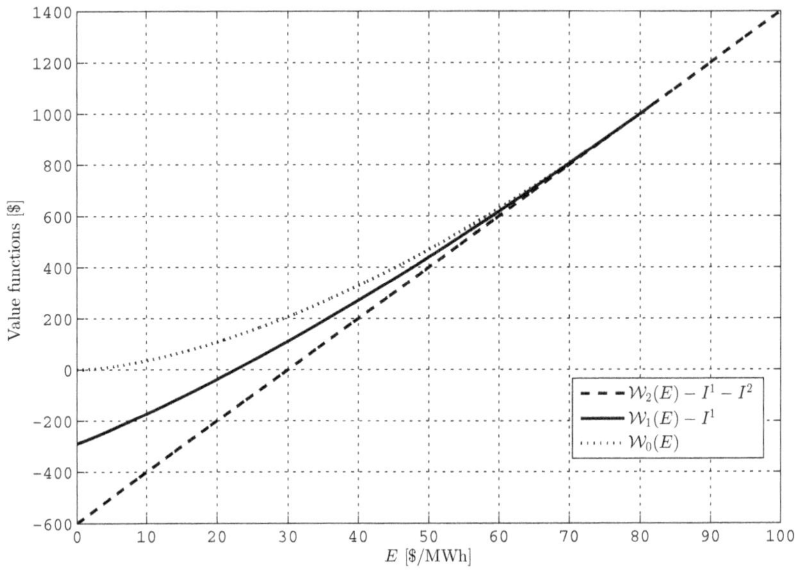

Fig. 7.16 Value functions with modular investment strategy for $\alpha = 0.05$, $\rho = 0.10$, $\sigma = 0.20$, $I^1 = 40$, $I^2 = 60$, $K^1 = 0.5$, $K^2 = 0.5$, $H = 2.5$, and $F = 20$

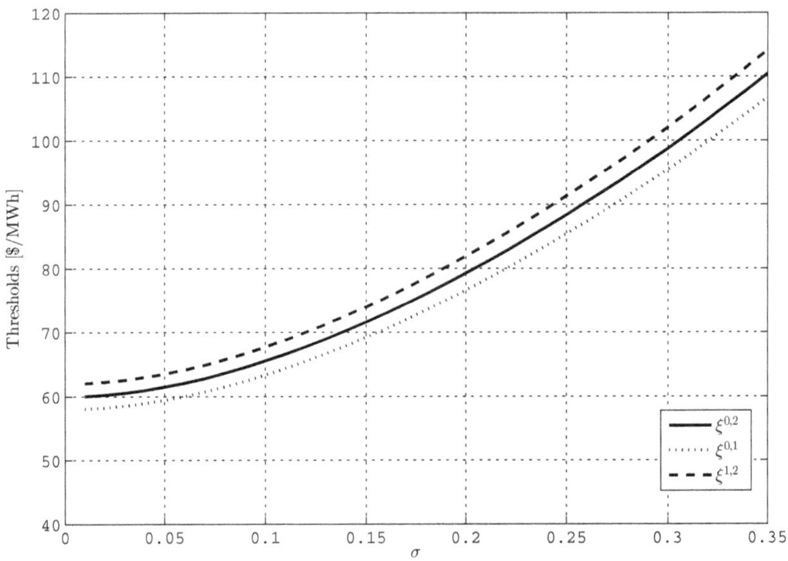

Fig. 7.17 Optimal thresholds under direct and modular investment strategies for $\alpha = 0.05$, $\rho = 0.10$, $I^1 = 40$, $I^2 = 60$, $K^1 = 0.5$, $K^2 = 0.5$, $H = 2.5$, and $F = 20$

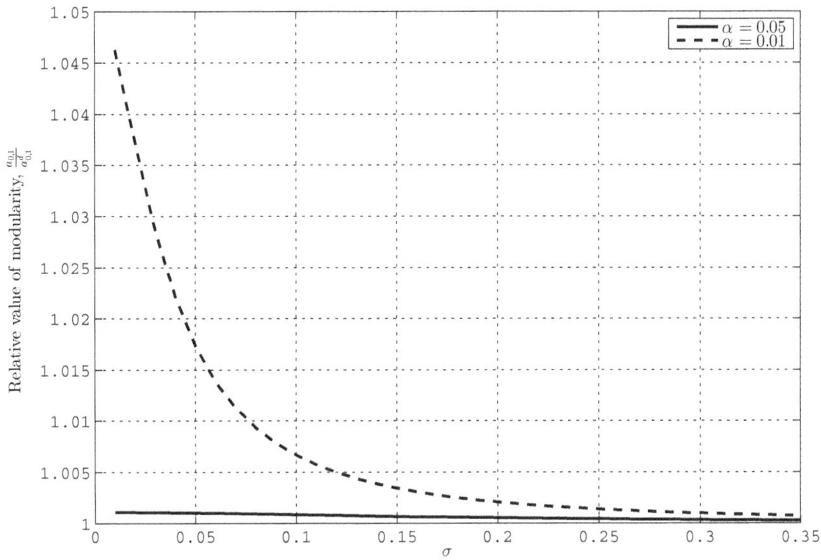

Fig. 7.18 Relative value of modular investment strategy for $\rho = 0.10$, $I^1 = 40$, $I^2 = 60$, $K^1 = 0.5$, $K^2 = 0.5$, $H = 2.5$, and $F = 20$

an increase in σ warrants delaying investment of any type. Hence, there is less discrepancy between the direct and modular investment strategies. See [31, 38] for applications of the modular investment approach to gas-fired power plants and distributed generation facilities, respectively. □

7.5 Continuous Capacity Sizing

Up to now, we have examined managerial discretion with respect to investment timing, operations, and modularity while assuming that the size of the completed power plant is simply a constant parameter, K. In reality, the size of the power plant itself may be a decision variable. Subject to land, permitting, and resource constraints, the power company may scale the plant's capacity in order to maximize profit. Depending on the type of plant, the sizing decision may be considered continuous or discrete. For example, in an analysis of distributed generation investment, [29] models gas-fired units as having discrete capacity sizes with batteries and solar photovoltaic panels having capacities that are continuous decision variables. Likewise, [2] examines the optimal investment timing and capacity sizing problem of a run-of-river hydropower plant in Norway by assuming that the scaling decision variable is continuous. By contrast, [14] treats the capacity sizing decision of a wind farm as a discrete one. Thus, either assumption may be valid depending on the characteristics of the technology and siting constraints. In this section, we assume that the endogenous sizing decision is continuous and follow in the spirit of [6]. A discrete treatment of the sizing decision is implemented in the next section.

As in previous sections, we assume that the power company has the discretion to invest in a power plant at a time of its choosing after which it will earn a profit flow that equals the stochastic revenue from electricity sales and a deterministic operating cost. Now, in addition, the power company may also determine the size of the plant, $\kappa(E)$, which depends on the electricity price and is the solution to the following now-or-never expected NPV maximization problem:

$$\kappa(E) \equiv \arg\max_{K} \left[\frac{E}{\rho - \alpha} - \frac{HF}{\rho} \right] K - \mathscr{I}(K). \tag{7.39}$$

We assume increasing marginal construction costs because of land and material restrictions, for example. Thus, the investment cost is:

$$\mathscr{I}(K) = AK^B, \tag{7.40}$$

where $A > 0$ and $B > 1$ are deterministic parameters. Consequently, with this convex investment cost, the optimal capacity size is obtained by differentiating the right-hand side of equation (7.39) with respect to K and setting it equal to zero:

$$\begin{aligned} &\tfrac{E}{\rho-\alpha} - \tfrac{HF}{\rho} - AB\kappa(E)^{B-1} = 0 \\ \Rightarrow \kappa(E) &= \max \left\{ \left[\tfrac{1}{AB} \left(\tfrac{E}{\rho-\alpha} - \tfrac{HF}{\rho} \right) \right]^{\frac{1}{B-1}}, 0 \right\}. \end{aligned} \tag{7.41}$$

Hence, the maximized expected now-or-never NPV is obtained via substitution of Eq. (7.41) into the right-hand side of Eq. (7.39):

$$\mathscr{W}_1(E;\kappa(E)) - \mathscr{I}(\kappa(E)) = \begin{cases} 0, & \text{if } \kappa(E) = 0 \\ \left[\tfrac{1}{AB} \left(\tfrac{E}{\rho-\alpha} - \tfrac{HF}{\rho} \right)^B \right]^{\frac{1}{B-1}} \left(\tfrac{B-1}{B}\right), & \text{otherwise.} \end{cases} \tag{7.42}$$

$\mathscr{W}_1(E;\kappa(E)) - \mathscr{I}(\kappa(E))$ indicates the maximized expected NPV of the power plant given that it is optimal to construct immediately. However, besides this sizing flexibility the power company also has discretion over the investment timing. As in previous sections, it is possible to show that the value of the option to invest is:

$$\mathscr{W}_0(E) = a_{0,1} E^{\beta_1}. \tag{7.43}$$

We next determine the optimal investment threshold via value-matching and smooth-pasting conditions between the functions in Eqs. (7.42) and (7.43):

$$a_{0,1}\xi^{\beta_1} = \left[\frac{\xi}{\rho - \alpha} - \frac{HF}{\rho}\right]^{\frac{B}{B-1}} \left(\frac{1}{AB}\right)^{\frac{1}{B-1}} \left(\frac{B-1}{B}\right) \tag{7.44a}$$

$$\beta_1 a_{0,1}\xi^{\beta_1 - 1} = \left(\frac{1}{\rho - \alpha}\right)\left[\frac{\xi}{\rho - \alpha} - \frac{HF}{\rho}\right]^{\frac{1}{B-1}} \left(\frac{1}{AB}\right)^{\frac{1}{B-1}}. \tag{7.44b}$$

Although Eqs. (7.44a) and (7.44b) are highly nonlinear, it is possible to solve them analytically for ξ:

$$\xi = \frac{(\rho - \alpha)HF\beta_1(B-1)}{\rho\,(\beta_1(B-1) - B)}. \tag{7.45}$$

Finally, by substituting ξ from Eq. (7.45) into Eq. (7.41), we obtain the optimal capacity size at the investment threshold price:

$$\kappa(\xi) = \left[\frac{1}{AB}\frac{HF}{\rho}\left(\frac{B}{\beta_1(B-1) - B}\right)\right]^{\frac{1}{B-1}}, \tag{7.46}$$

where we must ensure that $\beta_1(B-1) - B > 0$.

Illustrative Example 7.11 *Investment with continuous capacity sizing*

In order to gain more intuition about having flexibility over capacity sizing, we perform numerical examples with the following parameter values: $\alpha = 0.01$, $\rho = 0.10$, $A = 2.65 \times 10^{-5}$, $B = 2$, $H = 2.5$, and $F = 20$. We allow the volatility, σ, to vary between 0.01 and 0.10. The parameter A corresponds approximately to the investment cost of a typical gas-fired power plant. For example, the 430 MW CCGT plant built in Aghada [41] cost \$371 million. Since this is close to the capacity cost of \$876/kW assumed in Sect. 7.2, we use it to calculate a total investment cost for this plant to be \$377 million. By inserting this value into Eq. (7.40), we obtain $A = \frac{377\times10^6}{(430\times8760)^2} = 2.65 \times 10^{-5}$. Using these parameters, we obtain the value functions given in Fig. 7.19. Here, the function $\mathscr{W}_1(E;\kappa(E)) - \mathscr{I}(\kappa(E))$ represents the maximized expected NPV of the power plant from Eq. (7.42). In other words, this nonlinear function assumes that there is no discretion over the timing of the investment, but the capacity of the plant may be determined optimally as a function of the current electricity price, E. This now-or-never capacity size, $\frac{\kappa(E)}{8760}$, is illustrated in Fig. 7.23 for different values of E and is linearly increasing as long as the electricity price is high enough to cover the discounted operating costs. A doubling of A simply reduces the optimal now-or-never capacity level. For this reason, the maximized expected NPV in Fig. 7.19 is bounded by zero. Taking the value of waiting into account means that it is optimal to invest in the plant only when the price of electricity hits the threshold ξ, which is \$90/MWh in this case. Thus, the difference between the functions $\mathscr{W}_0(E)$ and $\mathscr{W}_1(E;\kappa(E)) - \mathscr{I}(\kappa(E))$ reflects the value of this deferral option. Finally, the linear function $\mathscr{W}_1(E;\kappa(\xi)) - \mathscr{I}(\kappa(\xi))$ is one in which

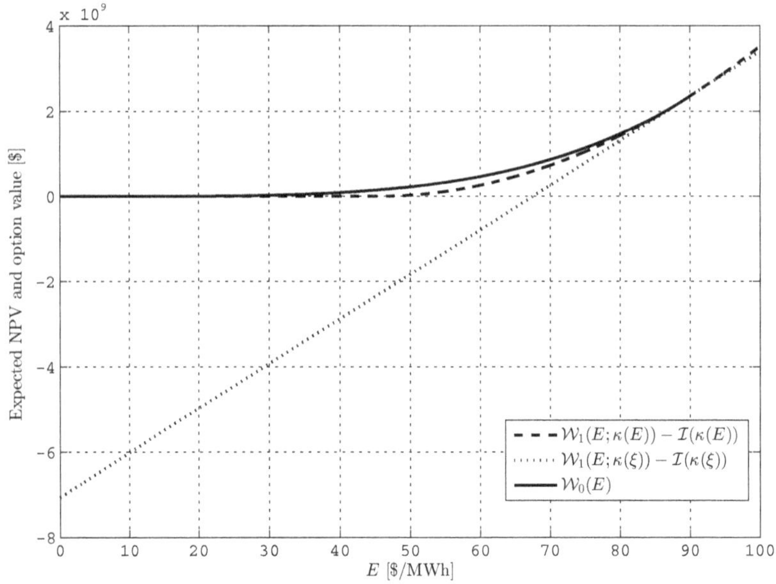

Fig. 7.19 Value functions with capacity sizing, $\sigma = 0.10$, and $A = 2.65 \times 10^{-5}$

there is no discretion over either investment timing or capacity sizing. As such, it reflects the now-or-never expected NPV of investing in a power plant of optimal capacity $\frac{\kappa(\xi)}{8760} = 1075$ MW immediately. Consequently, since the firm has no subsequent flexibility over its decision-making, it is exposed to losses if the electricity price decreases. Figure 7.20 repeats this figure for a doubled marginal cost of investment, i.e., $A = 5.31 \times 10^{-5}$. □

Illustrative Example 7.12 *Sensitivity analysis of investment with continuous capacity sizing with respect to volatility*

We next conduct sensitivity analysis with respect to the volatility, σ. In Fig. 7.21, we plot the optimal investment threshold price, ξ, and note that it increases monotonically. Interestingly, it is independent of the A parameter, i.e., a higher marginal cost of capacity will not affect the optimal timing of investment. This is also evident analytically from Eq. (7.45). The explanation for this result is provided by what happens to the optimal capacity size. In Fig. 7.22, we plot both the optimal capacity size, $\frac{\kappa(\xi)}{8760}$, and the now-or-never capacity size, $\frac{\kappa(E)}{8760}$, at the current electricity price of \$50/MWh as given in Eqs. (7.46) and (7.41), respectively, for two levels of A. Since the now-or-never decision is independent of the volatility, it is constant for all values of σ at 119.45 MW (and 59.72 MW for the higher value of A). By contrast, optimal capacity sizing is based on waiting until the electricity price hits ξ and building a power plant of the appropriate size. For $\sigma = 0.10$, this is 1075 MW (and 537.50 MW for the higher value of A). Hence, as uncertainty increases, it is optimal to

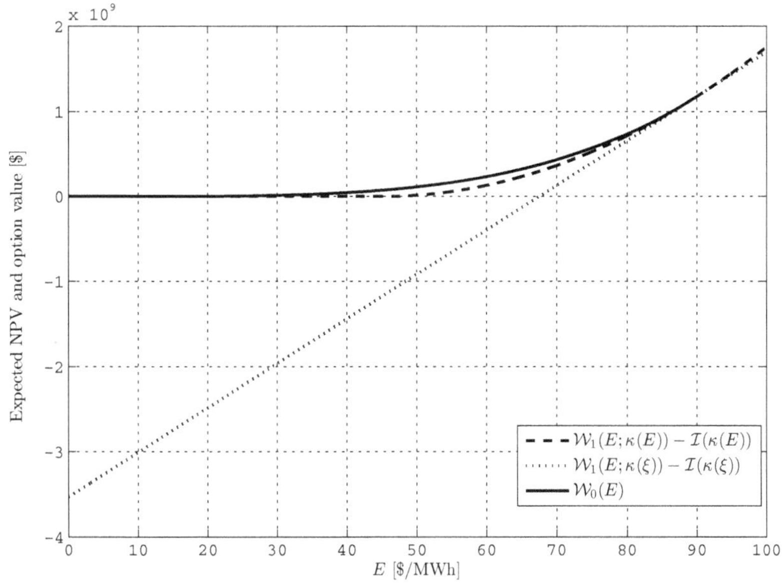

Fig. 7.20 Value functions with capacity sizing, $\sigma = 0.10$, and $A = 5.31 \times 10^{-5}$

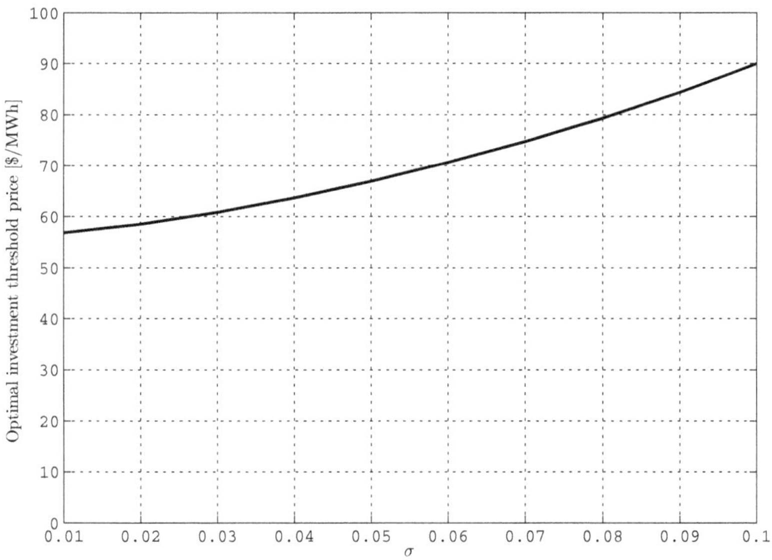

Fig. 7.21 Optimal investment threshold price with endogenous capacity sizing as a function of volatility, σ

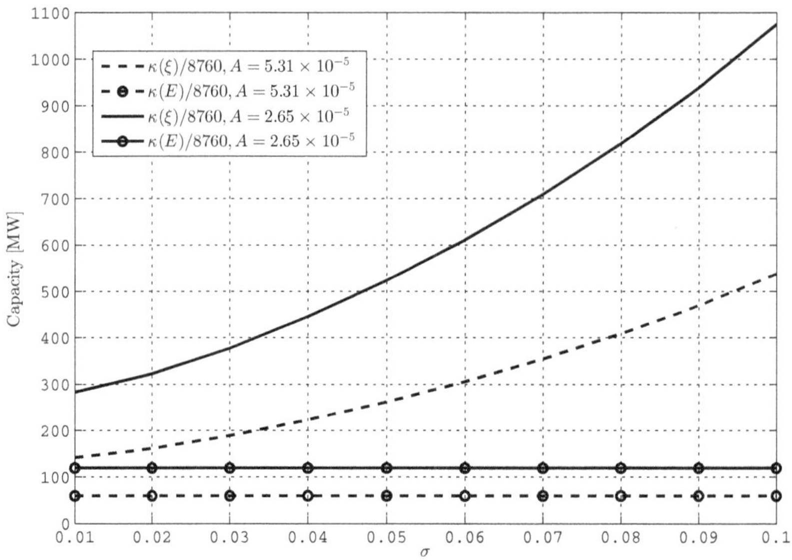

Fig. 7.22 Optimal capacity size as a function of volatility, σ

wait longer and to build a larger plant, but the impact of a higher marginal cost of capacity expansion is absorbed into the sizing decision only and leaves the optimal investment threshold unchanged. Finally, Fig. 7.23 plots the now-or-never capacity size, $\frac{\kappa(E)}{8760}$, from Eq. (7.41) as a function of the current electricity price to indicate the linear dependence as long as the price is high enough to cover operating costs. □

7.6 Mutually Exclusive Technologies

In the previous section, we assumed that it is possible to determine the size of the power plant endogenously as a continuous variable. While this supposition may be valid for certain types of facilities, it does not hold for those that are available only in discrete capacity sizes. For example, wind turbines and nuclear reactors cannot be scaled continuously. Likewise, even smaller gas-fired generators are typically optimized for performance and are available in discrete sizes [29]. Thus, in choosing capacities [14] or between different technologies [37, 43], it is also desirable to consider mutually exclusive discrete alternatives from the viewpoint of real options.

In this context, [11] proposes a simple adjustment to the standard real options treatment of investment under uncertainty when considering any finite number of discrete investment opportunities under uncertainty. For example, with two projects, $j = 1, 2$, of discrete size as given in Fig. 7.24, [11] would proceed as follows:

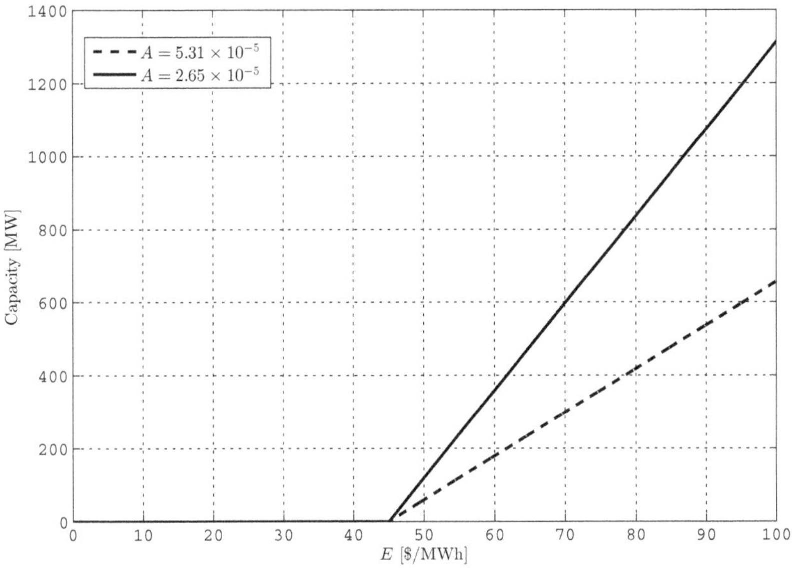

Fig. 7.23 Now-or-never capacity size as a function of initial electricity price, E

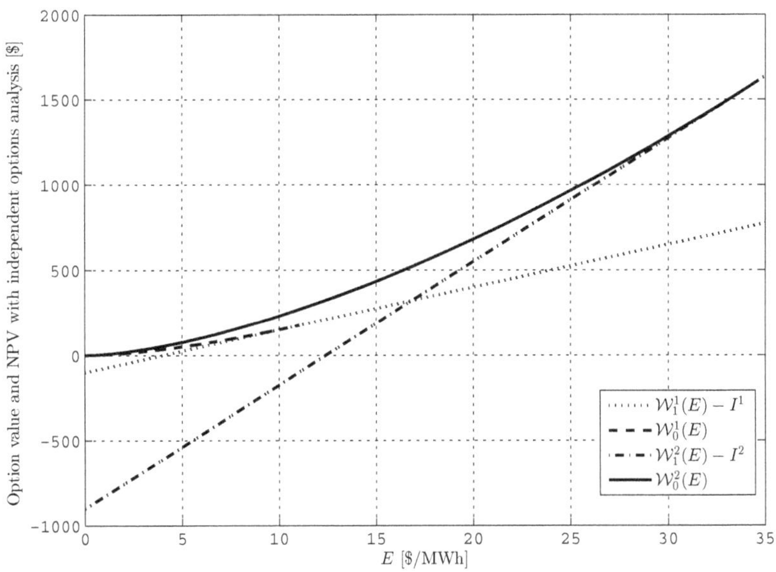

Fig. 7.24 Independent real options analyses of two discrete-sized projects, 1 and 2, for $\sigma = 0.30$

1. Find the optimal investment thresholds, ξ^j, along with the endogenous coefficients, $a_{0,1}^j$, from independent real options analysis of each alternative.
2. Let $j^* \equiv \arg\max_j \left\{a_{0,1}^j\right\}$ be the project with the higher option value coefficient.
3. If the current price, E, is less than project j^*'s threshold, ξ^{j^*}, then wait for the threshold ξ^{j^*} to be hit and invest in project j^*; otherwise, if $E > \xi^{j^*}$, then invest immediately in the project with the highest expected NPV, $\mathscr{W}_j(E) - I^j$.

This procedure seems sensible, but it can break down when the option value coefficient for the smaller project is higher than that for the larger project, i.e., $a_{0,1}^1 > a_{0,1}^2$, and the initial electricity price is equal to the indifference level between the two NPVs. In such a situation, [11] would suggest tossing a coin to break the tie. However, given the uncertainty in the electricity price, it seems intuitive that waiting for more information would be optimal in such a situation.

Following this line of reasoning, [7] allows for the value of the option to invest to be discontinuous, i.e., dichotomous with an upper branch that straddles the indifference point, $\tilde{E}$, at which the two projects' expected NPVs are equal. Specifically, if we let the now-or-never expected NPV of project $j = 1, 2$ be defined as:

$$\mathscr{W}_1^j(E) - I^j = \frac{K^j E}{\rho - \alpha} - \frac{K^j H^j F}{\rho} - I^j. \tag{7.47}$$

In order to have a tradeoff between the two projects, we assume that $K^2 > K^1$ and $I^2 > I^1$ such that $I^1/K^1 < I^2/K^2$. Without loss of generality, we set $H^1 = H^2$. Thus, the power company has a mutually exclusive choice between a smaller but relatively less costly (plant 1) or a large but relatively more costly (plant 2) option along with the right to determine the timing of the investment decision. By setting the expected NPVs of the two projects equal to each other, we find the indifference point:

$$\tilde{E} = \frac{(\rho - \alpha)}{\rho}\left[\frac{\rho\left(I^1 - I^2\right) + F\left(K^1H^1 - K^2H^2\right)}{K^1 - K^2}\right]. \tag{7.48}$$

If we do a real options analysis of each project j independently, i.e., assuming that the other project does not exist, then we obtain the usual optimal investment threshold prices and endogenous coefficients via value-matching and smooth-pasting conditions:

$$\xi^j = \left(\frac{\beta_1}{\beta_1 - 1}\right)\frac{(\rho - \alpha)}{\rho}\left[H^j F + \frac{I^j}{K^j}\right] \tag{7.49}$$

$$a_{0,1}^j = \frac{K^j\left(\xi^j\right)^{1-\beta_1}}{\beta_1(\rho - \alpha)}. \tag{7.50}$$

Hence, the independent value of the option to invest in project j is simply

$$\mathscr{W}_0^j(E) = a_{0,1}^j E^{\beta_1}. \tag{7.51}$$

The procedure in [7] for dealing with such mutually exclusive investment opportunities is as follows:

1. Order the projects by their capacities.
2. Find $a_{0,1}^j$ for each project j.
3. If $a_{0,1}^1 \leq a_{0,1}^2$, then the value of the option to invest will not be dichotomous because the larger project dominates the smaller one. In this case, the value of the investment opportunity is simply $\mathscr{W}_0(E) = \mathscr{W}_0^2(E)$, i.e., project 1 can effectively be ignored.
4. If $a_{0,1}^1 > a_{0,1}^2$, then the value of the option to invest will have two waiting regions:
   a. $E \in [0, \xi^1)$, which involves waiting for the price to increase until it is optimal to invest in project 1.
   b. $E \in (\xi^L, \xi^R)$, which involves waiting for the price to decrease (increase) until it is optimal to invest in project 1 (2).

   The value of the investment opportunity is thus:

$$\mathscr{W}_0(E) = \begin{cases} a_{0,1}^1 E^{\beta_1}, & \text{if } E \in [0, \xi^1) \\ a^R E^{\beta_1} + a^L E^{\beta_2}, & \text{if } E \in (\xi^L, \xi^R). \end{cases} \tag{7.52}$$

Thus, the dichotomous value function in Eq. (7.52) is defined over two ranges. Although ξ^1 and $a_{0,1}^1$ are known, the thresholds, ξ^L and ξ^R, as well as the coefficients, a^L and a^R, must be found endogenously via value-matching and smooth-pasting conditions between the second branch of $\mathscr{W}_0(E)$ in Eq. (7.52) and $\mathscr{W}_1^j(E) - I^j$ as follows:

$$\mathscr{W}_0(\xi^L) = \mathscr{W}_1^1(\xi^L) - I^1$$
$$\Rightarrow a^R \left(\xi^L\right)^{\beta_1} + a^L \left(\xi^L\right)^{\beta_2} = \frac{K^1 \xi^L}{\rho - \alpha} - \frac{K^1 H^1 F}{\rho} - I^1 \tag{7.53a}$$

$$\left.\frac{\mathrm{d}\mathscr{W}_0(E)}{\mathrm{d}E}\right|_{E=\xi^L} = \left.\frac{\mathrm{d}\mathscr{W}_1^1(E)}{\mathrm{d}E}\right|_{E=\xi^L}$$
$$\Rightarrow \beta_1 a^R \left(\xi^L\right)^{\beta_1 - 1} + \beta_2 a^L \left(\xi^L\right)^{\beta_2 - 1} = \frac{K^1}{\rho - \alpha} \tag{7.53b}$$

$$\mathscr{W}_0(\xi^R) = \mathscr{W}_1^2(\xi^R) - I^2$$
$$\Rightarrow a^R \left(\xi^R\right)^{\beta_1} + a^L \left(\xi^R\right)^{\beta_2} = \frac{K^2 \xi^R}{\rho - \alpha} - \frac{K^2 H^2 F}{\rho} - I^2 \tag{7.53c}$$

$$\left.\frac{\mathrm{d}\mathscr{W}_0(E)}{\mathrm{d}E}\right|_{E=\xi^R} = \left.\frac{\mathrm{d}\mathscr{W}_1^2(E)}{\mathrm{d}E}\right|_{E=\xi^R}$$

$$\Rightarrow \beta_1 a^R \left(\xi^R\right)^{\beta_1-1} + \beta_2 a^L \left(\xi^R\right)^{\beta_2-1} = \frac{K^2}{\rho-\alpha}. \tag{7.53d}$$

The system in Eqs. (7.53a)–(7.53d) is highly nonlinear and must be solved numerically. As in Sect. 7.3, guesses for the four unknowns are required. Reasonable guesses for ξ^L and ξ^R are $\frac{\xi^1+\tilde{E}}{2}$ and ξ^2, respectively. Likewise, a guess for a^R may be obtained by dropping the $\beta_2 a^L \left(\xi^R\right)^{\beta_2-1}$ term in smooth-pasting Eq. (7.53d) to solve explicitly for the remaining option value coefficient. This may be substituted into the remaining smooth-pasting Eq. (7.53b) to obtain a guess for a^L.

Illustrative Example 7.13 *Mutually exclusive investment with high volatility*

In order to illustrate how the waiting region may be dichotomous, we perform a numerical example with $\rho = 0.04$, $\alpha = 0$, $K^1 = 1$, $K^2 = 2.9$, $I^1 = 100$, and $I^2 = 900$. Without loss of generality, we set $F = 0$ and allow σ to range from 0.05 to 0.30. In other words, plant 2 is almost three times as large but has an investment cost that is nine times as high as that of plant 1. Note that these parameter values are different from those in Sect. 7.2 in order to obtain a nontrivial result with $a_{0,1}^1 > a_{0,1}^2$ for low values of volatility. Figure 7.24 illustrates that the two projects may be analyzed separately for a relatively high value of σ, i.e., 0.30. Here, it is clear that $a_{0,1}^1 < a_{0,1}^2$, because $\mathscr{W}_0^2(E) > \mathscr{W}_0^1(E)$. Thus, the optimal strategy is simple: the power company should disregard plant 1 and wait for the electricity price to hit the threshold $\xi^2 = 34.30$. The corresponding value functions are indicated in Fig. 7.25. □

Illustrative Example 7.14 *Mutually exclusive investment with low volatility*

With a relatively low level of volatility, e.g., $\sigma = 0.15$, we have $a_{0,1}^1 > a_{0,1}^2$. Consequently, the waiting region becomes dichotomous with an upper region around the indifference price, $\tilde{E} = 16.84$. This upper waiting region, reflected by $a^R E^{\beta_1} + a^L E^{\beta_2}$, extends from $\xi^L = 11.89$ to $\xi^R = 21.71$. For comparison, we have $\xi^1 = 6.76$ and $\xi^2 = 20.97$. Figure 7.26 shows that the lower portion of the $\mathscr{W}_0(E)$ function is precisely $\mathscr{W}_0^1(E)$. □

Illustrative Example 7.15 *Sensitivity analysis of mutually exclusive investment with respect to volatility*

As σ is varied, we obtain waiting and immediate investment regions for the two plants considered together in Fig. 7.27. For $\sigma \leq 0.21$, it is impossible to disregard plant 1, and the dichotomous value of waiting must be considered. For example, with $\sigma = 0.15$, there are lower and upper waiting regions. In the former, the power company should wait until the electricity price increases to ξ^1 before investing immediately in plant 1. By contrast, in the latter, the power company may end up investing in either plant 1 (if the price drops to ξ^L) or plant 2 (if the price increases to ξ^R).

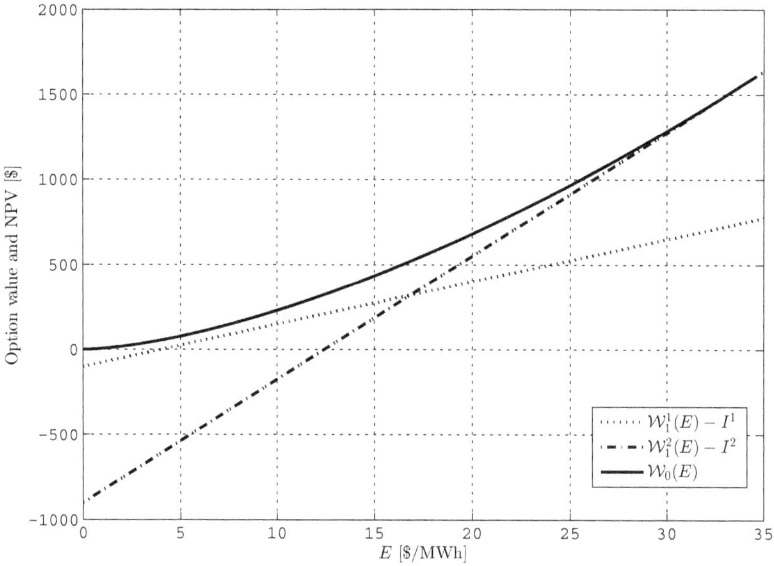

Fig. 7.25 Mutually exclusive investment opportunity in two discrete-sized projects, 1 and 2, for $\sigma = 0.30$

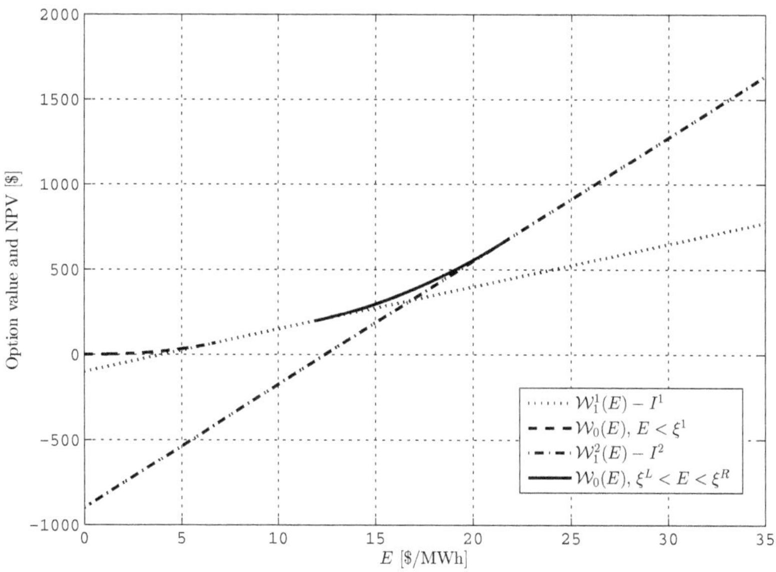

Fig. 7.26 Mutually exclusive investment opportunity in two discrete-sized projects, 1 and 2, for $\sigma = 0.15$

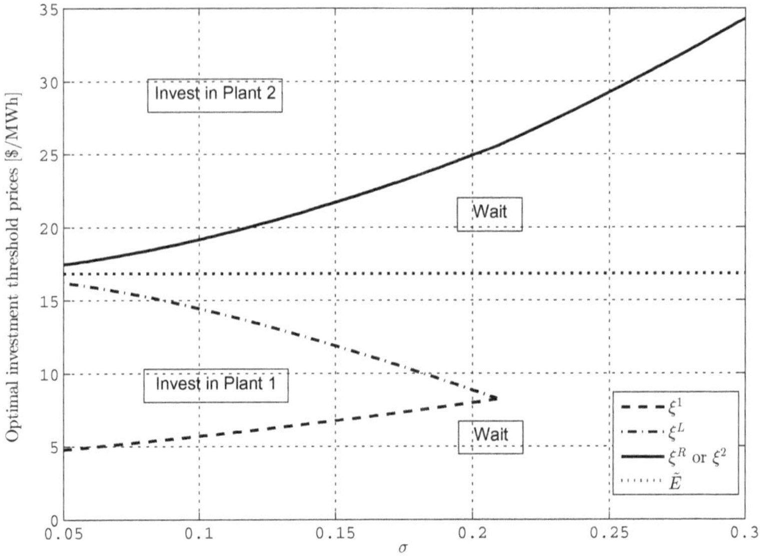

Fig. 7.27 Investment and waiting regions for a mutually exclusive investment opportunity in plants 1 and 2

However, if the current electricity price is in the range $[\xi^1, \xi^L]$ or $[\xi^R, \infty)$, then it is optimal to invest immediately in plant 1 or 2, respectively. The MATLAB code in Sect. 7.10 solves the nonlinear system in Eqs. (7.53a)–(7.53d) and generates Fig. 7.27. As [7] illustrates using a similar numerical example, the percentage gain from optimally delaying investment at the indifference price relative to investing immediately as suggested by [11] may be substantial. In our example, it will be 18.31 % for $\sigma = 0.15$. Finally, this mutually exclusive analysis may be extended to allow for switching options as in Sect. 7.4, i.e., having the right to switch from plant 1 to 2 [14], or allowing for subsequent improvement in the performance of one of the two plants [37]. □

7.7 Risk Aversion

In previous sections of this chapter, we assumed that the decision maker, i.e., typically a power company, is risk neutral because its objective is to maximize expected profit. While this may be justified if the standard assumptions of finance, viz., complete markets, hold and risk may be diversified by holding a portfolio of freely traded assets, such a supposition may not hold in practice. For example, besides market risk, power companies in the electric power industry may be exposed to heterogenous risk stemming from technological uncertainty associated with R&D in renewable energy technologies or the possibility of a change in policy support. Moreover, some municipally owned power companies may be inherently risk averse since they are answerable to a more conservative class of investor. Either way, it would be desirable to expand the framework for analysis to permit the decision maker to be risk averse when solving optimal timing or technology choice problems.

Taking the perspective of [19], we embed a utility function into the real options framework in order to examine how a risk-averse investor may make decisions under uncertainty with the deferral option. In the economics literature, constant relative risk aversion (CRRA) is a standard workhorse for both its analytical tractability and its desirable property that the fraction of wealth placed in a risky (as opposed to risk-free) asset by a decision-maker is independent of the initial level of wealth [33].[10] In particular, the CRRA utility function with relative risk aversion parameter $0 \leq \gamma \leq 1$ has the form:

$$\mathscr{U}(E) = \begin{cases} \frac{E^{1-\gamma}}{1-\gamma}, & \text{if } 0 \leq \gamma < 1 \\ \ln E, & \text{otherwise.} \end{cases} \tag{7.54}$$

Hence, as a concave function, $\mathscr{U}(E)$ captures the risk aversion of a conservative investor.

One difficulty with incorporating the CRRA utility function is the treatment of the operating costs. This is because the function is not separable in E and HF, i.e., $\mathscr{U}(E - HF) \neq \mathscr{U}(E) - \mathscr{U}(HF)$. Since we would like to avoid working with a function of the form $\frac{(E-HF)^{1-\gamma}}{1-\gamma}$, we decompose the cash flows using the approach of [4], in which operating costs are also included in a risk-averse analysis of real options decision making. Suppose that at the current time, i.e., $t = 0$, the power company has set aside all of the cash that it will need to pay for a power plant of nominal annual output 1 MWh that will cost I to build and will incur operational costs of HF per MWh of electricity generated. If the power plant operates forever after construction, then its discounted investment and operational costs are $I + \frac{HF}{\rho}$ (assuming a subjective discount rate ρ). This lump sum is assumed to be sitting in an interest-bearing account earning the same discount rate until the plant is constructed at optimal time τ, which implies an instantaneous cash flow of $\rho\left(I + \frac{HF}{\rho}\right)$. Consequently, the discounted (to time 0) utility of the cash flows from this lump sum is $\int_0^{\tau} e^{-\rho t}\mathscr{U}(HF + \rho I)\, dt$. Given that the power plant starts to earn revenues, E_t, at time τ that follow a GBM, the time-zero discounted expected utility of the cash flows is:

$$\int_0^{\tau} e^{-\rho t}\mathscr{U}(HF + \rho I)\, dt + \mathbb{E}_E\left[\int_{\tau}^{\infty} e^{-\rho t}\mathscr{U}(E_t)\, dt\right], \tag{7.55}$$

where E is the electricity price at $t = 0$. Now, since the first term in Eq. (7.55) may be reexpressed as $\int_0^{\infty} e^{-\rho t}\mathscr{U}(HF + \rho I)\, dt - \int_{\tau}^{\infty} e^{-\rho t}\mathscr{U}(HF + \rho I)\, dt$, we can

[10] The CRRA utility function is itself a special case of the hyperbolic absolute risk aversion (HARA) class employed in studies of investor behavior.

de facto decompose the cash flows as follows:

$$\int_0^\infty e^{-\rho t}\mathcal{U}\left(HF+\rho I\right)dt+\mathbb{E}_E\left[\int_\tau^\infty e^{-\rho t}\left\{\mathcal{U}\left(E_t\right)-\mathcal{U}\left(HF+\rho I\right)\right\}dt\right]. \tag{7.56}$$

From the law of iterated expectations and the strong Markov property of the GBM,[11] the conditional expectation in Eq. (7.56) may be rewritten as follows:

$$\begin{aligned}&\mathbb{E}_E\left[\int_\tau^\infty e^{-\rho t}\left\{\mathcal{U}\left(E_t\right)-\mathcal{U}\left(HF+\rho I\right)\right\}dt\right]\\&=\mathbb{E}_E\left[e^{-\rho\tau}\mathbb{E}_{E_\tau}\left[\int_0^\infty e^{-\rho t}\left\{\mathcal{U}\left(E_t\right)-\mathcal{U}\left(HF+\rho I\right)\right\}dt\right]\right].\end{aligned} \tag{7.57}$$

Since the first term in Eq. (7.56) is a constant, it may be ignored in determining the optimal time to invest. Thus, Eq. (7.57) is the discounted (to time $t=0$) expected utility of cash flows from a power plant that becomes active at τ and operates forever. Intuitively, the inner conditional expectation's independence from E means that the two expectations may be separated as follows:

$$\begin{aligned}&\mathbb{E}_E\left[e^{-\rho\tau}\mathbb{E}_{E_\tau}\left[\int_0^\infty e^{-\rho t}\left\{\mathcal{U}\left(E_t\right)-\mathcal{U}\left(HF+\rho I\right)\right\}dt\right]\right]\\=&\mathbb{E}_E\left[e^{-\rho\tau}\right]\mathbb{E}_{E_\tau}\left[\int_0^\infty e^{-\rho t}\left\{\mathcal{U}\left(E_t\right)-\mathcal{U}\left(HF+\rho I\right)\right\}dt\right]\\=&\quad\mathbb{E}_E\left[e^{-\rho\tau}\right]\left[\frac{\beta_1\beta_2\mathcal{U}(E_\tau)}{\rho\left(1-\beta_1-\gamma\right)\left(1-\beta_2-\gamma\right)}-\frac{\mathcal{U}(HF+\rho I)}{\rho}\right].\end{aligned} \tag{7.58}$$

In moving from the second to the third line of Eq. (7.58), we use Theorem 9.18 from [24], which finds a closed-form expression for the conditional expectation of an integral of a function of a Brownian motion. Here, $\beta_1>1$ and $\beta_2<0$ are again the positive and negative roots, respectively, of the characteristic quadratic function in Eq. (7.8). Hence, the value of the investment opportunity for a risk-averse decision-maker may be formulated as the solution to the following optimal stopping-time problem:

$$\mathcal{W}_0(E)=\sup_\tau\mathbb{E}_E\left[e^{-\rho\tau}\right]\left[\frac{\beta_1\beta_2\mathcal{U}(E_\tau)}{\rho\left(1-\beta_1-\gamma\right)\left(1-\beta_2-\gamma\right)}-\frac{\mathcal{U}(HF+\rho I)}{\rho}\right]. \tag{7.59}$$

Using the fact that the conditional expectation of the stochastic discount factor is of power form, i.e., $\mathbb{E}_E\left[e^{-\rho\tau}\right]=\left(\frac{E}{E_\tau}\right)^{\beta_1}$, as shown on page 315 of [12], and letting ξ denote the optimal threshold price, we can recast the optimal stopping-time problem in Eq. (7.59) as the following unconstrained nonlinear maximization problem:

[11] These effectively imply that price values after τ are independent of the values before τ and depend only on the value of the process at τ.

$$\mathscr{W}_0(E) = \max_{\xi \geq E} \left(\frac{E}{\xi}\right)^{\beta_1} \left[\frac{\beta_1 \beta_2 \mathscr{U}(\xi)}{\rho\,(1-\beta_1-\gamma)\,(1-\beta_2-\gamma)} - \frac{\mathscr{U}(HF+\rho I)}{\rho}\right]. \tag{7.60}$$

Taking the first-order necessary condition with respect to ξ, we obtain the following:

$$\left(\frac{E}{\xi}\right)^{\beta_1} \frac{\beta_1\beta_2\xi^{-\gamma}}{\rho(1-\beta_1-\gamma)(1-\beta_2-\gamma)}$$
$$-\beta_1\left(\frac{E}{\xi}\right)^{\beta_1}\left(\frac{1}{\xi}\right)\left[\frac{\beta_1\beta_2\xi^{1-\gamma}}{\rho(1-\beta_1-\gamma)(1-\beta_2-\gamma)(1-\gamma)} - \frac{(HF+\rho I)^{1-\gamma}}{\rho(1-\gamma)}\right] = 0. \tag{7.61}$$

Solving Eq. (7.61), we obtain the following optimal investment threshold price:

$$\xi = \left(\frac{\beta_2+\gamma-1}{\beta_2}\right)^{\frac{1}{1-\gamma}} (HF+\rho I). \tag{7.62}$$

Note that although investment thresholds are typically expressed in terms of β_1, e.g., as in Sect. 7.2, here it is more expedient to use β_2. Using the fact that $\beta_1\beta_2 = -\frac{2\rho}{\sigma^2}$, it can also be verified that ξ is the same as the investment threshold under risk neutrality from Eq. (7.13) for $\gamma = 0$. Although we have a closed-form expression for the optimal investment threshold under risk aversion, it is possible to prove analytically that the threshold increases with both volatility and risk aversion as one would expect. However, the proofs are tedious and are carried out in full in [4]. Intuitively, it is important to stress that both volatility and risk aversion increase the optimal investment threshold price but for vastly different reasons: greater uncertainty delays investment because the value of waiting for more information increases, thereby also increasing the opportunity cost of exercising the option to invest. By contrast, since greater risk aversion lowers the inherent payoff of an active power plant, it also reduces the marginal cost of delaying investment, which consists exclusively of stochastic cash flows, by more than the marginal benefit (see [4] for a rigorous proof).

Illustrative Example 7.16 *Investment under uncertainty with risk aversion*

In order to illustrate the properties of optimal investment with a deferral option from the perspective of a risk-averse power company, we perform numerical examples with the following parameter values: $\rho = 0.1$, $\alpha = 0.05$, $H = 2.5$, $F = 20$, $I = 100$, $\sigma = 0.20$, and $\gamma \in [0, 1)$. These are the same values as in Sect. 7.2 for ease of comparison. Indeed, when $\gamma = 0$, all results collapse to the risk-neutral ones. In Figs. 7.28 and 7.29, we draw the expected utility functions for $\gamma = 0.25$ and $\gamma = 0.75$, respectively, corresponding to the expression in Eq. (7.58). Notice that the curves denoted by $\frac{\beta_1\beta_2\mathscr{U}(E)}{\rho(1-\beta_1-\gamma)(1-\beta_2-\gamma)} - \frac{\mathscr{U}(HF+\rho I)}{\rho}$ are concave, as is to be expected for a risk-averse decision-maker. The value of the option to invest given this attitude toward risk is reflected by the curves $\mathscr{W}_0(E)$ as defined in Eq. (7.60). Although they are convex and always nonnegative, compared to the corresponding curve in Fig. 7.4, the value of the option to invest has been eroded. Furthermore, the investment threshold is slightly higher, i.e., $\xi = 80.04$ and $\xi = 81.76$ for $\gamma = 0.25$ and $\gamma = 0.75$, respectively, compared with $\xi = 79.30$ for the risk-neutral case. □

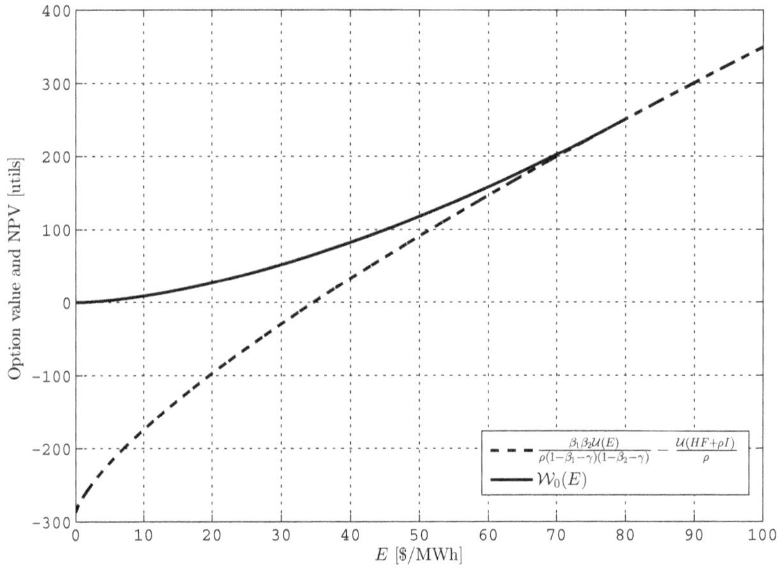

Fig. 7.28 Value of investment opportunity and expected utility of NPV for $\gamma = 0.25$

Illustrative Example 7.17 *Sensitivity analysis of investment thresholds with respect to relative risk aversion and volatility*

In Fig. 7.30, the optimal investment thresholds are indicated for various levels of relative risk aversion and volatility. As discussed, greater risk aversion increases the optimal investment threshold because of the lower valuation of the power plant. Thus, in order to justify investment, a higher trigger price is required. Although greater uncertainty also increases the optimal investment threshold price, it does so for a different reason, i.e., the higher value of waiting. These two parameters interact in order to increase the threshold yet further. □

Illustrative Example 7.18 *Sensitivity analysis of the value of the investment investment opportunity with respect to relative risk aversion and volatility*

Figure 7.31 assesses the extent to which risk aversion affects the valuation of the investment opportunity at some arbitrary common initial electricity price (here, $E = 50$). In particular, we plot the ratio of $\mathscr{W}_0(E)$ from Eq. (7.60) (with risk aversion) to that from Eq. (7.10). When $\gamma = 0$, the ratio is simply one, but it decreases rapidly as γ increases. For example, this ratio becomes just 0.25 for $\gamma = 0.25$ and drops thereafter to only 0.02 for $\gamma = 0.75$. A higher volatility exacerbates this reduction, whereas a lower volatility has the opposite effect. □

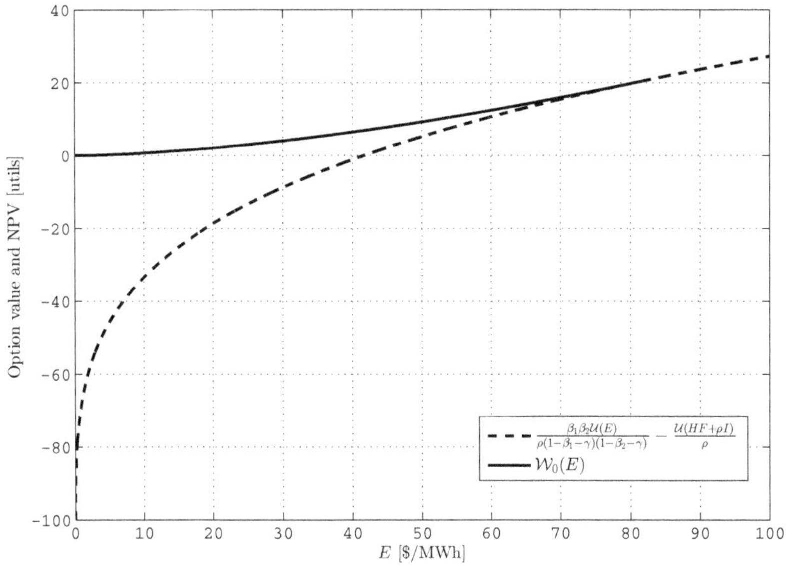

Fig. 7.29 Value of investment opportunity and expected utility of NPV for $\gamma = 0.75$

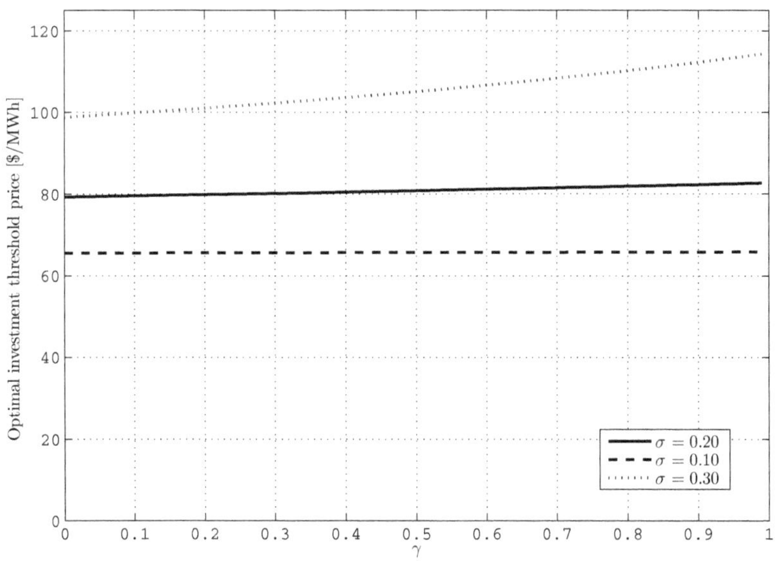

Fig. 7.30 Investment thresholds as functions of relative risk aversion, γ, and volatility, σ

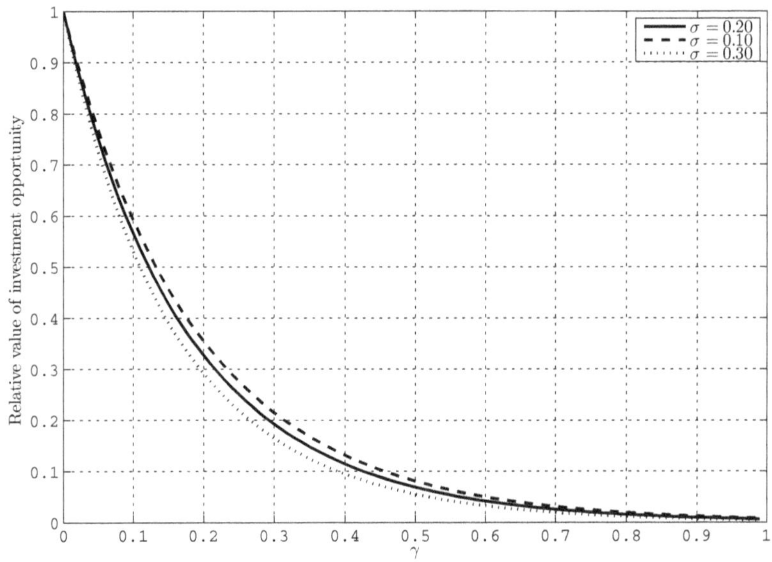

Fig. 7.31 Value of investment opportunity relative to a risk-neutral case as a function of relative risk aversion, γ, and volatility, σ, at initial electricity price, $E = 50$

7.8 Summary and Extensions

In this chapter, we address discretion over investment timing, technology choice, and capacity from the perspective of a single power company facing uncertainty in the electricity price. We also investigate how the optimal decision would change if the company were risk averse or could modularize its investment. Throughout, we model uncertainty using a GBM and assume a simple proposition for investment in order to facilitate analytical tractability from which we could formalize managerial insights. For example, we show how operational flexibility is used to increase the value of the investment opportunity while also lowering the optimal threshold price vis-à-vis the case without operational flexibility. Assessment of such flexibilities, and indeed others involving investment strategies or capacity sizing, would not be possible from the traditional now-or-never NPV approach to decision making.

Specifically, we begin by discussing the limitations of the now-or-never NPV approach in Sect. 7.1. When applied correctly, it can provide insights about optimal investment timing since the maximized NPV is the upper envelope of several mutually exclusive projects, each starting at a different point in time. However, since only discrete time points are considered, it is cumbersome to analyze every conceivable starting date for a given project. Furthermore, it is impossible to distinguish between two power plants of the same capacity that have different levels of operational flexibility. In order to overcome these limitations with the now-or-never NPV approach, the real options methodology is developed in Sect. 7.2. Using an elegant continuous-time approach, it solves the optimal investment timing problem as a stochastic dynamic program. Under the assumption of a GBM for the electricity price, a closed-form solution for the optimal investment threshold price is obtained, which facilitates comparison with the now-or-never NPV threshold to show that the real options approach recommends postponing investment precisely because of the value of waiting. Thus, unlike the now-or-never NPV approach, the real options approach detects the effect of increased uncertainty on the opportunity cost of killing the discretion to wait for more information. Somewhat paradoxically, the real options approach attributes a higher value to the investment opportunity even as it recommends more caution in pulling the investment trigger. This so-called hysteresis in making decisions involving fixed costs is present in many real-world decisions, e.g., farming and unemployment [5, 10]. Empirical work to detect its effects is relatively new, but a study of small run-of-river hydropower plants in Norway tries to examine how uncertainty in government subsidies for renewable energy technologies affected investment timing [26].

Besides tackling the issue of investment timing, we consider operational flexibility in Sect. 7.3. Such embedded options to change the operating status of the power plant in response to fluctuations in market conditions after the initial investment decision surely affects not only the value of the initial opportunity but also the optimal investment threshold. We show how to handle such operational flexibility by first using backward induction to value an active power plant that may switch between active and idle states. Next, we use the value of an active plant to determine the value of the investment opportunity and find that the added flexibility lowers the investment threshold while increasing the option value of investing. Furthermore, the value of this operational flexibility is shown to increase with volatility. Rather than having two operating states, it may be possible to have an arbitrary number or a continuous scale of output captured by a production function (see Chap. 6 of [12]). In a similar vein, Sect. 7.4 examines the possibility of building a power plant in stages. Although the modular investment strategy is always worth more and results in a lower initial threshold price for investment, somewhat surprisingly, the relative value of modularity decreases with uncertainty. This is because greater uncertainty makes it desirable to delay investment of any kind.

We also investigate the possibility of selecting the size of the power plant as an endogenous variable while also considering the deferral option. In Sect. 7.5, the capacity sizing decision is treated as continuous. Here, we again use backward induction to determine first the optimal size of the plant given that it was optimal to proceed with investment. The optimal size is a monotonic function of the electricity price and is subsequently substituted into the expected NPV function over which the optimal timing analysis is performed. We find that greater uncertainty leads to higher installed capacity simply because it is also beneficial to delay investment. By contrast, in Sect. 7.6, we consider mutually exclusive plants of discrete sizes. These could also be interpreted as competing technologies. Unlike the previous analyses, we show that it may be optimal to have a second waiting region around the indifference point of the two projects' expected NPVs, thereby rendering the option value dichotomous. This is particularly likely to occur for relatively low values of uncertainty. Finally, in Sect. 7.7, we relax the assumption of a risk-neutral power company to take the perspective of a risk-averse decision maker, e.g., a municipal authority. We recast the real options problem as an optimal stopping-time problem with a concave utility function driving the investment decision. As expected, greater risk aversion makes the investor more cautious and, thus, increases the investment threshold price.

Clearly, it is straightforward to develop real options models to analyze investment projects in the electricity industry that have two or more of the aforementioned characteristics. For example, [14] combines mutually exclusive wind turbine investment opportunities with the possibility of a follow-on project after the initial installation reaches the end of its lifetime. However, there are three areas for further analysis that have not been covered in this chapter. First, we have used investment in a single power plant (or mutually exclusive alternatives) as a motivating example. However, analysis of investment in transmission lines from a real options perspective would also be suitable as long as their peculiarities are taken into account. One such complication is the fact that transmission lines may take on the order of a decade to plan and construct as opposed to two to three years for most fossil-fueled power plants. Thus, neither the lead time nor the uncertainty involved in the planning process may be neglected. Given this background, [36] models uncertainty in the initial regulatory decision as a function of the market uncertainty. Their motivation is Hydro-Québec's proposal to construct an interconnection to the neighboring province of Ontario. In this work, although higher benefits from constructing the line will result in a higher probability of regulatory approval, there is still uncertainty in the authorization process. Another complication with assessing transmission investment is that the construction of an interconnection may change the very nature of the anticipated cash flows, i.e., the transmission owner's profit from collecting congestion rents on the new line [18, 23]. In considering the mutually exclusive option to build interconnections of various sizes between Norway and Germany, [13] addresses the impact that such a link would have on the nodal price differences in the connecting regions of the two countries. Consequently, these authors first run a bottom-up model to estimate the impact that such transmission links will have on the electricity price differences before incorporating them into the real options analysis.

A second facet that we have not covered in this chapter is real options analysis with two or more sources of uncertainty. Such a consideration may be important because in addition to the electricity, the cost of investment and the price of fuel may be stochastic along with other parameters, e.g., government subsidies for renewable energy technologies. Recall that with a single source of uncertainty, we end up solving an ODE. Analogously, with multiple uncertain factors, we can follow a similar valuation procedure using stochastic dynamic programming to obtain a partial differential equation (PDE). In general, PDEs are difficult to solve analytically because a free boundary rather than an optimal trigger must be obtained endogenously. Under certain conditions, e.g., when the payoff of the project is homogeneous in the underlying stochastic parameters, the resulting PDE may be reduced to an ODE through the use of a numéraire. For example, if the expected NPV of a power plant effectively depends on the relative value of the electricity price and investment cost, then their ratio is what ultimately drives the investment strategy. This dimension-reducing technique is outlined in [12, 30] (see Chap. 6). However, in general, homogeneity may not hold, e.g., when it is the electricity price and fuel price that are uncertain with a deterministic investment cost. In that case, the expected NPV of the plant cannot be expressed via a numéraire, and an analytical procedure to solve the resulting PDE is required. By guessing at a solution of the power form (when the underlying uncertainties follow GBMs), [1] shows rigorously how to obtain quasi-analytically the free boundary and related coefficients. This approach has been applied to problems in the energy sector, e.g., investment in carbon capture and sequestration technologies [17, 34].

A third and final main feature of real options analysis that we have not included in this chapter is game-theoretic interactions. Indeed, given that most liberalized electricity industries have a few large power companies, they can effectively exercise market power through their investment and operational decisions. How does this rivalry affect the timing of investment decisions? In a game-theoretic model without uncertainty, [15] finds an equilibrium between two firms in which there is a strategic incentive for one to preempt the other. This model is extended by [21] in the context of a duopoly to the case with uncertainty in which the decision-analytic incentive to delay investment interacts with the strategic incentive to preempt one's rival. These authors find that depending on parameter values, e.g., volatility, growth rate, and the relative market share of the firm that moves first, either collusive or preemptive equilibria may arise. In the context of electricity industries, [42] uses this framework to analyze a game between a nuclear and a gas-fired power plant. They find that the operational flexibility of the latter makes it more likely to be the leader, especially in cases of high price volatility. In order to gain policy insights for the United Kingdom electricity market, which has nearly 40 % of its installed capacity based on natural gas as of 2011, [3] sets up a similar duopoly model involving a renewable energy plant and a gas-fired power plant. These authors find that the "natural hedge" of the latter (in terms of being a price setter) gives it a built-in advantage over the former, thereby stymieing policy objectives to increase the share of renewables. However, they demonstrate that a policy measure such as a CO_2 price floor can reduce the value of the gas-fired power plant's operational flexibility and, thus, make it more likely for the renewable energy plant to be the leader.

7.9 End-of-Chapter Exercises

7.1 Prove that the optimal investment threshold in Eq. (7.13) increases with uncertainty. It may be easier to prove implicitly that β_1 decreases with uncertainty first, i.e., $\frac{\partial \beta_1}{\partial \sigma} < 0$. In order to do this, differentiate the characteristic quadratic function in Eq. (7.8) totally and evaluate it at $\beta = \beta_1$ to obtain $\left.\frac{\partial \mathcal{Q}}{\partial \beta}\frac{\partial \beta_1}{\partial \sigma} + \frac{\partial \mathcal{Q}}{\partial \sigma}\right|_{\beta=\beta_1} = 0$.

7.2 Re-work the real options analysis of Sect. 7.2 by assuming that the profit flow can be modeled directly as $X_t = E_t - HF$. Use a simple Brownian motion for the profit flow, i.e., $\mathrm{d}X_t = \alpha \mathrm{d}t + \sigma \mathrm{d}z_t$. Show that the value of the investment opportunity is $\mathscr{W}_0(X) = a_{0,1} e^{\eta_1 X}$, where η_1 is the positive root of $\frac{1}{2}\sigma^2\eta^2 + \alpha\eta - \rho = 0$, $\xi = \frac{1}{\eta_1} + I - \frac{\alpha}{\rho}$, and $a_{0,1} = \frac{e^{-\eta_1 \xi}}{\rho \eta_1}$.

7.3 In the model of operational flexibility from Sect. 7.3, prove that the optimal switching thresholds, $\xi^{1,2}$ and $\xi^{2,1}$, from Eqs. (7.21a)–(7.22b) converge to HF as $S^{1,2} \to 0$ and $S^{2,1} \to 0$.

7.4 After investment, an abandonment option can also be modeled. For example, in the modular investment model of Sect. 7.4, state 3 may represent the value of an abandoned plant, i.e., with $\mathscr{W}_3(E) = 0$ and fixed cost $I^3 > 0$. Set up and solve the problem in which it is possible to abandon the power plant only from state 2 with $I^3 = 10$. How does the solution change if it is possible to abandon the plant from either state 1 or state 2?

7.5 With continuous capacity sizing, operational flexibility may also affect the initial investment timing and scaling decisions. Re-work the model of Sect. 7.5 to allow for costless operational flexibility. In particular, assume that an active power plant may be shut down and restarted at any time. Thus, the expected NPV of an active power plant of size K is $\mathscr{W}_1(E;K) = K\left[\frac{E}{\rho-\alpha} - \frac{HF}{\rho}\right] + a_{1,2}(K)E^{\beta_2}$, whereas the expected NPV of a suspended plant of size K is $\mathscr{W}_2(E;K) = a_{2,1}(K)E^{\beta_1}$. Use value-matching and smooth-pasting conditions between these two functions at $E = HF$ to determine the coefficients $a_{1,2}(K)$ and $a_{2,1}(K)$. Next, determine the optimal capacity size by maximizing $\mathscr{W}_1(E;K) - \mathscr{I}(K)$ with respect to K. Finally, use this maximized expected NPV, $\mathscr{W}_1(E;\kappa(E)) - \mathscr{I}(\kappa(E))$, to determine the optimal investment threshold, ξ, and capacity, $\kappa(\xi)$.

7.6 In the model of mutually exclusive technology choice from Sect. 7.6, how would you modify the value function of technology 1 to allow for a subsequent switching option to technology 2? Show that its value function will be of the form $\mathscr{W}_1^1(E) = \frac{K^1 E}{\rho-\alpha} - \frac{K^1 H^1 F}{\rho} + a_{1,2}E^{\beta_1}$, where $\xi^{1,2}$, the optimal switching threshold from technology 1 to 2, and $a_{1,2}$ will be determined via value-matching and smooth-pasting conditions between $\mathscr{W}_1^1(E) - I^1 - I^2$ and $\mathscr{W}_1^2(E) - I^2$.

7.7 Show that the optimal investment threshold price under risk aversion from Eq. (7.62) converges to the risk-neutral threshold of Eq. (7.13) as $\gamma \to 0$.

7.10 MATLAB Codes

Here, we present MATLAB codes for solving selected examples numerically. The following code solves Illustrative Example 7.7 to analyze optimal investment timing with operational flexibility.

```
% Real options treatment of the investment timing problem
% with operational flexibility
rho=0.1;
alpha=0.05;
H=2.5;
F=20;
I=100;
sigmastart=0.15;
E=50;
S12=10; % Cost of switching from state 1 to state 2
S21=10; % Cost of switching from state 2 to state 1
interplot=zeros(1,5);

for kk=1:+1:21;
    sigma=sigmastart+0.01*(kk-1)
    beta1 = 0.5 - alpha/sigma^2 + ...
        sqrt((0.5-alpha/sigma^2)^2+2*rho/sigma^2);
    beta2 = 0.5 - alpha/sigma^2 - ...
        sqrt((0.5-alpha/sigma^2)^2+2*rho/sigma^2);
    e=0:+0.001:100;
    % Guesses for investment and switching thresholds
    xi010 = (beta1/(beta1-1))*(rho-alpha)*(H*F+rho*I)/rho;
    xi120 = (beta2/(beta2-1))*(rho-alpha)*(H*F/rho-S12);
    xi210 = (beta1/(beta1-1))*(rho-alpha)*(H*F/rho+S21);
    a010=xi010^(1-beta1)/(beta1*(rho-alpha));
    a120=-xi120^(1-beta2)/(beta2*(rho-alpha));
    a210=xi210^(1-beta1)/(beta1*(rho-alpha));
    % Initial guess for operational decisions
    vars0 = [a120 a210  xi120  xi210];
    % Specify options for fsolve
    opts=optimset('fsolve');
    opts=optimset(opts,'Maxiter',2000,'Tolx',...
        1e-6,'tolfun',1e-6);
    %*************************************
    %* next instruction calls funtion fffROinvof.m*
    %*************************************
    vars = fsolve(@fffROinvof,vars0,opts,...
        beta1,beta2,alpha,H,F,rho,S12,S21);
    a12 = vars(1);
    a21 = vars(2);
    xi12 = vars(3);
    xi21 = vars(4);
    z=@(y)(beta1-beta2)*a12*y.^beta2+...
        (beta1-1)*y/(rho-alpha)-beta1*(H*F/rho+I);
    xi01=fzero(z, xi010);
    a01=(beta2/beta1)*a12*xi01^(beta2-beta1)+...
        xi01^(1-beta1)/(beta1*(rho-alpha));
    e1=xi12:+0.001:100;
```

```
    e2=0:0.001:xi21;
    e0=0:0.001:xi01;
    W1=a12*e1.^beta2+e1/(rho-alpha)-H*F/rho;
    W2=a21*e2.^beta1;
    W0=a01*e0.^beta1;
    flexvalue=a01/a010;
    interplot=[interplot; sigma xi12 xi21 xi01
        flexvalue];
end

% delete first row leaving just calculated values
interplot(1,:) = [ ];

figure(1)
plot(e1, W1, '--k')
hold on
grid
plot(e2, W2, ':k')
plot(e0, W0, 'k')

figure(2)
plot(interplot(:,1), interplot(:,2), ':k')
hold on
grid
plot(interplot(:,1), interplot(:,3), 'k')
plot(interplot(:,1), interplot(:,4), '--k')

figure(3)
plot(interplot(:,1), interplot(:,5), 'k')
grid
```

```
% Function file: fffROinvof.m
% Called to solve non-linear system for operational
    decisions
function f=fffROinvof(vars,beta1,beta2,alpha,H,F,
   rho,S12,S21)
a12 = vars(1);
a21 = vars(2);
xi12 = vars(3);
xi21 = vars(4);
f=zeros(4,1);
f(1) = a12*xi12^beta2+xi12/(rho-alpha)-...
H*F/rho-a21*xi12^beta1+S12;
f(2)= beta2*a12*xi12^(beta2-1)+...
1/(rho-alpha)-beta1*a21*xi12^(beta1-1);
f(3)= -a21*xi21^beta1+...
xi21/(rho-alpha)-H*F/rho+a12*xi21^beta2-S21;
f(4) = -beta1*a21*xi21^(beta1-1)+...
1/(rho-alpha)+beta2*a12*xi21^(beta2-1);
```

The following MATLAB code solves Illustrative Example 7.15 numerically to determine optimal investment timing and sizing with mutually exclusive technology options.

```
% Real options treatment of mutually exclusive investment
% opportunity in discrete capacity sizes

rho=0.04;
alpha=0.0;
HA=2.5;
KA=1;
KB=2.9;
HB=2.5;
F=0;
IA=100;
IB=900;
sigmastart=0.05;
E=50;
num=251;


for kk=1:+1:num
    sigma(kk) = sigmastart + (kk-1)*0.001;

    % Option value parameters
    beta1(kk) = 0.5 - alpha/sigma(kk)^2 + ...
        sqrt((0.5-alpha/sigma(kk)^2)^2+2*rho/sigma(kk)^2);

    beta2(kk) = 0.5 - alpha/sigma(kk)^2 - ...
        sqrt((0.5-alpha/sigma(kk)^2)^2+2*rho/sigma(kk)^2);

    % Indifference point of NPVs
    xiind(kk)=(rho-alpha)*(rho*(IA-IB)+...
        F*(KA*HA-KB*HB))/(rho*(KA-KB));

    e=0:+0.001:35;

    % Independent options analysis
    WA1=KA*(e/(rho-alpha)-HA*F/rho);

    WB1=KB*(e/(rho-alpha)-HB*F/rho);


    xiA(kk) = (beta1(kk)/(beta1(kk)-1))*...
        (rho-alpha)*(HA*F+rho*IA/KA)/rho;
    xiAalt(kk) = xiA(kk);

    xiA0(kk) = (rho-alpha)*(HA*F+rho*IA/KA)/rho;

    xiB(kk) = (beta1(kk)/(beta1(kk)-1))*...
```

```
            (rho-alpha)*(HB*F+rho*IB/KB)/rho;

        xiB0(kk) = (rho-alpha)*(HB*F+rho*IB/KB)/rho;

        aA(kk)=KA*xiA(kk)^(1-beta1(kk))/(beta1(kk)*(rho-alpha));
        aB(kk)=KB*xiB(kk)^(1-beta1(kk))/(beta1(kk)*(rho-alpha));

        eA1=0:+0.001:xiA(kk);
        eB1=0:+0.001:xiB(kk);

        WA0=aA(kk)*eA1.^beta1(kk);
        WB0=aB(kk)*eB1.^beta1(kk);

        % Check whether project A is option-dominated
        if (aB(kk) > aA(kk))
            xiR(kk) = xiB(kk);
            xiL(kk) = NaN;
            xiAalt(kk) = NaN;
            W = WB0;
        % Otherwise, either project may be selected
        else
            % Form guesses as to the solution
            % to the non-linear system
            xiR0(kk) = xiB(kk);
            aR0(kk) = (KB*xiR0(kk)^(1-beta1(kk)))...
                /(beta1(kk)*(rho-alpha));
            xiL0(kk) = (xiind(kk)+xiA(kk))/2;
            aL0(kk) = (KA*xiL0(kk)^(1-beta2(kk)))...
                /(beta2(kk)*(rho-alpha)) - ...
                (beta1(kk)*aR0(kk)/beta2(kk))*...
                xiL0(kk)^(beta1(kk)-beta2(kk));
            % vars0is your guess
            vars0 = [aR0(kk) xiR0(kk) aL0(kk) xiL0(kk)];
            % OPTIMSET is recommended for setting options.
            opts=optimset('fsolve');
            opts=optimset(opts,'Maxiter',9000,...
                'Tolx',1e-4,'tolfun',1e-4);
            %*********************************************
            % next instruction calls fffdmv.m
            %*********************************************
            vars = fsolve(@fffdmv,vars0,opts,beta1(kk),...
                beta2(kk),rho,alpha,KA,KB,IA,IB,HA,HB,F);
            % The option value curve has three parts now
            aR(kk)=vars(1);
            xiR(kk)=vars(2);
            aL(kk)=vars(3);
            xiL(kk)=vars(4);
            W1 = aA(kk)*eA1.^beta1(kk);
            edi = xiL(kk):+0.001:xiR(kk);
            W2 = aL(kk)*edi.^beta2(kk) + aR(kk)*edi.^beta1(kk);
```

```
        end

    end

    figure(1)
    plot(sigma, xiAalt, '--k')
    hold on
    grid
    plot(sigma, xiL, '-.k')
    plot(sigma, xiR, 'k')
    plot(sigma, xiind, ':k')
    xlabel('$\sigma$')
    ylabel('Optimal investment threshold prices [(\$/MWh
        ]')
```

```
% Function file: fffdmv.m
% Called to solve non-linear system
% for mutually exclusive investment
function f=fffdmv(vars,beta1,beta2,rho,alpha,KA,KB,IA,IB,...
    HA,HB,F)
aR=vars(1);
xiR=vars(2);
aL=vars(3);
xiL=vars(4);
f=zeros(4,1);
f(1) = aR*xiL^beta1+aL*xiL^beta2-xiL*KA/(rho-alpha)+...
    KA*HA*F/rho+IA;
f(2) = beta1*aR*xiL^(beta1-1)+beta2*aL*xiL^(beta2-1)-...
    KA/(rho-alpha);
f(3) = aR*xiR^beta1+aL*xiR^beta2-xiR*KB/(rho-alpha)+...
    KB*HB*F/rho+IB;
f(4) = beta1*aR*xiR^(beta1-1)+beta2*aL*xiR^(beta2-1)-...
    KB/(rho-alpha);
```

References

1. Adkins, R., Paxson, D.: Renewing assets with uncertain revenues and operating costs. J. Financ. Quant. Anal. **46**, 785–813 (2011)
2. Bøckman, T., Fleten, S.-E., Juliussen, E., Langhammer, H., Revdal, I.: Investment timing and optimal capacity choice for small hydropower projects. Eur. J. Oper. Res. **190**, 255–267 (2008)
3. Chronopoulos, M., Bunn, D., Siddiqui, A.: Optionality and policymaking in re-transforming the British power market. Econ. Energy Environ. Policy **3**, 79–100 (2014)
4. Chronopoulos, M., De Reyck, B., Siddiqui, A.: Optimal investment under operational flexibility, risk aversion, and uncertainty. Eur. J. Oper. Res. **213**, 221–237 (2011)
5. Copeland, T.E., Antikarov, V.: Real Options: A Practitioner's Guide. Texere, Cheshire (2003)
6. Dangl, T.: Investment and capacity choice under uncertain demand. Eur. J. Oper. Res. **117**, 1–14 (1999)
7. Décamps, J.-P., Mariotti, T., Villeneuve, S.: Irreversible investment in alternative projects. Econ. Theory **28**, 425–448 (2006)
8. Deng, S.-J.: Valuation of investment and the opportunity to invest in power generation assets with spikes in power prices. Manag. Financ. **31**, 94–114 (2005)

9. Deng, S.-J., Johnson, B., Sogomonian, A.: Exotic electricity options and the valuation of electricity generation and transmission assets. Decis. Support Syst. **30**, 383–392 (2001)
10. Dixit, A.K.: Investment and hysteresis. J. Econ. Perspect. **6**, 107–132 (1992)
11. Dixit, A.K.: Choosing among alternative lumpy investment projects under uncertainty. Econ. Lett. **43**, 281–285 (1993)
12. Dixit, A.K., Pindyck, R.S.: Investment Under Uncertainty. Princeton University Press, Princeton (1994)
13. Fleten, S.-E., Heggedal, A.M., Siddiqui, A.: Transmission capacity between Norway and Germany—a real options analysis. J. Energy Markets **4**, 121–147 (2011)
14. Fleten, S.-E., Maribu, K., Wangensteen, I.: Optimal investment strategies in decentralized renewable power generation under uncertainty. Energy **32**, 803–813 (2007)
15. Fudenberg, D., Tirole, J.: Preemption and rent equalization in the adoption of new technology. Rev. Econ. Stud. **52**, 383–401 (1985)
16. Gollier, C., Proult, D., Thais, F., Walgenwitz, G.: Choice of nuclear power investments under price uncertainty: valuing modularity. Energy Econ. **27**, 667–685 (2005)
17. Heydari, S., Ovenden, N.C., Siddiqui, A.: Real options analysis of investment in carbon capture and sequestration technology. Comput. Manag. Sci. **9**, 109–138 (2012)
18. Hogan, W.W.: Contract networks for electric power transmission. J. Reg. Econ. **4**, 211–242 (1992)
19. Hugonnier, J., Morellec, E.: Real options and risk aversion. Working paper, HEC Lausanne, Switzerland (2007)
20. Hull, J.C.: Options, Futures, and Other Derivatives. Prentice Hall, Upper Saddle River (2014)
21. Huisman, K.J.M., Kort, P.M.: Effects of strategic interactions on the option value of waiting. Center Discussion Paper 1999-92, Tilburg University, The Netherlands (1999)
22. Huppmann, D., Gabriel, S.A., Leuthold, F.U.: A note on allowing negative energy prices in a discretely constrained MPEC. Energy Econ. **40**, 1023–1025 (2013)
23. Joskow, P., Tirole, J.: Merchant transmission investment. J. Ind. Econ. **53**, 233–264 (2005)
24. Karatzas, I., Shreve, S.: Methods of Mathematical Finance. Springer, New York (1999)
25. Kort, P., Murto, P., Pawlina, G.: Uncertainty and stepwise investment. Eur. J. Oper. Res. **202**, 196–203 (2010)
26. Linnerud, K., Andersson, A.-M., Fleten, S.-E.: Investment timing under uncertain renewable energy policy: an empirical study of small hydropower projects. Energy, (2014)
27. Lund, D.: How to analyze the investment-uncertainty relationship in real option models? Rev. Financ. Econ. **14**, 311–322 (2005)
28. Majd, S., Pindyck, R.S.: Time to build, option value, and investment decisions. J. Financ. Econ. **18**, 7–27 (1987)
29. Marnay, C., Venkatarmanan, G., Stadler, M., Siddiqui, A., Firestone, R., Chandran, B.: Optimal technology selection and operation of commercial-building microgrids. IEEE Trans. Power Syst. **23**, 975–982 (2008)
30. McDonald, R., Siegel, D.: Investment and the valuation of firms when there is an option to shut down. Int. Econ. Rev. **26**, 331–349 (1985)
31. Näsäkkälä, E., Fleten, S.-E.: Flexibility and technology choice in gas fired power plant investments. Rev. Financ. Econ. **14**, 371–393 (2005)
32. Pindyck, R.S.: The long-run evolution of energy prices. Energy J. **20**, 1–27 (1999)
33. Pratt, J.W.: Risk aversion in the small and in the large. Econometrica **32**, 122–136 (1964)
34. Rohlfs, W., Madlener, R.: Valuation of CCS-ready coal-fired power plants: a multi-dimensional real options approach. Ener. Syst. **2**, 243–261 (2011)
35. Ross, S.M.: Stochastic Processes. Wiley, Hoboken (1995)
36. Saphores, J.-D., Gravel, E., Bernard, J.-T.: Regulation and investment under uncertainty—an application to power grid interconnection. J. Reg. Econ. **25**, 169–186 (2004)
37. Siddiqui, A., Fleten, S.-E.: How to proceed with competing alternative energy technologies: a real options analysis. Energy Econ. **32**, 817–830 (2010)
38. Siddiqui, A., Maribu, K.: Investment and upgrade in distributed generation under uncertainty. Energy Econ. **31**, 25–37 (2009)

39. Siddiqui, A., Marnay, C.: Distributed generation investment by a microgrid under uncertainty. Energy **33**, 1729–1737 (2008)
40. Siddiqui, A., Takashima, R.: Capacity switching options under rivalry and uncertainty. Eur. J. Oper. Res. **222**, 583–595 (2012)
41. Šumbera, J.: Application of optimisation methods to electricity production problems. Ph.D. thesis, Department of Operational Research, Faculty of Informatics and Statistics, University of Economics, Prague, Czech Republic (2013)
42. Takashima, R., Goto, M., Kimura, H., Madarame, H.: Entry into the electricity market: uncertainty, competition, and mothballing options. Energy Econ. **30**, 1809–1830 (2008)
43. Wickart, M., Madlener, R.: Optimal technology choice and investment timing: a stochastic model of industrial cogeneration versus heat-only production. Energy Econ. **29**, 934–952 (2007)
44. World Nuclear Association: Small Nuclear Reactors. Available at http://www.world-nuclear.org/info/Nuclear-Fuel-Cycle/Power-Reactors (2014)

Appendix A
Engineering Economics

This appendix provides an overview of engineering economics. Its purpose is to familiarize the reader with the discounting of cash flows and the time value of money. Compounding periods are also treated, and the appendix concludes with examples of how to handle amortization of capital costs.

A.1 Introduction

In this appendix, we provide an overview of treating cash flows involved with real investments, e.g., in power plants or transmission lines, with respect to the time value of money. This field is known as engineering economics [2], and its main purpose is to develop a framework for economic analysis of real, rather than financial, investments. The complexity of investment projects in the electric power industry means that cash flows (both receipts and expenditures) may be uneven and occur at diverse points in time. From the perspective of investment appraisal, how are the cash flows from two different projects to be treated in order to facilitate comparison?

For example, in Figs. A.1 and A.2, we have projects 1 and 2, respectively, with cash flows of different magnitudes occurring at various points in time. Specifically, project 1 requires an expenditure of size P immediately followed by a revenue of size F received in year 8. Conversely, project 2 requires eight equal payments of size A each, after which the revenue in year 8 is received. By converting the cash flows of these two projects into either a single payment or a series of payments using an appropriate interest rate that reflects the time value of money, it is possible to compare their economic effects. Indeed, in general, any sequence of cash flows can be converted into another by accounting for the forgone opportunity to use the money profitably. In the next section, we discuss different types of interest rates.

A.J. Conejo et al., *Investment in Electricity Generation and Transmission*,
DOI 10.1007/978-3-319-29501-5

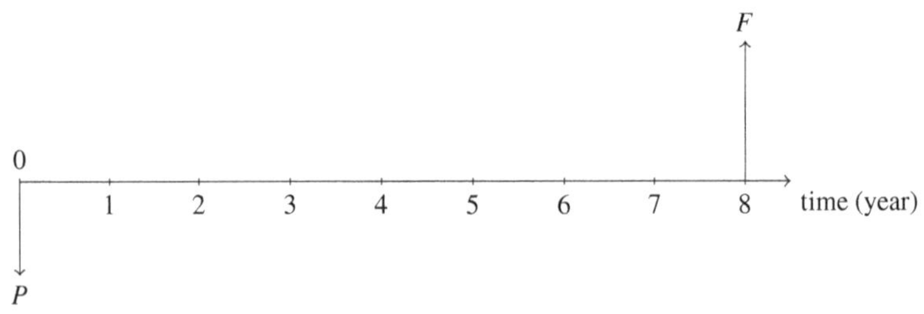

Fig. A.1 Cash flow diagram with a single payment (project 1)

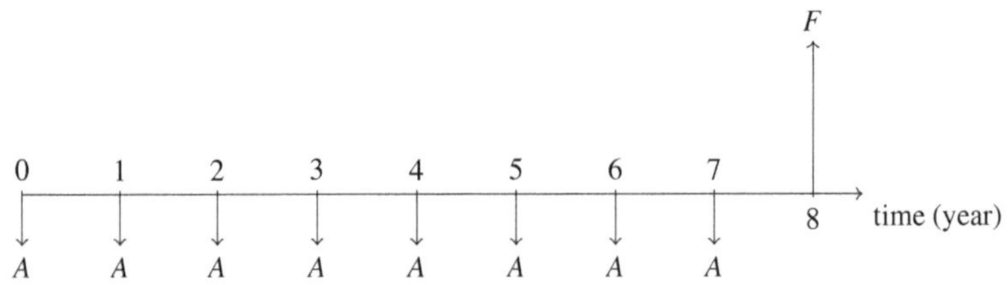

Fig. A.2 Cash flow diagram with multiple payments (project 2)

A.2 Interest Rates

The basic principle in engineering economics is that money today is generally worth more than the same amount available in the future. It is, thus, impossible to compare monetary amounts received at different points in time without having a common reference point. For this purpose, the future value is defined as the amount into which an initial investment will grow after earning interest for a given period of time. Likewise, the present value is the discounted value of some future payment or receipt. Thus, prior to doing cash flow calculations, it is useful to describe different types of interest rates.

Suppose that you put \$100 in the bank. If the annual interest rate is 5 %, then how much total money will you have in two years? The answer depends on the following factors:

1. Simple or compound interest.
2. Number of compounding periods (with continuous compounding in the limit).
3. Inflation.

With simple interest rates, only the principal earns interest. Thus, each year, the interest earned is $5\,\% \times \$100 = \$5$, which yields a future value after two years of \$110. In general, the formula for the future value, F_N, after N periods with an interest rate of i is:

$$F_N = P\,(1 + iN)\,. \tag{A.1}$$

However, it is more common to earn interest on the interest itself. This is referred to as compound interest and means that the interest earned after one year, $5\,\% \times \$100 = \$5$, will also earn interest in the second year. Consequently, the balance after two years will be the interest earned on the principal, $\$100 \times (1 + 0.05 \times 2) = \$110$, plus the interest earned on the interest, i.e., $5\,\% \times \$5 = \$0.25$. Hence, the future value after two years with compound interest will be \$110.25, or in general:

$$F_N = P\,(1+i)^N\,. \tag{A.2}$$

In Eq. (A.2), we assumed that compounding occurs annually. However, many financial instruments pay out interest every six months, e.g., United States federal treasury bonds, which means that the compounding period is not synchronous with the payment period. Consequently, since compounding occurs more frequently than the payment, the quoted interest rate will not be the effective annual interest rate. As an example, consider putting \$100 in a bond that will pay 10 % annual interest compounded semiannually. This effectively means that you will earn 5 % on the \$100 every six months. Thus, the balance after six months and twelve months will be $\$100 \times (1 + 0.05) = \$105$ and $\$105 \times (1 + 0.05) = \$110.25$, respectively. By contrast, the \$100 with a 10 % annual interest rate compounded annually will be worth only $\$100 \times (1 + 0.10) = \$110$ after one year. In general, the effective annual interest rate, i_a, with M compounding periods per year and an annual interest rate r will be:

$$i_a = \left(1 + \frac{r}{M}\right)^M - 1. \tag{A.3}$$

In our example, with semiannual compounding of 10 % annual interest, we obtain an effective annual interest rate of $i_a = \left(1 + \frac{0.10}{2}\right)^2 - 1 = 10.25\,\%$. As the number of compounding periods per year increase, we have in the limit the case of continuous compounding:

$$i_a = \lim_{M\to\infty}\left(1 + \frac{r}{M}\right)^M - 1 = e^r - 1. \tag{A.4}$$

Here, we use the definition of the mathematical constant e to obtain the result. Hence, for the same example, 10 % annual interest compounded continuously would correspond to an effective annual interest rate of $i_a = e^{0.10} - 1 = 10.52\,\%$.

The interest rates discussed so far have all ignored the effects of inflation, i.e., the rate at which prices in general are increasing. For example, in the United States, the Consumer Price Index (CPI) uses a basket of goods to determine the general price level in a given year, and the percentage change in successive CPIs is a measure of inflation. The consequence of inflation is that it reduces purchasing power and thus the real interest rate. For example, if the annual interest rate offered on an investment were $i = 5\,\%$ compounded annually, and the annual inflation rate were $f = 2.2\,\%$, then we could obtain the real interest rate, i', from the Fisher equation as follows:

$$1 + i' = \frac{1 + i}{1 + f}. \tag{A.5}$$

Multiplying out Eq. (A.5) yields $i = i' + f + fi'$. Since f and i' are "small," we have $i' \approx i - f$. Thus, in the example, we have $i' = \frac{1.05}{1.022} - 1 = 2.74\,\%$, or the approximation $i' \approx 0.05 - 0.022 = 2.8\,\%$. Finally, when it comes to discounting cash flows, the nominal (real) interest rate should be used with nominal (real) cash flows. For example, the present value of \$100 available in a year's time given a nominal interest rate $i = 10\,\%$ is simply $\frac{100}{1.1} = \$90.91$. If the inflation rate is $f = 7\,\%$ per annum, then the real amount available in a year's time is $\frac{100}{1.07} = \$93.46$. Discounting this by the real interest, i.e., $i' = \frac{1.1}{1.07} - 1 = 2.8\,\%$, yields the same answer, i.e., $\frac{93.46}{1.028} = \$90.91$.

A.3 Time Value of Money

Besides using the interest rate to discount single cash flows to obtain, e.g., present values, it is also straightforward to compare multiple cash flows occurring at various points in time. This is due to the principle of value additivity, i.e., the present value of the sum of two cash flows (say A in year ℓ and B in year m) is simply the sum of the cash flows' present values using the interest rate i:

$$PV\,(A + B) = \frac{A}{(1 + i)^{\ell}} + \frac{B}{(1 + i)^{m}}$$
$$\Rightarrow PV\,(A + B) = PV\,(A) + PV\,(B)\,. \tag{A.6}$$

The principle of value additivity in Eq. (A.6) can be generalized to any finite number of discrete cash flows, and it also applies to future values, i.e., $FV\,(A + B) = FV\,(A) + FV\,(B)$.

We illustrate this concept with the example in Fig. A.3 with cash flows of \$1,200 and \$1,500 received in years 4 and 6, respectively. The present value of this series of cash flows for an annual interest rate of 5 % is simply $PV = \frac{1200}{1.05^4} + \frac{1500}{1.05^6} = \$2{,}106.57$. Alternatively, we can recast everything in terms of the future value in year 8 as in

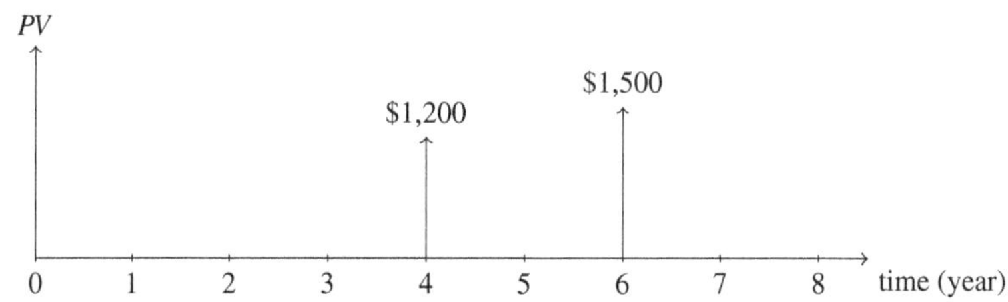

Fig. A.3 Principle of value additivity (present value)

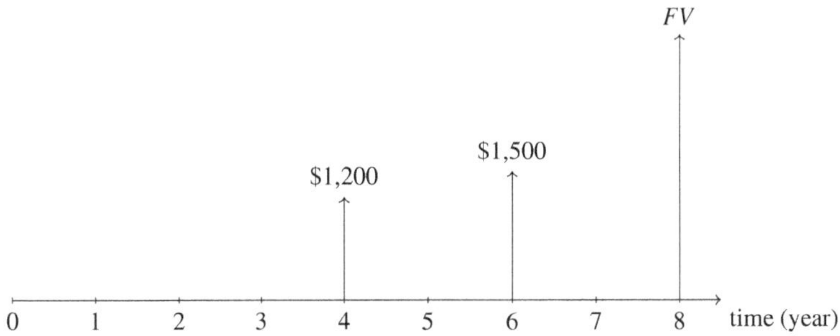

Fig. A.4 Principle of value additivity (future value)

Fig. A.4. Thus, we have $FV = 1200\,(1.05)^4 + 1500\,(1.05)^2 = \$3{,}112.36$. Moreover, the equivalence between the PV and FV can be established because $PV = \frac{FV}{1.05^8}$. Hence, in general, for any sequence of discrete cash flows and any sequence of annual interest rates, we have the following relationship:

$$PV_0 = \sum_n \frac{C_n}{(1+i_n)^n}, \tag{A.7}$$

where PV_0 is the present value today, C_n is the cash flow in year n, and i_n is the interest rate for year n.

A.4 Payment Schemes

Although the principle of value additivity could be used to deal with an arbitrary sequence of cash flows, it is often helpful to have formulas to handle frequently occurring annual payments. For example, some bonds issued by the British government in the eighteenth century to finance wars and infrastructure projects had perpetual coupon payments of equal amounts, C. Assuming a fixed interest rate, i, and the first payment starting in year 1, what is the present value of such a bond (known as a perpetuity) in year 0 (Fig. A.5)? Using the principle of value additivity,

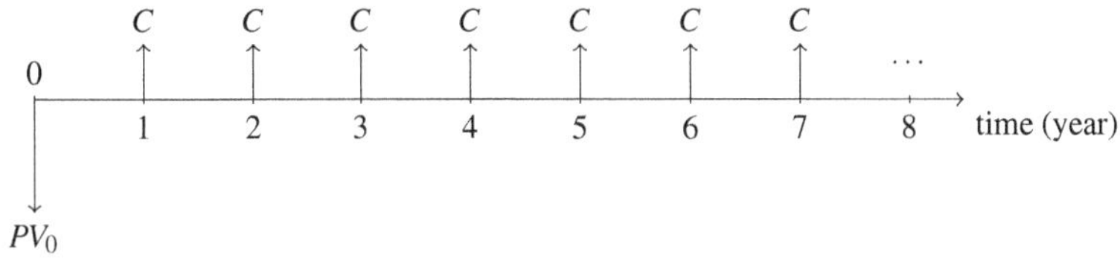

Fig. A.5 Cash flow diagram for a perpetuity

we can express the present value of a perpetuity in year 0 as:

$$
\begin{aligned}
PV_0 &= \sum_{n=1}^{\infty} \frac{C}{(1+i)^n} \\
\Rightarrow PV_0 &= C \times \left[\frac{1}{1 - \frac{1}{1+i}} - 1 \right] \\
\Rightarrow PV_0 &= C \times \left[\frac{1}{\frac{i}{1+i}} - 1 \right] \\
\Rightarrow PV_0 &= \frac{C}{i},
\end{aligned}
\tag{A.8}
$$

where we used the property of an infinite geometric series, i.e., $\sum_{n=0}^{\infty} x^n = \frac{1}{1-x}$ for $|x| < 1$, in going from the first to the second line. Hence, given an annual interest rate of 8 %, the present value today of a perpetual cash flow stream that pays \$150,000 per year forever is \$150,000 $\times \frac{1}{0.08}$ = \$1,875,000.

Although perpetuities are rare in both engineering-economic projects and financial transactions, they are, nevertheless, useful for valuing other payment schemes with level cash flows. Consider the annuity in Fig. A.6 with eight equal cash flows each of size A received in years 1 through 8. What is the present value of this stream of cash flows in year 0? Again, we can use the principle of value additivity to perform the calculation or use the property of a finite geometric series. However, it is less tedious to treat the present value of the cash flows in Fig. A.6 as the difference in the present values of two perpetual cash flows, the first of which begins in year 1 and the second in year 9. Since the present value of the former in year 0 is $\frac{A}{i}$, the present value of the latter in year 0 is $\frac{A}{i(1+i)^8}$, i.e., the discounted cash flows of $\frac{A}{i}$ from year 8. In general, the present value in year 0 of an annuity that starts in year 1 and goes to year N is:

$$
\begin{aligned}
PV_0 &= \frac{A}{i} - \frac{A}{i\,(1+i)^N} \\
\Rightarrow PV_0 &= A \left[\frac{(1+i)^N - 1}{i\,(1+i)^N} \right].
\end{aligned}
\tag{A.9}
$$

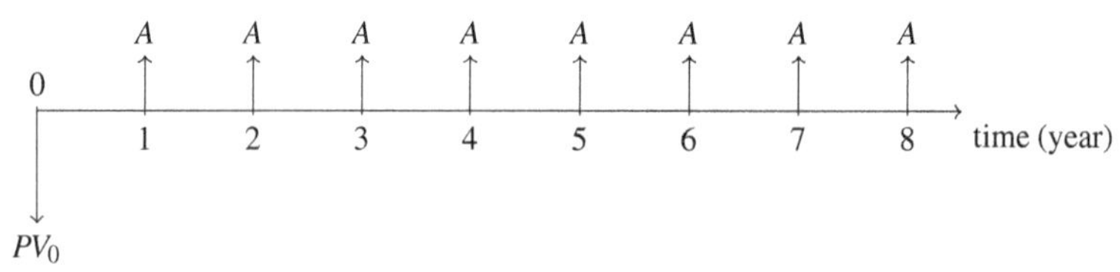

Fig. A.6 Cash flow diagram for an annuity

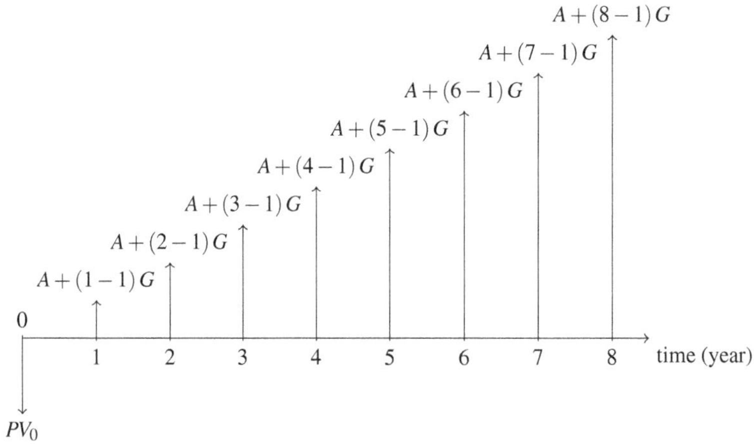

Fig. A.7 Cash flow diagram for a linear gradient series

Hence, the present value in year 0 of an eight-year annuity with annual cash flows of \$10,000 using an annual interest rate of 10 % is \$10,000 $\times \left[\frac{(1.1)^8-1}{0.10(1.1)^8}\right] = \$53{,}349$.

Besides constant annual cash flows, some engineering-economic applications may also involve periodic payments that increase or decrease by some common factor, G. Consider the cash flows in Fig. A.7, in which the payment received in year 1 is A, in year 2 is $A + G$, and so on until year 8, when the final payment of $A + 7G$ is received. Using the fact that the cash flows in Fig. A.7 may be reconfigured as two cash flow series, one of which is an eight-year annuity beginning in year 1 of A and the other is a gradient series beginning in year 2 of constant amount G, we can obtain the present value with a constant annual interest rate i as:

$$PV_0 = A\left[\frac{(1+i)^N - 1}{i\,(1+i)^N}\right] + \sum_{n=2}^{N} \frac{(n-1)\,G}{(1+i)^n}$$

$$\Rightarrow PV_0 = A\left[\frac{(1+i)^N - 1}{i\,(1+i)^N}\right] + G\left[\frac{(1+i)^N - iN - 1}{i^2\,(1+i)^N}\right], \tag{A.10}$$

where we use the property of an arithmetic–geometric series, i.e., $\sum_{n=0}^{N} nx^n = \frac{x\left[1-(N+1)x^N+Nx^{N+1}\right]}{(1-x)^2}$ for $x \neq 1$, in going from the first to the second line. Hence, the present value in year 0 of a cash-flow stream that starts in year 1 with \$100 and increases by annual amounts of \$25 with the last payment received in year 8 using an annual interest rate of 10 % is \$100 $\times \left[\frac{(1.1)^8-1}{0.10(1.1)^8}\right] + \$25 \times \left[\frac{1.1^8-0.1\times 8-1}{0.1^2\times 1.1^8}\right] = \$934.21$.

Using closed-form expressions for common types of payment schemes and the principle of value additivity, it is possible to account for the time value of money correctly in engineering-economic applications. We conclude this appendix with some worked examples and refer the reader to either [1] or [2] for a more in-depth treatment of the subject.

Illustrative Example A.1 *Doubling your money*

Suppose that you currently have an arbitrary amount of money, $P > 0$, and want to know how long it will take for it to double given an interest rate of r compounded continuously. In general, we know that the future value in N years with continuous compounding is $F_N = Pe^{rN}$ because the effective annual interest rate becomes $i_a = e^r - 1$ from Eq. (A.4). Thus, since we want to have $F_{N^*} = 2 \times P$, we obtain $e^{rN^*} = 2 \Rightarrow N^* = \frac{\ln 2}{r}$. As an approximation, it is convenient to divide 70 (since the natural logarithm of 2 is roughly 0.70) by the nominal interest rate (expressed in percentage terms) in order to determine N^*. □

Illustrative Example A.2 *Discounting continuous cash flows*

Rather than dealing with discrete cash flows, it is also possible to tackle continuous ones. Although cash flows may be discrete, they may occur so frequently, e.g., in the case of a power plant, that it is useful to treat them as if they were continuous. Consequently, the expressions for discounting discrete cash flows may be converted analogously. For example, consider the present value for a series of discrete cash flows, i.e., $PV_0 = \sum_{n=0}^{N} F_n (1+i)^{-n}$. With continuous cash flows and continuous compounding, the summation becomes an integral, the discrete periods, n, become continuous time, t, and the annual discount factor, $(1+i)^{-n}$, becomes the continuous discount factor, e^{-rt}. Hence, we have $PV_0 = \int_0^N F_t e^{-rt} \mathrm{d}t$. □

Illustrative Example A.3 *Amortization of capital costs with coincident payment and planning horizons*

Suppose that a power plant has a capital cost of $P = \$100$ million with an effective lifetime of $N = 30$ years. If the annual interest rate is 10 %, then how should the amortized cost of the plant be included in each year's cost calculations that will appear in the objective function of the corresponding optimization problem? If we restrict ourselves to investing only now, then the annuity to be paid in each year is obtained by inverting Eq. (A.9) to solve for A in terms of P and next discounting. Here, it is $A = \$100 \text{ million} \times \left[\frac{0.10 \times 1.1^{30}}{1.1^{30} - 1}\right] = \$10.61$ million, which is the amortized cost in each year. It is then discounted to yield the present value of the amortized cost in year n, i.e., $\frac{A}{(1+i)^n}$. For year 10, this is about \$4.09 million. As a check, if we add up all of the present values of the amortized costs in each year, then we obtain $\sum_{n=1}^{N} \frac{A}{(1+i)^n} = \$100$ million. □

Illustrative Example A.4 *Amortization of capital costs with noncoincident payment and planning horizons*

In many multiperiod investment problems, the lifetime of the equipment exceeds the planning horizon. For example, in Illustrative Example A.3, the effective lifetime of the power plant is $N = 30$ years, whereas the planning horizon may be only $M = 20$ years. Allowing for the possibility to invest in any of the next M years, we first define the decision variable x_m, which equals 1 if investment occurs in year m and is 0 otherwise. Thus, the mathematical formulation should specify $x_m \in \{0, 1\}$ for $m = 0, \ldots, M - 1$ and $\sum_{m=0}^{M-1} x_m \leq 1$ as constraints. If investment takes place in year m, then the present value of the amortized capital cost is discounted back from that year to obtain the expression for the present value to be included in the objective function: $\sum_{m=0}^{M-1} \frac{x_m}{(1+i)^m} \sum_{n=m+1}^{M} \frac{A}{(1+i)^{n-m}}$. □

References

1. Brealey, R.A., Myers, S.C., Marcus, A.J.: Fundamentals of Corporate Finance. McGraw-Hill, New York (2014)
2. Park, C.S., Sharp-Bette, G.P.: Advanced Engineering Economics. Wiley, New York (1990)

Appendix B
Optimization Under Uncertainty

This appendix provides a succinct tutorial introduction to optimization under uncertainty, spanning both two-stage stochastic programming and adaptive robust optimization. Two simple examples are used to illustrate the functioning of these two decision frameworks, and then, general problem formulations are provided. Next, observations on how to solve these problems are given. The appendix concludes by indicating how to extend these decision frameworks to a multistage setting.

B.1 Introduction

Two-stage stochastic programming is considered first, followed by adaptive robust optimization. These are the two most common frameworks for decision-making under uncertainty.

B.2 Two-Stage Stochastic Programming

A brief tutorial introduction to two-stage stochastic programming is provided in this section. Additional material can be found in the pioneering book by Birge and Louveaux [3], in the tutorial paper [9], and in [5], which has an electricity focus.

Illustrative Example B.1

Consider an electricity producer that has the option of building one of two production plants, A or B, as shown in Fig. B.1.

Plant A has a yearly investment cost of 4 and a yearly operating cost of 2 monetary units. On the other hand, Plant B has a yearly investment cost of 2 and a yearly operating cost of 3 monetary units. The capacities of units A and B are 20 and 30 units, repetitively.

A.J. Conejo et al., *Investment in Electricity Generation and Transmission*,
DOI 10.1007/978-3-319-29501-5

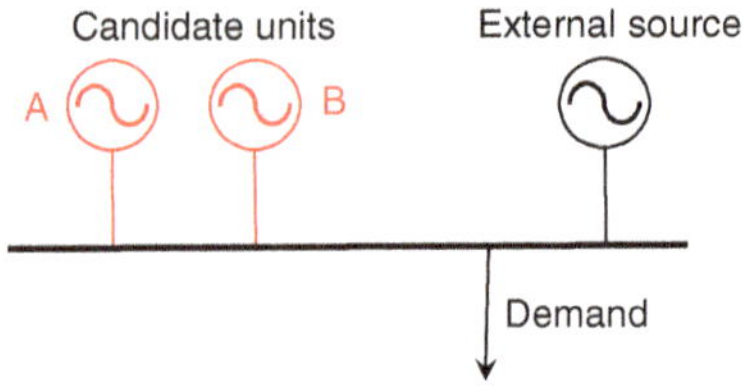

Fig. B.1 Stochastic programming example: investment in production facilities

The producer intends to build one of these two plants to supply next year's electricity demand, which is uncertain. It may also buy up to 40 units of energy from an expensive external source at a yearly cost of 4 monetary units per unit of energy.

The uncertain future demand is estimated to be 10 units with probability 0.5, 20 units with probability 0.25, or 30 units with probability 0.25 as well. We call these alternative realizations of the demand *scenarios*.

This producer needs to decide which plant to build without knowing next year's actual electricity demand. This decision is called a *first-stage* decision, since it is made at the initial stage, or a *here-and-now* decision, since a plant has to be built now to be ready for production next year.

Next year, with one of the two plants built, the producer will operate that plant and possibly buy energy from the expensive external source to supply the demand that finally materializes. This operating decision is a *second-stage* decision since it takes place once the first-stage decision is made, and a *wait-and-see* decision because it is made once the actual demand is known.

If the objective of this producer (investor) is to supply the demand minimizing the total expected cost (investment and operation), its decision-making problem is:

$$
\begin{aligned}
\min_{x_A, x_B; y_{A1}, y_{B1}, y_{E1}, y_{A2}, y_{B2} y_{E2}, y_{A3}, y_{B3}, y_{E3}} \quad & \\
4x_A + 2x_B + \quad & 0.50(2y_{A1} + 3y_{B1} + 4y_{E1}) + \\
& 0.25(2y_{A2} + 3y_{B2} + 4y_{E2}) + \\
& 0.25(2y_{A3} + 3y_{B3} + 4y_{E3}) \\
\text{s. t.} \quad & \\
x_A + x_B \le 1 \quad & \\
x_A, x_B \in \{0, 1\} \quad & \\
& y_{A1} + y_{B1} + y_{E1} = 10 \\
& y_{A2} + y_{B2} + y_{E2} = 20 \\
& y_{A3} + y_{B3} + y_{E3} = 30 \\
& 0 \le y_{A1}, y_{A2}, y_{A3} \le 20x_A \\
& 0 \le y_{B1}, y_{B2}, y_{B3} \le 30x_B \\
& 0 \le y_{E1}, y_{E2}, y_{E3} \le 40.
\end{aligned}
\tag{B.1}
$$

Binary variables x_A and x_A represent the build/not-build decisions pertaining to units A and B, respectively. These are here-and-now decisions. Variables y_{A1}, y_{A2}, and y_{A3} represent the production of unit A in scenarios 1, 2, and 3, respectively; variables y_{B1}, y_{B2}, and y_{B3} represent the production of unit B in scenarios 1, 2, and

Table B.1 Stochastic programming example: optimal operating decisions once plant A has been built

Unit/scenario	Scenario 1	Scenario 2	Scenario 3
Unit A	10	20	20
Unit B	0	0	0
External source E	0	0	10

3, respectively; and y_{E1}, y_{E2}, and y_{E3} represent the quantity bought from the external source in scenarios 1, 2, and 3, respectively. All these are wait-and-see decisions.

The objective function includes two parts. The left-hand-side part (i.e., $4x_A + 2x_B$) represents the investment cost and depends on here-and-now variables, while the right-hand-side part (i.e., $0.50(2y_{A1} + 3y_{B1} + 4y_{E1}) + 0.25(2y_{A2} + 3y_{B2} + 4y_{E2}) + 0.25(2y_{A3} + 3y_{B3} + 4y_{E3})$) represents the operation cost and depends on wait-and-see variables.

The first two constraints (on the left-hand side) involve here-and-now variables and state that only one plant can be built and that variables x_A and x_B are binary.

The following three constraints (on the right-hand side) enforce the supply of the demand in the three scenarios and involve just wait-and-see variables.

The following two double constraints (on the right-hand side as well) establish production limits for the two candidate plants. These constraints involve both here-and-now and wait-and-see variables.

Finally, the last double constraint (on the right-hand side) imposes a limit on the purchase of energy from the expensive external source and involves just wait-and-see variables.

Problem (B.1) is mixed-integer linear and can be solved using an appropriate solver. The optimal solution of this mixed-integer linear programming problem consists in building plant A, with the actual production per plant and scenario provided in Table B.1. The level of purchase per scenario is also provided in this table. The minimum expected total cost is 44 monetary units. A GAMS code to solve Problem (B.1) above is provided in Sect. B.2.2. □

B.2.1 Formulation

Problem (B.1) above has the following general form:

$$\begin{aligned}
&\min_{\mathbf{x};\mathbf{y}_w,\forall w} \\
&\quad f^{\mathrm{I}}(\mathbf{x}) + \mathscr{E}_w\{f^{\mathrm{O}}(\mathbf{y}_w)\} \\
&\text{s.t.} \\
&\quad \mathbf{h}^{\mathrm{I}}(\mathbf{x}) = \mathbf{0} \\
&\quad \mathbf{g}^{\mathrm{I}}(\mathbf{x}) \leq \mathbf{0} \\
&\quad \mathbf{x} \in \mathscr{X}
\end{aligned} \tag{B.2}$$

$$\mathbf{h}^{\mathrm{O}}_{w}(\mathbf{x}, \mathbf{y}_w) = \mathbf{0} \ \forall w \in \mathscr{W}$$
$$\mathbf{g}^{\mathrm{O}}_{w}(\mathbf{x}, \mathbf{y}_w) \leq \mathbf{0}, \ \forall w \in \mathscr{W}$$
$$\mathbf{y}_w \in \mathscr{Y}, \ \forall w \in \mathscr{W}.$$

In this problem, the superscript I denotes investment (first) stage, while the superscript O denotes operation (second) stage. The subscript w denotes scenario, and $\mathscr{W}$ is the set of all possible scenarios. Constraints $\mathbf{h}^{\mathrm{I}}(\mathbf{x}) = \mathbf{0}$ and $\mathbf{g}^{\mathrm{I}}(\mathbf{x}) \leq \mathbf{0}$ pertain to the investment decisions, while constraints $\mathbf{h}^{\mathrm{O}}_{w}(\mathbf{x}, \mathbf{y}_w) = \mathbf{0}$ and $\mathbf{g}^{\mathrm{O}}_{w}(\mathbf{x}, \mathbf{y}_w) \leq \mathbf{0}$ pertain to operation decisions in scenario w.

A more general version of problem (B.2) is:

$$\begin{aligned} &\min_{\mathbf{x}} \\ &\quad f^{\mathrm{I}}(\mathbf{x}) + \quad \mathscr{E}_w\{z^{\mathrm{O}}_{w}(\mathbf{x})\} \\ &\text{s.t.} \\ &\quad \mathbf{h}^{\mathrm{I}}(\mathbf{x}) = \mathbf{0} \\ &\quad \mathbf{g}^{\mathrm{I}}(\mathbf{x}) \leq \mathbf{0} \\ &\quad \mathbf{x} \in \mathscr{X}, \end{aligned} \tag{B.3}$$

where

$$\begin{aligned} z^{\mathrm{O}}_{w}(\mathbf{x}) = \{\min_{\mathbf{y}_w} \ & f^{\mathrm{O}}(\mathbf{y}_w) \\ \text{s.t.} \quad & \mathbf{h}^{\mathrm{O}}_{w}(\mathbf{x}, \mathbf{y}_w) = \mathbf{0} \\ & \mathbf{g}^{\mathrm{O}}_{w}(\mathbf{x}, \mathbf{y}_w) \leq \mathbf{0} \\ & \mathbf{y}_w \in \mathscr{Y}\} \ \forall w \in \mathscr{W}. \end{aligned} \tag{B.4}$$

Problem (B.3) seeks to minimize the total cost, including investment costs, $f^{\mathrm{I}}(\mathbf{x})$, involving investment variables $\mathbf{x}$, and expected operation costs, $\mathscr{E}_w\{z^{\mathrm{S}}_{w}\}$, involving operation variables $\mathbf{y}_w, \forall w$. Problem (B.4), which constrains problem (B.3), represents operation decisions regarding scenario $w \in \mathscr{W}$ and seeks to minimize the scenario cost.

Problem (B.3)–(B.4) is a general statement of a two-stage stochastic programming problem. Under mild mathematical assumptions, problems (B.3)–(B.4) and problem (B.2) are equivalent [3].

B.2.2 Solution

If problem (B.2) is linear, it is generally solvable up to millions of variables and constraints; if it is mixed-integer linear, solvability is directly linked to the number of integer variables, which should be kept in the thousands; if it is nonlinear, it is generally solvable up to thousands of variables and constraints; and if it is mixed-integer nonlinear, the off-the-shelf solution techniques are generally not available, and decomposition technique are advisable.

A GAMS code for solving problem (B.1), which is mixed-integer linear, is provided below:

```
1  sets g/g1,g2/, go/go1/, w/w1,w2,w3/;
2  variable z;
3  positive variables y(g,w), yo(go,w);
4  binary variables x(g);
5  parameter
6  i(g)       /g1   4, g2  2/,
7  c(g)       /g1   2, g2  3/,
8  cap(g)     /g1  20, g2 30/,
9  co(go)     /go1  4/,
10 capo(go)   /go1 40/,
11 d(w)       /w1 10, w2  20, w3  30/,
12 p(w)       /w1 .5, w2 .25, w3 .25/;
13 equation of, alt, lim(g,w), limo(go,w), bal(w);
14 of.. z =e= sum(g, i(g)*x(g)) +
15            sum((g,w), p(w)*c(g)*y(g,w)) + sum((go,w
               ), p(w)*co(go)*yo(go,w));
16 alt.. sum(g, x(g)) =l= 1;
17 lim(g,w).. y(g,w) =l= cap(g)*x(g);
18 limo(go,w).. yo(go,w) =l= capo(go);
19 bal(w).. sum(g, y(g,w)) + sum(go, yo(go,w)) =e= d(w
       );
20 model invgen /all/;
21 solve invgen using mip minimizing z;
22 display x.l, y.l, yo.l;
```

This GAMS code is briefly described in the following. Line 1 declares the sets (indexes) used, which include plants, external source (singleton), and scenarios. Lines 2–4 are variable declarations. Lines 5–12 specify the required data. Lines 13–19 state the equations (objective function and constraints) of the model. Line 20 defines the model, line 21 directs GAMS to solve the model, and line 22 indicates which output variables to display.

More often than not, problem (B.2) becomes large-scale as a result of considering a large number of scenarios to properly represent the uncertainty involved. In such a case, decomposition strategies are advisable [4] to achieve an efficient solution, or even tractability.

Benders decomposition is particularly appealing since it is adapted to attack problems involving *complicating variables*, which is generally the case with investment problems. Benders decomposition works as follows:

1. Fix investment decisions to given values by solving a so-called master problem, which approximately reproduces the original problem but involves a reduced number of variables and constraints.
2. For the given investment decisions, solve one operation problem per scenario and determine the sensitivity of the operation cost with respect to the investment decisions. Each scenario problem is generally called a subproblem.
3. Use this sensitivity information to enrich the master problem (making it closer to the original one) and to derive improved investment decisions by solving it.
4. This iterative procedure concludes if the investment decisions cannot be improved.

In cases in which the number of scenarios that need to be considered is very very large, decomposition techniques that build on scenario sampling might be appropriate, as proposed in [7]. Such techniques use a Monte Carlo type of sampling, and solve single-scenario problems within an iterative framework that, under some assumptions, allows for an asymptotic approximation of the original problem.

B.3 Adaptive Robust Optimization

A tutorial introduction to adaptive robust optimization is provided in this section. Further information can be found in the book by Ben-Tal et al. [1], the tutorial paper by Bertsimas et al. [2], and in [6], which has an application focus.

Illustrative Example B.2

A transmission operator needs to decide which of two alternative transmission lines to build, A or B, to connect a generation area and a demand area. One of these transmission lines needs to be built to supply next year's electricity demand, as illustrated in Fig. B.2.

The capacity of transmission lines A and B are 10 and 20 units, respectively, and their annualized building costs are 10 and 20 monetary units, respectively.

The transmission line to be built will allow shipping electricity from a generation area to a consumption one. The generation area includes a production unit with a yearly production cost of 2 monetary units and a capacity of 20 units, while the demand node includes an expensive production unit with a yearly production cost of 10 monetary units and a capacity of 20 units.

Next year's demand is uncertain, involving values of 10, 20, and 30 units, with probabilities 0.50, 0.25, and 0.25, respectively, with an average value 17.5 units.

For policy reasons, the operator needs to ensure that the demand is supplied in any case, i.e., it has a *robust* view.

In order to minimize building and operating costs and to enforce a robust view, the problem to be solved by the operator is:

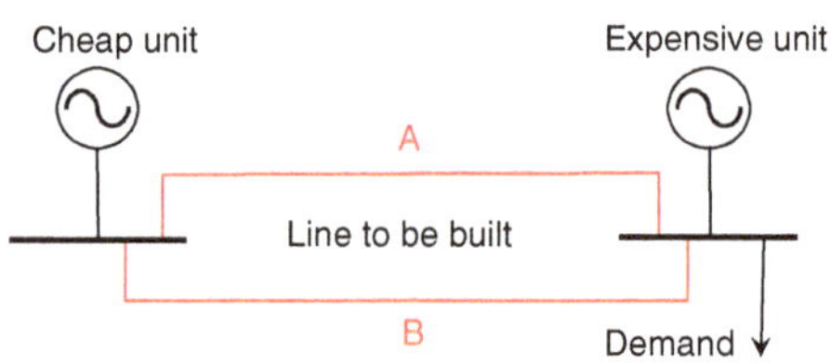

Fig. B.2 Robust optimization example: investment in transmission lines

$$\min_{x_A, x_B} \quad (10x_A + 20x_B) \; \max_{d} \quad \min_{y_1, y_2} \quad (2y_1 + 10y_2)$$
$$\text{s.t.}$$
$$y_1 + y_2 = d$$
$$y_1 \leq 10x_A + 20x_B$$
$$0 \leq y_1 \leq 20$$
$$0 \leq y_2 \leq 20$$
$$\text{s.t.}$$
$$d \in \{10, 20, 30\}$$
$$\text{s.t.}$$
$$x_A + x_B = 1$$
$$x_A, x_B \in \{0, 1\} \tag{B.5}$$

Binary variables x_A and x_B represent the build/not-build decisions pertaining to transmission lines A and B, respectively, variable d represents the alternative demand realizations, and variables y_1 and y_2 represent the actual production of the units in the generation and demand areas, respectively, for the robust operating condition.

Problem (B.5) is a trilevel problem involving three hierarchically interrelated problems. The left-hand-side problem allows deciding which of the two lines, A or B, to build (variables x_A and x_B) pursuing minimum total (investment and operating) cost. Once the investment decision has been made, the center problem pursues identifying the demand realization (variable d) that results in maximum operating cost. Finally, for a given investment decision and a demand realization, the right-hand-side problem decides the operation of the production units (variables y_1 and y_2) pursuing minimum operating cost.

Note that the above framework involves a *robust* view with respect to uncertainty (demand) since the decision of which line to build is made on the basis of worst uncertainty (demand) realization.

We solve this problem by enumeration as shown in Table B.2. That is, we consider both alternatives, i.e., building either line A or line B, and for each one of these two alternatives, we consider each of the three possible demand realizations. This renders six right-hand-side problems to be solved. Once the right-hand-side problems are solved, the solution of the middle problem for each investment alternative in the one

Table B.2 Adaptive robust optimization example: solution by enumeration

Investment		Demand	Operation		Cost
x_A	x_B	d	y_1	y_2	
1	0	10	10	0	30
1	0	20	10	10	130
1	0	30	10	20	230
0	1	10	10	0	40
0	1	20	20	0	60
0	1	30	20	10	160

resulting in the highest cost, i.e., 230 if line A is built, and 160 is line B is built instead. Finally, the left-hand-side problem picks the smallest of these two cost alternatives, i.e., 160. Thus, the optimal solution is building line B. □

B.3.1 Formulation

Trilevel problem (B.5) has the general form below:

$$
\begin{aligned}
\min_{\mathbf{x}} \quad & \max_{\mathbf{u}} \quad \min_{\mathbf{y}} \quad f(\mathbf{x}, \mathbf{u}, \mathbf{y}) \\
& \qquad\qquad\quad \text{s.t.} \\
& \qquad\qquad\quad \mathbf{h}^{\mathrm{O}}(\mathbf{x}, \mathbf{u}, \mathbf{y}) = \mathbf{0} \\
& \qquad\qquad\quad \mathbf{g}^{\mathrm{O}}(\mathbf{x}, \mathbf{u}, \mathbf{y}) \leq \mathbf{0} \\
& \qquad\qquad\quad \mathbf{y} \in \mathscr{Y} \\
& \quad \text{s.t.} \\
& \quad \mathbf{u} \in \mathscr{U} \\
\text{s.t.} \quad & \\
& \mathbf{h}^{\mathrm{I}}(\mathbf{x}) = \mathbf{0} \\
& \mathbf{g}^{\mathrm{I}}(\mathbf{x}) \leq \mathbf{0} \\
& \mathbf{x} \in \mathscr{X} .
\end{aligned}
\tag{B.6}
$$

Objective function $f(\mathbf{x}, \mathbf{u}, \mathbf{y})$ represents the minimization of the system's total cost, including investment and operating costs. The investment decision variables, gathered in vector $\mathbf{x}$, are binary variables representing the build/not-build nature of the investment decisions. The entries of vector $\mathbf{u}$ are variables describing the uncertain parameters. Similarly, the entries of vector $\mathbf{y}$ are the operating decision variables, which are considered to be continuous.

The worst case realization of the uncertainty and the successive adaptive actions are considered in the max–min right-hand-side problem, while the min left-hand-side problem seeks minimum total cost.

Constraints $\mathbf{h}^{\mathrm{I}}(\mathbf{x}) = \mathbf{0}$ and $\mathbf{g}^{\mathrm{I}}(\mathbf{x}) \leq \mathbf{0}$ represent investment requirements and limits, while constraints $\mathbf{h}^{\mathrm{O}}(\mathbf{x}, \mathbf{u}, \mathbf{y}) = \mathbf{0}$ and $\mathbf{g}^{\mathrm{O}}(\mathbf{x}, \mathbf{u}, \mathbf{y}) \leq \mathbf{0}$ include equality constraints related to the system operations, and inequality constraints related to system limits, respectively. Operation constraints depend on both $\mathbf{x}$ and $\mathbf{y}$, which implies that the investment decisions alter the configuration and the operation of the system. Finally, $\mathbf{u} \in \mathscr{U}$ defines the uncertainty set, a set representing the variability limits of the uncertain parameters.

B.3.2 Solution

Trilevel problem (B.6) cannot be solved directly. A generally effective solution strategy is using a *Benders decomposition* technique similar to the one explained in

Sect. B.2.2, [4]. In this case, a master problem proposes investment decisions (and thus plays the role of the left-hand-side problem), and a subproblem that involves both the central and right-hand-side problems evaluates such investment decisions and provides sensitivity information to enrich the master problem, which in turn provides better investment decisions. The procedure concludes if no improvement in investment decisions is achieved. The complication of this procedure is the joint solution of the the central and right-hand-side problems, which requires merging them into a single problem using duality theory, as explained in [1] or [2].

If operational (primal) information is transferred to the master problem instead of sensitivity (dual) information, the resulting decomposition technique is called *column-and-constraint generation* [10], which generally exhibits better computational efficiency than that of a dual Benders approach.

B.4 Multistage Decision-Making

A natural extension of the two-stage stochastic programming model is the *multistage stochastic programming* model, in which the sequence *decision–uncertainty realization–reaction* (recourse) is repeated a number of times. It is important to note that most investment problems are indeed multistage since investment decisions are made throughout a planning horizon at different points in time as uncertainty unfolds. Further information on multistage stochastic programming models is available in the textbook [5], which has an electricity focus.

Similarly, adaptive robust optimization models can be transformed into multistage decision models using linear decision rules, as explained, for instance, in [8]. This, however, requires significant simplifications.

B.5 End-of-Chapter Exercises

B.1 Formulate the two-stage stochastic programming problem (example) in Sect. B.1 as an adaptive robust optimization problem, solve it, and compare the results obtained with those in Sect. B.1.

B.2 Formulate the adaptive robust optimization problem (example) in Sect. B.2 as a two-stage stochastic programming problem, solve it, and compare the results obtained with those in Sect. B.2.

References

1. Ben-Tal, A., El Ghaoui, L., Nemirovski, A.: Robust Optimization. Princeton University Press, Princeton (2009)
2. Bertsimas, D., Brown, D., Caramanis, C.: Theory and applications of robust optimization. SIAM Rev. **53**(3), 464–501 (2011)
3. Birge, J.R., Louveaux, F.: Introduction to Stochastic Programming, 2nd edn. Springer, New York (2011)
4. Conejo, A.J., Castillo, E., Minguez, R., Garcia-Bertrand, R.: Decomposition Techniques in Mathematical Programming. Springer, Heidelberg (2006)
5. Conejo, A.J., Carrion, M., Morales, J.M.: Decision Making Under Uncertainty in Electricity Markets. Springer, New York (2010)
6. Gorissen, B., Yanikoglu, I., den Hertog, D.: Hints for practical robust optimization. European Banking CentER Discussion Paper No. 2013-065. Available at SSRN: http://ssrn.com/abstract=2361396 (2013)
7. Higle, J.L., Sen, S.: Stochastic Decomposition. Springer, New York (1996)
8. Kuhn, D., Wiesemann, W., Georghiou, A.: Primal and dual linear decision rules in stochastic and robust optimization. Math. Program **130**(1), 177–209 (2011)
9. Sen, S., Higle, J.L.: An introductory tutorial on stochastic linear programming models. Interfaces **29**(2), 33–61 (1999)
10. Zeng, B., Zhao, L.: Solving two-stage robust optimization problems using a column-and-constraint generation method. Oper. Res. Lett. **41**(5), 457–461 (2013)

Appendix C
Complementarity

This appendix provides an overview of complementarity modeling, including equilibrium models, mathematical programs with equilibrium constraints (MPECs), and equilibrium programs with equilibrium constraints (EPECs). These models are introduced and illustrated using simple investment examples.

C.1 Introduction

Equilibrium models are considered first, followed by hierarchical complementarity models, i.e., MPECs and EPECs. For the purposes of this appendix, an equilibrium model results from jointly considering the solution of a number of interrelated optimization problems. An MPEC involves a hierarchy and often results from considering an optimization problem that is constrained by a number of other optimization problems. Finally, an EPEC results from the joint consideration (in the equilibrium sense) of a number of interrelated MPECs.

C.2 Equilibria

The concept of equilibrium is first stated using a simple investment example, then illustrated through a numerical example. Finally, observations on solution algorithms are given.

C.2.1 Formulation

Consider an investor i willing to build an electricity production facility of capacity up to $x_i^{\max}$ units, with an annual investment cost I_i per production unit built, and an

A.J. Conejo et al., *Investment in Electricity Generation and Transmission*,
DOI 10.1007/978-3-319-29501-5

annual operating cost C_i per unit of energy produced. The objective of this investor is building the facility, producing electric energy using it, and selling the energy produced in the electricity market for a profit. For simplicity, we assume that the production unit can be built *instantaneously*, and that the electricity price depends linearly on the production of this investor and all other investors (Cournot viewpoint). The problem to be solved by this investor i is:

$$\begin{aligned} \min_{x_i, y_i} \quad & I_i x_i - (C_i - \lambda)\, y_i \\ \text{s.t.} \quad & y_i \leq x_i \\ & x_i \leq x_i^{\max} \\ & y_i \geq 0 \\ & x_i \geq 0, \end{aligned} \tag{C.1}$$

where x_i is the capacity to be built, y_i the production level, and λ the market price, which has the form:

$$\lambda = a - b \sum_j y_j, \tag{C.2}$$

where a and b are appropriate constants.

The objective function of Problem (C.1), to be minimized, is minus the profit of the investor and includes investment cost ($I_i x_i$) and operations profit ($(C_i - \lambda)\, y_i$). The first constraint enforces that the production level should be below the capacity built, the second constraint provides a cap on the capacity that can be built, and the third and fourth constraints are nonnegativity declarations.

Plugging (C.2) into the objective function of problem (C.1) and converting all constraints to the type less than or equal to yields:

$$\begin{aligned} \min_{x_i, y_i} \quad & I_i x_i - \left(C_i - (a - b \textstyle\sum_j y_j)\right) y_i \\ \text{s.t.} \quad & y_i \leq x_i : \quad \alpha_i \\ & x_i \leq x_i^{\max} : \quad \beta_i \\ & -y_i \leq 0 : \quad \gamma_i \\ & -x_i \leq 0 : \quad \delta_i, \end{aligned} \tag{C.3}$$

where α_i, β_i, γ_i, and δ_i are the dual variables associated with the four constraints of Problem (C.3).

The first-order optimality (KKT) conditions [4] of problem (C.3) are:

$$I_i - \alpha_i + \beta_i - \delta_i = 0 \tag{C.4a}$$

$$C_i - a + b y_i + b \sum_j y_j + \alpha_i - \gamma_i = 0 \tag{C.4b}$$

$$0 \leq \alpha_i \perp (y_i - x_i) \leq 0 \tag{C.4c}$$

$$0 \leq \beta_i \perp (x_i - x_i^{\max}) \leq 0 \tag{C.4d}$$

$$0 \leq \gamma_i \perp (-y_i) \leq 0 \tag{C.4e}$$

$$0 \leq \delta_i \perp (-x_i) \leq 0, \tag{C.4f}$$

where $u \perp v$ stands for $u \geq 0$, $v \geq 0$, and $u \cdot v = 0$, element by element. The operator $\perp$ is the so-called *perp* operator.

We use (C.4a) to solve for γ_i and plug the resulting expression into (C.4e), (C.4b) to solve for δ_i, and we plug the resulting expression into (C.4f). Also, we transform all inequalities to greater than or equal to constraints, which gives us:

$$0 \leq \alpha_i \perp (x_i - y_i) \geq 0 \tag{C.5a}$$

$$0 \leq \beta_i \perp (x_i^{\max} - x_i) \geq 0 \tag{C.5b}$$

$$0 \leq (I_i - \alpha_i + \beta_i) \perp x_i \geq 0 \tag{C.5c}$$

$$0 \leq (C_i - a + by_i + b\sum_j y_j + \alpha_i) \perp y_i \geq 0. \tag{C.5d}$$

The equilibrium for n investors behaving as indicated above is obtained by solving the following system of *complementarity* conditions:

$$\left.\begin{array}{l} 0 \leq \alpha_i \perp (x_i - y_i) \geq 0 \\ 0 \leq \beta_i \perp (x_i^{\max} - x_i) \geq 0 \\ 0 \leq (I_i - \alpha_i + \beta_i) \perp x_i \geq 0 \\ 0 \leq (C_i - a + by_i + b\sum_j y_j + \alpha_i) \perp y_i \geq 0 \end{array}\right\} \quad i = 1, \ldots, n. \tag{C.6}$$

The solution of this complementarity system provides the so-called Nash–Cournot equilibrium, as a result of representing the price demand dependency ($\lambda = a - b\sum_j y_j$) in the profit-maximization problem of each investor.

A problem of the form (C.6) is called a complementarity problem, CP, or mixed complementarity problem, MCP, if it also includes equality constraints.

Therefore, the general form of a CP is [2, 5]:

$$0 \leq x_i \perp g_i(x_1, \ldots, x_n) \geq 0, \quad \forall i = 1, \ldots, n, \tag{C.7}$$

while the general form of an MCP is [2]:

$$\left\{\begin{array}{ll} 0 \leq x_i \perp g_i(x_1, \ldots, x_n; y_1, \ldots, y_m) \geq 0, & \forall i = 1, \ldots, n \\ h_{n+j}(x_1, \ldots, x_n; y_1, \ldots, y_m) = 0, \; y_j \text{ free} & \forall j = 1, \ldots, m. \end{array}\right. \tag{C.8}$$

Note that equality constraints have associated free variables and that complementarity does not hold for these constraints.

Illustrative Example C.1

We consider three investors with investment costs 1, 1, and 1 units, operation costs 3, 4, and 5 units, and capacities 3, 4, and 2 units. The price slope and intercept are equal to 1 and 10, respectively. The complementarity system to be solved to find the equilibria of the three investors is:

Table C.1 Equilibrium example: capacity built and production level per investor

Investor i	Capacity built x_i	Production level y_i
1	2.250	2.250
2	1.250	1.250
3	0.250	0.250

$$\begin{cases} 0 \leq \alpha_1 \perp (x_1 - y_1) \geq 0 \\ 0 \leq \beta_1 \perp (3 - x_1) \geq 0 \\ 0 \leq (1 - \alpha_1 + \beta_1) \perp x_1 \geq 0 \\ 0 \leq (3 - 10 + y_1 + \sum_{j=1}^{3} y_j + \alpha_1) \perp y_1 \geq 0 \\ \\ 0 \leq \alpha_2 \perp (x_2 - y_2) \geq 0 \\ 0 \leq \beta_2 \perp (4 - x_2) \geq 0 \\ 0 \leq (1 - \alpha_2 + \beta_2) \perp x_2 \geq 0 \\ 0 \leq (4 - 10 + y_2 + \sum_{j=1}^{3} y_j + \alpha_2) \perp y_2 \geq 0 \\ \\ 0 \leq \alpha_3 \perp (x_3 - y_3) \geq 0 \\ 0 \leq \beta_3 \perp (2 - x_3) \geq 0 \\ 0 \leq (1 - \alpha_3 + \beta_3) \perp x_3 \geq 0 \\ 0 \leq (5 - 10 + y_3 + \sum_{j=1}^{3} y_j + \alpha_3) \perp y_3 \geq 0. \end{cases} \tag{C.9}$$

The equilibrium outcome resulting from solving the complementarity system above is provided in Table C.1.

Observe from Table C.1 that the capacity and the production level are identical for all producers; that is, each producer builds just the capacity that it can fully utilize. This is an expected result since building higher capacity than that to be used makes no economic sense.

A GAMS code [3] for solving this equilibrium problem is provided below:

```
OPTIONS mcp=path;

SETS i generators /g1*g3/;
ALIAS(i,j)

PARAMETERS
In(i) investment cost
/g1 1, g2 1, g3 1/
C(i) production cost
/g1 3, g2 4, g3 5/
X_max(i) capacity
/g1 3, g2 4, g3 2/
a price intercept /10/
b price slope      /1/;

POSITIVE VARIABLES
x(i) capacity built
```

```
y(i) production level
alpha(i) dual var. of maximum production
beta(i)  dual var. of maximum investment;

EQUATIONS
eq1(i), eq2(i), eq3(i), eq4(i);
eq1(i).. x(i) - y(i) =g= 0;
eq2(i).. X_max(i) - x(i) =g= 0;
eq3(i).. In(i) - alpha(i) + beta(i) =g= 0;
eq4(i).. C(i) - a + b*y(i) + b*sum(j,y(j)) + alpha(
   i) =g= 0;

MODEL Equilibrium /eq1.alpha, eq2.beta, eq3.x, eq4.
   y/;
SOLVE Equilibrium using mcp;
```

Note that the GAMS code above follows almost verbatim the form of the complementarity system (C.9). □

C.2.2 *Solution*

Further details on formulating CPs and MCPs, and on the available solution techniques to tackle them, are discussed, for instance, in [2]. PATH 3.0 [1] is a solver that embodies efficient solution techniques and allows solving CPs and MCPs of small and moderate size.

C.3 Mathematical Program with Equilibrium Constraints, MPEC

MPECs are considered next. An MPEC naturally arises in considering an optimization problem that is constrained by one or more optimization problems. Such problems are also called bilevel optimization problems since they consist of an upper level problem and a number of lower-level ones, the constraining optimization problems. If each constraining optimization problem is replaced by its optimality conditions, i.e., by its *equilibrium* conditions, the resulting problem is an optimization problem (or program) with *equilibrium* constraints.

C.3.1 Formulation

We consider that investor 1 is a leader and that investors 2 to n are followers. Thus, the problem to be solved by investor 1 is:

$$\begin{array}{ll}\min_{\Xi_1} & I_1 x_1 - \left(C_1 - (a - b\sum_j y_j)\right) y_1 \\ \text{s.t.} & y_1 \le x_1 \le x_1^{\max} \\ & y_1, x_1 \ge 0 \\ & \left.\begin{array}{l} 0 \le \alpha_i \perp (x_i - y_i) \ge 0 \\ 0 \le \beta_i \perp (x_i^{\max} - x_i) \ge 0 \\ 0 \le (I_i - \alpha_i + \beta_i) \perp x_i \ge 0 \\ 0 \le (C_i - a + b y_i + b\sum_j y_j + \alpha_i) \perp y_i \ge 0 \end{array}\right\} \quad i = 2, \ldots, n, \end{array} \tag{C.10}$$

where $\Xi_1 = \{x_1, y_1\} \cup \{x_i, y_i, \alpha_i, \beta_i, i = 2, \ldots, n\}$.

Problem (C.10) embodies the problem maximization of investor 1 and is subject to equilibrium conditions pertaining to investors 2 to n.

Considering the structure of (C.10), a general formulation of an MPEC is:

$$\begin{array}{ll}\min_{\Xi} & f(u_1, \ldots, u_{n^{\mathrm{U}}}; x_1, \ldots, x_n; y_1, \ldots, y_m) \\ \text{s.t.} & h_k^{\mathrm{U}}(u_1, \ldots, u_{n^{\mathrm{U}}}; x_1, \ldots, x_n; y_1, \ldots, y_m) = 0, \quad \forall k = 1, \ldots, p^{\mathrm{U}} \\ & g_l^{\mathrm{U}}(u_1, \ldots, u_{n^{\mathrm{U}}}; x_1, \ldots, x_n; y_1, \ldots, y_m) \le 0, \quad \forall l = 1, \ldots, q^{\mathrm{U}} \\ & 0 \le x_i \perp g_i(u_1, \ldots, u_{n^{\mathrm{U}}}; x_1, \ldots, x_n; y_1, \ldots, y_m) \ge 0, \quad \forall i = 1, \ldots, n \\ & h_{n+j}(u_1, \ldots, u_{n^{\mathrm{U}}}; x_1, \ldots, x_n; y_1, \ldots, y_m) = 0, \quad \forall j = 1, \ldots, m, \end{array} \tag{C.11}$$

where $\Xi = \{u_1, \ldots, u_{n^{\mathrm{U}}}\} \cup \{x_1, \ldots, x_n; y_1, \ldots, y_m\}$.

Problem (C.11) in compact form is [2, 5]:

$$\begin{array}{ll}\min_{\mathbf{u},\mathbf{x},\mathbf{y}} & f(\mathbf{u}, \mathbf{x}, \mathbf{y}) \\ \text{s.t.} & \mathbf{h}^{\mathrm{U}}(\mathbf{u}, \mathbf{x}, \mathbf{y}) = 0 \\ & \mathbf{g}^{\mathrm{U}}(\mathbf{u}, \mathbf{x}, \mathbf{y}) \le 0 \\ & \mathbf{h}(\mathbf{u}, \mathbf{x}, \mathbf{y}) = \mathbf{0} \\ & \mathbf{0} \le \mathbf{x} \perp \mathbf{g}(\mathbf{u}, \mathbf{x}, \mathbf{y}) \ge \mathbf{0}, \end{array} \tag{C.12}$$

where $\mathbf{u} \in \mathbb{R}^{n^{\mathrm{U}}}$, $\mathbf{x} \in \mathbb{R}^n$, $\mathbf{y} \in \mathbb{R}^m$, $f \in \mathbb{R}$, $\mathbf{h}^{\mathrm{U}} \in \mathbb{R}^{p^{\mathrm{U}}}$, $\mathbf{g}^{\mathrm{U}} \in \mathbb{R}^{q^{\mathrm{U}}}$, $\mathbf{h} \in \mathbb{R}^m$, and $\mathbf{g} \in \mathbb{R}^n$.

Illustrative Example C.2

Considering the three investors in the example of Sect. C.1 and assuming that investor 1 is a leader and that investors 2 and 3 are followers, the MPEC to be solved is:

Table C.2 MPEC example with investor 1 as the leader: capacity built and production level per investor

Investor i	Capacity built x_i	Production level y_i
1	3	3
2	2/3	2/3
3	2/3	2/3

$$
\begin{aligned}
\min_{\Xi_1}\ & x_1 - \left(3 - (10 - \textstyle\sum_{j=1}^{3} y_j)\right) y_1 \\
\text{s.t.}\quad & y_1 \le x_1 \le 3 \\
& y_1, x_1 \ge 0 \\
& \\
& 0 \le \alpha_2 \perp (x_2 - y_2) \ge 0 \\
& 0 \le \beta_2 \perp (4 - x_2) \ge 0 \\
& 0 \le (1 - \alpha_2 + \beta_2) \perp x_2 \ge 0 \\
& 0 \le (4 - 10 + y_2 + \textstyle\sum_{j=1}^{3} y_j + \alpha_2) \perp y_2 \ge 0 \\
& \\
& 0 \le \alpha_3 \perp (x_3 - y_3) \ge 0 \\
& 0 \le \beta_3 \perp (2 - x_3) \ge 0 \\
& 0 \le (1 - \alpha_3 + \beta_3) \perp x_3 \ge 0 \\
& 0 \le (5 - 10 + y_3 + \textstyle\sum_{j=1}^{3} y_j + \alpha_3) \perp y_3 \ge 0,
\end{aligned}
\tag{C.13}
$$

where $\Xi_1 = \{x_1, y_1\} \cup \{x_2, y_2, \alpha_2, \beta_2, x_3, y_3, \alpha_3, \beta_3\}$.

The outcome resulting from solving MPEC (C.13) is provided in Table C.2.

A GAMS code for solving MPEC (C.13) is provided below:

```
OPTIONS mpec=knitro;

PARAMETERS
I1 /1/, I2 /1/, I3 /1/
C1 /3/, C2 /4/, C3 /5/
X1max /3/, X2max /4/, X3max /2/
a /10/, b /1/;

FREE VARIABLES
Profit_leader;
POSITIVE VARIABLES
x1, x2, x3
y1, y2, y3
alpha2, alpha3  dual var. of maximum production
beta2,  beta3   dual var. of maximum capacity;

EQUATIONS
OF, plimit, climit,
eq1_1, eq1_2, eq1_3, eq1_4,
eq2_1, eq2_2, eq2_3, eq2_4;

OF..       Profit_leader =e= I1*x1 - (C1-a-b*(y1+y2+
    y3))*y1;
```

```
plimit..   y1 =l= x1;
climit..   x1 =l= X1max;

eq1_1..    x2 - y2 =g= 0;
eq1_2..    X2max - x2 =g= 0;
eq1_3..    I2 - alpha2 + beta2 =g= 0;
eq1_4..    C2 - a + b*y2 + b*(y1+y2+y3) + alpha2 =g=
     0;

eq2_1..    x3 - y3 =g= 0;
eq2_2..    X3max - x3 =g= 0;
eq2_3..    I3 - alpha3 + beta3 =g= 0;
eq2_4..    C2 - a + b*y3 + b*(y1+y2+y3) + alpha3 =g=
     0;

MODEL MPEC
/OF, plimit, climit,
eq1_1.alpha2, eq1_2.beta2, eq1_3.x2, eq1_4.y2,
eq2_1.alpha3, eq2_2.beta3, eq2_3.x3, eq2_4.y3/;
SOLVE MPEC using mpec max Profit_leader;
```

Observe that the GAMS code above follows almost verbatim MPEC (C.13). □

C.3.2 Solution

Solving large MPECs is possible, but it is generally not easy. The basic condition for solvability is that the constraining problem (the one producing the equilibrium constraints) be convex. In such a case, the optimality conditions of this constraining problem are both necessary and sufficient conditions for optimality. Therefore, replacing the constraining optimization problem by its optimality conditions yields a well-posed MPEC, which can be solved directly as a nonlinear programming problem or by linearizing the optimality conditions. The particular but common case consisting of a linear programming problem constrained by another linear programming problem is generally easy to tackle. Further details on solution techniques for MPECs are available in [2].

C.4 Equilibrium Program with Equilibrium Constraints, EPEC

We consider next EPECs. An EPEC arises from the joint consideration of a number of interrelated MPECs.

C.4.1 Formulation

We consider that both investor 1 and investor 2 are leaders and that investors 3 to n are followers. Thus, the problems of investors 1 and 2 have to be solved jointly to identify equilibria. That is:

$$\left\{\begin{array}{ll}
\min_{x_1,y_1} & I_1 x_1 - \left(C_1 - (a - b\sum_j y_j)\right) y_1 \\
\text{s.t.} & y_1 \le x_1 \le x_1^{\max} \\
& y_1, x_1 \ge 0 \\
& \left.\begin{array}{l}
0 \le \alpha_i \perp (x_i - y_i) \ge 0 \\
0 \le \beta_i \perp (x_i^{\max} - x_i) \ge 0 \\
0 \le (I_i - \alpha_i + \beta_i) \perp x_i \ge 0 \\
0 \le (C_i - a + b y_i + b\sum_j y_j + \alpha_i) \perp y_i \ge 0
\end{array}\right\} \quad i = 3, \ldots, n \\
& \\
\min_{x_2,y_2} & I_2 x_2 - \left(C_2 - (a - b\sum_j y_j)\right) y_2 \\
\text{s.t.} & y_2 \le x_2 \le x_2^{\max} \\
& y_2, x_2 \ge 0 \\
& \left.\begin{array}{l}
0 \le \alpha_i \perp (x_i - y_i) \ge 0 \\
0 \le \beta_i \perp (x_i^{\max} - x_i) \ge 0 \\
0 \le (I_i - \alpha_i + \beta_i) \perp x_i \ge 0 \\
0 \le (C_i - a + b y_i + b\sum_j y_j + \alpha_i) \perp y_i \ge 0
\end{array}\right\} \quad i = 3, \ldots, n,
\end{array}\right. \tag{C.14}$$

needs to be jointly solved.

The first problem in (C.14) represents the profit maximization of investor 1 and is constrained by the equilibrium conditions of investors 3 to n, while the second problem in (C.14) represents the profit maximization of investor 2 and is also constrained by the equilibrium conditions of investors 3 to n.

A general form of the EPEC (C.14) is [2, 5]:

$$\left\{\begin{array}{ll}
\min_{\mathbf{u}^{(1)},\mathbf{x}^{(1)},\mathbf{y}^{(1)}} & f^{(1)}\left(\mathbf{u}^{(1)}, \mathbf{x}^{(1)}, \mathbf{y}^{(1)}, \ldots, \mathbf{u}^{(o)}, \mathbf{x}^{(o)}, \mathbf{y}^{(o)}\right) \\
\text{s.t.} & \mathbf{h}^{\mathrm{U}(1)}\left(\mathbf{u}^{(1)}, \mathbf{x}^{(1)}, \mathbf{y}^{(1)}, \ldots, \mathbf{u}^{(o)}, \mathbf{x}^{(o)}, \mathbf{y}^{(o)}\right) = \mathbf{0} \\
& \mathbf{g}^{\mathrm{U}(1)}\left(\mathbf{u}^{(1)}, \mathbf{x}^{(1)}, \mathbf{y}^{(1)}, \ldots, \mathbf{u}^{(o)}, \mathbf{x}^{(o)}, \mathbf{y}^{(o)}\right) \le \mathbf{0} \\
& \mathbf{h}^{(1)}\left(\mathbf{u}^{(1)}, \mathbf{x}^{(1)}, \mathbf{y}^{(1)}, \ldots, \mathbf{u}^{(o)}, \mathbf{x}^{(o)}, \mathbf{y}^{(o)}\right) = \mathbf{0} \\
& \mathbf{0} \le \mathbf{x}^{(1)} \perp \mathbf{g}^{(1)}\left(\mathbf{u}^{(1)}, \mathbf{x}^{(1)}, \mathbf{y}^{(1)}, \ldots, \mathbf{u}^{(o)}, \mathbf{x}^{(o)}, \mathbf{y}^{(o)}\right) \ge \mathbf{0} \\
& \\
& \vdots \\
& \\
\min_{\mathbf{u}^{(o)},\mathbf{x}^{(o)},\mathbf{y}^{(o)}} & f^{(o)}\left(\mathbf{u}^{(1)}, \mathbf{x}^{(1)}, \mathbf{y}^{(1)}, \ldots, \mathbf{u}^{(o)}, \mathbf{x}^{(o)}, \mathbf{y}^{(o)}\right) \\
\text{s.t.} & \mathbf{h}^{\mathrm{U}(o)}\left(\mathbf{u}^{(1)}, \mathbf{x}^{(1)}, \mathbf{y}^{(1)}, \ldots, \mathbf{u}^{(o)}, \mathbf{x}^{(o)}, \mathbf{y}^{(o)}\right) = \mathbf{0} \\
& \mathbf{g}^{\mathrm{U}(o)}\left(\mathbf{u}^{(1)}, \mathbf{x}^{(1)}, \mathbf{y}^{(1)}, \ldots, \mathbf{u}^{(o)}, \mathbf{x}^{(o)}, \mathbf{y}^{(o)}\right) \le \mathbf{0} \\
& \mathbf{h}^{(o)}\left(\mathbf{u}^{(1)}, \mathbf{x}^{(1)}, \mathbf{y}^{(1)}, \ldots, \mathbf{u}^{(o)}, \mathbf{x}^{(o)}, \mathbf{y}^{(o)}\right) = \mathbf{0} \\
& \mathbf{0} \le \mathbf{x}^{(o)} \perp \mathbf{g}^{(o)}\left(\mathbf{u}^{(1)}, \mathbf{x}^{(1)}, \mathbf{y}^{(1)}, \ldots, \mathbf{u}^{(o)}, \mathbf{x}^{(o)}, \mathbf{y}^{(o)}\right) \ge \mathbf{0},
\end{array}\right. \tag{C.15}$$

where $\mathbf{u}^{(1)} \in \mathbb{R}^{n^{U(1)}}$, $\mathbf{x}^{(1)} \in \mathbb{R}^{n^{(1)}}$, $\mathbf{y}^{(1)} \in \mathbb{R}^{m^{(1)}}$, $f^{(1)} \in \mathbb{R}$, $\mathbf{h}^{U^{(1)}} \in \mathbb{R}^{p^{U(1)}}$, $\mathbf{g}^{U^{(1)}} \in \mathbb{R}^{q^{U(1)}}$, $\mathbf{h}^{(1)} \in \mathbb{R}^{m^{(1)}}$, $\mathbf{g}^{(1)} \in \mathbb{R}^{n^{(1)}}$, ..., $\mathbf{u}^{(o)} \in \mathbb{R}^{n^{U(o)}}$, $\mathbf{x}^{(o)} \in \mathbb{R}^{n^{(o)}}$, $\mathbf{y}^{(o)} \in \mathbb{R}^{m^{(o)}}$, $f^{(o)} \in \mathbb{R}$, $\mathbf{h}^{U^{(o)}} \in \mathbb{R}^{p^{U(o)}}$, $\mathbf{g}^{U^{(o)}} \in \mathbb{R}^{q^{U(o)}}$, $\mathbf{h}^{(o)} \in \mathbb{R}^{m^{(o)}}$, $\mathbf{g}^{(o)} \in \mathbb{R}^{n^{(o)}}$.

Illustrative Example C.3

We consider the example in Sect. C.1 and assume that investors 1 and 2 are both leaders, while investor 3 is a follower, the EPEC to be solved is:

$$\left\{\begin{array}{ll} \min_{\Xi_1} & x_1 - \left(3 - (10 - \sum_{j=1}^{3} y_j)\right) y_1 \\ \text{s.t.} & y_1 \le x_1 \le 3 \\ & y_1, x_1 \ge 0 \\ & 0 \le \alpha_3 \perp (x_3 - y_3) \ge 0 \\ & 0 \le \beta_3 \perp (2 - x_3) \ge 0 \\ & 0 \le (1 - \alpha_3 + \beta_3) \perp x_3 \ge 0 \\ & 0 \le (5 - 10 + y_3 + \sum_{j=1}^{3} y_j + \alpha_3) \perp y_3 \ge 0, \\ \\ \min_{\Xi_2} & x_2 - \left(4 - (10 - \sum_{j=1}^{3} y_j)\right) y_2 \\ \text{s.t.} & y_2 \le x_2 \le 4 \\ & y_2, x_2 \ge 0 \\ & 0 \le \alpha_3 \perp (x_3 - y_3) \ge 0 \\ & 0 \le \beta_3 \perp (2 - x_3) \ge 0 \\ & 0 \le (1 - \alpha_3 + \beta_3) \perp x_3 \ge 0 \\ & 0 \le (5 - 10 + y_3 + \sum_{j=1}^{3} y_j + \alpha_3) \perp y_3 \ge 0, \end{array}\right. \tag{C.16}$$

where $\Xi_1 = \{x_1, y_1\} \cup \{y_2\} \cup \{x_3, y_3, \alpha_3, \beta_3\}$ and $\Xi_2 = \{x_2, y_2\} \cup \{y_1\} \cup \{x_3, y_3, \alpha_3, \beta_3\}$.

This EPEC can be solved using a *diagonalization* procedure that relies on solving individual MPECs. It works as follows:

0. Variables x_2 and y_2 are fixed to initial values and the MPEC for investor 1 is solved to obtain $\tilde{x}_1$, $\tilde{y}_1$, $\tilde{x}_3$, and $\tilde{y}_3$.
1. Variables x_1 and y_1 are then fixed to the values obtained in the previous step, $\tilde{x}_1$ and $\tilde{y}_1$, and the MPEC for investor 2 is solved to obtain $\tilde{x}_2$, $\tilde{y}_2$, $\tilde{x}_3$, and $\tilde{y}_3$. Thus the MPEC to be solved is:

$$\begin{array}{ll} \min_{\Xi_2} & x_2 - \left(4 - (10 - \sum_{j=1}^{3} y_j)\right) y_2 \\ \text{s.t.} & y_2 \le x_2 \le 4 \\ & y_2, x_2 \ge 0 \\ & 0 \le \alpha_3 \perp (x_3 - y_3) \ge 0 \\ & 0 \le \beta_3 \perp (2 - x_3) \ge 0 \\ & 0 \le (1 - \alpha_3 + \beta_3) \perp x_3 \ge 0 \\ & 0 \le (5 - 10 + y_3 + \sum_{j=1}^{3} y_j + \alpha_3) \perp y_3 \ge 0 \\ & x_1 = \tilde{x}_1 \\ & y_1 = \tilde{y}_1, \end{array}$$

Table C.3 EPEC example with investors 1 and 2 as leaders: capacity built and production level per investor

Investor i	Capacity built x_i	Production level y_i
1	3	3
2	1	1
3	0.5	0.5

where $\Xi_2 = \{x_2, y_2\} \cup \{x_3, y_3, \alpha_3, \beta_3\}$.

2. Variables x_2 and y_2 are then fixed to the values obtained in the previous step, $\tilde{x}_2$ and $\tilde{y}_2$, and the MPEC for investor 1 is solved to obtain $\tilde{x}_1$, $\tilde{y}_1$, $\tilde{x}_3$, and $\tilde{y}_3$. Thus the MPEC to be solved is:

$$\begin{aligned}
\min_{\Xi_1} \; & x_1 - \left(3 - (10 - \textstyle\sum_{j=1}^{3} y_j)\right) y_1 \\
\text{s.t.} \; & y_1 \leq x_1 \leq 3 \\
& y_1, x_1 \geq 0 \\
& 0 \leq \alpha_3 \perp (x_3 - y_3) \geq 0 \\
& 0 \leq \beta_3 \perp (2 - x_3) \geq 0 \\
& 0 \leq (1 - \alpha_3 + \beta_3) \perp x_3 \geq 0 \\
& 0 \leq (5 - 10 + y_3 + \textstyle\sum_{j=1}^{3} y_j + \alpha_3) \perp y_3 \geq 0 \\
& x_2 = \tilde{x}_2 \\
& y_2 = \tilde{y}_2,
\end{aligned}$$

where $\Xi_1 = \{x_1, y_1\} \cup \{x_3, y_3, \alpha_3, \beta_3\}$.

3. Steps 1 and 2 are repeated until convergence is obtained, that is, until x_1, y_1, x_2, y_2, x_3, y_3 do not change in two consecutive iterations.

The outcome from solving EPEC (C.16) is provided in Table C.3.

The GAMS code for the MPEC of investor 1 is given below:

```
OPTIONS mpec=knitro;

PARAMETERS
I1 /1/, I2 /1/, I3 /1/
C1 /3/, C2 /4/, C3 /5/
X1max /3/, X2max /4/, X3max /2/
a /10/, b /1/
x2 /4/, y2 /4/;

FREE VARIABLES
Profit_leader;
POSITIVE VARIABLES
x1, x3
y1, y3
alpha3   dual var. of maximum production
beta3    dual var. of maximum capacity;

EQUATIONS
OF, plimit, climit,
```

```
eq2_1, eq2_2, eq2_3, eq2_4;

OF..        Profit_leader =e= -I1*x1 + ((a-b*(y1+y2+
    y3))-C1)*y1;
plimit..    y1 =l= x1;
climit..    x1 =l= X1max;

eq2_1..     x3 - y3 =g= 0;
eq2_2..     X3max - x3 =g= 0;
eq2_3..     I3 - alpha3 + beta3 =g= 0;
eq2_4..     C2 - a + b*y3 + b*(y1+y2+y3) + alpha3 =g=
      0;

MODEL MPEC
/OF, plimit, climit,
eq2_1.alpha3, eq2_2.beta3, eq2_3.x3, eq2_4.y3/;
SOLVE MPEC using mpec max Profit_leader;
```

Since the MPEC of investor 2 is similar to that of investor 1, it is not reproduced here. □

C.4.2 *Solution*

Solving EPECs is generally very complicated, due to the nonconvex nature of such problems. Two techniques are generally available to attack such problems. The first one, diagonalization, has already been described (Sect. C.3 above). The second one, which is valid only for certain type of EPECs, consists in deriving the optimality conditions of each MPEC (constituting the EPEC) and attempting to solve all these conditions together. Regarding this second technique, it is relevant to note that deriving the optimality conditions of an MPEC is not generally easy since the constraints of an MPEC do not meet regularity conditions [4].

C.5 Comparison

Investment and production outcomes for the equilibrium, MPEC, and EPEC models are summarized in Table C.4.

Resulting prices and total productions for the equilibrium, MPEC, and EPEC models are reported in Table C.5.

Resulting profits for the equilibrium, MPEC, and EPEC models are reported in Table C.6.

Considering the results reported in Tables C.4, C.5, and C.6, the observations below are in order:

Table C.4 Comparison across models: capacity built and production level per investor

	Investor 1		Investor 2		Investor 3	
	Investment	Production	Investment	Production	Investment	Production
	x_1	y_1	x_2	y_2	x_3	y_3
Equilibrium	2.25	2.25	1.25	1.25	0.25	0.25
MPEC	3.00	3.00	0.67	0.67	0.67	0.67
EPEC	3.00	3.00	1.00	1.00	0.50	0.50

Table C.5 Comparison across models: price and total production

	Price	Total production
Equilibrium	6.25	3.75
MPEC	5.67	4.33
EPEC	5.50	4.50

Table C.6 Comparison across models: profits

	Profit Producer 1	Profit Producer 2	Profit Producer 3	Total profit
Equilibrium	5.0625	1.5625	0.0625	6.6875
MPEC	5.0000	0.4444	−0.2222	5.2222
EPEC	4.5000	0.5000	−0.2500	4.7500

1. The Nash–Cournot equilibrium provides the most favorable outcome for all investors in terms of profit. Nevertheless, investor 3 experiences losses for the MPEC and EPEC solutions.
2. The price is lowest for the EPEC equilibrium, and thus this is the best option for consumers.
3. In terms of production level, the EPEC equilibrium is the best option for investor 1, while the Nash–Cournot equilibrium is best for investor 2, and the MPEC solution is best for investor 3.

C.6 End-of-Chapter Exercises

C.1 Consider two interrelated linear programming problems and derive the equilibrium resulting from the joint consideration of their respective optimality conditions.

C.2 Write a linear programming problem constrained by another (related) linear programming problem and derive the resulting MPEC.

C.3 Write two interrelated MPECs similar to the one in the previous exercise and identify their equilibrium (or equilibria) using a diagonalization technique.

C.4 Expand the example in Sect. C.1 considering six investors, the three in that example and three additional ones, identical to the original three. Compute the equilibrium outcomes in this new market condition and compare them with the outcomes of the example in Sect. C.1.

C.5 Repeat Exercise C.4 considering investor 1 as the leader and solving an MPEC.

C.6 Repeat Exercise C.4 considering investors 1 and 2 as leaders and solving an EPEC.

References

1. Ferris, M. C., Munson, T. S.: Interfaces to PATH 3.0: Design, Implementation and Usage. In: Pang, J.-S. (ed.) Computational Optimization. Springer, New York (1999)
2. Gabriel, S., Conejo, A. J., Hobbs, B., Fuller D., Ruiz C.: Complementarity Modeling in Energy Markets. Springer, New York (2012)
3. GAMS. Available at www.gams.com (2015)
4. Luenberger, D. G., Ye, Y.: Linear and Nonlinear Programming, 3rd edn. Springer, New York (2008)
5. Ruiz, C., Conejo, A. J., Fuller, D., Gabriel, S., Hobbs, B.: A tutorial review of complementarity models for decision-making in energy markets. EURO J. Decis. Process. **2(1–2)**, 91–120 (2014)

Appendix D
Risk Management

This appendix provides an overview of risk management and the corresponding risk metrics, especially the conditional value-at-risk (CVaR) that is used in Chap. 4. This metric is introduced and illustrated using a simple example corresponding to the generation expansion planning (GEP) problem.

D.1 Introduction

In the investment decision-making optimization problems, the objective function is either maximizing profit, e.g., in the case in which a single producer invests in new generating units, or minimizing system cost, e.g., in the centralized GEP problem. However, those problems are generally subject to diverse sources of uncertainty, e.g., the expansion costs and the future demand growth. One relevant approach to modeling those uncertainties is *stochastic programming*, whereby uncertain data are represented through a set of plausible *scenarios*. Each scenario contains a realization for uncertain parameters and a probability. For example, the uncertainty of future investment cost of a specific generating unit can be represented by two scenarios, i.e., \$600,000/MW and \$800,000/MW, whose probabilities are 0.6 and 0.4, respectively. This provides a distribution of the objective function (profit or system cost), and therefore, it is needed to optimize a function characterizing such a distribution, e.g., its *expected value*. The main disadvantage of maximizing *expected profit* or minimizing *expected cost* is that the remaining parameters characterizing the profit or cost distribution are neglected. For example, the expected profit maximization may result in a significant loss if a specific scenario with a nonnegligible probability is realized.

In order to control the risk of experiencing distributions with undesirable properties, several risk management tools have been developed in the literature. Those tools mostly define risk metrics that can be incorporated into the optimization problems as an additional term in the objective function and/or as additional constraints. The most relevant risk metrics in the technical literature are listed below [1]:

A.J. Conejo et al., *Investment in Electricity Generation and Transmission*,
DOI 10.1007/978-3-319-29501-5

- Variance,
- Shortfall probability,
- Expected shortage,
- Value-at-risk (VaR),
- Conditional value-at-risk (CVaR) [2, 3].

Among the risk metrics above, the last one is the most advantageous, for the following reasons:

1. In contrast to the first metric, no quadratic (or nonlinear) term needs to be included in this one. Therefore, it maintains the linearity of linear programming (LP) and mixed-integer LP (MILP) problems.
2. Unlike the second and the fourth metrics, the CVaR does not add any binary variables to the problem. Therefore, it maintains the convexity of LP problems and does not increase the number of binary variables in the MILP problems.
3. Unlike all other metrics, the CVaR provides information about the profit (or system cost) throughout the distribution. For example, it is able to detect and quantify a *fat tail* that may appear in the profit (or system cost) distribution.
4. Unlike all other metrics, the CVaR is a *coherent* risk metric, i.e., it satisfies all desirable properties for a risk metric. These properties are (i) translation invariance, (ii) subadditivity, (iii) positive homogeneity, and (iv) monotonicity. Further information about these properties are available in [1].

Considering the context above, the CVaR risk metric is considered in this book. In the rest of this appendix, first a risk-neutral GEP example is provided. Then, the CVaR risk metric is described and applied to an illustrative example.

Illustrative Example D.1 *Risk-neutral GEP problem*

This illustrative example addresses a centralized GEP problem in which the system operator makes the optimal generation expansion decisions, whose objective is to minimize the *expected system cost*. The existing generating portfolio includes two conventional units, e_1 and e_2. The capacity of unit e_1 is 1000 MW with a production cost of \$14/MWh. In addition, the capacity of unit e_2 is 800 MW with a production cost of \$20/MWh. The only candidate unit to be built is conventional unit c_1, whose production cost is \$15/MWh. The transmission system is not considered in this example.

For expansion purposes, two future time periods, t_1 and t_2, are considered (dynamic expansion approach). The generation expansion decisions are made at the beginning of each time period. For the sake of simplicity, the capacity expansion options are assumed continuous, so that the maximum capacity to be built in each time period is 1500 MW. The amortization rates are equal to 30 % and 15 % in the first and second time periods, respectively.

The expansion cost at the beginning of the first time period is known, which is \$300,000/MW. However, such cost in the second time period is uncertain and represented by five equiprobable scenarios. According to these scenarios, the expansion cost in the second time period can be 20 % lower than, 10 % lower than, equal to,

20 % higher than, or 50 % higher than that cost in the first time period. Note that once the first time period concludes, the system operator knows which scenario is actually realized. As another source of uncertainty, we consider an inelastic consumer with uncertain consumption level. This uncertainty in the first time period is represented by three equiprobable scenarios: 2000 MW, 2250 MW, and 2500 MW. For the sake of simplicity, we assume that the demand level in the second time period is known at the beginning of that time period, which is identical to the realized demand level in the first time period raised by 20 %. Based on the two sources of uncertainty considered, 15 (i.e., 5 × 3) equiprobable scenarios can be generated, as given in Table D.1. The risk-neutral GEP problem is formulated by (D.1) below:

$$\min_{\Delta_\omega=\{x^{\text{C}}_{t_1c_1\omega},\, x^{\text{C}}_{t_2c_1\omega},\, p^{\text{E}}_{t_1e_1\omega},\, p^{\text{E}}_{t_2e_1\omega},\, p^{\text{E}}_{t_1e_2\omega},\, p^{\text{E}}_{t_2e_2\omega},\, p^{\text{C}}_{t_1c_1\omega},\, p^{\text{C}}_{t_2c_1\omega}\}\ \forall\omega=\omega_1,\omega_2,\ldots,\omega_{15}}$$

$$\sum_{\omega=\omega_1}^{\omega_{15}} \phi_\omega \Bigg\{ 8760 \Big[14\,(p^{\text{E}}_{t_1e_1\omega} + p^{\text{E}}_{t_2e_1\omega}) + 20\,(p^{\text{E}}_{t_1e_2\omega} + p^{\text{E}}_{t_2e_2\omega})$$

$$+ 15\,(p^{\text{C}}_{t_1c_1\omega} + p^{\text{C}}_{t_2c_1\omega}) \Big] + (0.30 \times 300000)\, x^{\text{C}}_{t_1c_1\omega} + 0.15\, C_{t_2\omega}\, x^{\text{C}}_{t_2c_1\omega} \Bigg\} \quad \text{(D.1a)}$$

subject to

$$0 \le x^{\text{C}}_{t_1c_1\omega} \le 1500 \qquad \forall\omega=\omega_1,\omega_2,\ldots,\omega_{15} \quad \text{(D.1b)}$$

$$0 \le x^{\text{C}}_{t_2c_1\omega} \le 1500 \qquad \forall\omega=\omega_1,\omega_2,\ldots,\omega_{15} \quad \text{(D.1c)}$$

Table D.1 Illustrative Example D.1: uncertain data across different scenarios

Scenario (ω)	Expansion cost in t_2 [\$/MW] ($C_{t_2\omega}$)	Demand level in t_1 [MW] ($P^{\text{D}}_{t_1\omega}$)	Demand level in t_2 [MW] ($P^{\text{D}}_{t_2\omega}$)
ω_1	240,000	2000	2400
ω_2	270,000	2000	2400
ω_3	300,000	2000	2400
ω_4	360,000	2000	2400
ω_5	450,000	2000	2400
ω_6	240,000	2250	2700
ω_7	270,000	2250	2700
ω_8	300,000	2250	2700
ω_9	360,000	2250	2700
ω_{10}	450,000	2250	2700
ω_{11}	240,000	2500	3000
ω_{12}	270,000	2500	3000
ω_{13}	300,000	2500	3000
ω_{14}	360,000	2500	3000
ω_{15}	450,000	2500	3000

$$x^{\mathrm{C}}_{t_1c_1\omega} = x^{\mathrm{C}}_{t_1c_1\tilde{\omega}} \quad \forall \omega = \omega_1, \omega_2, \ldots, \omega_{15};\ \forall \tilde{\omega} = \omega_1, \omega_2, \ldots, \omega_{15} \tag{D.1d}$$

$$p^{\mathrm{E}}_{t_1e_1\omega} + p^{\mathrm{E}}_{t_1e_2\omega} + p^{\mathrm{C}}_{t_1c_1\omega} = P^{\mathrm{D}}_{t_1\omega} \quad \forall \omega = \omega_1, \omega_2, \ldots, \omega_{15} \tag{D.1e}$$

$$p^{\mathrm{E}}_{t_2e_1\omega} + p^{\mathrm{E}}_{t_2e_2\omega} + p^{\mathrm{C}}_{t_2c_1\omega} = P^{\mathrm{D}}_{t_2\omega} \quad \forall \omega = \omega_1, \omega_2, \ldots, \omega_{15} \tag{D.1f}$$

$$0 \le p^{\mathrm{E}}_{t_1e_1\omega} \le 1000 \quad \forall \omega = \omega_1, \omega_2, \ldots, \omega_{15} \tag{D.1g}$$

$$0 \le p^{\mathrm{E}}_{t_2e_1\omega} \le 1000 \quad \forall \omega = \omega_1, \omega_2, \ldots, \omega_{15} \tag{D.1h}$$

$$0 \le p^{\mathrm{E}}_{t_1e_2\omega} \le 800 \quad \forall \omega = \omega_1, \omega_2, \ldots, \omega_{15} \tag{D.1i}$$

$$0 \le p^{\mathrm{E}}_{t_2e_2\omega} \le 800 \quad \forall \omega = \omega_1, \omega_2, \ldots, \omega_{15} \tag{D.1j}$$

$$0 \le p^{\mathrm{C}}_{t_1c_1\omega} \le x^{\mathrm{C}}_{t_1c_1\omega} \quad \forall \omega = \omega_1, \omega_2, \ldots, \omega_{15} \tag{D.1k}$$

$$0 \le p^{\mathrm{C}}_{t_2c_1\omega} \le \left(x^{\mathrm{C}}_{t_1c_1\omega} + x^{\mathrm{C}}_{t_2c_1\omega}\right) \quad \forall \omega = \omega_1, \omega_2, \ldots, \omega_{15}. \tag{D.1l}$$

The objective function (D.1a) minimizes the expected system cost including the expected production cost and the expected expansion cost. Note that ϕ_ω represents the probability of scenario ω. Constraints (D.1b) and (D.1c) bound the capacity of candidate unit c_1 to be built in time periods t_1 and t_2, respectively. The nonanticipativity constraints are enforced by (D.1d). Constraints (D.1e) and (D.1f) enforce the power balance between production and consumption levels in time periods t_1 and t_2, respectively. The remaining constraints (D.1g)–(D.1l) restrict the production levels. Note that problem (D.1) is linear.

Table D.2 Illustrative Example D.1: risk-neutral generation expansion decisions

Scenario	Generation expansion decision in the first time period [MW] ($x^{\mathrm{C}}_{t_1c_1\omega}$)	Generation expansion decision in the second time period [MW] ($x^{\mathrm{C}}_{t_2c_1\omega}$)
ω_1	900	500
ω_2		500
ω_3		0
ω_4		0
ω_5		0
ω_6		800
ω_7		800
ω_8		0
ω_9		0
ω_{10}		0
ω_{11}		1100
ω_{12}		1100
ω_{13}		300
ω_{14}		300
ω_{15}		300

Table D.3 Illustrative Example D.1: system cost distribution over scenarios

Scenario	System cost [$ million]
ω_1	664.02
ω_2	666.27
ω_3	667.92
ω_4	667.92
ω_5	667.92
ω_6	758.04
ω_7	761.64
ω_8	764.28
ω_9	764.28
ω_{10}	764.28
ω_{11}	852.06
ω_{12}	857.01
ω_{13}	861.00
ω_{14}	863.70
ω_{15}	867.75

The optimal solution of the GEP problem (D.1) is presented in Table D.2. Accordingly, a new 900-MW unit is built at the beginning of the first time period. However, the generation expansion decision in the second time period depends on scenario realization at the end of the first time period. In the case of scenarios ω_1 and ω_2, an additional 500-MW unit is built, i.e., the capacity of the candidate unit increases up to 1400 MW (900 + 500). An additional 800-MW capacity is built in the second time period if either scenario ω_6 or scenario ω_7 is realized. This additional capacity in the case of scenarios ω_{11} and ω_{12} is 1100 MW. The capacity of the candidate unit reaches 1200 MW if one of scenarios ω_{13} to ω_{15} is realized at the end of the first time period. Finally, the generation portfolio is not expanded in the second time period if one of the remaining scenarios is realized.

The expected system cost in the optimal solution is \$763.21 million. Table D.3 gives the system cost distribution, i.e., the optimal system cost for each scenario. Note that the system cost exhibits high volatility varying from \$664.02 million to \$867.75 million. □

D.2 Conditional Value-at-Risk

In a minimization problem with discrete scenarios, e.g., the GEP problem (D.1), the CVaR for a given confidence level $\alpha \in (0,1)$ is defined as the expected value of the system cost higher than the $(1-\alpha)$-quantile of the cost distribution over scenarios. If all scenarios are equiprobable, the CVaR is computed as the expected system cost in the $(1-\alpha) \times 100\,\%$ worst scenarios. Similarly, in a maximization problem with discrete scenarios, e.g., the generation investment problem of a producer maximizing its expected profit, the CVaR is defined as the expected value of the profit smaller than the $(1 - \alpha)$-quantile of the profit distribution over scenarios.

The CVaR of a discrete distribution function can be incorporated into the minimization problem, e.g., the GEP problem (D.1), as follows:

$$\min_{\Delta_\omega,\ \zeta,\ \eta_\omega} \quad \mathbb{E}_\omega\left\{f(\omega)\right\} - \beta\left(\zeta - \frac{1}{1-\alpha}\mathbb{E}_\omega\left\{\eta_\omega\right\}\right) \tag{D.2a}$$

subject to

$$\text{GEP constraints (D.1b)} - \text{(D.1l)} \tag{D.2b}$$

$$\zeta + f(\omega) \leq \eta_\omega \quad \forall \omega \tag{D.2c}$$

$$\eta_\omega \geq 0 \quad \forall \omega, \tag{D.2d}$$

where function $f(\omega)$ represents the system cost under scenario ω. In addition, $\mathbb{E}_\omega$ is the expectation operator.

The term $\left(\zeta - \frac{1}{1-\alpha}\mathbb{E}_\omega\left\{\eta_\omega\right\}\right)$ in the objective function (D.2a) computes the CVaR risk metric. Note that ζ is an auxiliary continuous variable that computes the value-at-risk (VaR). The optimal value of the VaR, i.e., ζ^*, for a given confidence level α is the minimum system cost value such that the probability of the system cost being higher than this value is less than or equal to $1 - \alpha$. Based on the VaR, the CVaR is defined as the expected value of those cost values that are greater than or equal to the VaR. In the objective function (D.2a), the nonnegative weighting parameter β makes a tradeoff between the expected system cost and the risk, so that a higher value of β implies that the system operator is more risk averse. Note that $\beta = 0$ makes the problem risk-neutral.

Constraints (D.2b) include all constraints of the original risk-neutral GEP problem. Finally, constraints (D.2c) and (D.2d) allow incorporating the CVaR risk metric. Note that η_ω is an auxiliary nonnegative continuous variable equal to the summation of the VaR, ζ, and the system cost under scenario ω if this summation is nonnegative, and equal to zero if such a summation is negative.

Similarly, the CVaR risk metric can be incorporated into the maximization problem with discrete scenarios, e.g., the generation investment problem of a producer maximizing its expected profit, as follows:

$$\max_{\Gamma,\ \zeta,\ \eta_\omega} \quad \mathbb{E}_\omega\left\{p(\omega)\right\} + \beta\left(\zeta - \frac{1}{1-\alpha}\mathbb{E}_\omega\left\{\eta_\omega\right\}\right) \tag{D.3a}$$

subject to

$$\text{Original constraints of the risk-neutral problem} \tag{D.3b}$$

$$\zeta - p(\omega) \leq \eta_\omega \quad \forall \omega \tag{D.3c}$$

$$\eta_\omega \geq 0 \quad \forall \omega, \tag{D.3d}$$

where function $p(\omega)$ represents the producer's profit under scenario ω. In addition, variable set Γ includes all variables of the original risk-neutral problem. The optimal value of the VaR, i.e., ζ^*, for a given confidence level α is the maximum profit value such that the probability of the profit being lower than this value is less than or equal to $1-\alpha$. Based on the VaR, the CVaR is defined as the expected value of those profit values that are less than or equal to the VaR.

Illustrative Example D.2 *Risk-constrained GEP problem*

In this section, the CVaR risk metric is incorporated into Illustrative Example D.1. We assume a confidence level α=0.80 and a weighting parameter β=10. The corresponding risk-constrained GEP model is formulated by linear problem (D.4) below:

$$\min_{x^{\mathrm{C}}_{t_1c_1\omega},\, x^{\mathrm{C}}_{t_2c_1\omega},\, p^{\mathrm{E}}_{t_1e_1\omega},\, p^{\mathrm{E}}_{t_2e_1\omega},\, p^{\mathrm{E}}_{t_1e_2\omega},\, p^{\mathrm{E}}_{t_2e_2\omega},\, p^{\mathrm{C}}_{t_1c_1\omega},\, p^{\mathrm{C}}_{t_2c_1\omega},\, \eta_\omega;\ \forall\omega=\omega_1,\omega_2,\ldots,\omega_{15},\ \zeta}$$

$$\sum_{\omega=\omega_1}^{\omega_{15}} \phi_\omega \left\{ 8760 \left[14\,(p^{\mathrm{E}}_{t_1e_1\omega} + p^{\mathrm{E}}_{t_2e_1\omega}) + 20\,(p^{\mathrm{E}}_{t_1e_2\omega} + p^{\mathrm{E}}_{t_2e_2\omega}) \right.\right.$$

$$\left.\left. + 15\,(p^{\mathrm{C}}_{t_1c_1\omega} + p^{\mathrm{C}}_{t_2c_1\omega}) \right] + (0.30 \times 300000)\, x^{\mathrm{C}}_{t_1c_1\omega} + 0.15\, C_{t_2\omega}\, x^{\mathrm{C}}_{t_2c_1\omega} \right\}$$

$$- 10 \left(\zeta - \frac{1}{1-0.80} \sum_{\omega=\omega_1}^{\omega_{15}} \phi_\omega \eta_\omega \right) \tag{D.4a}$$

subject to

Constraints (D.1b)–(D.1l) (D.4b)

$$\zeta + 8760 \left[14\,(p^{\mathrm{E}}_{t_1e_1\omega} + p^{\mathrm{E}}_{t_2e_1\omega}) + 20\,(p^{\mathrm{E}}_{t_1e_2\omega} + p^{\mathrm{E}}_{t_2e_2\omega}) + 15\,(p^{\mathrm{C}}_{t_1c_1\omega} + p^{\mathrm{C}}_{t_2c_1\omega}) \right]$$

$$+ (0.30 \times 300000)\, x^{\mathrm{C}}_{t_1c_1\omega} + 0.15\, C_{t_2\omega}\, x^{\mathrm{C}}_{t_2c_1\omega} \le \eta_\omega \qquad \forall\omega = \omega_1, \omega_2, \ldots, \omega_{15} \tag{D.4c}$$

$$\eta_\omega \ge 0 \qquad \forall\omega = \omega_1, \omega_2, \ldots, \omega_{15}. \tag{D.4d}$$

The optimal solution of the risk-constrained GEP problem (D.4) is provided in Table D.4. In this case, the risk-averse system operator decides to build a higher capacity in the first time period (1200 MW) with respect to that in the risk-neutral GEP problem (900 MW). In addition, a smaller capacity is built in the second time period for different scenarios with respect to that in the risk-neutral GEP problem. In this way, a higher capacity is available in the first time period, which increases the ability of the risk-averse system operator to cope with future uncertainties.

The expected system cost in the risk-constrained case is \$766.81 million. As expected, it is higher than that in the risk-neutral case (\$763.21 million). The system cost and the optimal value of η_ω per scenario in the risk-constrained case are provided in Table D.5. Note that the system cost varies from \$675.84 million to \$861.36 million, which exhibits in this example a comparatively lower volatility with respect to the risk-neutral case.

The optimal value for the VaR (ζ^*) at confidence level α=0.80 is −\$858.72 million, which is in fact minus the value of the system cost for scenario ω_{12}. This implies

Table D.4 Illustrative Example D.2: risk-constrained generation expansion decisions

Scenario	Generation expansion decision in the first time period [MW] ($x^{C}_{t_1 c_1 \omega}$)	Generation expansion decision in the second time period [MW] ($x^{C}_{t_2 c_1 \omega}$)
ω_1	1200	200
ω_2		200
ω_3		0
ω_4		0
ω_5		0
ω_6		500
ω_7		500
ω_8		0
ω_9		0
ω_{10}		0
ω_{11}		800
ω_{12}		800
ω_{13}		0
ω_{14}		0
ω_{15}		0

that \$858.72 million is the minimum system cost value such that the probability of the system cost being higher than that value is less than or equal to 0.20. As given in Table D.5, the system costs in the last three scenarios (\$861.36 million) with identical probabilities ($\frac{1}{15}$) are less than or equal to the VaR.

Since all scenarios in Illustrative Example D.2 are equiprobable, the CVaR at confidence level $\alpha = 0.80$ is equal to the expected system cost in the 20 % worst scenarios. Therefore, it determines the mean of the system cost value in the last three scenarios, i.e., \$861.36 million. Another general technique for computing the CVaR even in cases with different probabilities is to derive the optimal value of the term $-\left(\zeta - \frac{1}{1-\alpha}\sum_{\omega}\phi_{\omega}\eta_{\omega}\right)$ in the objective function. In this example, such a term is computed as follows:

$$
\begin{aligned}
-\left(\zeta - \frac{1}{1-\alpha}\sum_{\omega=\omega_1}^{\omega_{15}}\phi_{\omega}\eta_{\omega}\right) &= 858.72 \times 10^6 + \frac{1}{1-0.80}\left[\left(\frac{1}{15} \times 2.64 \times 10^6\right)\right.\\
&\quad \left. + \left(\frac{1}{15} \times 2.64 \times 10^6\right) + \left(\frac{1}{15} \times 2.64 \times 10^6\right)\right]\\
&= 861.36 \times 10^6.
\end{aligned}
$$

It is worth mentioning that the value of CVaR in the risk-neutral GEP problem, i.e., Illustrative Example D.1, is \$867.75 million, which is comparatively higher than that in the risk-constrained GEP problem (\$861.36 million).

Table D.5 Illustrative Example D.2: system cost and η_ω for different scenarios

	System cost [$ million]	η_ω [$ million]
ω_1	675.84	0
ω_2	676.74	0
ω_3	677.40	0
ω_4	677.40	0
ω_5	677.40	0
ω_6	761.10	0
ω_7	763.35	0
ω_8	765.00	0
ω_9	765.00	0
ω_{10}	765.00	0
ω_{11}	855.12	0
ω_{12}	858.72	0
ω_{13}	861.36	2.64
ω_{14}	861.36	2.64
ω_{15}	861.36	2.64

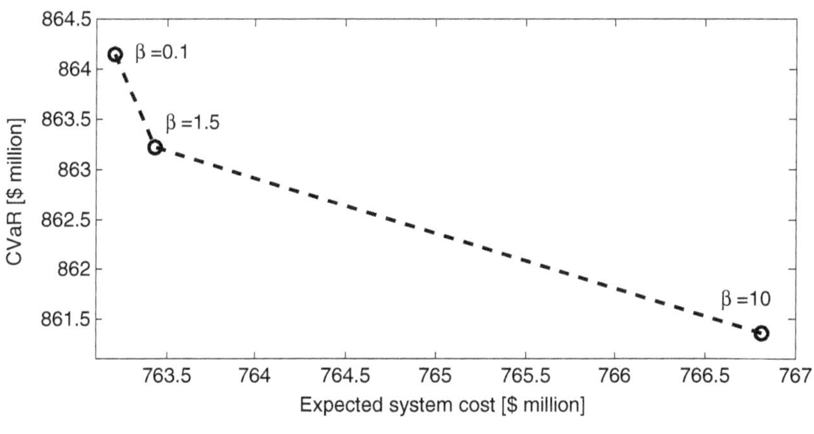

Fig. D.1 Illustrative Example D.2: expected system cost versus CVaR (efficient frontier)

Finally, the risk-constrained GEP problem (D.4) is solved considering different values of the weighting parameter β. Figure D.1 depicts the so-called efficient frontier for this case, which shows how the expected system cost increases as the CVaR decreases. □

References

1. Conejo, A. J., Carrión, M., Morales, J. M.: Decision Making Under Uncertainty in Electricity Markets. Springer, New York (2010)
2. Rockafellar, R.T., Uryasev, S.: Optimization of conditional value-at-risk. J. Risk **2**, 21–41 (2000)
3. Rockafellar, R.T., Uryasev, S.: Conditional value-at-risk for general loss distributions. J. Bank. and Finance **26**(7), 1443–1471 (2002)

Appendix E
Dynamic Programming

This appendix presents a high-level introduction to dynamic programming. First, a discrete-time approach is taken in which decisions may be made only at specific points in time. Next, a continuous-time framework is developed and applied to the problem of optimal stopping. Examples illustrate all concepts numerically.

E.1 Introduction

In this appendix, we will survey dynamic programming concepts that are used to analyze dynamic decision making under uncertainty. These are relevant for any problem in which decisions are made or updated sequentially in time as new information is revealed. As a concrete example, consider the power company in Chap. 7 that holds the perpetual option to invest in a power plant of a given capacity. Facing uncertainty in the underlying value of the plant, the company can choose to delay its investment decision in order to maximize the conditional expected net present value (NPV) of the power plant. For example, it may invest in the power plant either immediately, in five years, or in ten years. Given these discrete choices, the power company's optimal investment timing decision is the one that maximizes the conditional expected NPV. To that end, the power company compares the expected NPV from immediate investment with the continuation function, which contains the value of all subsequent decisions to invest or to wait. Conceptually, discrete-time dynamic programming handles dynamic optimization under uncertainty by decomposing the main problem into smaller subproblems and exploiting their recursive nature. Hence, this so-called Bellman principle of optimality formalizes the notion that the optimal decision policy based on some initial choices and conditions is such that the subsequent decisions for the resulting subproblems are also optimal [1].

However, it is plausible to imagine that the power company makes such comparisons at more frequent intervals. In the limit, we can convert the decision-making problem into a continuous-time one by letting discrete time steps over which the underlying stochastic process is defined go to zero. Consequently, instead of

A.J. Conejo et al., *Investment in Electricity Generation and Transmission*,
DOI 10.1007/978-3-319-29501-5

comparing the value functions from decisions at discrete points in time, the resulting Bellman equation is cast in terms of rates, i.e., the instantaneous rate of return on the value of the investment equals its expected appreciation per unit time. If the underlying stochastic process is an Itô process [3], then the Bellman equation may be rearranged by applying Itô's lemma to yield a partial differential equation (PDE). In the case of a perpetual option, the PDE becomes an ordinary differential equation (ODE) that can be solved analytically subject to boundary conditions.

In the final section of this appendix, we will consider optimal stopping-time problems. This approach makes the formal link between solving an investment problem and finding the optimal threshold price. Illustrative examples for a typical investment decision-making problem are used to convey the main concepts.

E.2 Discrete-Time Dynamic Programming

We will demonstrate discrete-time dynamic programming using a simple setup in which the investor faces uncertainty in the price of the underlying asset, i.e., a power plant. The expected value of the plant, V_t (in \$), changes at discrete points in time $t = 0, \Delta t, 2\Delta t, \ldots, T\Delta t$, where Δt is the length of each time increment and $T+1$ is the finite number of time periods over which the investment decision may be made. Over any time increment, Δt, the value of the plant can either jump up or down by a factor of $U > 1$ or $D < 1$, respectively. We let $0 \leq p \leq 1$ be the probability associated with an upward jump and assume that successive changes in the price are independent of all previous movements in the plant's value. Thus, given V_t, the value in period $t + \Delta t$ is UV_t with probability p and DV_t with probability $1-p$. Figure E.1 illustrates the discrete movements in the plant's value.

Given that the power company has discretion over investment timing, which time period is the optimal one in which to initiate investment? Assuming no operating costs and that the investment cost is I (in \$), the expected NPV from immediate investment in any time period t is simply $\max\{V_t - I, 0\}$. This expected NPV must now be compared with the discounted expected value (conditional on the current plant value) of waiting to invest in a subsequent period, where ρ is the periodic discount rate. If $\mathscr{W}(V_t, t)$ is the value of the investment opportunity in period t when the plant value is V_t, then we have the recursive relationship in Eq. (E.1) for $t = 0, \ldots, (T-1)\Delta t$ [4]:

$$\mathscr{W}(V_t, t) = \max\left\{V_t - I, \frac{1}{1+\rho\Delta t}\mathbb{E}_{V_t}\left[\mathscr{W}(V_{t+\Delta t}, t+\Delta t)\right]\right\}. \tag{E.1}$$

The boundary condition with respect to which the recurrence is solved is in Eq. (E.2):

$$\mathscr{W}(V_{T\Delta t}, T\Delta t) = \max\{V_{T\Delta t} - I, 0\}. \tag{E.2}$$

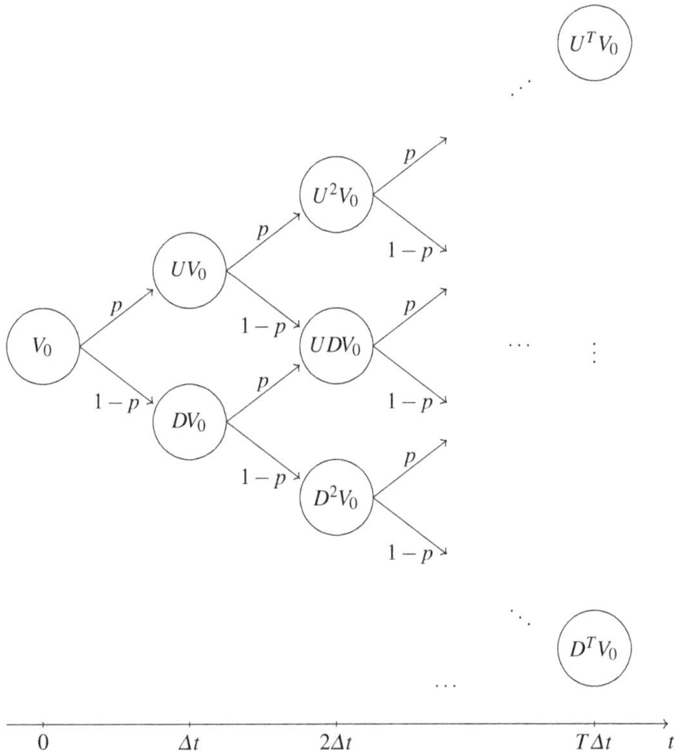

Fig. E.1 Discrete movements in the plant's value over time

Illustrative Example E.1 *Discrete-time dynamic programming*

We now apply the discrete-time dynamic programming approach in Eqs. (E.1)–(E.2) to a specific example. Consider the movements of the plant value given in Fig. E.2. Here, the parameters are $T = 2$, $\Delta t = 1$, $U = 1.2$, $D = 0.9$, $p = \frac{1}{2}$, $\rho = 0.10$, $V_0 = 100$, and $I = 100$. Thus, the power company has two years over which to observe the value of the underlying plant and to make its investment decision. The expected annual percentage change in the value of the power plant is $\frac{1}{2} \times 0.20 + \frac{1}{2} \times -0.10 = 5\,\%$, and the annualized volatility in the percentage changes is $\sqrt{\frac{1}{2} \times (0.20 - 0.05)^2 + \frac{1}{2} \times (-0.10 - 0.05)^2} = 15\,\%$. We solve for the value of the option to invest and obtain the optimal investment policy by solving equations (E.1)–(E.2) using backward induction starting with year 2:

- In year 2, the power company's investment decision is a simple now-or-never one, which means that Eq. (E.2) is evaluated for each value of V_2:

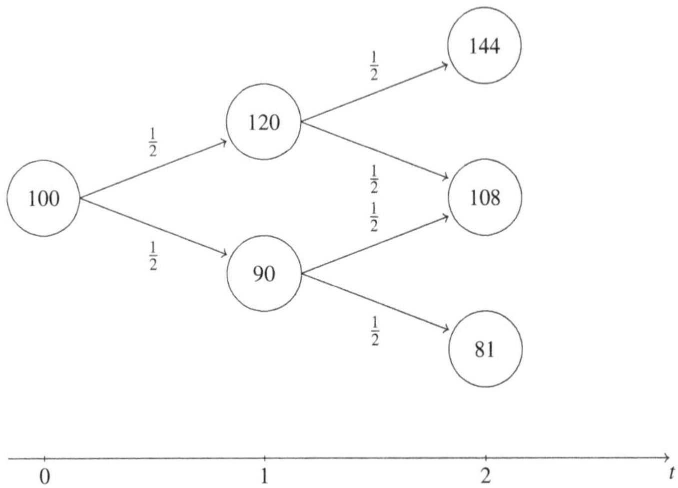

Fig. E.2 Discrete movements in the plant's value for two time periods

$$\mathscr{W}(V_2, 2) = \max\{V_2 - I, 0\} = \begin{cases} 44, & \text{if } V_2 = 144 \\ 8, & \text{if } V_2 = 108 \\ 0, & \text{if } V_2 = 81. \end{cases}$$

- In year 1, the expected NPV from immediate investment must be compared with the expected value of waiting conditional on the current plant value using Eq. (E.1):

$$\begin{aligned} \mathscr{W}(V_1, 1) &= \max\left\{V_1 - I, \frac{1}{1+\rho}\mathbb{E}_{V_1}\left[\mathscr{W}(V_2, 2)\right]\right\} \\ &= \begin{cases} \max\{20, \frac{1}{1.1}\left[\frac{1}{2} \times 44 + \frac{1}{2} \times 8\right]\}, & \text{if } V_1 = 120 \\ \max\{-10, \frac{1}{1.1}\left[\frac{1}{2} \times 8 + \frac{1}{2} \times 0\right]\}, & \text{if } V_1 = 90 \end{cases} \\ &= \begin{cases} 23.636, & \text{if } V_1 = 120 \\ 3.636, & \text{if } V_1 = 90. \end{cases} \end{aligned}$$

- Finally, in year 0, we have:

$$\begin{aligned} \mathscr{W}(V_0, 0) &= \max\{V_0 - I, \frac{1}{1+\rho}\mathbb{E}_{V_0}\left[\mathscr{W}(V_1, 1)\right]\} \\ &= \max\left\{0, \frac{1}{1.1}\left[\frac{1}{2} \times 23.636 + \frac{1}{2} \times 3.636\right]\right\} \\ &= 12.397. \end{aligned}$$

Hence, the value of the investment opportunity is worth \$12.397 initially even though the expected now-or-never NPV is precisely zero. The optimal investment policy is to wait until year 2 and to invest only if the value of the plant is 144 or 108. □

E.3 Continuous-Time Dynamic Programming

Rearranging Eq. (E.1), we obtain that the Bellman equation in the continuation region is:

$$\begin{aligned} \mathscr{W}(V_t, t) &= \frac{1}{1+\rho\Delta t}\mathbb{E}_{V_t}\left[\mathscr{W}(V_{t+\Delta t}, t+\Delta t)\right] \\ \Rightarrow (1+\rho\Delta t)\,\mathscr{W}(V_t, t) &= \mathbb{E}_{V_t}\left[\mathscr{W}(V_{t+\Delta t}, t+\Delta t)\right] \\ \Rightarrow \rho\Delta t\mathscr{W}(V_t, t) &= \mathbb{E}_{V_t}\left[\mathscr{W}(V_{t+\Delta t}, t+\Delta t) - \mathscr{W}(V_t, t)\right]. \end{aligned} \tag{E.3}$$

By letting $\Delta\mathscr{W} \equiv \mathscr{W}(V_{t+\Delta t}, t+\Delta t) - \mathscr{W}(V_t, t)$ and dividing Eq. (E.3) by Δt before taking the limit as Δt goes to zero, we obtain:

$$\begin{aligned} \rho\mathscr{W}(V_t, t) &= \frac{1}{\Delta t}\mathbb{E}_{V_t}\left[\Delta\mathscr{W}\right] \\ \Rightarrow \lim_{\Delta t\to 0}\rho\mathscr{W}(V_t, t) &= \lim_{\Delta t\to 0}\frac{1}{\Delta t}\mathbb{E}_{V_t}\left[\Delta\mathscr{W}\right] \\ \Rightarrow \rho\mathscr{W}(V_t, t) &= \frac{1}{\mathrm{d}t}\mathbb{E}_{V_t}\left[\mathrm{d}\mathscr{W}\right] \\ \Rightarrow \rho\mathscr{W}(V_t, t)\,\mathrm{d}t &= \mathbb{E}_{V_t}\left[\mathrm{d}\mathscr{W}\right]. \end{aligned} \tag{E.4}$$

The last line of Eq. (E.4) reflects the return-equilibrium condition that the instantaneous rate of return from owning the investment opportunity is equal to the expected appreciation as if it were operated optimally.

In the limit as $\Delta t \to 0$, the underlying plant value also converges to a continuous-time stochastic process rather than a discrete one as in Sect. E.2. Now, if the underlying stochastic processes, $\{V_t, t \geq 0\}$, is an Itô process, i.e., of the form $\mathrm{d}V_t = a(\cdot)\,\mathrm{d}t + b(\cdot)\,\mathrm{d}z_t$, where $a(\cdot)$ and $b(\cdot)$ are the drift and diffusion parameters depending on V_t as well as t, and $\mathrm{d}z_t$ is the increment to a Wiener process, then the continuous-time Bellman equation in Eq. (E.4) may be simplified as a result of applying Itô's lemma. At its essence, Itô's lemma enables the calculation of the total differential of a function that depends on an Itô process and time. Intuitively, it is related to a Taylor series expansion for a deterministic function of several variables, e.g., $f(x, y)$, which yields $\mathrm{d}f = \frac{\partial f}{\partial x}\mathrm{d}x + \frac{\partial f}{\partial y}\mathrm{d}y + \frac{1}{2}\frac{\partial^2 f}{\partial x^2}(\mathrm{d}x)^2 + \frac{1}{2}\frac{\partial^2 f}{\partial y^2}(\mathrm{d}y)^2 + \cdots$. In contrast to the deterministic setting, second-order terms of an Itô process do not vanish but are of the order of $\mathrm{d}t$ since $(\mathrm{d}V_t)^2 = a(\cdot)^2(\mathrm{d}t)^2 + b(\cdot)^2(\mathrm{d}z_t)^2 + 2a(\cdot)\,b(\cdot)\,\mathrm{d}t\mathrm{d}z_t$. Here, $(\mathrm{d}z_t)^2 = \mathrm{d}t$, whereas $\mathrm{d}t\mathrm{d}z_t = (\mathrm{d}t)^{\frac{3}{2}}$ and $(\mathrm{d}t)^2$ are both "small" compared to $\mathrm{d}t$. Hence, we have for $\mathscr{W}(V_t, t)$, where $\{V_t, t \geq 0\}$ is an Itô process:

$$\begin{aligned} \mathrm{d}\mathscr{W} &= \frac{\partial \mathscr{W}}{\partial V_t}\mathrm{d}V_t + \frac{\partial \mathscr{W}}{\partial t}\mathrm{d}t + \frac{1}{2}\frac{\partial^2 \mathscr{W}}{\partial V_t^2}(\mathrm{d}V_t)^2 \\ &= \frac{\partial \mathscr{W}}{\partial V_t}a(\cdot)\,\mathrm{d}t + \frac{\partial \mathscr{W}}{\partial V_t}b(\cdot)\,\mathrm{d}z_t + \frac{\partial \mathscr{W}}{\partial t}\mathrm{d}t + \frac{1}{2}\frac{\partial^2 \mathscr{W}}{\partial V_t^2}b(\cdot)^2\,\mathrm{d}t \\ &= \left(\frac{\partial \mathscr{W}}{\partial V_t}a(\cdot) + \frac{\partial \mathscr{W}}{\partial t} + \frac{1}{2}\frac{\partial^2 \mathscr{W}}{\partial V_t^2}b(\cdot)^2\right)\mathrm{d}t + \frac{\partial \mathscr{W}}{\partial V_t}b(\cdot)\,\mathrm{d}z_t. \end{aligned} \tag{E.5}$$

Applying Itô's lemma to the return-equilibrium condition in Eq. (E.4) yields the following PDE:

$$\begin{aligned} \rho\mathscr{W}(V_t,t)\,\mathrm{d}t &= \mathbb{E}_{V_t}\left[\left(\frac{\partial \mathscr{W}}{\partial V_t}a(\cdot) + \frac{\partial \mathscr{W}}{\partial t} + \frac{1}{2}\frac{\partial^2 \mathscr{W}}{\partial V_t^2}b(\cdot)^2\right)\mathrm{d}t + \frac{\partial \mathscr{W}}{\partial V_t}b(\cdot)\,\mathrm{d}z_t\right] \\ &\Rightarrow \rho\mathscr{W}(V_t,t)\,\mathrm{d}t = \left(\frac{\partial \mathscr{W}}{\partial V_t}a(\cdot) + \frac{\partial \mathscr{W}}{\partial t} + \frac{1}{2}\frac{\partial^2 \mathscr{W}}{\partial V_t^2}b(\cdot)^2\right)\mathrm{d}t \\ &\Rightarrow \frac{1}{2}b(\cdot)^2\frac{\partial^2 \mathscr{W}}{\partial V_t^2} + a(\cdot)\frac{\partial \mathscr{W}}{\partial V_t} + \frac{\partial \mathscr{W}}{\partial t} - \rho\mathscr{W}(V_t,t) = 0. \end{aligned} \tag{E.6}$$

The PDE in Eq. (E.6) may be solved numerically, e.g., by starting at the terminal time T and working backward through a grid based on the finite-differencing method subject to boundary conditions [2]. Besides finding the value of the option to invest, the solution should also comprise the "free boundary," υ_t, i.e., the optimal threshold plant value at which to invest for every t. The corresponding boundary conditions are the value-matching and smooth-pasting conditions as follows:

$$\mathscr{W}(V_t,t)|_{V_t=\upsilon_t} = V_t - I|_{V_t=\upsilon_t} \tag{E.7}$$

$$\left.\frac{\partial \mathscr{W}(V_t,t)}{\partial V_t}\right|_{V_t=\upsilon_t} = \left.\frac{\partial (V_t - I)}{\partial V_t}\right|_{V_t=\upsilon_t}. \tag{E.8}$$

Intuitively, the value-matching condition in Eq. (E.7) states that at the free boundary, the value of the option to invest equals the expected NPV of the active plant, whereas the smooth-pasting condition in Eq. (E.8) is a first-order condition for optimality [3].

If the option to invest is perpetual or the time to expiry is long, then the value function no longer depends directly on time, i.e., it is simply $\mathscr{W}(V)$. Consequently, the PDE in Eq. (E.6) becomes an ODE as the time derivative vanishes:

$$\frac{1}{2}b(\cdot)^2\mathscr{W}''(V) + a(\cdot)\mathscr{W}'(V) - \rho\mathscr{W}(V) = 0. \tag{E.9}$$

Now, instead of solving for a free boundary, we solve for an optimal threshold, υ, which is independent of time. Thus, Eq. (E.9) is solved together with the following value-matching and smooth-pasting conditions:

$$\mathscr{W}(V)|_{V=\upsilon} = V - I|_{V=\upsilon} \tag{E.10}$$

$$\left.\frac{\mathrm{d}\mathscr{W}(V)}{\mathrm{d}V}\right|_{V=\upsilon} = \left.\frac{\mathrm{d}(V-I)}{\mathrm{d}V}\right|_{V=\upsilon}. \tag{E.11}$$

Illustrative Example E.2 *Continuous-time dynamic programming*

Suppose that a power company has the perpetual option to invest in a power plant with underlying value $\{V_t, t \geq 0\}$, which follows a GBM, i.e., $\mathrm{d}V = \alpha V \mathrm{d}t + \sigma V \mathrm{d}z_t$. As in Illustrative Example E.1, we let $\rho = 0.10$ be the annual discount rate and set the investment cost to be $I = 100$. Since the expected appreciation in V_t in Illustrative Example E.1 is 0.05 with a volatility of 0.15, we set $\alpha = 0.05$ and let the volatility parameter, σ, be equal to 0.15. Using Eq. (E.9), we obtain the following ODE:

$$\frac{1}{2}\sigma^2 V^2 \mathscr{W}''(V) + \alpha V \mathscr{W}'(V) - \rho \mathscr{W}(V) = 0. \tag{E.12}$$

The solution to this second-order ODE is of the power form, i.e.:

$$\mathscr{W}(V) = a_{0,1} V^{\beta_1} + a_{0,2} V^{\beta_2}, \tag{E.13}$$

where $\beta_1 > 1$ and $\beta_2 < 0$ are the roots of the characteristic quadratic equation $\mathscr{Q}(\beta) = \frac{1}{2}\sigma^2\beta(\beta - 1) + \alpha\beta - \rho = 0$ defined by the insertion of Eq. (E.13) into Eq. (E.12). Furthermore, as the value of the plant goes to zero, the value of the option to invest in it should also become worthless, i.e., $lim_{V\to 0}\mathscr{W}(V) = 0$ is a boundary condition, which implies that $a_{0,2} = 0$. Thus, the solution to the value of the option to invest is:

$$\mathscr{W}(V) = a_{0,1} V^{\beta_1}. \tag{E.14}$$

Using the value-matching and smooth-pasting conditions in Eqs. (E.10) and (E.11), respectively, we can now solve for the optimal investment threshold, υ, and the endogenous constant, $a_{0,1}$:

$$\upsilon = \left(\frac{\beta_1}{\beta_1 - 1}\right) I \tag{E.15}$$

$$a_{0,1} = \frac{\upsilon^{1-\beta_1}}{\beta_1}. \tag{E.16}$$

From the parameters, we first obtain that $\beta_1 = 1.7209$, which leads to $\upsilon = 238.7198$ and $a_{0,1} = 0.0112$ from Eqs. (E.15) and (E.16), respectively. Hence, it is optimal to invest when the value of the plant hits υ, and the value of the option to invest when $V = 100$ is $a_{0,1}V^{\beta_1} = 31.0341$, which is considerably higher than the expected now-or-never NPV of zero. □

As an exercise, the reader is encouraged to extend Illustrative Example E.1 to $T = 20$ to facilitate comparison with the results of Illustrative Example E.2. With the discrete approximation, the value of the option to invest is 30.4767, and in the next-to-last year, the lowest possible plant value that triggers immediate investment is 239.8806. Hence, both values with the discrete approximation are similar to the continuous-time solution.

E.4 Optimal Stopping Time Problems

Another way to approach the problem of investment under uncertainty is to recast it in terms of the optimal stopping time. Specifically, suppose that the same power company as in Illustrative Example E.2 is trying to determine the optimal time, $\tau \geq 0$, at which to invest assuming a perpetual option and that $\{V_t, t \geq 0\}$ follows a GBM:

$$\mathscr{W}(V) = \sup_{\tau \geq 0} \mathbb{E}_V \left[e^{-\rho\tau} \{V_\tau - I\} \right]. \tag{E.17}$$

Using the law of iterated expectations and the strong Markov property of the GBM, we may express the objective in the right-hand side of Eq. (E.17) as follows:

$$\begin{aligned} &\mathbb{E}_V \left[\mathbb{E}_{V_\tau} \left[e^{-\rho\tau} \{V_\tau - I\} \right] \right] \\ &\quad = \mathbb{E}_V \left[e^{-\rho\tau} \mathbb{E}_{V_\tau} [V_\tau - I] \right] \\ &\quad = \mathbb{E}_V \left[e^{-\rho\tau} \right] \{V_\tau - I\}. \end{aligned} \tag{E.18}$$

Thus, the independence of V_τ from V enables the inner expectation to be separated from the outer one. As a result, the objective function in Eq. (E.17) becomes:

$$\mathscr{W}(V) = \sup_{\tau \geq 0} \mathbb{E}_V \left[e^{-\rho\tau} \right] \{V_\tau - I\}. \tag{E.19}$$

In order to solve the optimal stopping-time problem in Eq. (E.19), it is expedient to convert it into a nonlinear unconstrained optimization problem by finding an expression for the conditional expectation of the stochastic discount factor, $\mathbb{E}_V \left[e^{-\rho\tau} \right]$, in terms of the optimal investment threshold value, υ. To do so, we let $g(V) \equiv \mathbb{E}_V \left[e^{-\rho\tau} \right]$ and condition on the first thing that happens in the next $\mathrm{d}t$ time units:

$$\begin{aligned} g(V) &= o(\mathrm{d}t)\, e^{-\rho \mathrm{d}t} + (1 - o(\mathrm{d}t))\, e^{-\rho \mathrm{d}t} \mathbb{E}_V [g(V + \mathrm{d}V)] \\ &\Rightarrow g(V) = o(\mathrm{d}t)\, e^{-\rho \mathrm{d}t} + (1 - o(\mathrm{d}t))\, e^{-\rho \mathrm{d}t} \mathbb{E}_V [g(V) + \mathrm{d}g + o(\mathrm{d}t)] \\ &\Rightarrow g(V) = o(\mathrm{d}t)\, e^{-\rho \mathrm{d}t} + (1 - o(\mathrm{d}t))\, e^{-\rho \mathrm{d}t} \mathbb{E}_V \Big[g(V) + \mathrm{d}V g'(V) \end{aligned}$$

$$+\frac{1}{2}(\mathrm{d}V)^2 g''(V) + o(\mathrm{d}t)\Big]$$

$$\Rightarrow g(V) = o(\mathrm{d}t) + e^{-\rho \mathrm{d}t} g(V) + e^{-\rho \mathrm{d}t}\alpha V g'(V)\,\mathrm{d}t + e^{-\rho \mathrm{d}t}\frac{1}{2}\sigma^2 V^2 g''(V)\,\mathrm{d}t$$

$$\Rightarrow g(V) = o(\mathrm{d}t) + (1 - \rho \mathrm{d}t)\, g(V) + (1 - \rho \mathrm{d}t)\,\alpha V g'(V)\,\mathrm{d}t$$

$$+ (1 - \rho \mathrm{d}t)\,\frac{1}{2}\sigma^2 V^2 g''(V)\,\mathrm{d}t$$

$$\Rightarrow \frac{1}{2}\sigma^2 V^2 g''(V) + \alpha V g'(V) - \rho g(V) = \frac{o(\mathrm{d}t)}{\mathrm{d}t}. \tag{E.20}$$

The first line of Eq. (E.20) is obtained by conditioning on whether the threshold $\upsilon \geq V$ is reached in the next $\mathrm{d}t$ time units, which is presumed to occur with probability $o(\mathrm{d}t)$. If it does, then the discount factor is simply $e^{-\rho \mathrm{d}t}$. Otherwise, with probability $(1 - o(\mathrm{d}t))$, the problem starts all over again with time displacement $\mathrm{d}t$, which leads to discounting of $e^{-\rho \mathrm{d}t}$ and a change of $\mathrm{d}V$ in the underlying plant value. Next, we move to the second line of Eq. (E.20) by arguing heuristically that $g(V + \mathrm{d}V)$ is $g(V) + \mathrm{d}g$ plus some terms of order $o(\mathrm{d}t)$. In order to get to the third line, we apply Itô's lemma to the $\mathrm{d}g$ term. The fourth line of Eq. (E.20) arises from the fact that terms of order $\mathrm{d}t$ become $o(\mathrm{d}t)$ when multiplied by other terms of order $o(\mathrm{d}t)$. In the fifth line, we use the relationship $e^{-\rho \mathrm{d}t} = (1 - \rho \mathrm{d}t) + o(\mathrm{d}t)$, and finally, the sixth line of Eq. (E.20) is a consequence of rearrangement.

If we take the limit as $\mathrm{d}t$ goes to zero, then the last line of Eq. (E.20) becomes the following familiar second-order ODE:

$$\frac{1}{2}\sigma^2 V^2 g''(V) + \alpha V g'(V) - \rho g(V) = 0. \tag{E.21}$$

The general solution to the ODE in Eq. (E.21) is $g(V) = a_1 V^{\beta_1} + a_2 V^{\beta_2}$, where β_1 and β_2 are the positive and negative roots of the characteristic quadratic equation $\mathcal{Q}(\beta) = \frac{1}{2}\sigma^2 \beta(\beta - 1) + \alpha\beta - \rho = 0$. In order to obtain the endogenous constants a_1 and a_2, we use the following boundary conditions:

$$\lim_{V \to 0} g(V) = 0, \tag{E.22}$$

$$g(\upsilon) = 1. \tag{E.23}$$

Equation (E.22) states that as the initial plant value goes to zero, which is an absorbing state for the GBM, it becomes improbable for the threshold υ to be hit. Thus, τ should go to infinity, which means that the expected discount factor becomes zero. The only way for this to hold is if $a_2 = 0$; otherwise, the V^{β_2} term explodes to infinity. At the other extreme, if the initial plant value is already very close to the investment threshold, υ, then the so-called hitting time should be approaching zero, which implies that the expected discount factor is as reflected by Eq. (E.23). Consequently, we have $a_1 = \frac{1}{\upsilon^{\beta_1}}$. Hence, the optimal stopping-time problem in Eq. (E.17) becomes the following nonlinear optimization problem:

$$\mathscr{W}(V) = \max_{\upsilon \geq V} \left(\frac{V}{\upsilon}\right)^{\beta_1} (\upsilon - I). \tag{E.24}$$

Illustrative Example E.3 *Optimal stopping time*

Taking the first-order necessary condition for the problem in Eq. (E.24) yields the following:

$$\left(\frac{V}{\upsilon}\right)^{\beta_1} - \beta_1 (\upsilon - I) \left(\frac{V}{\upsilon}\right)^{\beta_1 - 1} \frac{V}{\upsilon^2} = 0. \tag{E.25}$$

Solving this for υ yields the expression in Eq. (E.15). The second-order sufficiency condition may be verified as well:

$$\begin{aligned} &-\beta_1 \left(\frac{V}{\upsilon}\right)^{\beta_1 - 1} \frac{V}{\upsilon^2} - \beta_1 \left(\frac{V}{\upsilon}\right)^{\beta_1 - 1} \frac{V}{\upsilon^2} + \beta_1 (\beta_1 + 1)(\upsilon - I) \left(\frac{V}{\upsilon}\right)^{\beta_1} \frac{1}{\upsilon^2} \\ &= \beta_1 \left(\frac{V}{\upsilon}\right)^{\beta_1} \frac{1}{\upsilon} \left[\frac{(\upsilon - I)}{\upsilon} (\beta_1 + 1) - 2\right] \\ &< 0. \end{aligned} \tag{E.26}$$

The result in the last line of Eq. (E.26) follows because (i) $\beta_1 \left(\frac{V}{\upsilon}\right)^{\beta_1} \frac{1}{\upsilon} > 0$ and (ii) $\frac{(\upsilon - I)}{\upsilon} (\beta_1 + 1) - 2 < 0$ if the definition of υ from Eq. (E.15) is employed. □

References

1. Bellman, R.E.: Dynamic Programming. Princeton University Press, Princeton (1957)
2. Brennan, M., Schwartz, E.: Finite difference methods and jump processes arising in the pricing of contingent claims: a synthesis. J. Fin. Quant. Anal. **13**, 461–474 (1978)
3. Dixit, A.K., Pindyck, R.S.: Investment under Uncertainty. Princeton University Press, Princeton (1994)
4. Dreyfus, S.E.: Dynamic Programming and the Calculus of Variations. Academic Press, New York (1965)

Index

A.J. Conejo et al., *Investment in Electricity Generation and Transmission*,
DOI 10.1007/978-3-319-29501-5

GPSR Compliance
The European Union's (EU) General Product Safety Regulation (GPSR) is a set of rules that requires consumer products to be safe and our obligations to ensure this.

If you have any concerns about our products, you can contact us on

ProductSafety@springernature.com

In case Publisher is established outside the EU, the EU authorized representative is:

Springer Nature Customer Service Center GmbH
Europaplatz 3
69115 Heidelberg, Germany

www.ingramcontent.com/pod-product-compliance
Lightning Source LLC
Chambersburg PA
CBHW070800080926
51093CB00003B/63
* 9 7 8 3 3 1 9 2 9 4 9 9 5 *